Introductory Statistics for the Behavioral Sciences

Introductory Statistics for the Behavioral Sciences

THIRD EDITION

Robert K. Young
University of Texas

Donald J. Veldman
University of Texas

HOLT, RINEHART AND WINSTON   New York   Chicago   San Francisco   Atlanta
Dallas   Montreal   Toronto   London   Sydney

Library of Congress Cataloging in Publication Data

Young, Robert K.
 Introductory statistics for the behavioral sciences.

 Includes index.
 1. Statistics. I. Veldman, Donald J., joint author.
II. Title. [DNLM: 1. Behavioral sciences—Programmed
texts. 2. Statistics—Programmed texts. HA29 Y75i]
HA29.Y65 1976 519.5′07′7 76-49552

ISBN 0-03-089677-0

Preface to the Third Edition

Changes in the third edition reflect our continuing assessment of the needs of students in the social sciences and related areas. These students tend to be people with a limited background in mathematics who also tend to be intimidated by a course in introductory statistics. We have tried to make the material presented as interesting, meaningful, and understandable as possible; at the same time we have tried to provide those statistical tools necessary for understanding and/or doing research. Basically, the book remains what it was in the first two editions with some changes, additions, and deletions. In keeping with our attempt to make the material presented meaningful, we have included exercises in most chapters which require both the gathering of data and the use of those statistical techniques currently under study. Use of these exercises, we believe, will tend to make the material presented less abstract and make it more apparent how useful the techniques of descriptive and inferential statistics are.

Several other major changes have been made in the text. We have added a chapter at the end of the text dealing with those changes we see taking place in the field. As a consequence, we cover nonparametric statistics, Bayesian statistics, and consider the possible impact of computer and calculator technology on the teaching of statistics. In response to those multitudes of students who have become confused and frustrated over the process of interpolation, the computation of the median and semi-interquartile range in Chapters Four and Five has been simplified. Students may now use the midpoint of the interval within which the median (as well as Q_1 and Q_3) falls as the median. We hope the reduction in student frustration more than compensates for the apparent lack of elegance. In addition, we have deleted the chapter on testing and prediction. Recent court decisions regarding the use of selection tests make our coverage less than adequate. Adequate coverage would require much more space than can be given in an introductory statistics book. To reflect the wider use our text has had in such diverse areas as nursing and library science, as well as in social science and education courses, we have attempted to draw examples and illustrations from as broad a range of topics as possible. We hope they do not miss the appropriate mark.

After twelve years we have to admit defeat on one issue. The t test used to compare two means appears to be solidly entrenched in research literature. We have found that students are sometimes puzzled about not being able to find this test in our text. Consequently, the t test, for the comparison of two independent groups as well as for two correlated groups, has been briefly included. The relationship of these tests to the corresponding analysis of variance techniques has been emphasized.

We wish to acknowledge the help, guidance, and suggestions provided us by our reviewers: Dr. Joe W. Darnall, of Hardin-Simmons University and Dr. Carolyn H. Denham of California State University at Long Beach. We would also like to thank Dr. Ron Wyllys of the University of Texas at Austin for his continued help and support.

Austin, Texas   R.K.Y.
November 1976   D.J.V.

Preface to the Second Edition

The changes made for the second edition of *Introductory Statistics for the Behavioral Sciences* are based on our experiences in using the book as a text. Basically the book remains what it has been, a combination of programmed exercises and text, but with many additions. The programs were intended to provide drill and/or practice, and therefore relatively few questions were provided at the ends of the chapters. However, the need for more general questions was reported by many students; for this reason we have increased the number of questions at the ends of the chapters. As a further aid to self-study, extended discussions of the answers are also provided in the back of the book to give the student a better understanding of why the specific answers are correct. What we believe to be a unique aspect of the book is the final chapter in which we have attempted to provide for the student some idea as to which statistical tests to use under particular conditions. We have found that students learn *how* to use the various tests, but often do not know *when* to use them. The final chapter is designed to give the student some help with this problem. Most of the other changes in the text were prompted by student needs for more elaboration or clarification of certain topics.

Nothing has occurred since the publication of the first edition to convince us that inclusion of the *t* test between means would be warranted. In one sense this procedure is an anachronism in an introductory statistics text, since it is only a special case (two-groups) of one-way analysis of variance. We have, however, changed our ideas about the use of the programmed material. Students prefer to study these sections as homework rather than during scheduled classes or laboratory settings.

An interesting use of the text has been in the development of a self-paced course in introductory statistics, a course in which students work through the book at their own rate, taking a test on each chapter. Although the project is still in the developmental stage, the proportion of students completing the self-paced course has approximated the proportion of students completing the regular lecture course. This indicates that well-motivated students can work their way through the book without the aid of an instructor or formal course meetings.

Austin, Texas
February 1972

R.K.Y.
D.J.V.

Preface to the First Edition

This book is divided into two major sections—text and programs. The text portion of the book was truly a joint effort. Each chapter was written and rewritten by both authors until a final version had been reached that both could agree on. The programs, on the other hand, are the sole responsibility of the first author.

It is with pleasure that we acknowledge the invaluable aid of students in our classes who suffered through the use of mimeographed materials during our several revisions of text and programs. We also acknowledge the permissions granted to us by the following authors and publishers: Appendix A is included through the courtesy of Holt, Rinehart and Winston, Inc. and Professors Helen Walker and Joseph Lev; Appendix B is included through the courtesy of Cambridge University Press and Houghton Mifflin Company; we are indebted to the Literary Executor of the late Sir Ronald A. Fisher, F.R.S., Cambridge, to Dr. Frank Yates, F.R.S., Rothamsted, and to Oliver & Boyd Ltd., Edinburgh, for their permission to abridge Tables Nos. III, IV, and VI from their book *Statistical Tables for Biological, Agricultural and Medical Research*—data from these tables make up Appendixes C, G, and a portion of E; the remainder of Appendix E and all of Appendix D are included through the courtesy of the Iowa State University Press and Professor George Snedecor; Appendix F is included through the courtesy of the Institute of Mathematical Statistics, Professor E. G. Olds, Appleton-Century-Crofts, Inc., and Professors Benton Underwood, Carl Duncan, Janet Taylor, and John Cotton; finally, Appendix H was prepared through the courtesy of Dr. Paul C. Jennings and The University of Texas Computation Center.

We would like to acknowledge the invaluable help of Mesdames Faye Gibson, Priscilla Williams, Laura Schwenk, and Peggy Hampton who typed and retyped the manuscript.

TO THE INSTRUCTOR

Introductory Statistics for the Behavioral Sciences represents a considerable departure from "traditional" textbooks in this field. The most obvious innovation is the use of the programmed learning technique to present a considerable amount of drill material. In accordance with accepted procedures for developing programmed learning materials, we have determined error rates in trial classes and have revised many frames on this basis.

Another obvious departure from tradition lies in the selection of topics. These changes, we believe, reflect the more stable developments in statistics during the past fifteen years. For this reason we present single- and double-classification analysis of variance, while omitting the t test of differences between sample means. This particular departure can be further justified in that the analysis of variance is easier to learn, is a more general method than the t test, and is, of course, algebraically equivalent, since $F = t^2$ in the two-group case.

We did not intend this text to be an encyclopedia of statistical techniques. Rather, we have emphasized a few widely used methods for the analysis of experimental and observational data. Also, while recognizing the importance of statistical description, we have oriented the book toward statistical inference.

We recommend that the instructor integrate the use of both textual and programmed materials in a laboratory setting. Our experience has been that the use of the programs in class is more successful than assigning them as out-of-class homework. We have also found that when only the textual materials were used, students felt that the presentation was too condensed. On the other hand, students who were given only the programs felt that statistics was an endless series of steps with no central themes. Use of both written and programmed materials tends to minimize both of these reactions.

As noted in the section addressed to the student, we advise a careful reading of the text in a given chapter before *and* after working through the programmed learning materials.

TO THE STUDENT

Introductory Statistics for the Behavioral Sciences may be a considerable change from textbooks you have encountered in other courses. As you will see, a large portion of the book consists of sequential exercises or "programs," each of which contains a series of questions or "frames." Each of these frames is designed to give you a little more information about the topic at hand, as well as a question or two which you should be able to answer easily from knowledge you have accumulated from previous frames. As you work through the program, your fund of information about the topic will gradually but systematically grow. By the time you finish a program, you will probably be surprised at the amount of material you have acquired in this manner.

The usual frame will have one or more blanks in it. For example: $2 + 2 = \_\_$ and $2 + \_\_ = 5$. The correct answer to the question(s) will be placed immediately below a heavy STOP rule which looks this this:

═══════════════════════════════════════

If you place a sheet of paper or an envelope against this STOP rule, you will shield the answer to the problem you are attempting to solve.

After you have deciced on a response and written it down on a separate sheet of paper, slide the paper shield down to expose the printed answer.

We strongly recommend that you write down your responses to every frame, since experience with programmed learning techniques indicates that you will probably understand and remember the material better if you do so. Although spaces are provided in the book itself, the use of scratch paper for this purpose will permit you to review the programmed material as often as you wish without the distraction of previous answers to the questions. One of the main benefits of writing down your answers is to keep your pace in completing the programs slow enough for you to derive a maximum amount of learning from the materials.

The programs should *not* be considered as a kind of test, but instead as an aid to learning. The answers are made immediately available so that you can check your progress continually. If you make a mistake, be sure you understand completely the nature of your error before continuing on to the next frame. If the mistake was not simply in computation, you should review previous frames to clear up the misunderstanding before going on.

The procedure we recommend for studying each chapter is first to read over the text material rather quickly to get an idea of the organization of the material and of the various topics to be covered. Next, work through the accompanying program. Finally, return to the textual part of the chapter and read it carefully and thoroughly. While the programs tend to make the learning of formulas, terms, and notation less troublesome, the text gives you an overview and organization that you would not get from the programs alone. Numerous examples are presented in the programs, and therefore few exercises are provided at the ends of the chapters. You will find the answers to these problems at the end of the book. Working out these problems would be a good way to test your own understanding of the material just completed before you move on to another chapter.

You will find that some topics that are covered in the text materials are not covered in the programs; some chapters, in fact, do not have accompanying programs. Some concepts are easily understood without the aid of a program, and others can be presented more clearly with a few lines of text than they can by means of a program.

Finally, beware of the feeling that you can short-cut the programmed learning sections without losing much. Our experience with these materials has been that students who skim through the programs often become very confused around the middle of the semester when they discover that their understanding of the concepts and methods lacks both permanence and depth. A slower, more systematic acquisition usually results in a saving of effort in the long run.

Austin, Texas   *R.K.Y.*
February 1965   *D.J.V.*

Contents

One

Introduction

The human organism is extremely complex, and so is its behavior. Human behavior is the product of many influences acting together and in competition. Some of these influences are typically strong, others are weak, and some fluctuate periodically in intensity. Some influences are present only as memories, others stem from internal functions, while still others originate in perceptions of external stimuli. The patterning of these myriad influences varies from time to time and from person to person to produce a bewildering array of individual differences in behavior.

The goal of the behavioral sciences is to abstract from this variability the most essential consistencies. The fewer conceptual categories and essential relationships that are required to account for, predict, or explain human behavior, the closer the scientist is to his goal.

Statistical procedures are of particular importance in the behavioral sciences because of the need for data reduction and abstraction. Because of the variability of human behavior, the study of single individuals contributes little to the search for general consistencies. Groups of people must be studied in order to avoid the danger of generalizing from an atypical person. Suppose, for instance, that you believed men to be taller than women, in general. In presenting evidence to support your belief, you could hardly expect anyone to be very much convinced if you simply pointed out that Zane was taller than his sister Jane. What alternative procedure is possible? You could measure the heights of a number of men and women, determine the average height in each group, and then point out that the average for the measured men exceeded that for the measured women. When you describe the groups in terms of their average heights, you are making use of *descriptive statistics*.

Still, a critic might object that you just happened to choose a group of tall men, or short women, or both. This objection points up the fact that you were drawing a conclusion about all men and all women, even though you only measured a small sample of each. Statistical procedures explained in this book will allow you to estimate how likely it is that the difference in height you observed would have been found if indeed the average height of all men was the same as that of all women. In the form of a definition, then, *statistics is a set of techniques for describing groups of data and for making decisions in the absence of complete information.* The procedures for determining the probability that particular samples of observations are only the result of chance variation are called *inferential statistics* because we use them to decide what may be inferred about the larger groups from which our samples were obtained.

COMMUNICATION AND MEASUREMENT

Science may be defined as the search for consistencies in the variation of natural phenomena. The primary criterion of a scientific finding is its replicability. The consistency must be capable of reproduction or repeated demonstration. If you were the only person who could find that men are taller than women, your finding would have no scientific validity. The need to demonstrate a consistency repeatedly leads directly to a second essential aspect of scientific work: *Procedures and results must be communicated with a high degree of objectivity.* Another scientist must be able to carry out the necessary operations to gather data in the same manner as did the original discoverer of the consistency. He must also be able to compare his findings directly with those obtained previously. Communication is essential among scientists, and objectivity is the measure of its adequacy. Ambiguous or personalistic communication in science may be worse than no communication at all.

Measurement may be considered simply as a refined method of objective communication. You may object to this, thinking perhaps of communicating the color of an apple to someone else. The word *red*, however, is only a gross measure of the wavelength of light. Considerably greater objectivity is afforded through the use of more refined measures, although at the expense of much more effort. Well, then, what about communicating facts, such as the sex of a particular person? To say that X is a male would certainly be difficult to translate into any particular measurement, as we typically use the term. As will be explained more fully in the next section of this chapter when certain conditions are met, however, even specification of group membership may be considered a form of measurement.

MEASUREMENT SCALES

A controversy arose among statisticians a few years ago about the relationships between measurement scales and the various statistical techniques for manipulating data. One side argued that the treatment of the data at hand was uniquely determined by the scale of measurement employed, while the other side argued that this was not the case.

Rather than ignore the topic of measurement altogether, we feel that an overview of measurement scales would be helpful in providing a convenient framework for organizing the various statistical techniques to be presented. We also believe that the topic of measurement is important in its own right.

The student should recognize that any statistical method can be applied to all data, no matter how they were obtained. She should also recognize, on the other hand, that the goal of statistics is meaningful summary statements. Statistical indices obtained without regard to this goal may give results that have little, if any, meaning. Such would be the case if we surveyed a hospital and found the "average sickness," or surveyed a town to find the "average religion." We have tried to avoid dogmatic statements concerning the methods that may or may not be used with the various measurement scales, using this topic only as a convenient organizational framework for the material to be presented.

There are four degrees of measurement that may be identified among the various methods used to communicate observations objectively. These levels of measurement are called *scales,* in the sense that an individual object or event is measured by placement in some category or at some point along a continuum. The essential differences among these four levels of measurement are in the amount of information which can be communicated by their use.

NOMINAL (CATEGORY) SCALES The labeling of X as a male in the preceding section of this chapter was an implicit use of a *nominal* scale (sex). Any time we describe something by giving it the name of one or another of a particular set of mutually exclusive categories or classes, we are using a nominal scale of measurement. By *mutually exclusive* we mean that it should be impossible to classify a single observation in two different categories at once. The set of categories should also be complete, in the sense of including all possible relevant objects or events.

Male and female form a 2-category scale of measurement. The 50 United States might form a 50-category scale if one wished to describe an individual's place of residence, for instance. These 50 categories could be changed to a 4-category scale—East, Midwest, South, West (provided the boundary lines were unambiguously specified). Although the exact method for determining group membership might be difficult to specify, hair color could form a category scale—blonde, brunette, red, black. If indeterminate shades were encountered, the category "other" could be added. The need for such a category suggests that the principle of classification should be more carefully defined. Several other examples are given in Table 1-1.

The categories on a nominal scale do not have to be labeled with words or letters; numbers can sometimes be used in this manner. Perhaps the most familiar use of numerals for a nominal scale is the numbering of football jerseys. Any player with a jersey numbered from 10 to 19 is known to be a quarterback, 20 to 29 is a center, 30 to 39 is a guard, and so on. Nothing is implied by the difference between 20 and 30 except that the wearers play different positions. The letters A, Q, R, D, X would have served just as well

Table 1-1   Some common scales we use every day

Nominal	Ordinal	Interval	Ratio
Male–female	Freshman–sophomore	One o'clock	10 inches
Single–married	junior–senior	two o'clock	12 inches
Yes–no	First runnerup	three o'clock	14 inches
Group 1	second runnerup	Sunday	5 pounds
group 2	First place	Monday	10 pounds
group 3	second place	Tuesday	15 pounds
Urban–rural	third place	20°, 40°, 60°	50 cents
Quarterback	Blue ribbon	1920, 1930, 1940	75 cents
halfback	red ribbon		one dollar
fullback	green ribbon		
520–529 = astronomy	Good–better–best		
530–539 = physics	BA, Ma, Ph.D.		
540–549 = chemistry	Chapter 1		
Open–close	chapter 2		
	chapter 3		

to identify the various categories. The important thing to understand here is that the numbers themselves do not indicate the type of measurement; the use of the numbers in describing objects or events determines the kind of measurement involved.

The use of nominal scales is so much a part of our everyday lives that it is sometimes difficult to see clearly the essential differences between this type of descriptive measurement and others that afford information beyond the mere classification of objects and events. *Nominal* is derived from the same root as the word *name,* and the use of a nominal scale is nothing more than the consistent use of a set of unambiguous names as labels for classifying objects or events. The categories of classification are not ordered in any way; placement of an observation in a particular category simply indicates that it is *different* from observations in other categories, and does not imply in any sense at all that it is more or less than other observations.

ORDINAL (RANK) SCALES   Big, medium, and small form a scale of measurement useful in describing the relative size of objects. When objects are described on ordinal scales, we are able to make statements such as: "Apples are more red than bananas"; or "*A* is bigger than *B*"; or "He was the last of a long line of tycoons." With this type of measurement we can talk about the *order* of a set of objects or events with regard to some specific attribute like size, but we cannot say anything about how much bigger one thing is than another. The use of an ordinal scale is involved in the description of events when we make statements such as: "He was the first to finish the test"; or "That's the fifth phone call she's had this evening."

The term *rank order* is often used to signify the placement of objects or events on an ordinal scale. As an example, consider the conversion of a nominal scale to an ordinal scale, which is accomplished when we put the names of the 50 United States in the order of the size of their populations. Such an operation would be rank ordering the states according to size of population. In this example we assign numbers to each of the states; the numbers are their ranks, or scores, on the ordinal scale of population size. We could, of course, rank-order the 50 states according to many other characteristics. Each different rank ordering would constitute the use of a different ordinal scale, even though the same numbers (1 to 50) were used each time.

The numbers assigned to objects or events when they are placed on an ordinal scale signify only their place in a continuous arrangement; they do not signify anything about the distance from one datum to the next with regard to the characteristic involved. If we rank-order the three cars Volkswagen, Chevrolet, and Cadillac according to price, we ignore all of the information contained in the differences between the dollar costs of the three cars, and assign the ranks 1, 2, and 3 to the three makes. Although the difference in cost between cars 1 and 2 is much less than that between cars 2 and 3, the rank numbers do not represent this information. For this reason, the use of an ordinal scale is inappropriate when reliable information is available about the relative distances between the objects or events that are to be measured with regard to some particular attribute.

An ordinal scale is distinguished from a nominal scale by the additional property of *order* among the categories included on the scale. It is important to recognize that the assumption of order among a set of categories is not always obviously justified. An example of this is found in the measurement of social class. The labels given the various categories—upper, middle, and lower, for instance—imply order. In some kinds of research, however, the treatment of the three "levels" as an ordered continuum would contradict certain other assumptions regarding the unique character of the social groups represented.

INTERVAL SCALES   An interval type of scale is sometimes called *equal intervals* to emphasize the characteristic that distinguishes it from an ordinal scale. If we know only that Mary scored higher on a quiz than did Roberta, who in turn scored higher than Alice, we do not have as much information as is provided by the scores 50, 40, and 10 for the three girls. The ranks 1, 2, and 3 do not convey any information regarding the *distances* among the scores, since the interval between any two ranks is unspecified. However, the interval between two scores on an interval scale can be determined by arithmetic manipulation of the scores.

An important characteristic of interval scales is that they can be added, subtracted, multiplied, and divided without affecting the relative distances among the scores. For instance, Fahrenheit and Celsius measures of temperature are both interval scales. Celsius temperatures can be converted to Fahrenheit by means of the formula $C = 5(F - 32)/9$. You can demonstrate to yourself that the difference between 20°F and 60°F is twice that between 60°F and 80°F, and that this same ratio (2:1) holds after the temperatures are converted to Celsius.

Another important characteristic of interval scales is that the zero point is arbitrary, which means that we cannot interpret meaningfully the size of particular score ratios. We cannot say, for instance, that 100°F is twice as hot as 50°F. Most psychological tests are scored on interval scales, particularly those which are based on the performance of large normative groups. The so-called IQ measure is an example of this; a zero IQ score almost certainly does not represent a complete absence of mental functioning or total lack of capacity for learning.

RATIO SCALES   Ratio scales have all the properties of interval scales, with the addition of an absolute zero point. Distance and time measures have genuine zero points. The distance 2 feet is twice the distance 1 foot, and the same kind of statement may be made with measures such as minutes, days, years, and so on. Note, however, that year measurement on the scale 1900, 1901, 1902, and so on does not meet the requirements of a ratio scale.

Ratio scales are rarely encountered among psychological measures. As we mentioned previously, measures based on the performance of persons in some normative group are only interval in nature. It would be a mistake, however, to conclude that the attributes measured on interval scales are any less "real" or any less meaningful than those measured on ratio scales. It is often possible to measure the same attribute on scales with different degrees of refinement. Consider the problem of measuring the time it takes each of two children to put together a puzzle. We could measure their times on (1) a ratio scale—2 minutes versus 3 minutes; (2) an interval scale—faster than 60 percent of children this age versus faster than 45 percent of children this age; (3) an ordinal scale—first versus second; or (4) a nominal scale—both children finished. Obviously a nominal scale has little meaning when the data can be quantified.

THE USES OF STATISTICS

As we will see more clearly in later chapters, the level of measurement determines to some extent which statistical procedures we may wish to select for descriptive or inferential purposes. Upon this foundation we will gradually build an understanding of two general kinds of statistical measures. First, we will deal with descriptive statistics, which allow us to summarize the characteristics of large groups of individuals. These summary measures will allow us to ascribe meaning to individual scores, in terms of some reference group. Once we have gained an understanding of these descriptive indices and their computation from raw data, we will begin to explore the major techniques for drawing inferences about larger populations on the basis of descriptive statistics derived from small groups. It is inferential statistics that behavioral scientists find so powerful. For example, we are not ultimately interested in knowing whether the 30 sixth-grade pupils in Fairmount School in Grand Rapids, Michigan, learn fractions faster with programmed instruction than with traditional teaching methods. The significance of finding a consistent difference between the two methods would lie in the extent to which we could infer that such differences might be expected from *other* sixth-grade classes throughout the country. From another point of view, the purpose of seeking such consistencies would be to test the predictive power of a particular theory of learning.

SAMPLES AND POPULATIONS   In statistical theory, the term *population* is used to signify the entire class of objects or events to which generalizations are to be referred, and a *sample* is some part of a particular population. For the purposes of this book we will consider all populations to be of indeterminate size, and assume that an infinite number of different samples could be drawn from any given population. We will differentiate, however, among populations in terms of the scope of their definitions.

The population of interest in a particular scientific study may be relatively restricted in scope. If an experimenter is exploring a newly discovered phenomenon, he may wish to restrict his inferences to a narrow population, such as female sophomore college students at the University of Montana, in order to insure greater homogeneity of subjects within the samples he studies. Such restriction of population is not always clearly recognized by those who conduct or read about experiments. Even though we might have little reason to challenge the notion that a sample of Montana coeds was not representative of the more general population of all female college sophomores, the fact is that generalizing to the larger population does involve an untested assumption. If an experimenter in Delaware was unable to duplicate the Montana results, the assumption would suddenly become very questionable.

Descriptive statistics may be used to characterize either samples or populations. Because statistical description of a population is possible only in a purely theoretical sense, samples are typically described and inferential statistical methods are used to draw conclusions about the nature of the populations concerned. Although this topic will be discussed in greater detail in Chapter Seven, we may note here that inferences about the nature of populations, which are based on the characteristics of samples, are possible only when the samples have been drawn in such a fashion that the population is represented without systematic bias. This is accomplished by means of *random selection* of sample members, in which every object or event in the population has an equal probability of being included in a sample.

TYPES OF PROBLEMS ENCOUNTERED   All of the material we have been presenting sounds very abstract, and it is. However, if you can see beyond the abstractions, you will find statistics to be an extremely practical set of procedures. Use of statistics will enable you to make decisions concerning a wide variety of things about which you could otherwise only guess. For example, you may notice that people tend to turn right at a choice point in a store. While some turn left, many more turn right. Is this due just to random chance, or is there a tendency for people to have a bias to turn right? This question, you will find, can be easily answered by using the statistical procedures presented in this book. For another example, suppose you notice that men tend to be taller than women. You measure heights in your statistics class and you find that, as you thought, the men are indeed taller than the women. Your neighbor says that the difference is due just to chance, just as flipping two heads out of three tries is due to chance. The procedures you will learn in this course will allow you to test the validity of your findings and also your neighbor's claim.

These questions are, of course, only a few of the wide variety of questions that can be answered after learning the procedures in this text. Other questions you will be able to answer would be ones such as these:

1. I've flipped for coffee and lost five times in a row. Is there something wrong here?
2. The quality control for this kind of cola doesn't seem as good as the other kind. Sometimes I get a bad drink, and sometimes I get a good one. Does one vary more in quality than the other?
3. The intelligence of twins seems to be more alike than does the intelligence of siblings. Is this true?
4. Women seem to take more books out of the library than do men. Is this true in general?
5. People seem to have learned a lot in this course. Did they really?
6. Even people who are not sick seem to have a fever. Suppose we take at random ten people who say they are well. I'll bet that their temperature will be above normal.

SUMMARY

Because of the complexity and variability of human behavior, statistical procedures are especially important in the work of researchers who attempt to isolate consistencies from observations of such behavior. Descriptive statistics are summary measures derived from raw data, and typically represent the characteristic behavior of groups of individuals. Inferential statistics are derived from descriptive statistics in order to aid decision making in the absence of complete information. Descriptions of samples drawn from populations serve as the basis for inferences regarding the nature of the populations from which the samples were randomly selected.

Objective measurements may be derived from the direct experience of observers at several levels of information value, called *scales of measurement*. A *nominal scale* consists of a group of unordered, mutually exclusive categories. Addition of the property of order among the categories yields an *ordinal scale*. Quantitative measurement on dimensions with equal distance intervals constitutes use of an *interval scale*. An interval scale with an absolute rather than an arbitrary zero point is called a *ratio scale*. The level of precision of measurement determines to some extent the particular statistical indices that are useful for description and inference, and also provides a framework for organizing the various statistical indices to be discussed in subsequent chapters.

Two

Distributions and Graphs

Statisticians use various methods to present data in an organized fashion. In subsequent chapters we will consider techniques for abstracting some of the primary characteristics of sets of data in the form of descriptive statistics, or summary measures. In this chapter, however, we wish only to present data in a systematic, orderly way.

Suppose that you measured the height of each of 28 male students in your college, and recorded the scores (rounded to the nearest inch) as they were obtained. Your record sheet might look like this: 68, 70, 64, 74, 71, 68, 66, 69, 70, 70, 67, 74, 67, 69, 71, 67, 69, 72, 68, 71, 69, 70, 67, 68, 70, 69, 68, 70.

It is quite difficult to get a very adequate idea of the characteristics of this group of data from such a list. Since you know that height is a variable with characteristics of a ratio scale, it makes sense to put these 28 scores in the order of their size: 64, 66, 67, 67, 67, 67, 68, 68, 68, 68, 68, 69, 69, 69, 69, 69, 70, 70, 70, 70, 70, 70, 71, 71, 71, 72, 74, 74.

This is a considerable improvement, but there is a good deal of redundancy in such a presentation; most of the scores occur more than once.

THE FREQUENCY DISTRIBUTION

The frequency distribution can be used to present a set of data in an economical fashion—a way that avoids the redundancy of listing identical scores. This tabular presentation makes use of two columns, one for the score values and one for the frequencies with which the values occur. Table 2-1 shows a frequency distribution for the 28 heights previously listed.

Why did we put in score values for 73 and 65 inches? We put in these values to make it more apparent that there were no heights recorded with these values, and to make the list of heights consecutive.

LINEAR SCALES

Another way of presenting the distribution in Table 2-1 is by means of a segmented line or linear scale, as shown in Figure 2-1. This measure is

Table 2-1   Frequency distribution of 28 heights

Height	Frequency
74	2
73	0
72	1
71	3
70	6
69	5
68	5
67	4
66	1
65	0
64	1

analogous to a yardstick, with the frequencies written in at the appropriate points.

Note in Figure 2-1 that the particular height scores are represented by particular points on the line. Since we rounded the heights to the nearest whole inch, the line segment from 67.5 to 68.5 actually represents the location of the five original scores that were rounded to 68 inches.

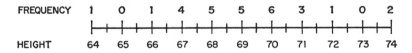

Figure 2-1   *Linear scale frequency distribution of 28 heights.*

INTERVALS LARGER THAN 1.00

With many kinds of measurement, we encounter far more score values than can reasonably be included in a frequency distribution or graph. One way of dealing with this problem is to round the scores, as we did with the height measurements. Another way of looking at this procedure is in terms of the grouping that is accomplished. If all scores between 71.5 and 72.5 are rounded to 72.00, we are in effect grouping together all individuals whose actual heights lie between these limits.

A general rule of thumb for such grouping is to aim for approximately a dozen categories. Fewer categories are too gross to adequately represent most distributions, while many more than this are difficult to interpret.

These categories are called *intervals*, and all cases (that is, an individual's score) falling within the boundaries or limits of an interval are included within it. When scores are grouped in this way the value of an individual case is lost, and all that is known about it is that someone (or something) got a score that falls within the upper and lower limits of the interval. If the width of the interval is set too narrow, then too many intervals result; if it is set too wide, then too few intervals result.

The following example will help illustrate this. Suppose you have a large group of entrance exam scores that range from a low of 100 to a high of 480. A simple listing of these scores, in order from lowest to highest, would be so difficult to read that the reader could get relatively little information from it. Because of this you decide that a frequency distribution would be most appropriate. Your frequency distribution might be like Table 2-2, which illustrates what would happen if you list every score by itself so

Table 2-2   Frequency distribution with an interval size of 1.00

X	f
480	1
479	1
478	2
477	0
476	1
.	.
.	.
.	.
104	2
103	1
102	0
101	1
100	1

that the interval size equals 1.00. On the other hand, it might be like Table 2-3, which groups the data into intervals of 200.

Table 2-3   Frequency distribution with an interval size of 200

X	f
300–499	575
100–299	480

In these two tables, X refers to the value of the variable, which in this case is the test score, and f refers to the frequency or number of cases (that is, test scores) falling within each interval.

It is obvious that both frequency distributions fail at the single, most important consideration in tabular and graphic presentation—that is, useful communication. In both cases the basic characteristics of the distribution of scores are obscured. In the first case, so much information is presented in such a cumbersome manner that the reader does not have the time to digest all information. In the second case, the information is lumped together in such a gross fashion that relatively little information is provided. Because of these faults—a concern for too much data presented on the one hand, and a concern for too gross a lumping of information on the other—a set of conventions or rules built up over the years deals with how a frequency distribution should be constructed to best achieve the goal of maximal communication.

Initially, there should be about a dozen intervals in any frequency distribution. This is not a hard and fast rule, since as few as 7 or as many as 20 would not usually be considered awkward. Fewer than 7 intervals encounter the problem mentioned before (too gross a lumping) and more than 20 get to the point where too much information is being presented.

If we divide 12 into the difference between the largest and smallest score in our distribution, we arrive at the interval size that would give us exactly 12 intervals. Thus, if $480 - 100 = 380$ were divided by 12, then $380/12 = 31.67$. That is, if we were to use an interval size of 31.67, then the 100–480 range of scores would have exactly 12 intervals.

Unfortunately, our search for effective communication of information has not yet reached its goal because it is usually the case that intervals which are not whole numbers are not easily understood by the reader. That is, the goal of 12 intervals should be modified to get an interval size that is at least a whole number. Furthermore, the interval size itself should be what you would call a *round* number. With large numbers, interval sizes of 500 or 1000 would be preferable to interval sizes of 777 or 863. And with intermediate interval sizes, intervals of 25, 30, 40, or 50 would all be preferable to 31.67. If we were to choose 25 as the interval size, we would end up with 16 intervals in our frequency distribution (16 intervals $\times$ 25 interval size $= 400$, which is approximately the difference between 480 and 100, the largest and smallest scores), and if we were to choose 50, we would end up with 8 intervals. Hence, either one of these interval sizes falls within the acceptable number of intervals. Similarly, interval sizes of 1, 2, 3, 4, 5, 10, 20 are all considered appropriate interval sizes if this results in around 12 intervals in the distribution.

Suppose we chose an interval size of 50 because it divides easily into 100. Have we settled all our problems? No, because now we have to decide on the way to set up the interval. That is, should the lowest interval stretch from 100 to 149, from 101 to 150, from 75 to 125, from 92 to 142, or what? The convention generally adopted is to adjust the interval bounds so that the apparent lower limit is divisible by the interval size. Of the several sets of limits on the lowest interval presented above, only the 100 to 149 interval meets this criterion. The reason why the first number should be divisible by the interval size is that we read from left to right, and the number on the left, the lower limit of the interval, is the first one seen and read. If this number is easily seen and understood, then the data of the frequency distribution are also more likely to be understood. Now our frequency distribution would look like that presented in Table 2-4.

In summary, the frequency distribution in Table 2-4 was settled on be-

**Table 2-4   Frequency distribution
with an interval size of 50**

X	f
450–499	40
400–449	120
350–399	170
300–349	245
250–299	200
200–249	140
150–199	110
100–149	30

cause the difference between the highest and lowest scores when divided by 12 yields a number which was (very) approximately 50. For this reason 50 was used as the interval size. Next, for reasons of ease of reading and ready understanding, the lower limit of each interval was set so as to be divisible by 50, the size of the interval. This procedure is appropriate for any frequency distribution you may want to construct. Notice that the heading over the interval column is X, referring to the various values the test scores can take on, while the heading over the right-hand column is f, which refers to the frequency or number of cases that fall within each interval. Hence, it can be seen from Table 2-4 that 245 people had scores between 300 and 349.

Notice that the difference between 300 and 349 is not 50 but 49; yet we have said that the interval size, the width of the interval, was equal to 50. How can this be? The reason for this is that these are the *apparent* limits of each interval—the numbers that appear in the frequency distribution—and not the *real* limits of each interval. In this case the apparent limits of the interval in question are 300 and 349. The real limits of this interval stretch from 299.5 to 349.5, a difference of 50. This means that a test score of 349.3 would be assigned to the 300–349 interval, while a test score of 349.8 would be assigned to the 350–399 interval. In this way all cases could be assigned to an interval, no matter what their test-score value was.

THE HISTOGRAM

Return now to Table 2-1, in which a frequency distribution of heights was presented. We can convert into vertical distances the frequencies corresponding to each height. The resulting two-dimensional representation of the distribution is the familiar bar graph, or histogram, which is drawn in Figure 2-2.

The vertical dimension in Figure 2-2 represents the frequency of occur-

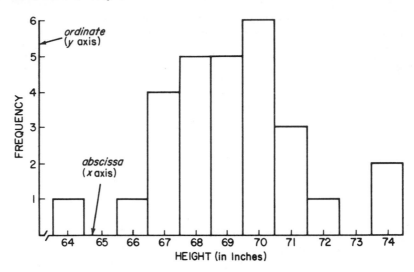

Figure 2-2   *Histogram of 28 heights.*

rence, while the horizontal dimension is the segmented linear scale shown in Figure 2-1. The scale has been marked off, however, at points corresponding to scores halfway between the whole-number points. A score of **71.0**, for instance, falls in the center of the baseline segment marked **71**. This segment is bounded by points corresponding to **70.5** and **71.5**. These points are known as the *lower real limit* and the *upper real limit* of the interval for this case, **71** inches.

Graphs are usually constructed so that the smaller frequencies are lower on the vertical dimension (called the *ordinate*), and the smaller score values are to the left on the horizontal dimension (called the *abscissa*), as in Figure 2-2. The ordinate is also known as the *y* axis, and the abscissa is correspondingly called the *x* axis.

THE FREQUENCY POLYGON

The height of a vertical bar in a histogram represents the frequency of occurrence for a particular score. In a similar type of figure, called a *frequency polygon*, the center of the top of each histogram bar becomes a point that is used to represent score frequency, and each point is connected by a line. When the frequency is 0, the point is located on the baseline of the figure. Figure 2-3 shows a frequency polygon for the height data given in Table 2-1.

In this type of graph, as in the linear scale, the point along the baseline that is used to represent a given score value is located in the center of a line segment running from —0.5 to +0.5 on either side of the particular score value (that is, from 71.5 to 72.5 for the score value 72.0).

Note that the points for the various score frequencies represented in Figure 2-3 are located at the intersections of imaginary reference lines

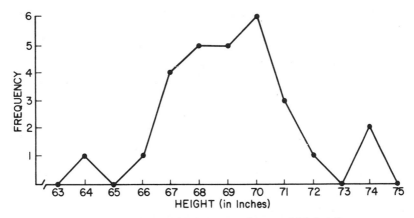

Figure 2-3   *Frequency polygon of 28 heights.*

drawn across from the ordinate (frequency) and up from the abscissa (score value).

CONTINUOUS AND DISCRETE DATA

The term *continuous data* refers to measurements that can attain any value, while the term *discrete data* implies that only certain values (usually whole numbers) are possible. In the measurement of height used previously as an example, we rounded each height to its nearest whole inch. We were treating continuous data as if it were possible to measure height only in whole numbers. Not all characteristics of objects and events are measurable along continuous dimensions. For instance, it is really quite meaningless to talk about 1.5 families, or to say that the typical family has 3.5 members. Nevertheless, it is often quite useful to treat discrete data as if they were continuous, just as we previously found it convenient to treat continuous data as if they were discrete.

We can represent discrete units of measurement along the continuous baseline of a graph and draw histograms or frequency polygons just as we did with height measurements. Strictly speaking, however, the histogram is appropriate only for discrete data such as size of family, while the frequency polygon is intended for continuous measures such as height.

Often it is of interest to construct graphs which allow visual comparison between two different groups. For example, one of the authors does not require students to take the three 1-hour exams which are given during the course of the semester. The students are encouraged to take all the 1-hour exams (as well as the final exam), but many choose not to take one or more of the exams. Although the course grades are assigned in such a way as to give everyone the same chance to make a good grade, the students who take all the hour exams seem to do better than those students who fail to take one or more exams. These data are presented in Figure 2-4.

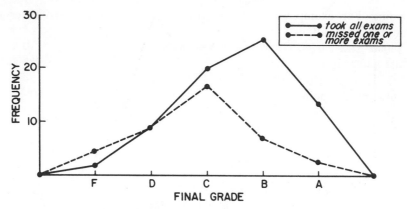

Figure 2-4   *Final grades of students who took all possible exams, and grades of those who missed one or more exams.*

INTERVALS LARGER THAN 1.00

As in our discussion of frequency distributions, the number of intervals appropriate for a frequency polygon or histogram is also about 12. And again the same reasons hold for the graphs as for the frequency distributions. Fewer than 8 intervals often result in too gross a lumping together of the data, while greater than 20 results in the presentation of too much data, to the detriment of a readily understandable graph. As an example, consider the frequency distribution of new car costs shown in Figure 2-5.

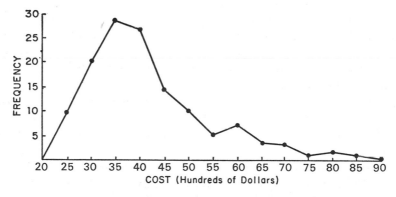

Figure 2-5   *Distribution by price level of new cars purchased.*

Clarity of presentation is the whole point of grouping data in this fashion. Fifteen intervals were chosen for Figure 2-5 because of the additional clarity afforded by representing cost in hundreds of dollars.

The interval size of the data presented in Figure 2-5 is $500. The data in a frequency polygon will usually be presented so that the upper and lower limits of an interval will extend half an interval above and below the plotted point. For example, 20 cars are represented in the $3000 interval. This means that all cars costing from $2750–3250 are represented as costing $3000. If a car cost $3255, it would have cost more than the upper limit of the $3000 interval, and it would then be represented in the $3500 interval.

Look carefully at the frequency polygon in Figure 2-5. In a frequency polygon the points are raised over the midpoint of each interval. Thus if the data of Table 2-4 were to be presented either in a frequency polygon or a histogram, the *midpoint* of each interval rather than the lower apparent limit would have been chosen so that the midpoints would be divisible by the interval size. The reason for this, of course, is that only the value of the midpoint of each interval is placed along the abscissa (x axis) in presenting data in either the frequency polygon or the histogram. If the midpoint is not divisible by the interval size, then a set of awkward or not especially convenient numbers will be used. Such a situation would be encountered if a frequency polygon were constructed from the frequency distribution presented in Table 2-4. The midpoints of these intervals are 124.5, 174.5, 224.5, and so on, numbers that appear to be somewhat awkward to use. While there is certainly nothing incorrect about using these numbers, other midpoints such as 100, 150, 200, and so on, would have made the data presented in the frequency polygon more readily understandable.

GRAPHING OF NOMINAL DATA

Certain problems are encountered when nominal data are presented graphically. Since nominal data are unordered, and hence do not lie along a dimension, it could be misleading to the reader to present each datum as if it did lie along a dimension, the abscissa. One solution to this dilemna is to use a bar graph with some distance between successive bars to emphasize the nominal character of the data. This is illustrated in the graph presented in Figure 2-6. Another way to present nominal data is shown in Figure 2-7, which illustrates the various types of vehicle accidents occurring in North Dakota in 1974.

GRAPHING RELATIONSHIPS

In the graphs we have discussed so far, the ordinate always represented the same variable—frequency of occurrence. It is important to recognize that these graphs represented relationships between (1) height, cost, or grade, and (2) frequency of occurrence in the group measured. We will also consider the graphing of relationships between any two variables. The graphs to be presented are sometimes called *line graphs*. They may resemble frequency polygons, but they are not, since neither of the two dimensions will be used to represent the variable of frequency of occurrence.

DEPENDENT AND INDEPENDENT VARIABLES Suppose we decide to conduct an experiment to determine the relationship between the rate of

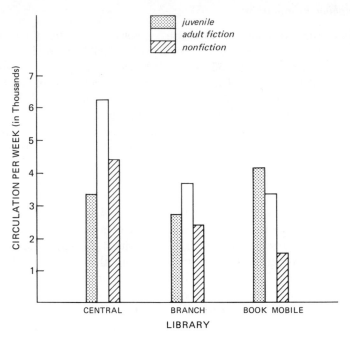

Figure 2-6 *Circulation of books in River City's library system.*

PERCENTAGE AND TYPE OF MOTOR VEHICLE COLLISIONS

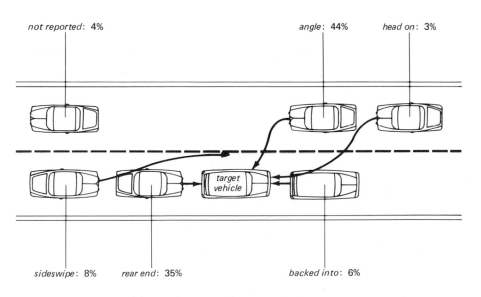

Source: North Dakota vehicular accident facts, 1974.

Figure 2-7 *Nominal data can be presented in a variety of interesting ways.*

learning and the rate at which the material to be learned is presented. We select five groups of five people each and ask them to memorize different lists of 20 words.

In this example we are trying to determine whether speed of learning depends on the rated imagery of the material. Words that do not have images associated with them, such as *method, knowledge,* and *thought,* would be given a low rating, and words that have strong images associated with them, such as *tripod, hurdle,* and *ladder,* would be given a high rating. We present the list of words, one at a time, to the people serving in the experiment, and count the number of words each person remembers correctly. Thus, the number of correct responses is a measure of the degree of learning—how well the material has been learned—which is known as the *dependent variable.* The other variable, rated imagery value, is called the *independent variable,* and this variable is under the control of the experimenter, who varies it systematically from one to another of the five groups.

After collecting the data (the number of words remembered by each person serving in the experiment), we can construct the line graph shown in Figure 2-8.

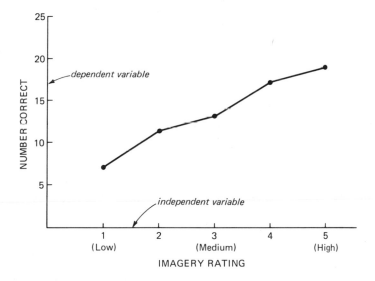

Figure 2-8   *Relationship between imagery rating and ease of learning.*

The independent variable (the one controlled by the experimenter) is typically placed on the abscissa, and the dependent (response) variable is represented by the ordinate. The points in Figure 2-8 represent the average scores on the dependent variable for each of the five groups that represent the variation of the independent variable.

Figure 2-8 shows that there is a definite relationship between number of correct responses and degree of rated imagery—the higher the rated imagery, the easier to learn the list.

GRAPHS WITH TWO INDEPENDENT VARIABLES   Line graphs are also valuable in comparing the relationships between two variables that occur in two or more groups of data, thereby constituting a third variable. In Figure 2-9 the dependent variable is the number of trials to learn, one independent

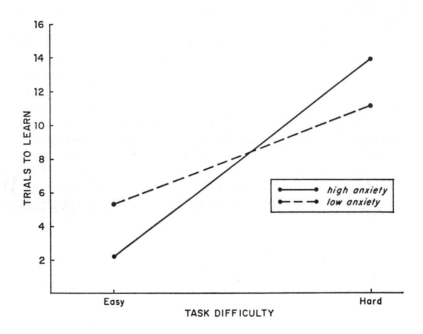

Figure 2-9   *Relationship between trials to learn and task difficulty as a function of subjects' anxiety levels.*

variable is the difficulty of the material to be learned, and the high and low anxiety of the subjects chosen for the two groups constitutes measurement on a second independent variable (anxiety).

Figure 2-9 shows more clearly than could a purely verbal description the fact that the relationship between task difficulty and trials to learn depends on the level of anxiety of the people in the experiment. This interaction of anxiety and task difficulty indicates that the score on the dependent variable, or the rate of learning, depends on the *combination* of the two independent variables. We apply the term *independent* to the anxiety variable in this example because it is under the experimenter's control to the extent that she can select two groups of people (high and low anxious) for comparison. She might also have introduced some experimental treatment to increase the anxiety of one of the groups, but this would not be a necessary condition for calling the anxiety variable *independent* in this case.

The graph in Figure 2-9 could have been drawn in another way with equal logic. Figure 2-10 graphs the same data. This time, however, we have used the abscissa to represent the level of anxiety, with two separate graph

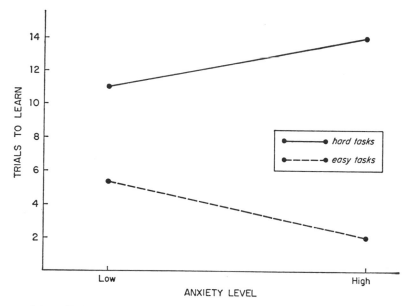

Figure 2-10 *Relationship between trials to learn and subjects' anxiety levels as a function of task difficulty.*

lines for the levels of task difficulty. Both graphs represent the same experimental results; the choice of either one is purely arbitrary.

UNBIASED PRESENTATION   The purpose of graphic presentation of data is, of course, communication of complex sets of data. It is easy to distort inadvertently the data you wish to present in graphic form, and some unscrupulous characters do this purposefully to bias the impression of their graph. For example, one common method is failure to label the axes. By doing this, the same data can be made to appear quite different, depending upon the choice of size of abscissa or ordinate. This is illustrated in Figure 2-11. Figure 2-11A shows the index of economic gain in the country of Belgravia as illustrated by the party in power. Figure 2-11B shows the same data interpreted by the party not in power. It is in the best interests of the party in power to show that the index is increasing rapidly, and for this reason the graph in Figure 2-11A is drawn to show this. On the other hand, the party out of power would find it to its advantage to make it appear that the index is sluggish and is increasing very slowly, if at all. Although both graphs present the same data, the information conveyed by Figures 2-11A and 2-11B is quite different. Failure to label both axes allows even greater distortion of information. When this happens, the person drawing the graph can convey virtually any impression he wishes to give.

   Even when both axes are properly labeled, a rapid change can be made to look like a slow change, or vice versa, simply by varying the size of the unit of measure on the abscissa or ordinate. This is illustrated in Figure 2-12.

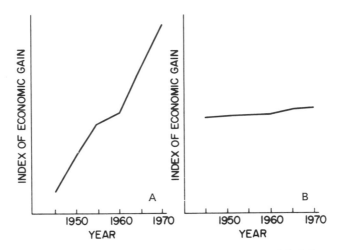

Figure 2-11   *When either the abscissa or ordinate is not labeled properly, the impression given by a graph can vary considerably.*

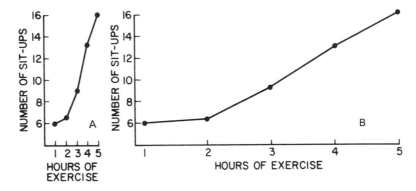

Figure 2-12   *The impression of the data contained in a graph can be easily changed by varying the ratio of ordinate height to abscissa length.*

Both graphs represent the average number of sit-ups a group of housewives can do as a result of a number of hours spent in a well-advertised exercise course. Both graphs present exactly the same data. However, Figure 2-12A makes it appear that the number of sit-ups a person can do increases quite rapidly, whereas Figure 2-12B makes it appear that the increase is much more gradual. To avoid the possibility of giving a biased impression with graphic presentation, most statisticians have adopted the convention that the height of the ordinate should be about three-fourths the length of the abscissa. This is, of course, arbitrary, and any other ratio of ordinate to abscissa length could be used. This ratio seems to give a better impression than other ratios, and for this reason has been adopted.

Another convention sometimes used is illustrated in Figures **2-2** and **2-3**.

Ideally, the point where the abscissa and ordinate intersect should be the point where the values of both graphs are equal to zero. However, sometimes the values of the ordinate or abscissa are such that a graph cannot be drawn this way. To inform the reader of this, two slash marks, cutting the axis, are sometimes used to warn the reader that a portion of the abscissa or ordinate has been omitted. (See Figures 2-2 and 2-3.)

Data can easily be presented in a manner that gives a false impression. For this reason, great care should be taken to present as clear and unbiased a figure as possible. It is quite easy to vary the impression that the same group of data conveys. Probably because of this, great wits remark that, "Figures don't lie but liars figure." Of the same genre is, "There are liars, damn liars, and statisticians."

SUMMARY

This chapter discussed two general ways of presenting data: graphic and tabular. The intent of both methods is to communicate as much information as possible, while at the same time reducing the possibility of distortion or misunderstanding. With tabular presentation, the frequency distribution is used. A set of conventions or rules was outlined—12 intervals should be used, the lower limit should be divisible by the interval size, and so on—to increase ease of communication. Similarly, a set of conventions was also outlined for graphic presentation—either a frequency polygon, histogram, or line graph—so as to reduce misunderstanding or distortion. While these methods are important, subsequent chapters will show you how similar information can be communicated through the use of descriptive statistics, such as the average.

PROBLEMS*

1. Scores of 40, 90, 43, 87, 78, 32, 54, 60, 70, 65, 80, 45, and 50 are found on a test. *E* wishes to construct a frequency polygon from these data. What interval size should be employed?
2. From Problem 1, the interval containing the lowest score would have what limits?
3. If the interval size were 20, a score of 68 would have what values for upper and lower limits?
4. A distinction should be made between the apparent limits and the real limits of an interval. The apparent limits of an interval are 50–59; what are the real limits?
5. A city census grouped the inhabitants by age. The limits of one interval were given as 20–24; what are the real limits of that interval?
6. An agricultural research worker uses four different types of fertilizer on four different corn fields. What would be the independent variable and the dependent variable?

* The answers to these and other questions can be found in the back of the book.

7. With graphic presentation of data the independent variable is usually placed along the _____ while the dependent variable is placed along the _____.

8. Construct a frequency polygon from these data: 12, 15, 14, 10, 16, 11, 16, 14, 15, 10, 16, 12, 15, 14.

9. Construct a histogram from the data presented in this frequency distribution.

X	f
12–14	2
9–11	6
6–8	1
3–5	4
0–2	3

10. Construct a frequency polygon from these data.

X	f
45–49	1
40–44	4
35–39	12
30–34	16
25–29	10
20–24	12
15–19	6
10–14	2

11. Construct a frequency distribution, linear scale, histogram, and frequency polygon for the following group of test scores: 12, 10, 12, 11, 9, 10, 11, 12, 13.

12. Graph the relationship between age and hours of sleep, given the following data:

Age	Average hours of sleep
10	8
5	9
3	10
2	11
1	12

13. Label the ordinate, abscissa, x axis, y axis, dependent variable, and independent variable on the graph completed for Problem 12.

EXERCISES

In this chapter and in most subsequent ones, we will present some exercises. On the one hand, they will attempt to illustrate some of the concepts presented in the chapter, and on the other, they are perhaps more realistic than are the problems presented at the end of each chapter.

1. Have each member of the class estimate the height, weight, or age of some well-known person, or of a member of the class, or of the instructor. From these data construct a frequency distribution and a frequency polygon. Save these data for use in exercises in subsequent chapters.
2. Take 100 single-digit numbers from the random number table in the back of the book (Appendix H). Construct a frequency distribution and a histogram from these data.

1   Graphic representation is a convenient way to present data. A picture that represents data is called a _____.

══════ STOP ══════════════════════ STOP ══════

graph

---

2   Although there are several ways to present data in graphs we will be primarily concerned with those which have an independent variable and a _____ variable.

══════ STOP ══════════════════════ STOP ══════

dependent

---

3   The _____ variable is commonly placed on the x axis (or abscissa or baseline) of a graph while the dependent variable is placed on the y axis or ordinate.

══════ STOP ══════════════════════ STOP ══════

independent

---

4   Another name for x axis (or baseline) is _____, and another name for y axis is _____.

══════ STOP ══════════════════════ STOP ══════

abscissa, ordinate

---

5   The _____ variable is placed on the abscissa and the _____ variable is placed on the ordinate.

══════ STOP ══════════════════════ STOP ══════

independent, dependent

---

6   We usually think of the _____ variable as being varied by the experimenter and the dependent variable as the response given by the subject.

══════ STOP ══════════════════════ STOP ══════

independent

---

7   The line marked x is called the abscissa and the line marked y is called the _____

y |
    |
    |_____
       x

══════ STOP ══════════════════════ STOP ══════

ordinate

8

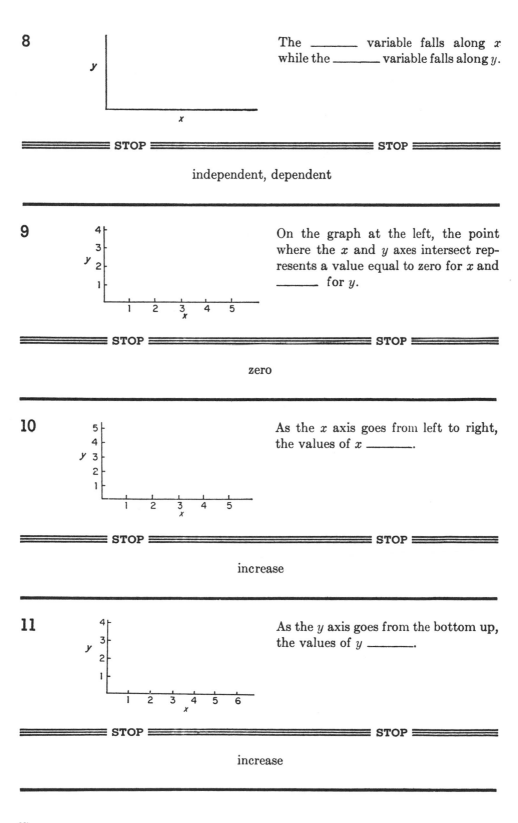

The _____ variable falls along x while the _____ variable falls along y.

═══ STOP ═══════════════════════ STOP ═══

independent, dependent

─────────────────────────────────────

9

On the graph at the left, the point where the x and y axes intersect represents a value equal to zero for x and _____ for y.

═══ STOP ═══════════════════════ STOP ═══

zero

─────────────────────────────────────

10

As the x axis goes from left to right, the values of x _____.

═══ STOP ═══════════════════════ STOP ═══

increase

─────────────────────────────────────

11

As the y axis goes from the bottom up, the values of y _____.

═══ STOP ═══════════════════════ STOP ═══

increase

─────────────────────────────────────

12

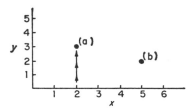

If we extend a line up from the x axis we can read the x value of a point. For example the x value of point (a) is 2 while for point (b) $x =$ _____.

═══════ STOP ═══════════════════════ STOP ═══════

5

13

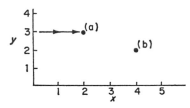

Similarly, if we extend a line over from the y axis we can read the y value of a point. For point (a) $y = 3$, for point (b) $y =$ _____.

═══════ STOP ═══════════════════════ STOP ═══════

2

14

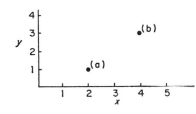

For point (a) $x = 2$ and $y = 1$. For point (b) $x =$ _____ and $y =$ _____

═══════ STOP ═══════════════════════ STOP ═══════

4, 3

15

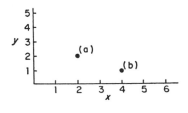

For point (a) $x =$ _____ and $y =$ _____. For point (b) $x =$ _____ and $y =$ _____.

═══════ STOP ═══════════════════════ STOP ═══════

2, 2; 4, 1

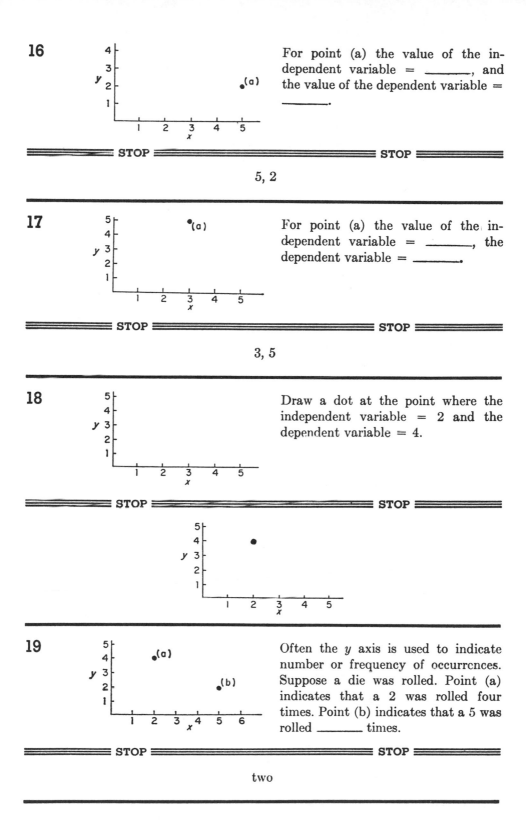

16

For point (a) the value of the independent variable = _____, and the value of the dependent variable = _____.

═══ STOP ═══ ═══ STOP ═══

5, 2

17

For point (a) the value of the independent variable = _____, the dependent variable = _____.

═══ STOP ═══ ═══ STOP ═══

3, 5

18

Draw a dot at the point where the independent variable = 2 and the dependent variable = 4.

═══ STOP ═══ ═══ STOP ═══

19

Often the y axis is used to indicate number or frequency of occurrences. Suppose a die was rolled. Point (a) indicates that a 2 was rolled four times. Point (b) indicates that a 5 was rolled _____ times.

═══ STOP ═══ ═══ STOP ═══

two

20

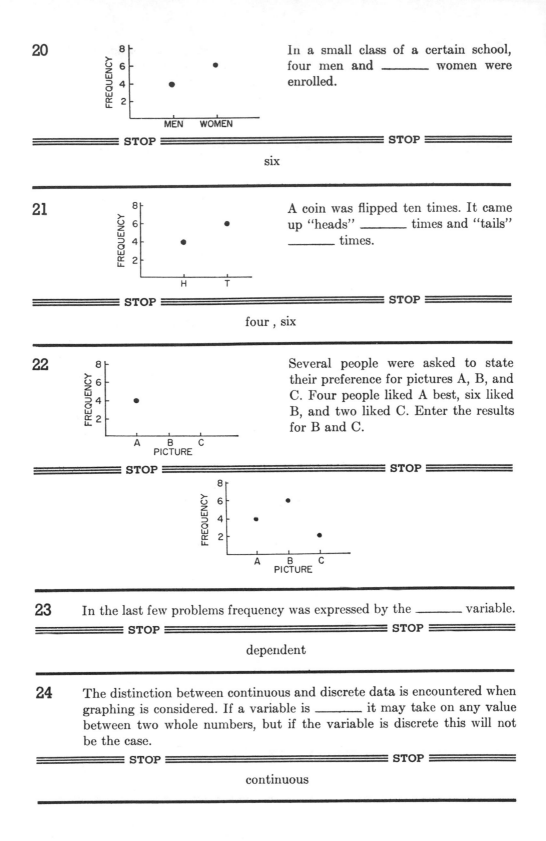

FREQUENCY
8
6
4
2

MEN   WOMEN

In a small class of a certain school, four men and _____ women were enrolled.

═══ STOP ═══════════════════ STOP ═══

six

21

FREQUENCY
8
6
4
2

H   T

A coin was flipped ten times. It came up "heads" _____ times and "tails" _____ times.

═══ STOP ═══════════════════ STOP ═══

four , six

22

FREQUENCY
8
6
4
2

A   B   C
PICTURE

Several people were asked to state their preference for pictures A, B, and C. Four people liked A best, six liked B, and two liked C. Enter the results for B and C.

═══ STOP ═══════════════════ STOP ═══

FREQUENCY
8
6
4
2

A   B   C
PICTURE

---

23   In the last few problems frequency was expressed by the _____ variable.

═══ STOP ═══════════════════ STOP ═══

dependent

---

24   The distinction between continuous and discrete data is encountered when graphing is considered. If a variable is _____ it may take on any value between two whole numbers, but if the variable is discrete this will not be the case.

═══ STOP ═══════════════════ STOP ═══

continuous

25   Age is a _____ variable because a person can be any age between, for example, 10 and 11 years.

═══════ STOP ════════════════════════ STOP ═══════

continuous

---

26   On the other hand, the number of holes dug in the ground is commonly given as an example of a _____ variable because a person can dig 2 holes or 3 holes, but he cannot, for example, dig 2½ holes.

═══════ STOP ════════════════════════ STOP ═══════

discrete

---

27   The number of children a family has is another example of a _____ variable.

═══════ STOP ════════════════════════ STOP ═══════

discrete

---

28   Temperature, on the other hand, would be a _____ variable.

═══════ STOP ════════════════════════ STOP ═══════

continuous

---

29   The problem in graphic presentation comes from the fact that we often must treat a continuous variable as if it were a _____ variable.

═══════ STOP ════════════════════════ STOP ═══════

discrete

---

30   Suppose we wanted to measure height. We would need to summarize the data and would then call people 70 inches tall when they were 70.13 inches tall, 65 inches when they were 64.82, and so on. It is obvious that height is a _____ variable.

═══════ STOP ════════════════════════ STOP ═══════

continuous

---

31   By changing every height measure to a whole number we are then treating height as a _____ variable.

═══════ STOP ════════════════════════ STOP ═══════

discrete

---

32   Thus in order to collect and summarize our data we often must treat a _____ variable as if it were a _____ variable.

═══════ STOP ═══════════════════════════ STOP ═══════

continuous, discrete

---

33   In the previous example, every person between 69.5–70.5 inches tall was recorded as 70 inches. In this case 69.5 was the lower limit of the 70-inch interval. If we looked at the 67-inch interval 66.5 would be the _____ _____.

═══════ STOP ═══════════════════════════ STOP ═══════

lower limit

---

34   On the other hand the upper limit of the 72-inch interval would be _____.

═══════ STOP ═══════════════════════════ STOP ═══════

72.5

---

35   Finally we note that the lower and upper limits for the 64-inch interval are _____ and _____.

═══════ STOP ═══════════════════════════ STOP ═══════

63.5 and 64.5

---

36   Not all intervals are one inch wide. For example the 3-inch interval 65–67 has a lower limit of 64.5 and an upper limit of _____.

═══════ STOP ═══════════════════════════ STOP ═══════

67.5

---

37   The 5-inch interval 50–54 has lower and upper limits of _____ and _____. Subtract the lower limit from the upper limit and we see that the width of this interval is _____.

═══════ STOP ═══════════════════════════ STOP ═══════

49.5 and 54.5; 5 inches

---

38   An interval has upper and lower limits of 17.5 and 11.5. The width of this interval is _____.

═══════ STOP ═══════════════════════════ STOP ═══════

6

39 Not all intervals extend from 0.5 to 0.5. For example, those people reporting themselves to be 10 years old have a lower limit of _____ and an upper limit of _____. (The reason for this is that people don't say they are 11 until their 11th birthday.)

═══════ **STOP** ═══════════════════════════════ **STOP** ═══════

10.00, 10.99

───

40 In summary, problems in graphing tend to occur because we treat _____ variables as _____ variables.

═══════ **STOP** ═══════════════════════════════ **STOP** ═══════

continuous, discrete

───

41 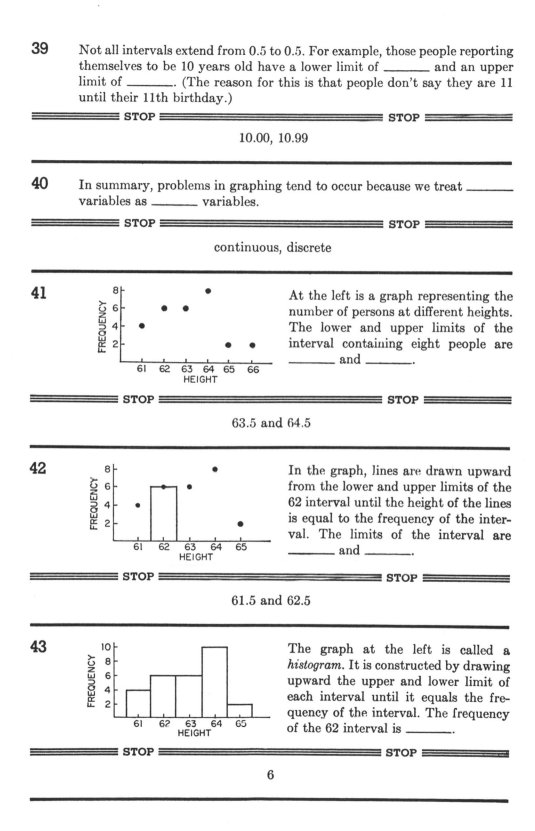 At the left is a graph representing the number of persons at different heights. The lower and upper limits of the interval containing eight people are _____ and _____.

═══════ **STOP** ═══════════════════════════════ **STOP** ═══════

63.5 and 64.5

───

42 In the graph, lines are drawn upward from the lower and upper limits of the 62 interval until the height of the lines is equal to the frequency of the interval. The limits of the interval are _____ and _____.

═══════ **STOP** ═══════════════════════════════ **STOP** ═══════

61.5 and 62.5

───

43 The graph at the left is called a *histogram*. It is constructed by drawing upward the upper and lower limit of each interval until it equals the frequency of the interval. The frequency of the 62 interval is _____.

═══════ **STOP** ═══════════════════════════════ **STOP** ═══════

6

───

44

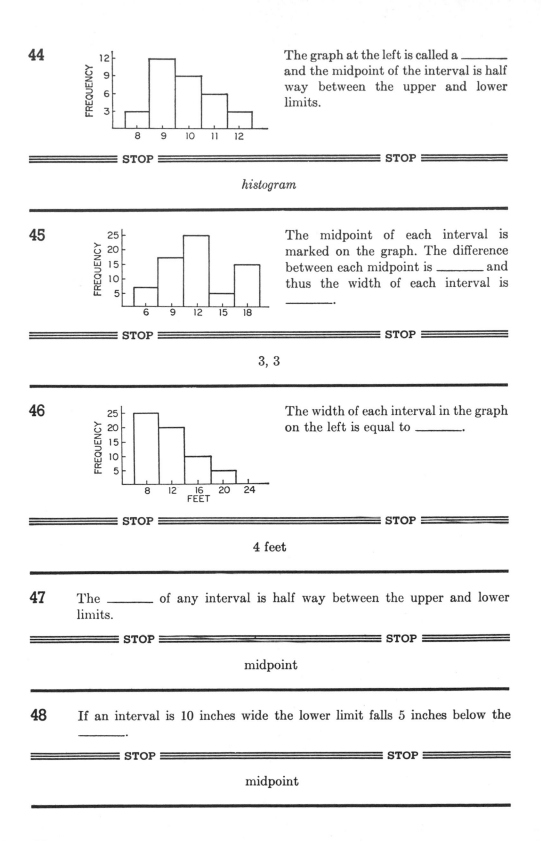

The graph at the left is called a _____ and the midpoint of the interval is half way between the upper and lower limits.

════ STOP ═══════════════════ STOP ════

histogram

45

The midpoint of each interval is marked on the graph. The difference between each midpoint is _____ and thus the width of each interval is _____.

════ STOP ═══════════════════ STOP ════

3, 3

46

The width of each interval in the graph on the left is equal to _____.

════ STOP ═══════════════════ STOP ════

4 feet

---

47   The _____ of any interval is half way between the upper and lower limits.

════ STOP ═══════════════════ STOP ════

midpoint

---

48   If an interval is 10 inches wide the lower limit falls 5 inches below the _____.

════ STOP ═══════════════════ STOP ════

midpoint

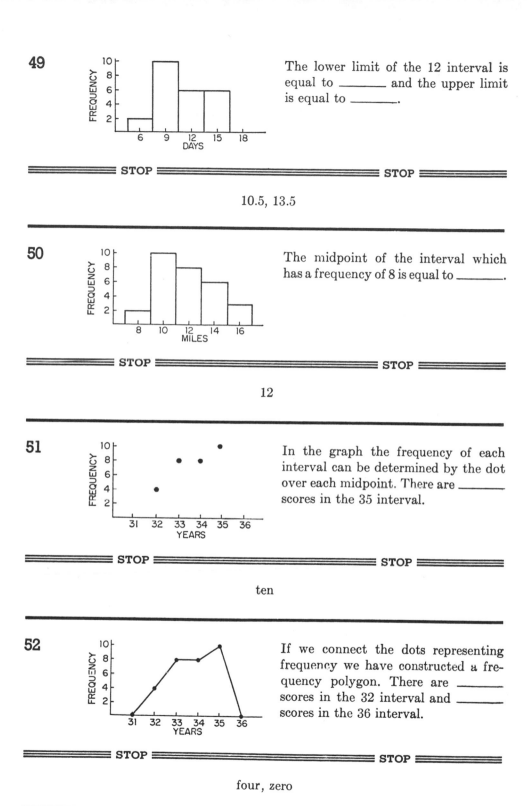

49

The lower limit of the 12 interval is equal to _____ and the upper limit is equal to _____.

═══ STOP ═══════════════════ STOP ═══

10.5, 13.5

50

The midpoint of the interval which has a frequency of 8 is equal to _____.

═══ STOP ═══════════════════ STOP ═══

12

51

In the graph the frequency of each interval can be determined by the dot over each midpoint. There are _____ scores in the 35 interval.

═══ STOP ═══════════════════ STOP ═══

ten

52

If we connect the dots representing frequency we have constructed a frequency polygon. There are _____ scores in the 32 interval and _____ scores in the 36 interval.

═══ STOP ═══════════════════ STOP ═══

four, zero

53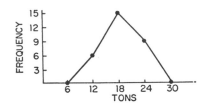

The graph at the left is called a
_____ _____ and there are
_____ scores in the interval with a
midpoint of 12.

━━━━━ STOP ━━━━━━━━━━━━━━━ STOP ━━━━━

frequency polygon, six

54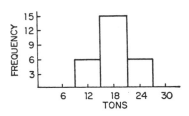

The graph at the left is called a
_____ and there are _____ scores
in the 12 interval.

━━━━━ STOP ━━━━━━━━━━━━━━━ STOP ━━━━━

histogram, six

55 In the construction of the frequency polygon dots representing frequency
are placed over the _____ of each interval. In the construction of the
histogram lines are drawn upward from the _____ of each interval.

━━━━━ STOP ━━━━━━━━━━━━━━━ STOP ━━━━━

midpoint; limits

56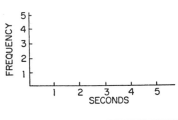

Three rats ran a maze in 2 seconds,
five rats ran it in 3 seconds, and two
rats ran it in 4 seconds. Draw a frequency polygon to illustrate these data.

━━━━━ STOP ━━━━━━━━━━━━━━━ STOP ━━━━━

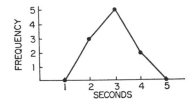

57

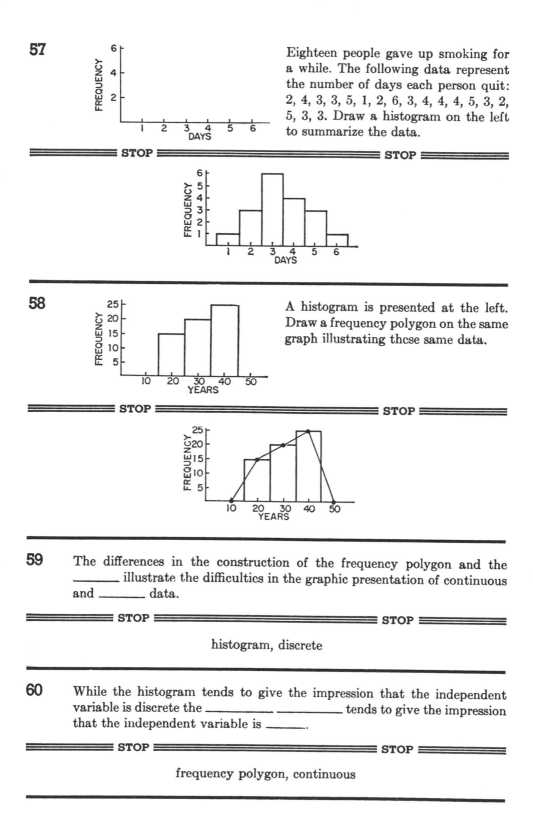

Eighteen people gave up smoking for a while. The following data represent the number of days each person quit: 2, 4, 3, 3, 5, 1, 2, 6, 3, 4, 4, 4, 5, 3, 2, 5, 3, 3. Draw a histogram on the left to summarize the data.

≡ STOP ≡≡≡≡≡≡≡≡≡≡≡≡≡≡≡≡≡ STOP ≡

58

A histogram is presented at the left. Draw a frequency polygon on the same graph illustrating these same data.

≡ STOP ≡≡≡≡≡≡≡≡≡≡≡≡≡≡≡≡≡ STOP ≡

59 The differences in the construction of the frequency polygon and the _____ illustrate the difficulties in the graphic presentation of continuous and _____ data.

≡ STOP ≡≡≡≡≡≡≡≡≡≡≡≡≡≡≡≡≡ STOP ≡

histogram, discrete

60 While the histogram tends to give the impression that the independent variable is discrete the _____ _____ tends to give the impression that the independent variable is _____.

≡ STOP ≡≡≡≡≡≡≡≡≡≡≡≡≡≡≡≡≡ STOP ≡

frequency polygon, continuous

61   Thus when the independent variable is thought to be _____, a histogram is preferred; but when the independent variable is thought to be _____, a frequency polygon is preferred.

════ STOP ════════════════ STOP ════

discrete; continuous

62   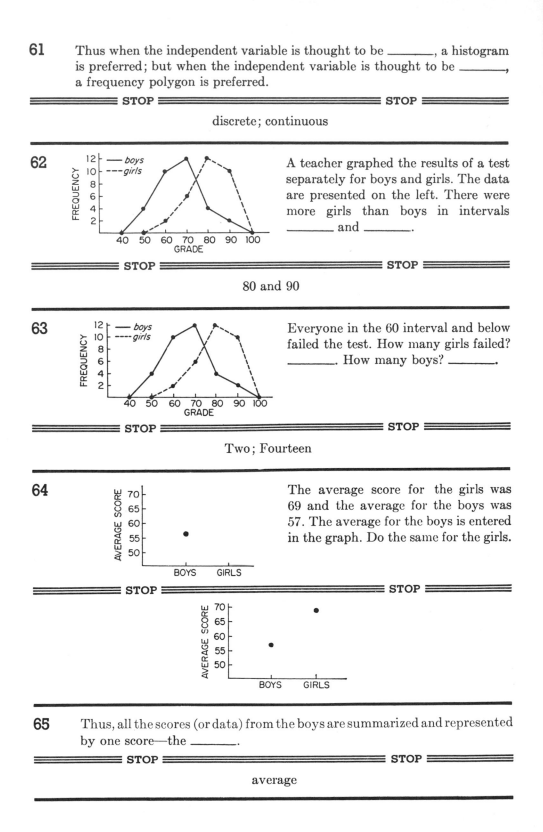 A teacher graphed the results of a test separately for boys and girls. The data are presented on the left. There were more girls than boys in intervals _____ and _____.

════ STOP ════════════════ STOP ════

80 and 90

63   Everyone in the 60 interval and below failed the test. How many girls failed? _____. How many boys? _____.

════ STOP ════════════════ STOP ════

Two; Fourteen

64   The average score for the girls was 69 and the average for the boys was 57. The average for the boys is entered in the graph. Do the same for the girls.

════ STOP ════════════════ STOP ════

65   Thus, all the scores (or data) from the boys are summarized and represented by one score—the _____.

════ STOP ════════════════ STOP ════

average

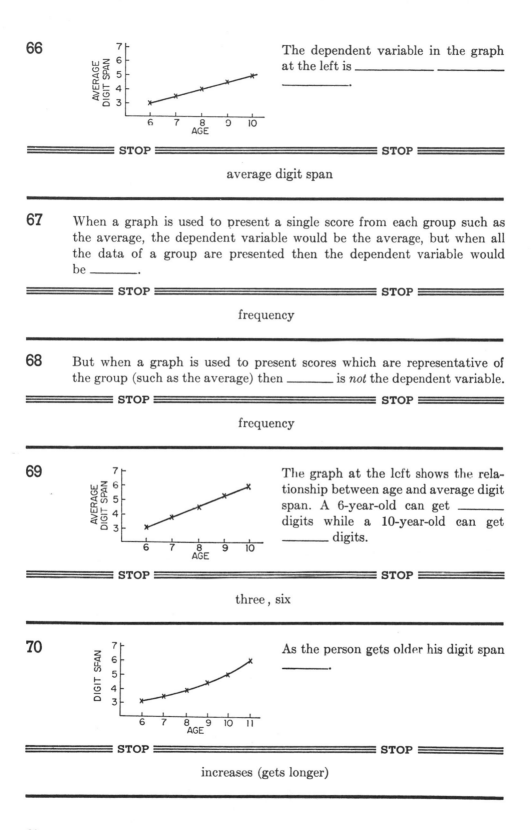

66

The dependent variable in the graph at the left is _____ _____ _____.

≡≡≡≡≡ STOP ≡≡≡≡≡≡≡≡≡≡≡≡≡≡≡≡≡≡≡≡≡ STOP ≡≡≡≡≡

average digit span

---

67 When a graph is used to present a single score from each group such as the average, the dependent variable would be the average, but when all the data of a group are presented then the dependent variable would be _____.

≡≡≡≡≡ STOP ≡≡≡≡≡≡≡≡≡≡≡≡≡≡≡≡≡≡≡≡≡ STOP ≡≡≡≡≡

frequency

---

68 But when a graph is used to present scores which are representative of the group (such as the average) then _____ is *not* the dependent variable.

≡≡≡≡≡ STOP ≡≡≡≡≡≡≡≡≡≡≡≡≡≡≡≡≡≡≡≡≡ STOP ≡≡≡≡≡

frequency

69

The graph at the left shows the relationship between age and average digit span. A 6-year-old can get _____ digits while a 10-year-old can get _____ digits.

≡≡≡≡≡ STOP ≡≡≡≡≡≡≡≡≡≡≡≡≡≡≡≡≡≡≡≡≡ STOP ≡≡≡≡≡

three , six

70

As the person gets older his digit span _____.

≡≡≡≡≡ STOP ≡≡≡≡≡≡≡≡≡≡≡≡≡≡≡≡≡≡≡≡≡ STOP ≡≡≡≡≡

increases (gets longer)

71

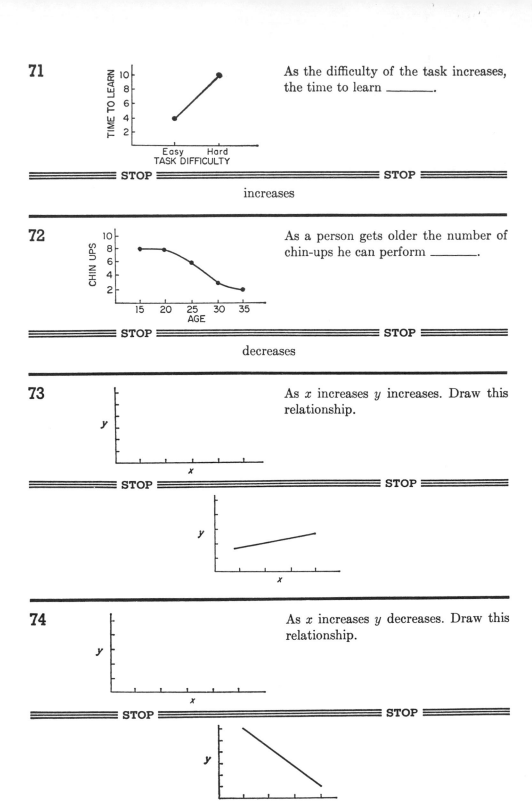

As the difficulty of the task increases, the time to learn _____.

═══════ STOP ═══════════════════════════ STOP ═══════

increases

72

As a person gets older the number of chin-ups he can perform _____.

═══════ STOP ═══════════════════════════ STOP ═══════

decreases

73

As *x* increases *y* increases. Draw this relationship.

═══════ STOP ═══════════════════════════ STOP ═══════

74

As *x* increases *y* decreases. Draw this relationship.

═══════ STOP ═══════════════════════════ STOP ═══════

75

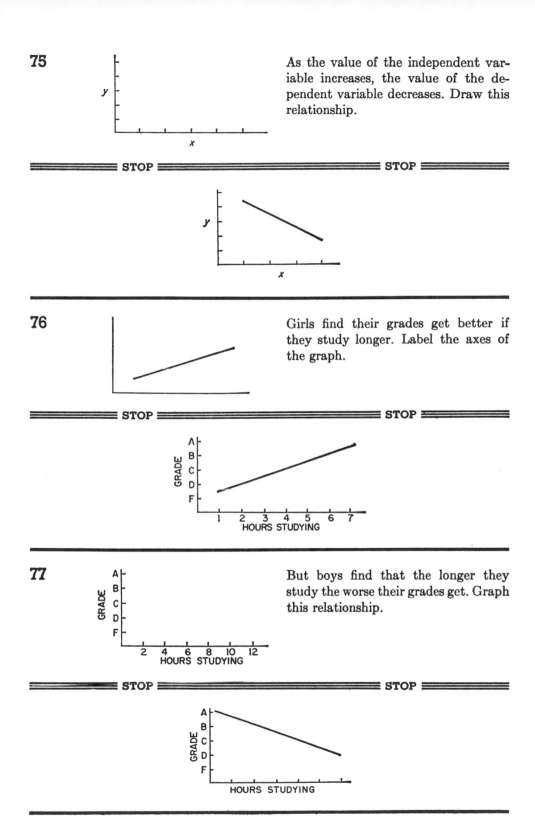

As the value of the independent variable increases, the value of the dependent variable decreases. Draw this relationship.

═══ STOP ═══ STOP ═══

76

Girls find their grades get better if they study longer. Label the axes of the graph.

═══ STOP ═══ STOP ═══

77

But boys find that the longer they study the worse their grades get. Graph this relationship.

═══ STOP ═══ STOP ═══

78

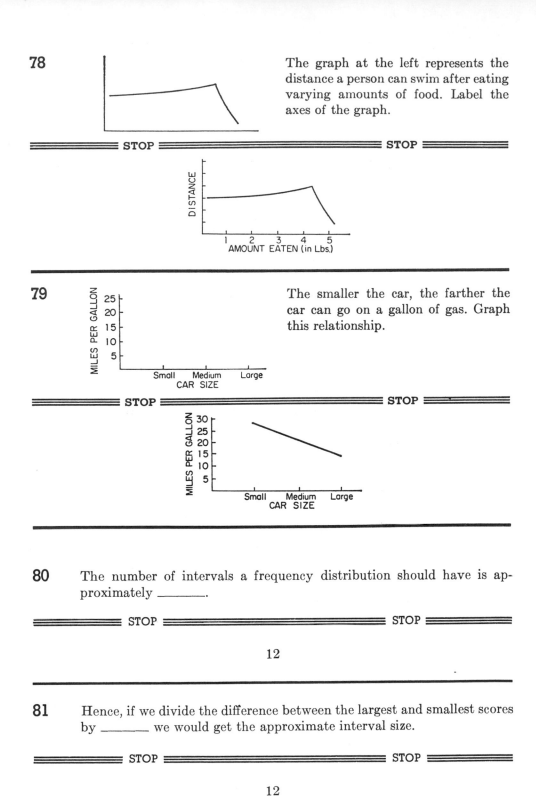

The graph at the left represents the distance a person can swim after eating varying amounts of food. Label the axes of the graph.

══════ STOP ═══════════════════════════ STOP ══════

79

The smaller the car, the farther the car can go on a gallon of gas. Graph this relationship.

══════ STOP ═══════════════════════════ STOP ══════

80   The number of intervals a frequency distribution should have is approximately _____.

══════ STOP ═══════════════════════════ STOP ══════

12

81   Hence, if we divide the difference between the largest and smallest scores by _____ we would get the approximate interval size.

══════ STOP ═══════════════════════════ STOP ══════

12

82   The difference between the largest and smallest score in a certain distribution is 100. The approximate interval size would be _____.

══════════ STOP ══════════════════════════ STOP ══════════

8.33

---

83   However, 8.33 is too cumbersome a number for an interval size. A more convenient interval size would be _____.

══════════ STOP ══════════════════════════ STOP ══════════

10

---

84   The lower apparent limit should be divisible by the interval size. If 32 is the smallest score in the distribution and the interval size is 10, then what are the apparent limits of the lowest interval?

══════════ STOP ══════════════════════════ STOP ══════════

30–39

---

85   With an interval size of 20, the lowest score in a certain distribution of scores is 57. What are the apparent limits of the lowest interval?

══════════ STOP ══════════════════════════ STOP ══════════

40–59

---

86   If 40–59 are the apparent limits, what are the real limits?

══════════ STOP ══════════════════════════ STOP ══════════

39.5–59.5

---

87   The apparent limits of another interval are 90–94. What are the real limits?

══════════ STOP ══════════════════════════ STOP ══════════

89.5–94.5

---

88 What is the midpoint of the 90–94 interval?

═══════ STOP ═══════════════════════ STOP ═══════

92

---

89 The highest and lowest scores in a certain distribution are 77 and 12. The appropriate interval size would be _____.

═══════ STOP ═══════════════════════ STOP ═══════

$^{65}\!/_{12}$ is about 5

---

90 The apparent limits of this lowest interval would be _____.

═══════ STOP ═══════════════════════ STOP ═══════

10–14

---

91 The real limits of that same interval would be _____.

═══════ STOP ═══════════════════════ STOP ═══════

9.5–14.5

---

92 The midpoint of that interval would be _____.

═══════ STOP ═══════════════════════ STOP ═══════

12

Three

Notation

Single Summation

Statistics requires the use of many numbers. To facilitate the use of these numbers, which may represent almost anything, it is necessary to have a system of notation that enables the statistician to indicate that a certain set of operations is to be performed. For example, you are probably well acquainted with the operations when you see 3^4. (This, of course, simply stands for the expression $3 \times 3 \times 3 \times 3 = 81$.) This is an example from a mathematical system of notation.

A clear understanding of the symbols and formulas presented in this book will greatly enhance your understanding of the concepts to be discussed. Of particular importance are the formulas. A formula may be considered a type of "recipe," since it states in a particular kind of shorthand form the sequence of steps necessary to obtain a particular statistic. Let us borrow the formula for the mean from the next chapter to use as an example:

$$M = \frac{\Sigma X}{N} \quad \text{or} \quad M = \Sigma X / N$$

As you will see in the program shortly, this formula may be translated as follows: To obtain the mean (M), find the sum of (Σ) all the values of the variable X and then divide this sum (ΣX) by the number of values (N). Note that the bar in the first formula and the slash in the second both indicate division.

The symbols X and Y are both used in statistics to indicate a variable. While a constant is always equal to the same value, a variable may take on different values. If we were to measure the heights of a group of six-year-old children, for example, height would be a variable and the value of the variable would change from individual to individual.

If X is used to indicate a variable or a group of scores, then we may use a subscript to indicate a specific score. For example, X_5 means the fifth score in the X group. In some situations we may wish to indicate that we are referring to a specific value of a variable, without stating which one; in such cases we use a symbolic subscript. For instance, X_i indicates that we are referring to a single value (the "*i*th") of the variable X and not to the entire group of scores.

In the formulas presented in this book you will frequently encounter a symbol that you may not have seen before in mathematics courses. This symbol Σ is the Greek capital letter sigma and is used to indicate summation. This symbol indicates an operation to be performed just as $+$, $-$, $/$, and $\times$ indicate operations. When there is no danger of ambiguity about the number of values to be summed, the symbol is used as shown in the previous example. However, when there might be the possibility of some ambiguity, the following notation is used:

$$ M_x = \frac{\sum\limits_{1}^{N} X}{N} $$

The 1 appearing below the Σ and the N above it indicate that the values in the group of X scores are to be summed from 1 to N. That is, all the X scores from X_1 to X_N are to be summed.

Note that we have also added a subscript to M. Just as we use $\sum\limits_{1}^{N} X$ instead of ΣX when there may be some ambiguity, we will use M_x instead of M when we wish to indicate where M came from. Thus, M_x would indicate the mean of the X scores, and M_y would indicate the mean of the Y scores.

PROBLEMS

1. Given 2, 4, 8, 5, 3, find $\sum\limits_{1}^{N} X$.

2. Given 12, 0, 1, 3, 18 find (a) $\sum\limits_{1}^{4} X$; (b) $\sum\limits_{2}^{3} X$; (c) $\sum\limits_{1}^{N} X$.

3. Given 0, 1, 3, 2, find $\sum\limits_{1}^{N} X^2$.

4. Given 4, 5, 8, 4, 1, 6, find (a) $\sum\limits_{1}^{3} X^2$; (b) $\sum\limits_{1}^{N} X^2$.

5. Given 1, 3, 5, 2, find (a) $\sum\limits_{1}^{N} X$; (b) $\sum\limits_{1}^{N} X^2$; (c) $\sum\limits_{1}^{N} X^2 - \sum\limits_{1}^{N} X$.

6. Given 2, 0, 4, 6, 1, find (a) $\sum\limits_{1}^{N} X^2$; (b) $\left(\sum\limits_{1}^{N} X\right)^2$; (c) $\sum\limits_{1}^{N} X^2 - \left(\sum\limits_{1}^{N} X\right)^2$.

7. Given 0, 1, 2, 3, 5, 7, find (a) $\sum\limits_{1}^{N} X^2$; (b) $\dfrac{\left(\sum\limits_{1}^{N} X\right)^2}{N}$; (c) $\sum\limits_{1}^{N} X^2 - \dfrac{\left(\sum\limits_{1}^{N} X\right)^2}{N}$.

8. Given 1, 8, 6, 1, 7, 7, find (a) $\sum\limits_{1}^{N} X^2$; (b) $\dfrac{\left(\sum\limits_{1}^{N} X\right)^2}{N}$; (c) $\sum\limits_{1}^{N} X^2 - \dfrac{\left(\sum\limits_{1}^{N} X\right)^2}{N}$.

9. The mean is defined as $\dfrac{\sum\limits_{1}^{N} X}{N}$. Find the mean of the following numbers: 12, 6, 4, 8, 9, 1, 9.

10. The variance is defined as

$$\frac{\sum\limits_{1}^{N} X^2 - \dfrac{\left(\sum\limits_{1}^{N} X\right)^2}{N}}{N-1}.$$

Find the variance of the following numbers: 0, 6, 2, 4, 3.

11. Given 4, 3, 8, 7, 2, 6, find the mean.
12. Given 2, 6, 6, 6, find the variance.
13. Given 0, 2, 5, 7, 8, find the mean.
14. Given 0, 5, 7, 2, 1, find the variance.

1   Height, weight, income, and so on, are all variables. Age would be an example of a _____.

═════════ STOP ═══════════════════════════════ STOP ═════════

variable

---

2   The statistician collects data from variables and uses numbers to represent such data. For example, 10 inches, 14 inches, and 18 inches would be called _____.

═════════ STOP ═══════════════════════════════ STOP ═════════

data

---

3   If the statistician were interested in the variable of age, his data would be collected in terms of _____.

═════════ STOP ═══════════════════════════════ STOP ═════════

years (months, and so on)

---

4   Three months, 12 years, 42 days would all be examples of scores from the variable of _____.

═════════ STOP ═══════════════════════════════ STOP ═════════

time (age)

---

5   Suppose the statistician has collected five scores concerned with the variable of height (in inches): 72, 60, 67, 62, 74. These scores would be called _____.

═════════ STOP ═══════════════════════════════ STOP ═════════

data

---

6   Thus, numbers of trials to learn a list, times to run a maze, IQ scores, would all be _____.

═════════ STOP ═══════════════════════════════ STOP ═════════

data

---

7   A convenient way of indicating a variable is by use of a letter such as X or Y. For example, the variable of age could be denoted _____ and the height variable _____.

═════════ STOP ═══════════════════════════════ STOP ═════════

X, Y(or Y, X)

---

8 The five height scores mentioned before could be called X scores since they come from a variable which was arbitrarily given the notation _____.

═══════ STOP ═══════════════════════ STOP ═══════

X

---

9 Subscripts are commonly used to specify a certain score. Using the five height scores 72, 60, 67, 62, and 74, X_2 would refer to the second score 60. X_4 would refer to _____.

═══════ STOP ═══════════════════════ STOP ═══════

the fourth score (62)

---

10 If there are N scores in a distribution then the first score would be denoted X_1 while the last score would be denoted _____.

═══════ STOP ═══════════════════════ STOP ═══════

X_N

---

11 A statistician may wish to indicate the score of an individual i but does not know where he is in the group of scores. If such is the case then X_i is used to denote the ith individual. The score of individual j could be denoted as _____.

═══════ STOP ═══════════════════════ STOP ═══════

X_j

---

12 A group of X_i scores is commonly called a *distribution of scores*. Thus if we collected a group of data concerned with the heights of fifth grade school children we would have a _____ of scores.

═══════ STOP ═══════════════════════ STOP ═══════

distribution

---

13 The statistician may wish to perform certain operations on these scores. He might wish to find the sum of these scores. Thus, $72 + 60 + 67 + 74 + 62 =$ _____.

═══════ STOP ═══════════════════════ STOP ═══════

335

---

14 The method by which the summation of these scores is indicated is called *notation*. Thus $72 + 60 + 67 + 62 + 74$ is one form of _____.

════════ STOP ═══════════════════════════ STOP ═══════

notation

---

15 But, when there are many scores this form of notation is inconvenient. Thus the statistician would look for other forms of _____ which would be more convenient.

════════ STOP ═══════════════════════════ STOP ═══════

notation

---

16 A second form of notation to indicate the sum of the same group of scores (72, 60, 67, 62, 74) would be $X_1 + X_2 + X_3 + X_4 + X_5$. X_1 would be equal to _____.

════════ STOP ═══════════════════════════ STOP ═══════

72

---

17 If $X_1 = 72$, $X_2 = 60$, $X_3 = 67$, $X_4 = 62$, and $X_5 = 74$, then $X_1 + X_2 + X_3 + X_4 + X_5 =$ _____.

════════ STOP ═══════════════════════════ STOP ═══════

335

---

18 One form of notation, indicating the sum of a group of numbers, is the capital Greek letter sigma (Σ). Σ means _____ _____.

════════ STOP ═══════════════════════════ STOP ═══════

sum of

---

19 When Σ comes before a variable notation (such as X) then Σ means the sum of the scores. Thus ΣX means the _____ _____ the X scores.

════════ STOP ═══════════════════════════ STOP ═══════

sum of

---

20 In the previous example the scores were 72, 60, 67, 62, and 74. In that case, $\Sigma X =$ _____.

════════ STOP ═══════════════════════════ STOP ═══════

335

---

21 If the scores were 0, 2, 4, 0, 6 then $\sum X = $ _____.

══════ STOP ══════════════════════ STOP ══════

12

---

22 Numbers below and above the $\sum$ indicate that all numbers between (and including) the two are to be summed. For example $\sum\limits_{3}^{6}$ would mean that the numbers between (and including) the third and the _____ numbers would be added.

══════ STOP ══════════════════════ STOP ══════

sixth

---

23 Suppose eight scores are available: 0, 0, 3, 2, 1, 4, 6, 2. Then $\sum\limits_{1}^{4} X = $ _____.

══════ STOP ══════════════════════ STOP ══════

5

---

24 Suppose six scores are available: 0, 0, 2, 3, 0, 5. Then $\sum\limits_{1}^{N} X = \sum\limits_{1}^{6} X = $ _____.

══════ STOP ══════════════════════ STOP ══════

10

---

25 Remember, N means number of scores and $\sum\limits_{1}^{N} X$ means the sum of all the X scores. Suppose ten scores are available. Then $N = $ _____.

══════ STOP ══════════════════════ STOP ══════

10

---

26 If five scores are available then $\sum\limits_{1}^{N} X$ means the sum of _____ _____ _____.

══════ STOP ══════════════════════ STOP ══════

all five scores

---

27 Here are scores 0, 3, 2, 1, 0, 4, 3. $\sum\limits_{1}^{N} X = $ _____.

══════ STOP ══════════════════════ STOP ══════

13

28 Here are some other scores: 0, 3, 2, 1, 3, 2. $N =$ _____, $\sum_{1}^{5} X =$ _____.

══════ STOP ══════════════════ STOP ══════

6, 9

---

29 Here are still other scores: 0, 3, 1, 4, 0, 2, 6. $\sum_{3}^{5} X =$ _____.

══════ STOP ══════════════════ STOP ══════

5

---

30 The statistician needs to square his scores. To square means to multiply a number by itself. The square of 4 is equal to _____.

══════ STOP ══════════════════ STOP ══════

$4 \times 4 = 16$

---

31 The notation for a square is called a *superscript*. For example if one wished to indicate that 4 is to be squared then 4^2 would be the notation employed. The notation indicating 8 is to be squared would then be _____.

══════ STOP ══════════════════ STOP ══════

8^2

---

32 If we selected a score X then the square of X could be indicated by _____.

══════ STOP ══════════════════ STOP ══════

X^2

---

33 Suppose we have a group of scores: 0, 1, 3, 2, 2. We could indicate the square of each of these by the notation _____, _____, _____, _____, and _____.

══════ STOP ══════════════════ STOP ══════

0^2, 1^2, 3^2, 2^2, and 2^2

---

34 If we had a group of numbers 0, 1, 3, 2, 2 and we wished to find the sum of the squared numbers, then one type of notation would be $0^2 + 1^2 + 3^2 + 2^2 + 2^2$. Or if we had a group of scores X_1, X_2, X_3 we could indicate the sum of these squares by the notation _____ + _____ + _____.

══════ STOP ══════════════════ STOP ══════

$X_1^2 + X_2^2 + X_3^2$

---

35 However, a more convenient notation would be $\sum_1^5 X^2$ since $\sum$, as we know, means *sum of*, then the notation $\sum_1^5 X^2$ means the sum of the _____ of the five scores.

══════ STOP ══════════════ STOP ══════

squares

───────────────────────────────────────

36 Suppose we have a group of numbers 0, 1, 2, 2, 3, 1. The notation $\sum_1^6 X^2$ means that you must _____ each of the six scores and *then* sum them.

══════ STOP ══════════════ STOP ══════

square

───────────────────────────────────────

37 Suppose we have a second group of numbers: 3, 1, 1, 2, 2, 4. Then $\sum_1^N X^2 =$ _____, $\sum_1^3 X^2 =$ _____.

══════ STOP ══════════════ STOP ══════

35, 11

───────────────────────────────────────

38 A third group of numbers: 0, 1, 2, 3, 1, 3. Then $\sum_1^N X^2 =$ _____.

══════ STOP ══════════════ STOP ══════

24

───────────────────────────────────────

39 However there are other operations which the statistician must perform. One of these is finding the square of a sum. If the sum of 0, 1, 3, 2 is equal to 6 then the square of this *sum* is equal to _____.

══════ STOP ══════════════ STOP ══════

36

───────────────────────────────────────

40 To find the square of the sum of 0, 2, 1, 4, 3 you must first sum the scores, which equals _____, then you must square this sum, and this equals _____.

══════ STOP ══════════════ STOP ══════

10, 100

───────────────────────────────────────

41 $(\sum\limits_{1}^{N} X)^2$ is used to indicate the square of a sum. For example, with scores 0, 1, 1, 2, 1, 1, $\sum\limits_{1}^{N} X =$ _____, and $(\sum\limits_{1}^{N} X)^2 =$ _____.

$\equiv$ STOP $\equiv$ STOP $\equiv$

6, 6 × 6 = 36

---

42 You should note that $\sum\limits_{1}^{N} X^2$ and $(\sum\limits_{1}^{N} X)^2$ are quite different. $\sum\limits_{1}^{N} X^2$ (read sum X^2) means to find the sum of a group of squares while $(\sum\limits_{1}^{N} X)^2$ (read sum X, the quantity squared) means _____ _____ _____ _____ _____ _____ _____ _____ _____.

$\equiv$ STOP $\equiv$ STOP $\equiv$

to square the sum of a group of scores

---

43 If we have a group of scores 0, 1, 3, 2, 3, 1. Then $\sum\limits_{1}^{N} X = 10$, $\sum\limits_{1}^{N} X^2 = 24$, and $(\sum\limits_{1}^{N} X)^2 =$ _____.

$\equiv$ STOP $\equiv$ STOP $\equiv$

100

---

44 If we have another group of scores 1, 1, 3, 2, 0, we find $\sum\limits_{1}^{N} X = 7$, $\sum\limits_{1}^{N} X^2 =$ _____, and $(\sum\limits_{1}^{N} X)^2 =$ _____.

$\equiv$ STOP $\equiv$ STOP $\equiv$

15, 49

---

45 In another group of scores 0, 0, 1, 2, 2, 2. $\sum\limits_{1}^{N} X =$ _____, $\sum\limits_{1}^{N} X^2 =$ _____, and $(\sum\limits_{1}^{N} X)^2 =$ _____.

$\equiv$ STOP $\equiv$ STOP $\equiv$

7, 13, 49

---

46 Often the statistician finds it necessary to perform other operations on his data. For example he may wish to divide. Suppose he has scores 0, 1, 2, 2, 4, 1 and wishes to divide by N. N in this case = _____.

$\equiv$ STOP $\equiv$ STOP $\equiv$

6

47 If he has a group of scores and he wishes to divide the sum of these scores by the number of scores he would then divide $\sum\limits_{1}^{N} X$ by _____.

═══════ STOP ═══════════════════════════ STOP ═══════

$$N$$

---

48 10/5 means that 5 is divided into 10. Thus, $\sum\limits_{1}^{N} X/N$ means that N is divided into _____.

═══════ STOP ═══════════════════════════ STOP ═══════

$$\sum\limits_{1}^{N} X$$

---

49 If you have scores 1, 2, 3, 0, 2, 4 then $\sum\limits_{1}^{N} X = 12$ and $\sum\limits_{1}^{N} X/N =$ _____.

═══════ STOP ═══════════════════════════ STOP ═══════

$$12/6 = 2$$

---

50 If you have scores 1, 2, 2, 5, 0 then $\sum\limits_{1}^{N} X/N =$ _____.

═══════ STOP ═══════════════════════════ STOP ═══════

$$10/5 = 2$$

---

51 With scores 3, 0, 1, 4. Then $\sum\limits_{1}^{N} X/N =$ _____.

═══════ STOP ═══════════════════════════ STOP ═══════

$$8/4 = 2$$

---

52 With scores 3, 0, 5, 2, 1, 4. Then $\sum\limits_{1}^{N} X/N =$ _____.

═══════ STOP ═══════════════════════════ STOP ═══════

$$15/6 = 2.50$$

---

53 Finally, with scores 3, 0, 1, 2, 4, 8. Then $\sum\limits_{1}^{N} X/N =$ _____.

═══════ STOP ═══════════════════════════ STOP ═══════

$$18/6 = 3$$

---

54 Suppose you had scores 3, 0, 1, 2, 4. Then $\sum_1^N X = 10$, $(\sum_1^N X)^2 = $ _____.

═══ STOP ═══════════════════════════ STOP ═══

$$100$$

---

55 If $N = 5$ and if $(\sum_1^N X)^2 = 100$, then $(\sum_1^N X)^2/N = $ _____.

═══ STOP ═══════════════════════════ STOP ═══

$$100/5 = 20$$

---

56 If you have scores 2, 1, 3, 2 then $\sum_1^N X = 8$, $(\sum_1^N X)^2 = 64$, and $(\sum_1^N X)^2/N = $

_____.

═══ STOP ═══════════════════════════ STOP ═══

$$64/4 = 16$$

---

57 Remember $(\sum_1^N X)^2$ and $\sum_1^N X^2$ are quite different. If you have scores 0, 1, 3, 2, 4 then $(\sum_1^N X)^2 = 100$, but $\sum_1^N X^2 = $ _____.

═══ STOP ═══════════════════════════ STOP ═══

$$30$$

---

58 So, too, do $(\sum_1^N X)^2/N$ and $\sum_1^N X^2/N$ have much different meanings. If you have scores 3, 0, 1, 4 then $(\sum_1^N X)^2/N = 64/4 = 16$, while $\sum_1^N X^2/N = $

_____.

═══ STOP ═══════════════════════════ STOP ═══

$$26/4 = 6.50$$

---

59 With scores of 0, 1, 1, 4, 4, 2 then $\sum_1^N X = 12$, $\sum_1^N X/N = 2$, $\sum_1^N X^2/N = $ $38/6 = 6.33$, and $(\sum_1^N X)^2/N = $ _____.

═══ STOP ═══════════════════════════ STOP ═══

$$144/6 = 24$$

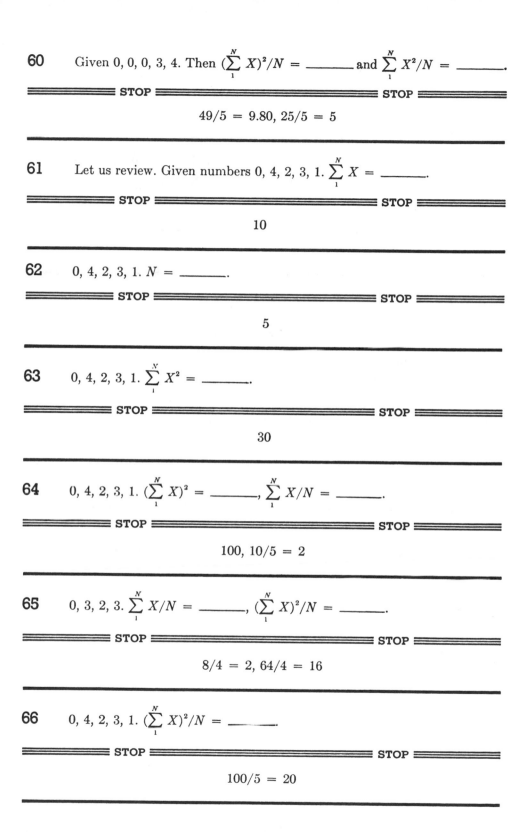

60 Given 0, 0, 0, 3, 4. Then $(\sum_1^N X)^2/N =$ _____ and $\sum_1^N X^2/N =$ _____.

═══ STOP ═══════════════════════════ STOP ═══

$$49/5 = 9.80, \; 25/5 = 5$$

---

61 Let us review. Given numbers 0, 4, 2, 3, 1. $\sum_1^N X =$ _____.

═══ STOP ═══════════════════════════ STOP ═══

10

---

62 0, 4, 2, 3, 1. $N =$ _____.

═══ STOP ═══════════════════════════ STOP ═══

5

---

63 0, 4, 2, 3, 1. $\sum_1^N X^2 =$ _____.

═══ STOP ═══════════════════════════ STOP ═══

30

---

64 0, 4, 2, 3, 1. $(\sum_1^N X)^2 =$ _____, $\sum_1^N X/N =$ _____.

═══ STOP ═══════════════════════════ STOP ═══

$$100, \; 10/5 = 2$$

---

65 0, 3, 2, 3. $\sum_1^N X/N =$ _____, $(\sum_1^N X)^2/N =$ _____.

═══ STOP ═══════════════════════════ STOP ═══

$$8/4 = 2, \; 64/4 = 16$$

---

66 0, 4, 2, 3, 1. $(\sum_1^N X)^2/N =$ _____.

═══ STOP ═══════════════════════════ STOP ═══

$$100/5 = 20$$

---

67   0, 4, 2, 3, 1. $\sum\limits_{1}^{N} X^2/N =$ _____.

≡≡≡≡≡ STOP ≡≡≡≡≡≡≡≡≡≡≡≡≡≡≡≡≡≡≡≡≡≡≡≡≡≡ STOP ≡≡≡≡≡

$$30/5 = 6$$

---

68   Simplicity in notation tends to be preferred. For example, when there will be no confusion, we would prefer to use $\sum X$ instead of $\sum\limits_{1}^{N} X$. Given scores 0, 1, 3, 5. $\sum X =$ _____.

≡≡≡≡≡ STOP ≡≡≡≡≡≡≡≡≡≡≡≡≡≡≡≡≡≡≡≡≡≡≡≡≡≡ STOP ≡≡≡≡≡

9

---

69   With scores 1, 2, 4. $\sum\limits_{1}^{N} X^2 =$ _____ and $\sum X^2 =$ _____.

≡≡≡≡≡ STOP ≡≡≡≡≡≡≡≡≡≡≡≡≡≡≡≡≡≡≡≡≡≡≡≡≡≡ STOP ≡≡≡≡≡

21, 21

---

70   With scores 2, 3, 1. $\sum X^2 =$ _____ and $(\sum X)^2 =$ _____.

≡≡≡≡≡ STOP ≡≡≡≡≡≡≡≡≡≡≡≡≡≡≡≡≡≡≡≡≡≡≡≡≡≡ STOP ≡≡≡≡≡

14, 36

---

71   With scores 0, 1, 5, 2. $\sum X^2 =$ _____ and $(\sum X)^2 =$ _____.

≡≡≡≡≡ STOP ≡≡≡≡≡≡≡≡≡≡≡≡≡≡≡≡≡≡≡≡≡≡≡≡≡≡ STOP ≡≡≡≡≡

30, 64

Four

Measures of Central Tendency

A middle-aged man needing a haircut accidentally visited a hair stylist. A woman, instead of a male barber, cut and styled his hair, and he was unsure about what to expect. But he was very pleased with his new hairstyle, which completely covered his bald spot. After paying the fee and realizing he was expected to give her a tip, he inquired as to the amount of the average tip. The hair stylist replied, "five dollars." Conforming as best he could, he tipped her the five dollars. Then she replied, "Thanks, mister, that's the first time anyone ever came up to my average."

Clearly, his definition of *average* differed from her definition just as statisticians differ in their definition of average. The term itself is sufficiently unclear, as illustrated above, that statisticians prefer to use the phrase *measure of central tendency.*

We mentioned earlier that the two major types of statistics are *descriptive* and *inferential.* Before we can begin the study of inferential statistics, we must learn the techniques used to describe the characteristics of the data from which we wish to draw inferences. In Chapter two we explained and illustrated how graphs may be used to describe distributions of data. In this chapter we will discuss one of two very important summary measures for describing a set of data.

In practice, groups of scores obtained from such dependent variables as the size of fish in a certain lake or the number of auto accidents over a holiday weekend can be adequately described by two summary scores. That is, once we know the value of these two summary scores, we have sufficient information to provide a fairly accurate description of the whole distribution of individual scores. The first of these summary scores is a measure of central tendency, and the second is a measure of variability.

There are additional summary scores that are used when an increasingly accurate description of a distribution of scores is desired. (For example, one summary measure describes the degree of symmetry in a distribution of scores.) However, these are rather infrequently used, and for this reason they are beyond the scope of this book. In this chapter we will discuss the measures of central tendency most commonly used in descriptive and inferential statistics, and in the following chapter we will present the analogous measures of variability.

While each measure of central tendency is defined in its own way, all of them may be considered different methods of finding that which is commonly called the *average*. Once we have computed the average, we may use it to describe the set of data from which it was taken, or, if the set of data is thought to be representative of a larger group of scores, the average may be used to infer the characteristic score of the larger group. For example, suppose we select a group of 20 from a large class of 400 students by a coin-flipping procedure. Then we have each of the 20 students take a test. We score the test and compute the average. If our interest is only in the 20 students, we will use the average to *describe* their performance on the test; but if our interest were to go beyond the 20 students to the 400 students in the class, then we would use the average of the 20 scores to *infer* what the average score of all 400 students would be if we gave each one the test. Thus, the same descriptive statistic can be used to describe a set of data or to make inferences about a larger group. Usually, however, behavioral scientists do not wish to make inferences to a restricted larger group, but to a general population such as "all college students." Often we find that the distinction between a descriptive and an inferential statistic lies in the use to which it is put.

We have been using the word *average* in our discussion with no specific definition because it is a well-known word meaning much the same thing as *measure of central tendency*. At this point we will be somewhat more precise and define measure of central tendency as an attempt to identify the most characteristic score in a group of scores. It answers the question, "What is the single score that best describes all of the scores in the distribution?" There are three common measures of central tendency: the *mode*, the *median*, and the *mean*. Although each of these measures of central tendency is an average, each is defined in a different way and each is used under different circumstances.

THE MODE

The mode is the easiest of the three measures to obtain, and it is also most subject to fluctuation when the values of a few scores are changed. For this reason the mode is typically used only for a quick estimate of the central tendency of a distribution. The mode is defined as *the most frequent score in the distribution*. In Table 4-1 each mode is the score that appears more often than any other score in the distribution.

Suppose that a ten-question quiz was given to a class and that the distribution of scores for the 17 students was as shown in Table 4-1, column A. The instructor used the mode as a measure of central tendency, and announced that all scores below this point would be given a grade of *C*. You were one of the two people with a score of 5, and breathed a sigh of relief. Then the teacher found two errors in scoring and added one point to each

Table 4-1 Modes in frequency distributions ($N = 17$)

(A)		(B)	
Score	Frequency	Score	Frequency
9	1	9	1
8	1	8	1
7	3	7 (mode)	4
6	1	6	0
5	2	5	2
4	1	4	1
3	1	3	2
2 (*mode*)	4	2	3
1	2	1	2
0	1	0	1

of these scores, changing one score from 2 to 3 and another score from 6 to 7. Table 4-1, column B, shows the drastic effect these small changes had on the mode; it rose from 2 to 7. It also had a drastic effect upon your grade, which was then in the C range. The mode may be easy to obtain, but it is also subject to extreme fluctuation with small changes in a few scores.

THE MEDIAN

The median is defined as *that point on a scale below which 50 percent of the cases fall*. The median is not a score, but a point on a scale dividing the upper and lower half of the cases.

Figure 4-1 is a histogram that represents the distribution of number of aggressive play responses found in a group of closely related puppies. Each puppy was observed for a 3-minute period, and the number of aggressive responses made during this time was counted. Computation of the median simply involves finding the middle score; that is, the score below which 50

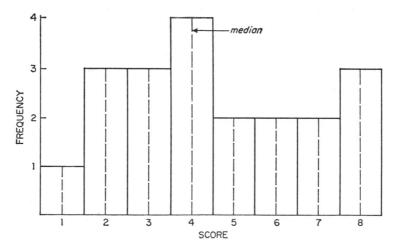

Figure 4-1 *Frequency distribution with median.*

percent of the cases fall. In Figure 4-1 it can be seen by counting and summing the frequencies at each interval that there were a total of 20 puppies (that is, $N = 20$). Since the median is the point below which 50 percent of the cases fall, then $20 \times 0.50 = 10$; thus, ten cases fall below the median. Counting up from the lowest scores, we find we have $1 + 3 + 3 = 7$ cases below a score value of 4, and $7 + 4 = 11$ cases below 5. Hence, the tenth score falls in the 3.5–4.5 interval and the median is 4. If desired, interpolation can be used to determine a single point within the interval, which can be called the *median,* although in this context we will define the median as the *midpoint* of the interval containing the middle score or, similarly, the midpoint of the interval containing that point below which 50 percent of the cases fall. In our example the median number of aggressive play responses in a 3-minute period has been found to be 4.

Alternately, if exactly 50 percent of the cases fall below a certain point, then that point is the median. For example, suppose six students got the following scores on a test: 8, 10, 5, 7, 9, 10; arranging the scores in rank order 5, 7, 8, 9, 10, 10, we find that exactly half (three) fall below 8.50 (the upper limit of the 8 interval). Hence, 8.50 is the median. Similarly, if cases are widely dispersed and there is no middle score (as would be the situation with an even number of cases), another problem arises. For example, in another test the scores of 75, 92, 83, 100, 65, and 89 were found; when ranked as 65, 75, 83, 89, 92, 100, it can be seen that 50 percent of the cases fall at 83.5 and below, while another 50 percent fall at 88.5 and above. All the points between 83.5 and 88.5 meet the definition of a median. Therefore, the convention that has been adopted is to take the midpoint of all those scores which could be the median. In this case that would be 86.0, which can be found by adding the two scores in question and dividing by 2:
$(88.5 + 83.5)/2 = 86.0$

The median can be easily computed when the interval size is greater than 1.00. For example, suppose a research worker was interested in determining the cephalic index (the width of the head expressed as a percentage of the length of the head). People with an index under 80 are called *long headed;* over 80, *broad* or *round headed.* These data are presented in Table 4-2 for a group of people living in a mountain village. It can be seen by summing the frequency column (f) in the table that 146 adults were measured. The median then is defined as the midpoint of the interval below which 50 percent (that is, $146 \times 0.50 = 73$ cases) fall. Counting up from the bottom, we find that 71 cases fall below the upper limit of the 74–75 interval, and 91 cases fall below the 76–77 interval. Hence, the median falls in the 76–77 interval. Since we have said that the midpoint of that interval is the median, then 76.5 is defined as the median.

THE MEAN

The mean is the measure of central tendency that most closely approximates what is commonly meant by *average.* The arithmetic mean, usually called

Table 4-2   Frequency distribution of adult head sizes in a mountain village

X	f
88–89	1
86–87	5
84–85	8
82–83	10
80–81	14
78–79	17
76–77	20
74–75	18
72–73	15
70–71	10
68–69	9
66–67	7
64–65	7
62–63	4
60–61	1

simply *the mean,* is the most widely used measure of central tendency, and it is also the most laborious to compute because the value of every score in the distribution must be considered. The mean may be defined as *the sum of all the scores in the distribution divided by the total number of scores.* This definition is represented by the formula

$$M = \frac{\Sigma X}{N}$$

where M is the mean of the distribution of X scores, Σ means *sum of,* X signifies a score in the X distribution, and N is the number of scores in the distribution.

Suppose a therapist had been working with a group of eight patients, and she wished to determine the mean number of hours of therapy the patients had received. To find the mean, she first found the number of hours of therapy for each patient, as shown in Table 4-3.

The sum of the scores indicates that the eight patients were given a total of 53 hours of therapy (that is $\Sigma X = 53$). By dividing 8 into 53, she found that the mean number of hours of therapy given the patients within the group was 6.63.

The mean possesses an interesting characteristic, which is sometimes useful in checking its computation. This has to do with the sum of the deviation scores, where a deviation score is the difference between the "raw" score and the mean, or

$$x = X - M$$

where x is the deviation score, X is the raw score, and M is the mean of the distribution of X scores.

Table 4-3   Number of hours of therapy
(X) received by each of eight patients

$$X$$

4
12
5
3
9
3
10
7

$$\Sigma X = \overline{53}$$
$$N = 8$$

$$M = \frac{\Sigma X}{N} = \frac{53}{8} = 6.63$$

Suppose a research worker wished to determine the grade level achieved by the members of a small farm community. He interviewed the members of this community and obtained the data presented in Table 4-4.

Table 4-4   Mean and deviation scores
for grade level achieved in school

X	M	x
4	5	-1
5	5	0
3	5	-2
7	5	$+2$
4	5	-1
9	5	$+4$
3	5	-2
5	5	0
6	5	$+1$
4	5	-1
$\Sigma X = \overline{50}$		$\Sigma x = \overline{0}$
$N = 10$		

$$M = \frac{\Sigma X}{N} = \frac{50}{10} = 5$$

Table 4-4 contains three columns of numbers. In the column on the left (X) the scores (grade level achieved) are presented. In the middle column (M), 5 has been entered, representing the mean of the scores on the left. The scores in the right-hand column (x) were obtained by subtracting 5 (the mean) from each of the scores in the left column. You will find that the sum of the deviation scores (x) equals zero. The sum of the deviation scores about the mean will always equal zero. Thus, in addition to the mean being the sum of the scores divided by N, it is also the point where the sum of the deviation scores will equal zero. This characteristic of the mean will be used again in a later chapter.

COMPARISON OF MODE, MEDIAN, AND MEAN

The three measures of central tendency which we have been discussing— the mode, the median, and the mean—are *not* all equally useful alternatives for arriving at a measure of central tendency. On the contrary, each of these three measures has characteristics which make it advantageous to use under some conditions and most unfortunate if used under others. In this section we will contrast and compare the three measures so that you can get a better understanding of their characteristics.

Let us look at a couple of examples. These illustrate that the choice of mean, median, or mode depends in part on the level of measurement which has been used, as well as on the question being asked.

Suppose a political scientist looked at the different types of governments used in the world today. He classified them as belonging to one of five general types A, B, C, D, and E. On a certain continent, more type D governments were found than any other type. Hence, the modal type of government on that continent is *D*, and is the most characteristic. Notice that the various governments could be ranked as to degree of freedom accorded the individual, and the median could be found. However, such a ranking might reflect the bias of the rater, and therefore the political scientist felt more comfortable in simply using the mode.

A linguist interested in regional dialects identifies ten different ones, ranging from type Q spoken by 5 million people to type E spoken by 47 million people. The various dialects are spoken by 5, 7, 8, 10, 15, 19, 22, 23, 35, and 47 million people, respectively. Several questions can be asked about the data. What is the most characteristic dialect? The answer to this is the mode, that is, the most frequent dialect, *E*. Next, what number best describes the number of people speaking each dialect? Here the answer would be the median, 17 million, both because the data are skewed and because the numbers are estimates, which are probably more representative of an ordinal scale than an interval scale.

THE MODE   There are several conditions under which the mode would be the proper measure of central tendency to use. We mentioned before that the mode is the quickest of the three measures to compute. Therefore, when speed is the most important consideration, the mode would be the preferred measure of central tendency. We gave an illustration in Table 4-1 which showed that the mode can fluctuate wildly with small changes in a few scores. Hence, only when accuracy is not an especially important consideration should the mode be used.

The mode should be used when nominal data are being described. In our discussion of the nominal scale we used the numbers on football jerseys as an example. If the coach of some team wanted to describe the most characteristic type of player, he would use the mode. In this case, computation of the mean or median from the numerals on the backs of the football jerseys would yield a meaningless number.

THE MEDIAN   The median is the preferred measure of central tendency when an ordinal scale is used. In such a situation the score assigned to each case has no meaning other than to indicate that one score is greater than (or less than) another score. Under these conditions we would be most reluctant to compute the mean, and would prefer to use the median as a measure of central tendency. We would encounter just such a situation in an athletic league in which several teams participate. If we wish to identify those teams in the first division or second division, we would look at those teams which are above (or below) the median position. That is, we would be interested in those teams in the top half or in the bottom half of the team standings. We would not, on the other hand, be interested in a team's number of games won because we cannot determine a team's position in the league from knowledge of the number of games won.

Under certain conditions we may want to estimate central tendency when certain scores are indeterminant or missing. An example of this situation is occasionally encountered in learning experiments that use children as subjects. If the task to be learned is too difficult, many of the slower learners will become discouraged and refuse to continue the task. If the experimenter is interested in the time it takes to learn the task, he has two alternative methods for handling the data for children who quit. He may arbitrarily assign a value based on the number of trials already completed, or he may consider only those children who completed the task. If he computed the mean from his data, he would erroneously underestimate the value of the central tendency in either case. For example, if he computes the mean using the first procedure (assigning values based on trials completed), he has permitted low scores (fast learners) but not high scores (slow learners who become discouraged and quit). Since scores that are far from the mean are given much more weight than scores that are close to the mean, the low scores would tend to be given more weight than the high scores. The consequence of this is to lower the value of the mean.

On the other hand, if the experimenter simply does not consider the scores of the subjects who quit before they complete the task, the error becomes even greater. That is, the high scores are eliminated rather than underestimated, and an even lower estimate of the mean is given. The median does not give additional weight to a score simply because it is more extreme. Hence, the median would give a more accurate estimate of the central tendency of the distribution. Since the time it takes the slow learners is probably above the median before they quit without learning, it makes no difference if they continue until they learn. Hence, the median would give a more accurate estimate of the central tendency of a distribution in which extreme scores are missing for one reason or another.

A similar situation would be encountered, but with opposite consequences, if an experimenter used a group of very cooperative mentally retarded adults in a learning experiment. Even if only one of the adults could not learn the task, use of the mean would inflate the value of the measure of central tendency. The reason is that the score for an adult who could not learn is indeterminant, and the use of any score would underestimate the number of trials it would take him to learn the task. Again the proper measure of central tendency would be the median because it can be correctly computed without knowing the exact value of every score.

THE MEAN   The mean is the most difficult measure of central tendency to compute. It is used normally with interval- and ratio-scaled data. We noted in previous sections that the computation of the mean gives more weight to extreme scores than to scores near the mean. Hence, the mean should be used as a measure of central tendency only when the distribution of scores has approximately the same number of extremely high and low scores. That is, the mean should be used only when the distribution is symmetrical. However, when the distribution is symmetrical, both the median and mean will be approximately the same. Why, then, should we ever use the mean as a measure of central tendency? There are at least two answers to this question. First, the mean is the most stable measure of central tendency; changes of a few scores have little influence on its value. Second, the formula for the computation of the mean allows for subsequent algebraic manipulations of the data, which cannot be carried out with either the median or the mode. The most powerful methods of inferential statistics are based on algebraic manipulations utilizing the mean. Hence, when the interest of the research worker goes beyond description to statistical inference, the mean would be the preferred measure of central tendency.

The impression should not be given that once the scale of measurement—nominal, ordinal, interval, or ratio—has been determined, the proper measure of central tendency is also determined. While knowing which scale of measurement is being used may help to determine the most appropriate measure of central tendency, other considerations are also important.

The mode can be used with any of the measurement scales and is the only measure of central tendency which can be used meaningfully with a nominal scale. The median can be used with ordinal, interval, and ratio scales, while the mean can be used with interval and ratio scales. Thus all three, the mode, median, and mean, can be used with an interval scale. The proper measure of central tendency in this case is determined by considerations such as the ease of computation, the accuracy desired, and the necessity of drawing inferences from the data.

The important point to be understood here is that any statistic can be computed from any set of numerical data; the purpose of such computation, however, is to obtain a meaningful result, and this is what determines the choice of the statistic.

THE EFFECTS OF DISTRIBUTION SYMMETRY

Most distributions of scores commonly encountered are symmetrical, such that the number of extremely high scores is roughly equal to the number of extremely low scores. That is, the distribution of scores is evenly balanced around the middle of the distribution. For any distribution that is symmetrical, the median and mean will have the same value. If the symmetrical distribution has a single peak, the mode will also equal the median and mean. Such a distribution is shown in Figure 4-2A.

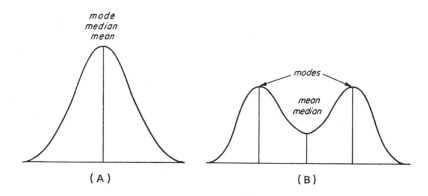

Figure 4-2   (A) *A unimodal symmetrical distribution, and* **(B)** *a bimodal symmetrical distribution.*

A distribution with a single peak has a single mode and is called *unimodal,* while one with two peaks is called *bimodal.* As is shown in Figure 4-2B, it is possible to have a bimodal symmetrical distribution. In this case, neither of the modes is equal to the mean and the median.

When a distribution is not symmetrical, it is called a *skewed* distribution, and it may be either positively or negatively skewed. If you remember that scores on the abscissa are typically lower on the left and higher on the right, then the "tail" of a positively skewed distribution points toward the right (positive) end of the distribution. Similarly, the "tail" of a negatively skewed distribution points toward the left (negative) side of the scale.

Figures 4-3A and 4-3B show why the skewness of a distribution is important in selecting the proper measure of central tendency. The mode, median, and mean all have different values if the distribution is skewed. Notice that the mean tends to be located toward the long-tailed end of the distribution. The reason for this is that the formula for the mean will weight extreme scores more heavily than it does scores nearer the mean. Thus, the few extreme scores have the effect of pulling the mean toward this end of the distribution. The mode is less influenced in this manner, and is located more toward the end of the distribution where the scores tend to pile up. It would not be an unusual case to find the mode equal to the highest (or lowest) score in the distribution. For example, on a moderately easy test,

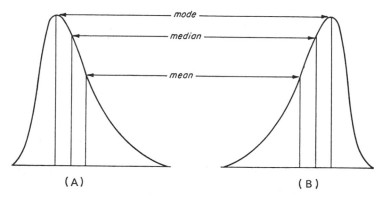

Figure 4-3   **(A)** *A positively skewed distribution, and* **(B)** *a negatively skewed distribution.*

probably more 100s would be earned than any other single score, although the individual scores might go as low as 40 or 50. The median is located between the point where the scores pile up the highest (the mode) and the point where the sum of the deviations is equal to zero (the mean). The reason is that the value of the median is influenced neither by the piling up of scores, as in the case of the mode, nor by the extremity of the scores, as in the case of the mean. The median is simply the point below which 50 percent of the cases fall.

COMPUTATION OF A MEAN FROM A FREQUENCY DISTRIBUTION
Just as we are able to compute the mode and median from a group of scores arranged in a frequency distribution, we can also compute the mean. Suppose, an anthropologist discovers a large number of human-like fossil skulls in a certain valley in Africa. She estimates the brain size from the skull parts for each fossil skull to the nearest 100 cubic centimeters. These data are presented in Table 4-5.

The mean of a group of scores arranged in a frequency distribution

**Table 4-5   Frequency distribution
of brain sizes**

X	Frequency (f)	fX
8	2	16
7	6	42
6	5	30
5	7	35
4	4	16
3	1	3
		$\Sigma fX = \overline{142}$

$$M = \frac{\Sigma fX}{N} = \frac{142}{25} = 5.68$$

may be found by multiplying the frequency (f) of each score by that score (X). This product is denoted fX. Next, the fX values are summed to yield ΣfX. Finally, ΣfX is divided by the total number of cases (N) to yield the mean. The only new idea involved in the computation of the mean from a frequency distribution is the multiplication of X (the score) by f (the frequency) to yield fX. For example, in the data from the experiment presented in Table 4-5 we see that there were six skulls that were 700 cubic centimeters in size. We then would multiply $6 \times 7 = 42$ to find fX. This would give us exactly the same result as if we summed all those scores of 7, that is, $7 + 7 + 7 + 7 + 7 + 7 = 42$. When the frequency of a score is very large, it is obvious that multiplying the frequency by the score is considerably easier than finding the sum of all identical scores.

   While it is apparent that computation of the mean is relatively easy when a frequency distribution with a large number of cases is considered, it does not necessarily follow that this method is preferable to simply summing all scores and dividing by N. If your only interest is in the computation of the mean, then construction of a frequency distribution and computation of the mean from the fX values would probably require more work than the direct method of calculating the mean. This is especially true if you have a calculator available.

COMPUTATION OF THE MEAN IN A FREQUENCY DISTRIBUTION WITH INTERVALS GREATER THAN 1.00   Just as computation of the median with intervals greater than 1.00 required only a slight change in procedure, computation of the mean under similar conditions requires only a slight adjustment. Consider the frequency distribution presented in Table 4-6, which presents data as to the number of soft drinks consumed by a group of spectators at a football game on a hot Saturday afternoon. The best guess as to the value of a case within an interval is the midpoint. (Notice that this same assumption was made in the computation of the mean from a frequency distribution when the interval size was 1.00.) Hence, if we assume that all cases within an interval fall at the midpoint, the mean can easily be computed. This is shown in Table 4-6. It should be stressed

Table 4-6   Computation of the mean from grouped data

Number of drinks	Midpoint (X)	f	fX
12–13	12.50	1	12.50
10–11	10.50	1	10.50
8–9	8.50	3	25.50
6–7	6.50	5	32.50
4–5	4.50	2	9.00
2–3	2.50	4	10.00
			$\Sigma fX = 100.00$

$$M = \frac{\Sigma fX}{N} = \frac{100.00}{16} = 6.25$$

that computation of the mean from a frequency distribution with the data grouped into intervals is a mixed blessing. The assumption made about the values of the cases within an interval must be made because the original values of the cases are lost after grouping. In most cases it would seem that the added convenience of grouping of data does not compensate for the loss of information and consequent loss of accuracy in the computation of the mean.

COMPUTATION OF THE MEAN OF LARGE SCORES   The use of the fX method can save unnecessary work when many of the scores in a distribution are identical. Another method of computing the mean can save considerable effort when all scores are large. Suppose you want to compute the mean of these five scores: 312, 314, 310, 313, 311. The direct method would be to add them and divide by N:

X
312
314
310
313
311
$\Sigma X = \overline{1560}$

$$M = \frac{\Sigma X}{N} = \frac{1560}{5} = 312$$

An alternative method can be derived from the fact that the mean of a set of scores, all of which have been reduced by a constant C, is equal to the mean of the original scores minus the constant. This equation can be proved algebraically:

$$\frac{\Sigma(X - C)}{N} = \frac{\Sigma X}{N} - \frac{\Sigma C}{N} = \frac{\Sigma X}{N} - C$$

Applying this procedure to the previous data, we subtract a constant (310) from each of the X scores to get a series of $(X - C)$ scores, compute the mean of the reduced scores, and add the constant $(C = 310)$ to this mean.

X	C	$(X - C)$
312	310	2
314	310	4
310	310	0
313	310	3
311	310	1
	$\Sigma(X - C) = $	$\overline{10}$

$$M_{(X-C)} = \frac{\Sigma(X - C)}{N} = \frac{10}{5} = 2$$

$$M_x = M_{(X-C)} + C = 2 + 310 = 312$$

Any constant may be chosen. Usually, the most convenient one is the smallest score in the distribution, since negative $(X - C)$ values are thus avoided. The practical implication of this example is that if you add a constant to each score in a distribution, the value of the mean is increased by a value equal to that constant. That is, if the mean of a group of scores is 32 and you add 10 to each score, the new mean becomes 42.

SUMMARY

This chapter has presented some of the techniques for determining the single score which best characterizes or describes a whole distribution of scores. While most people are quite familiar with the term *average,* the concepts of mean, median, and mode, and their appropriate uses, become somewhat more difficult. It will be seen that the mean is the most useful definition of central tendency. The reason for this is that the mean can be manipulated algebraically, something which cannot be done with the other measures. This characteristic of the mean will lead to its continued use in inferential statistics, whereas we will rarely encounter the median or mode hereafter.

PROBLEMS

1. Compute the mean of the following scores: 4, 8, 5, 3, 4, 6.
2. Compute the mean of the following scores: 1,123,425; 1,123,420; 1,123,426; 1,123,423; 1,123,426.
3. Compute the mean of the scores 46, 56, 58, 48 by first subtracting a constant of 50.
4. Compute the median of each of the following groups of scores:
   (a) 4, 2, 7, 11, 5.
   (b) 4, 2, 7, 5.
   (c) 4, 2, 11, 8.
5. Compute the median of each of the following frequency distributions of scores:

(a) X	f	(b) X	f
5	3	8	3
4	6	7	6
3	2	6	4
2	4	5	3
1	1	4	8

6. The mean, median, and mode of a certain distribution of scores are 60, 75, and 78. From this information draw a graph of the distribution.
7. One section of the public library had the following number of books checked out in one week: 105, 185, 175, 165, 140, 190, 216. What is the mean (from Sunday through Saturday) number of books checked out? The median?
8. The following frequency distribution represents annual income (of heads of households, in $ thousand) in a rural village. What measure of central tendency would be most appropriate? Find it.

X	f
25–29	1
20–24	1
15–19	2
10–14	7
5–9	25
0–4	14

9. The Hutterites, a religious group living in the northern great plains of North America, tend to have large families. One investigator found that eight adult women past child-bearing age in one community had had 8, 11, 15, 12, 6, 14, 13, and 9 children, respectively. What is the mean number of children in each family?

10. Find the mean and median for each of the following sets of scores:
   (a) 0, 1, 2, 5, 7.
   (b) 1, 3, 4, 7.
   (c) 4, 12, 1, 9.
   (d) 2, 3, 4, 5, 3.
   (e) 301, 315, 292, 312.

(f) X	f		(g) X	f		(h) X	f
5	8		5	2		5	8
4	4		4	4		4	6
3	1		3	11		3	0
2	2		2	5		2	2
1	1		1	8		1	8

EXERCISES

1. Compute the appropriate measure of central tendency from the data in Exercise 1 of Chapter Two. What considerations determined the appropriate measure of central tendency?

2. Select 20 single-digit numbers from the table of random numbers (Appendix H), and compute the mean, median, and mode from these data.

1   When we wish to get an indication of how well a baseball player has been hitting we look at his batting _____.

══════════ STOP ══════════════════════ STOP ══════════

average

---

2   The batting _____ is simply the player's number of hits divided by the total times he has been at bat.

══════════ STOP ══════════════════════ STOP ══════════

average

---

3   For example, if a player has made thirty hits in one hundred times at bat his average number of hits is equal to 30/100 or _____.

══════════ STOP ══════════════════════ STOP ══════════

.300

---

4   Thus .300 is a convenient way of describing a batter's hitting ability. Similarly if we wish to get an indication of how well a bowler has been bowling we look at his _____.

══════════ STOP ══════════════════════ STOP ══════════

average

---

5   A single score which is used to describe (or characterize) a large number of scores is called the _____.

══════════ STOP ══════════════════════ STOP ══════════

average

---

6   More generally a measure of central tendency is used to describe a group of scores. Thus an average is a _____ _____ _____ _____.

══════════ STOP ══════════════════════ STOP ══════════

measure of central tendency

---

7   Although the average seems simple to compute, the statistician has three different ways to compute an average or _____ _____ _____ _____.

══════════ STOP ══════════════════════ STOP ══════════

measure of central tendency

---

8   Three common measures of central tendency (or average) are the mean, median, and mode. The mode is defined as the score that appears most often (or the score with the greatest frequency). With scores 0, 1, 3, 3, 4, 6, the most frequent score is 3. Hence the mode would be equal to _____.

═══════ STOP ═══════════════════════════════ STOP ═══════

3

---

9   The score with the greatest frequency defines the _____.

═══════ STOP ═══════════════════════════════ STOP ═══════

mode

---

10   With scores 0, 1, 5, 7, 7, the mode is equal to _____.

═══════ STOP ═══════════════════════════════ STOP ═══════

7

---

11   With scores 0, 1, 1, 2, 3, 5, 5, 5, 7, 9, 9, the mode is equal to _____.

═══════ STOP ═══════════════════════════════ STOP ═══════

5

---

12   The median may be defined as the point (or score) below which 50 percent of the cases (or scores) fall. More simply, it is the middle score. Hence with scores 0, 0, 2, 3, 8, the median would be equal to 2 because there are 50 percent above and _____ below this score.

═══════ STOP ═══════════════════════════════ STOP ═══════

50 percent

---

13   While the median is the point below which 50 percent of the scores fall, the *mode* is defined as the most frequent _____.

═══════ STOP ═══════════════════════════════ STOP ═══════

score

---

14   The _____ is defined as that point below which 50 percent of the cases fall.

═══════ STOP ═══════════════════════════════ STOP ═══════

median

15   The median is that point below which _____ _____ _____ _____
_____ _____.

═════════ STOP ═══════════════════════════ STOP ═════════

50 percent of the cases fall

---

16   With scores 1, 1, 3, 4, 7, 8, 12 the median would equal _____ while the
mode would be equal to _____.

═════════ STOP ═══════════════════════════ STOP ═════════

4, 1

---

17   With scores of 15, 17, 19, 22, 22, 22, 24, 31, 49, both the median and the
mode would equal _____.

═════════ STOP ═══════════════════════════ STOP ═════════

22

---

18   The measure of central tendency which most closely approximates the
meaning of "average" is the *mean*. The formula for the mean is $\sum X/N$,
that is, all the scores are added together and then divided by _____
_____ _____ _____ _____.

═════════ STOP ═══════════════════════════ STOP ═════════

the total number of scores (N)

---

19   When the meaning of $\sum X$ is not exactly clear, we would use the formula
$M = \sum_{1}^{N} X/(?)$.

═════════ STOP ═══════════════════════════ STOP ═════════

N

---

20 $M = (?)/N$

═════════ STOP ═══════════════════════════ STOP ═════════

$\sum X \text{ (or } \sum_{1}^{N} X)$

---

21   Given the following scores: 2, 6, 8, 12, $M = \sum X/N = $ _____.

═════════ STOP ═══════════════════════════ STOP ═════════

$28/4 = 7$

---

22 We have defined three measures of central tendency: (1) the mode—the score with the greatest _____; (2) the median—the score below which _____ of the cases fall; and (3) the mean—the sum of all the scores divided by _____.

═══════ STOP ═══════════════════════════ STOP ═══════

frequency (number); 50 percent; N (the number of scores or cases)

---

23 Thus the score with the greatest frequency defines the _____; the sum of all the scores divided by N is equal to the _____; and the point below which 50 percent of the cases (scores) fall is equal to the _____.

═══════ STOP ═══════════════════════════ STOP ═══════

mode; mean; median

---

24 Given scores 0, 1, 1, 2, 3, 5, 9, the mode = _____, the median = _____, and the mean = _____.

═══════ STOP ═══════════════════════════ STOP ═══════

1, 2, 3

---

25 We have five scores 0, 1, 3, 7, 9. The median is equal to _____.

═══════ STOP ═══════════════════════════ STOP ═══════

3

---

26 If we have an odd number of scores the median is the _____ score.

═══════ STOP ═══════════════════════════ STOP ═══════

middle

---

27 If on the other hand we have four scores, the median would be the point half way between the second and _____ score.

═══════ STOP ═══════════════════════════ STOP ═══════

third

---

28 Thus if we had scores 0, 1, 3, 5, the median would be the point half way between _____ and _____.

═══════ STOP ═══════════════════════════ STOP ═══════

1 and 3

---

29 With scores 0, 1, 4, 7, the median would be _____.

════════ STOP ════════════════════════ STOP ════════

2.5

---

30 Again the median would be the point below which _____ percent of the cases fall.

════════ STOP ════════════════════════ STOP ════════

50

---

31 With scores 0, 1, 4, 7, the median would equal 2.5 while the mean $\sum X/N =$ _____.

════════ STOP ════════════════════════ STOP ════════

3

---

32 Much of the data in psychology are taken from populations that fall into a normal curve or normal distribution, which looks like the figure. For example, if we measured the heights of hundreds of males the distribution of heights would be _____.

════════ STOP ════════════════════════ STOP ════════

normal

---

33 If heights of males are normally distributed, this means that while some people would be very tall and some very short, most would be about _____ height.

════════ STOP ════════════════════════ STOP ════════

average

---

34 If we should wish to graph the distribution of heights, we would place height along the x axis (or abscissa or baseline) and the frequency along the _____ or ordinate.

════════ STOP ════════════════════════ STOP ════════

y axis

35 Thus if we found fifty people who were 72 inches tall the height of the normal curve (that is, the y axis value) at 72 inches would be equal to _____.

══════ STOP ══════════════════════ STOP ══════

50

---

36 If we found thirty people who were 75 inches tall, then the height of the ordinate at 75 inches would equal _____.

══════ STOP ══════════════════════ STOP ══════

30

---

37 When plotting a distribution we place the large values of the x axis on the right and the small values on the _____.

══════ STOP ══════════════════════ STOP ══════

left

38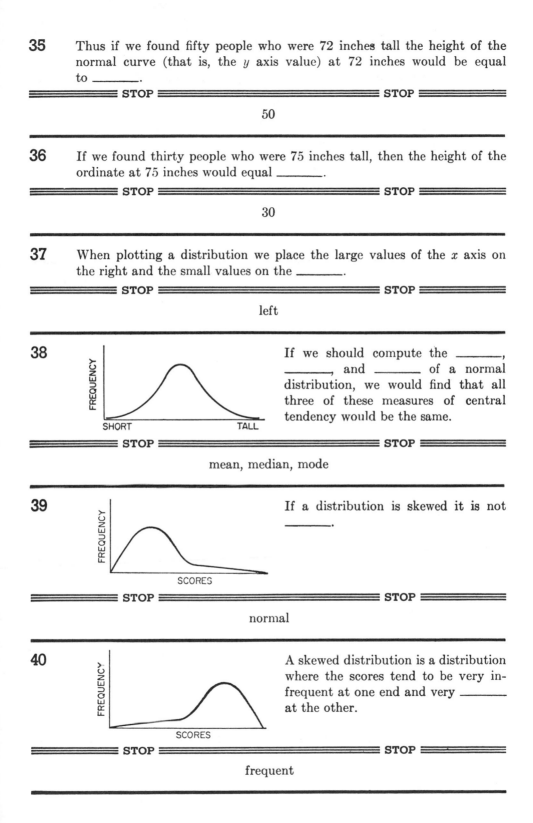

If we should compute the _____, _____, and _____ of a normal distribution, we would find that all three of these measures of central tendency would be the same.

══════ STOP ══════════════════════ STOP ══════

mean, median, mode

39

If a distribution is skewed it is not _____.

══════ STOP ══════════════════════ STOP ══════

normal

40

A skewed distribution is a distribution where the scores tend to be very infrequent at one end and very _____ at the other.

══════ STOP ══════════════════════ STOP ══════

frequent

41   If a distribution is positively skewed this means that there are few high scores and a large number of _____ scores.

═══════ STOP ═══════════════════════ STOP ═══════

low

---

42   If a distribution is negatively skewed the most frequent scores would be the high scores and the _____ _____ scores would be the low scores.

═══════ STOP ═══════════════════════ STOP ═══════

least frequent

43

We have said that a positively skewed distribution has a large number of low scores and that a negatively skewed distribution has a large number of high scores. The distribution on the left is _____ skewed.

═══════ STOP ═══════════════════════ STOP ═══════

positively

44

The distribution on the left has many high scores. This distribution is _____ skewed.

═══════ STOP ═══════════════════════ STOP ═══════

negatively

45

The distribution on the left is _____ skewed.

═══════ STOP ═══════════════════════ STOP ═══════

positively

---

46 A normal distribution is not skewed and we find that the mean, median, and mode are all equal. However, with a skewed distribution the mean, median, and mode would not all be _____.

═══════ STOP ═══════════════════════════ STOP ═══════

equal

───

47 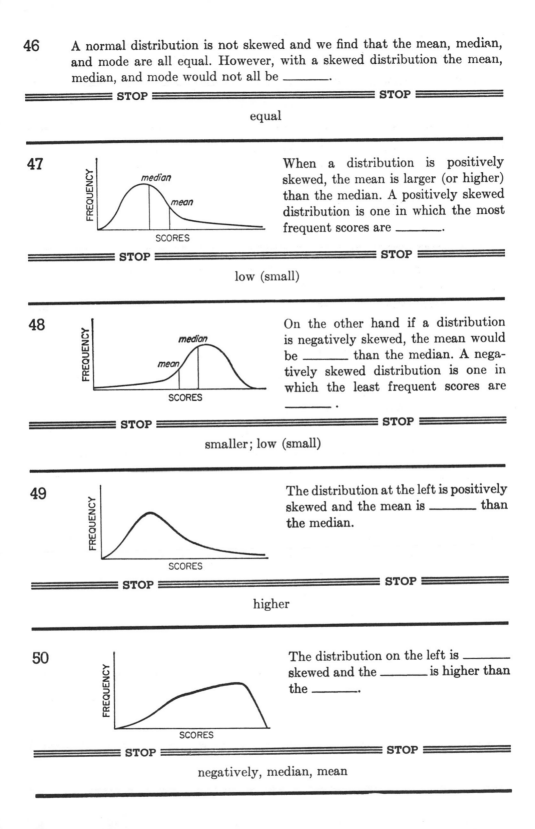 When a distribution is positively skewed, the mean is larger (or higher) than the median. A positively skewed distribution is one in which the most frequent scores are _____.

═══════ STOP ═══════════════════════════ STOP ═══════

low (small)

───

48 On the other hand if a distribution is negatively skewed, the mean would be _____ than the median. A negatively skewed distribution is one in which the least frequent scores are _____ .

═══════ STOP ═══════════════════════════ STOP ═══════

smaller; low (small)

───

49 The distribution at the left is positively skewed and the mean is _____ than the median.

═══════ STOP ═══════════════════════════ STOP ═══════

higher

───

50 The distribution on the left is _____ skewed and the _____ is higher than the _____.

═══════ STOP ═══════════════════════════ STOP ═══════

negatively, median, mean

───

51

O 1 2 3 SCORES 94

Given numbers 0, 1, 2, 3, 94. This distribution of scores is _____ skewed.

═══════ STOP ═══════════════════ STOP ═══════

positively

---

52 Given scores 0, 1, 2, 3, 94. This distribution is positively skewed, in which case the mean should be higher than the median. $M = \sum X/N =$ _____, and the median = _____.

═══════ STOP ═══════════════════ STOP ═══════

20, 2

---

53 $M = 20$ and the median = 2. Hence with this positively skewed distribution, the mean is _____ than the median.

═══════ STOP ═══════════════════ STOP ═══════

higher

---

54 $M = 5$, median = 25. We would suspect that the distribution from which the mean and median were computed was _____ skewed.

═══════ STOP ═══════════════════ STOP ═══════

negatively

---

55 However, with a normal distribution the mean would be _____ _____ the median.

═══════ STOP ═══════════════════ STOP ═══════

equal to

---

56 Because computation of the mean can be manipulated algebraically (and for many other reasons) the mean is most commonly used in inferential statistics. Given scores 0, 1, 5, 6, 3, $M =$ _____.

═══════ STOP ═══════════════════ STOP ═══════

3

---

57 If we let X stand for the raw score of a single individual and M stand for the mean, then $(X - M)$ is the _____ between the raw score and the mean.

═══════ STOP ═══════════════════ STOP ═══════

difference

58 Given scores 0, 1, 5, 6, 3, $M = 3$. The difference between the mean and a raw score may be given the notation x (lower case or small x). Thus if $x = 4$, this means that the difference between the mean and the raw score is equal to _____.

═════════ STOP ═══════════════════════════ STOP ═════════

4

---

59 _____ $= X - M$ and is called a *deviation score*.

═════════ STOP ═══════════════════════════ STOP ═════════

x

---

60 $x = X -$ _____, a deviation score, indicates the difference between a raw score and the _____.

═════════ STOP ═══════════════════════════ STOP ═════════

M, mean

---

61 $$x = \text{_____} - M$$

═════════ STOP ═══════════════════════════ STOP ═════════

X

---

62 and $x =$ _____.

═════════ STOP ═══════════════════════════ STOP ═════════

$X - M$

---

63 If a raw score is below the mean, then the deviation score will be negative. For example, if $X = 10$ and $M = 16$, then $x =$ _____.

═════════ STOP ═══════════════════════════ STOP ═════════

-6

---

64 With scores 1, 7, $M = 4$, $x_1 = 1 - 4 = -3$, $x_2 = 7 - 4 = 3$, and $\sum x$ (the sum of the deviation scores) $= (-3) + (+3) =$ _____.

═════════ STOP ═══════════════════════════ STOP ═════════

0

65 Given scores 0, 1, 5, 6, 3. $M = 3$, $x_1 = X_1 - M = 0 - 3 = -3$, $x_2 =$ _____.

════════ STOP ════════════════════════════ STOP ════════

$$-2$$

---

66 Given scores 0, 1, 5, 6, 3, $M = 3$. The five xs are -3, -2, _____, _____, and _____.

════════ STOP ════════════════════════════ STOP ════════

$$+2, +3, \text{ and } 0$$

---

67 Given scores 1, 1, 3, 7. $M =$ _____ and the four deviation (x) scores are _____, _____, _____, and _____.

════════ STOP ════════════════════════════ STOP ════════

$$3, -2, -2, 0, \text{ and } +4$$

---

68 With scores 1, 1, 3, 7, the deviation scores are -2, -2, 0, $+4$. $\sum x =$ _____.

════════ STOP ════════════════════════════ STOP ════════

$$0$$

---

69 Given scores 3, 7, 4, 6, $M =$ _____, the deviation scores are _____, _____, _____, and _____. Finally, $\sum x =$ _____.

════════ STOP ════════════════════════════ STOP ════════

$$5, -2, +2, -1, \text{ and } +1; 0$$

---

70 Write down any four different scores. Compute M and the deviation scores and then find the sum of the deviation scores. $\sum x =$ _____.

════════ STOP ════════════════════════════ STOP ════════

$$0$$

---

71 Thus the mean may be defined in two ways: (1) $\sum X/N$ and (2) the point where the summed deviations equal _____.

════════ STOP ════════════════════════════ STOP ════════

$$0$$

---

72 Given scores 7, 1, 2, 4, 3, 0, 0, 1, 0. Arrange the scores from smallest to largest.

═══════ STOP ═══════════════════════════════════ STOP ═══════

0, 0, 0, 1, 1, 2, 3, 4, 7

---

73 From frame 72, the mode = _____, the mean = _____.

═══════ STOP ═══════════════════════════════════ STOP ═══════

0, 2

---

74 From frame 72, the median = _____.

═══════ STOP ═══════════════════════════════════ STOP ═══════

1.00

---

75 Four witnesses saw a robbery. Three reported seeing two robbers, one reported seeing eight robbers. What is the appropriate measure of central tendency?

═══════ STOP ═══════════════════════════════════ STOP ═══════

mode

---

76 Five themes were ranked by four judges. Probably the most appropriate measure of central tendency would be _____.

═══════ STOP ═══════════════════════════════════ STOP ═══════

median

---

77 An educator gave a group of children a puzzle to solve and measured the length of time it took each. Three could not solve the puzzle, but five could. Here are their times (in minutes): 3, 5, 7, 5, 6. What is the appropriate measure of central tendency here? _____

═══════ STOP ═══════════════════════════════════ STOP ═══════

median

---

78 Arrange the scores from frame 77, in size from smallest to largest.

═══════ STOP ═══════════════════════════════════ STOP ═══════

3, 5, 5, 6, 7, ∞, ∞, ∞

---

79 The median of this group of scores is _____.

═══════ STOP ═══════════════════════════════════ STOP ═══════

6.50

---

80   The symbol f is used to indicate the frequency of a score. Given the scores 3, 5, 5, 2, 3, 3, 0, 3. For $X = 5$, $f = 2$, and for $X = 3$, $f =$ _____.

═══════ STOP ═══════════════════════════ STOP ═══════

4

──

81     $X = 3$, $f = 4$, and if f is multiplied by X then $fX =$ _____.

═══════ STOP ═══════════════════════════ STOP ═══════

$3 \times 4 = 12$

──

82   Given the scores 1, 2, 5, 4, 2, 2, 3, 2. For $X = 2$, $f =$ _____ and $fX =$ _____.

═══════ STOP ═══════════════════════════ STOP ═══════

$4, 2 \times 4 = 8$

──

83   Given the scores 3, 2, 3, 5, 5, 3, 3, 2, 3. $X = 3$, $f =$ _____, and $fX =$ _____. The sum of the scores equal to 3 = _____.

═══════ STOP ═══════════════════════════ STOP ═══════

$5, 15; 15$

──

84   In general, the sum of all the cases with the same score equals _____.

═══════ STOP ═══════════════════════════ STOP ═══════

fX

──

85   Given the scores 0, 2, 1, 3, 3, 2, 1, 0, 1, 2.

X	f	fX
3	2	?
2	3	
1	?	
0	2	

═══════ STOP ═══════════════════════════ STOP ═══════

$f = 3, fX = 3 \times 2 = 6$

──

86

X	f	fX
3	1	_?_
2	2	_?_
1	4	_?_
0	3	_?_

The sum of all cases equal to 2 = _____.

══════ STOP ══════════════════════ STOP ══════

fX	
3	
4	4
4	
0	

87

X	f	fX
10	2	_?_
9	1	_?_
8	3	_?_
7	4	_?_

With $X = 8$, the sum of these scores = _____.

══════ STOP ══════════════════════ STOP ══════

fX	
20	
9	
24	24
28	

88

fX
20
5
8

$\sum fX$ is the sum of all the fX values. The value of $\sum fX =$ _____.

══════ STOP ══════════════════════ STOP ══════

33

89

X	f	fX
3	4	_?_
2	2	_?_
1	5	_?_

$\sum fX =$ _____.

══════ STOP ══════════════════════ STOP ══════

fX	
12	21
4	
5	

90

X	f
3	4
2	2
1	5

The total number of cases is equal to the sum of the frequencies at each score. $N =$ _____.

══════ STOP ══════════════════════════ STOP ══════

$$4 + 2 + 5 = 11$$

91

X	f	fX
3	2	?
2	2	?
1	1	?

Given the scores 1, 2, 2, 3, 3, $\sum X =$ _____. $\sum fX =$ _____.

══════ STOP ══════════════════════════ STOP ══════

fX	
6	11; 11
4	
1	

---

92 $\sum X = \sum fX$. Then in addition to $M = \sum X/N$, another formula for the mean is $M = (?)/N$.

══════ STOP ══════════════════════════ STOP ══════

$$\sum fX$$

---

93 $N = 8, \sum fX = 32, M = \sum fX/N =$ _____.

══════ STOP ══════════════════════════ STOP ══════

$$32/8 = 4$$

94

X	f	fX
3	4	12
2	2	4
1	4	4

$\sum fX =$ _____, $M = \sum fX/N =$ _____, and $N =$ _____.

══════ STOP ══════════════════════════ STOP ══════

20, 2, 10

95

X	f	fX
12	6	?
11	4	?
10	6	?

$\sum fX =$ _____, $M =$ _____.

══════ STOP ══════════════════════════ STOP ══════

fX	
72	176, 176/16 = 11
44	
60	

96　The median can be easily computed from grouped data. The median is defined as the point below which _____ percent of the cases fall.

═══════ STOP ═══════════════════ STOP ═══════

50

97

X	f
5	4
4	7
3	8
2	4
1	1

In the frequency distribution at the left, $N =$ _____ and _____ cases fall below the median.

═══════ STOP ═══════════════════ STOP ═══════

24, twelve

98　Thus in frame 97, the median is the point below which _____ cases fall.

═══════ STOP ═══════════════════ STOP ═══════

twelve

99　How many cases fall below 1.50? _____. How many below 2.50? _____. How many below 3.50? _____.

═══════ STOP ═══════════════════ STOP ═══════

one, five, thirteen

100　Since five cases fall below 2.50 and thirteen cases fall below 3.50, we know that the median falls in the interval with lower and upper real limits of _____ and _____, respectively.

═══════ STOP ═══════════════════ STOP ═══════

2.50 and 3.50

---

101 The median is the point below which 12 cases fall. Five cases fall below 2.50 and eight cases fall into the 2.50–3.50 interval. The midpoint of the 2.50–3.50 interval is ———.

════ STOP ════════════════════ STOP ════

3.00

102 From frame 101, the median is equal to ———.

════ STOP ════════════════════ STOP ════

3.00

103

X	f
4	1
3	4
2	5
1	4
0	2

In the frequency distribution at the left, N = ———, and ——— cases fall below the median.

════ STOP ════════════════════ STOP ════

16, 8

104 The lower and upper limits of the interval in which the median falls are ——— and ———, respectively.

════ STOP ════════════════════ STOP ════

1.50, 2.50

105 And the median is equal to ———.

════ STOP ════════════════════ STOP ════

2.00

106

X	f
24–27	1
20–23	4
16–19	12
12–15	8
8–11	2
4–7	3

In the frequency distribution at the left, N = ———, and ——— cases fall below the median.

════ STOP ════════════════════ STOP ════

30, 15

107 The upper and lower real limits of the interval in which the median falls are _____ and _____, respectively.

===== STOP ===== STOP =====

15.5, 19.5

---

108 The midpoint of the 15.5–19.5 interval is _____.

===== STOP ===== STOP =====

17.5

---

109 The median falls in the 15.5–19.5 interval. The median is equal to _____.

===== STOP ===== STOP =====

17.5

---

110 The median of another frequency distribution falls in the 110–119 interval. The median is equal to _____.

===== STOP ===== STOP =====

114.5

---

111 The median of still another frequency distribution falls in the 150–199 interval. The median is equal to _____.

===== STOP ===== STOP =====

174.5

Measures of Variability

Two boys were standing on opposite sides of a river. John told Steven to come across the river. "I can't," said Steven, "because I can't swim and there is no boat." "It doesn't make any difference if you can't swim," John replied. "If you cross the river here, on the average it is only 4 feet deep." But Steven still refused to cross, for luckily he was aware of the fact that the most characteristic score—the measure of central tendency—does not necessarily give information about all the scores (depths in this case) in the distribution to be considered. That is, in some places the river may have been very shallow and in some places the river may have been very deep. It would only take a few deep spots to prevent Steven from crossing.

Chapter Four was concerned with methods for locating the most characteristic score in a distribution. The three measures of central tendency represent the first of two essential types of descriptive statistics. This chapter concerns the second major group: measures of variability.

Variability is so much a part of our experience that the abstract concept needs to be carefully considered and should be clearly distinguished from the centrality concept discussed in the preceding chapter. If a group of observations resulted in identical measurements, neither centrality nor variability indices would be necessary; description of any single score would suffice. The presence of any degree of variation among the measures, however, necessitates the use of *both* concepts for an adequate descriptive summary of the characteristics of the group of scores.

In Figure 5-1 a frequency polygon shows each of two sets of test scores. In Class A the scores tend to cluster close to the mean; the variability of the scores is small. In Class B the scores spread out to a much greater degree around the mean. Note that the means for the two classes are identical, although the variabilities are considerably different.

Important differences in variability are not always as obvious as those in Figure 5-1. In any case, however, it would be desirable to have an index of variability that we could use to describe and compare distributions, just as we are now able to describe and compare central tendencies using

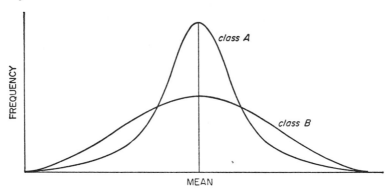

Figure 5-1 *Two distributions with identical means but with different variabilities.*

the indices discussed in Chapter Four. We will describe three such indices in this chapter: the *range*, the *semi-interquartile range*, and the *standard deviation*.

With two summary statistics—a measure of central tendency and a measure of variability—we can describe almost any distribution of scores with a reasonable degree of accuracy. There are additional measures that can be used to obtain even greater descriptive precision, but the use of these measures in the behavioral sciences is too rare to warrant discussion in this book.

THE RANGE

The range is simply the difference between the values of the largest and smallest scores in a distribution. The range is the easiest measure of variability to compute and, like the mode, is also the least stable. Changes in only a single score—if it is the largest or smallest—can affect its value greatly. For this reason the usefulness of the range as a measure of variability is confined to quick estimation or to provision of a rough check on the computation of more refined measures.

Although the range and the mode are both simple to compute and are subject to large fluctuations with small changes in the distribution, there is a very important difference between them. The mode can be used with nominal, ordinal, or interval data, and it is the only measure of central tendency that can be used with nominal data. The range, however, cannot be used meaningfully with nominal or ordinal data. Its use should be restricted to interval data, where it is meaningful to talk about the largest and smallest scores. Categories on a nominal scale, of course, have no order; no one class is any "more" or "less" than another, in the sense of an ordered continuum.

Consider these two groups of five scores: 4, 2, 8, 1, 5, and 4, 5, 3, 3, 5. The mean of both groups is 20/5 = 4, but the range of the first group is 8 − 1 = 7, while the range for the second group is 5 − 3 = 2. Since the range of the first group exceeds that of the second group, we conclude that the first group of scores is more variable than the second, even though the means are identical.

THE SEMI-INTERQUARTILE RANGE

Just as the mode and the range are measures of comparable precision, so also are the median and the semi-interquartile range. The semi-interquartile range (SIQR) is defined as half the distance on the scale between Q_1 (the point below which 25 percent of the cases fall) and Q_3 (the point below which 75 percent of the cases fall). The following formula expresses this definition in condensed form:

$$SIQR = \frac{Q_3 - Q_1}{2}$$

In Figure 5-2 the locations of Q_1, Q_2, and Q_3 are shown, as well as the

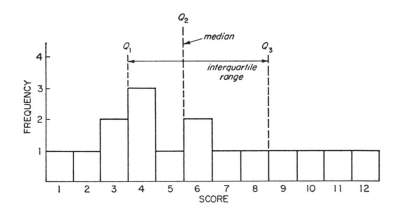

Figure 5-2   *Locations of quartile points.*

interquartile range. The SIQR is half the interquartile range. Note that Q_2 is the median of the distribution.

One thing about the definition of the semi-interquartile range (and of the median for that matter) is often confusing: Although the Q points are located by reference to the ordinal arrangement of the scores, the Q points are expressed in terms of an interval scale of measurement along the baseline. In Figure 5-2, for instance, Q_1 represents the point separating the lower 25 percent of the cases from the upper 75 percent, but it is expressed as a score of 3.5. Similarly, the SIQR was found to be $(8.5 - 3.5)/2 = 5/2 = 2.5$ points on the scale of the baseline.

To illustrate one advantage of the semi-interquartile range over the range as a measure of variability, consider the following fictitious example:

A group of judges was asked to express opinions about two paintings. The first was entitled "Winter Landscape" and the second was entitled "Ankle Ascending an Elbow." The judges were asked to rate each painting by checking one of the seven scale points shown on the baseline of Figure 5-3.

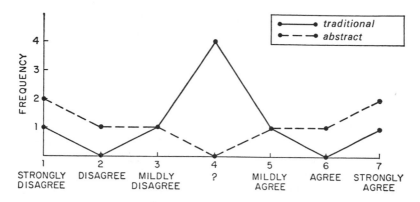

Figure 5-3 *Responses of judges to an abstract and to a traditional painting.*

Both groups of scores have the same range: $7 - 1 = 6^1$, but the difference in variability is reflected by the semi-interquartile range. For the painting "Winter Landscape," $Q_3 = 4.5$, $Q_1 = 3.5$, and SIQR $= (4.5 - 3.5)/2 = 0.50$. For the painting "Ankle Ascending an Elbow," $Q_3 = 6.5$, $Q_1 = 1.5$, and SIQR $= (6.5 - 1.5)/2 = 2.50$. The ratings of the Ankle painting have a considerably larger SIQR, although the range fails to indicate the obvious difference in variability. Note that the median $(Q_2) = 4.0$ for both groups.

Computation of the semi-interquartile range is no more difficult than computation of the median. In fact, the locations of the Q_1 and Q_3 points are determined in exactly the same manner as is Q_2, the median, except that 25 percent and 75 percent of the cases locate the points rather than 50 percent.

Suppose that a sample of 48 residents in a middle-class suburb yielded the distribution of scores shown in Table 5-1 when they were asked to indicate their annual income to the nearest thousand.

To find the value of the semi-interquartile range, we must first find Q_1 and Q_3. Multiplying the number of cases (48) by the proportion of cases below Q_1 (0.25) and by Q_3 (0.75), we find that Q_1 and Q_3 are the points below which 12 and 36 cases fall, respectively. Counting up from the bottom of the distribution, we find that 12 cases fall in interval 12 or below. Therefore, $Q_1 = 12.50$, which is the upper limit of interval 12. Again counting up from the bottom of the distribution, we find that 33 cases fall in interval 16 or

---

[1] Some statisticians prefer to define the range as the distance between the lower real limit of the lowest score and the upper real limit of the highest score. In our examples this would produce ranges 1.0 point larger than those described.

Table 5-1   Frequency distribution of incomes

X	f	X	f
20	1	13	3
19	2	12	3
18	5	11	4
17	7	10	2
16	8	9	1
15	6	8	1
14	4	7	1

below (that is, 33 cases fall below 16.5) and that 40 cases fall in interval 17 or below. Hence, Q_3 is equal to 17. Substituting into the equation, we find that SIQR $= (17.00 - 12.50)/2 = 4.50/2 = 2.25$.

An anthropologist wanted a scale to determine whether a tribe was Plains Indians. Therefore he looked at behaviors characteristic of Plains Indians (for example, dependence on the buffalo, use of a movable teepee, and so on) and counted the number of such characteristics possessed by each of several Plains Indian tribes. The data are given below.

Each score represents the number of characteristics possessed by a specific tribe: 15, 16, 12, 22, 20, 18, 13, 17, 17, 15, 18, 14. The scale of measurement seems to be more characteristic of an ordinal scale than an interval scale, and therefore computation of the SIQR as a measure of variability seems appropriate. By arranging the data in sequence—12, 13, 14, 15, 15, 16, 17, 17, 18, 18, 20, 22—it can be seen that $Q_1 = 14.5$ and $Q_3 = 18$. Hence, SIQR $= (18 - 14.5)/2 = 1.75$. The median for these data is 16.5. Thus, if the anthropologist were to encounter a different tribe with a new set of behaviors, it would be expected to get a score similar to those found above before being counted as a Plains Indian tribe.

THE STANDARD DEVIATION

Just as the range and semi-interquartile range are similar to the mode and median, respectively, so too is the standard deviation similar to the mean. The standard deviation is probably the most widely used measure of variability and since the value of the standard deviation depends upon all the scores in the distribution and not on just a few, it is the most stable measure of variability.

Before we describe methods for calculating the standard deviation, let us be sure we understand a few important symbols and terms. The symbol for the standard deviation of a group of scores is S. The square of the standard deviation is called the *variance*, and its symbol is S^2. The variance will become particularly important in later chapters.

The *standard deviation* is defined as *the square root of the sum of the squared deviations about the mean divided by the number of cases minus* one.[2] This definition can be symbolized by the deviation-score formula for the standard deviation

$$S = \sqrt{\frac{\Sigma x^2}{N-1}}$$

where S is the standard deviation, $x = X - M$, $x^2 = (X - M)^2$, and $N =$ number of cases.

We will initially work through a simple example (Table 5-2), using the definition of the standard deviation given above, and then describe formulas that allow more rapid calculation.

Table 5-2   Calculation of the variance and standard deviation by the deviation-score method

X	M	x	x²
2	4	−2	4
5	4	1	1
3	4	−1	1
7	4	3	9
3	4	−1	1
$\Sigma X = 20$		$\Sigma x = 0$	$\Sigma x^2 = 16$

$$S^2 = \frac{\Sigma x^2}{N-1} = \frac{16}{4} = 4$$

$$S = \sqrt{\frac{\Sigma x^2}{N-1}} = \sqrt{\frac{16}{4}} = \sqrt{4} = 2$$

Suppose we have a group of five scores: 2, 5, 3, 7, 3. The mean of these five scores is $M = \Sigma X/N = 20/5 = 4$. The first step in finding the standard deviation is to calculate the deviation of each score from the mean. The procedure for finding deviation scores should be familiar to you; it was described in the previous chapter and is illustrated in Table 5-2. We have computed the deviation scores in column x by subtracting the value of the mean from each raw score in column X. Thus, if $x = X - M$, then the first score is $2 - 4 = -2$. Next, each of the deviation scores is squared and entered in column x^2. The value of Σx^2 is found by summing the scores in the x^2 column. Thus, $\Sigma x^2 = 16$, and this sum is next divided by $N - 1$, or 4. Dividing Σx^2 by $N - 1$ yields the variance, which is equal to 4, and the square root of the variance is the standard deviation 2.

Unfortunately, calculation of the variance and standard deviation by the deviation-score formula is usually laborious because the value of the

---

[2] If our only interest in the standard deviation were for description of our data, we would use the formula $\sigma = \sqrt{\Sigma x^2/N}$, dividing by N instead of $N - 1$. This formula results in a slight error when we wish to make inferences to populations from samples. Since the orientation of this book is more toward inference than description, we will normally use the formula for S as a measure of variability.

mean is rarely a whole number. By means of some algebraic manipulation, which you may wish to carry out for your own satisfaction, it is possible to derive a much more practical formula for calculating the variance and standard deviation. This formula is exactly equal to the deviation-score formula, but it is called the *raw-score formula for the variance:*

$$S^2 = \frac{\Sigma X^2 - \dfrac{(\Sigma X)^2}{N}}{N - 1} \quad \text{and} \quad S = \sqrt{S^2}$$

To demonstrate the equivalence of the two formulas, Table 5-3 shows the application of the raw-score formula to the data of the preceding example.

Table 5-3   Calculation of the variance and standard deviation by the raw-score method

X	X^2
2	4
5	25
3	9
7	49
3	9
$\Sigma X = 20$	$\Sigma X^2 = 96$

$$S^2 = \frac{\Sigma X^2 - \dfrac{(\Sigma X)^2}{N}}{N - 1} = \frac{96 - \dfrac{(20)(20)}{5}}{5 - 1} = \frac{96 - 80}{4} = \frac{16}{4} = 4$$

The sum of the scores and the sum of the squares of the scores are found and entered into the formula. The results of the computations are exactly the same as those found by the deviation-score formula. The primary advantage of the raw-score formula is that it is only necessary to square the original scores.

MEANING OF THE STANDARD DEVIATION

While the interpretation of the range and semi-interquartile range is relatively straightforward, the interpretation of the standard deviation is somewhat more difficult. Initially we might note that, like the other two measures of variability, the larger the value of the standard deviation, the greater the degree of variability. Furthermore, while it may not be immediately apparent, the value of the standard deviation (or the variance) can never be less than zero. If you ever compute a standard deviation and get a negative value, you can be sure that you have made an error in your calculations. In addition, the absolute value of the standard deviation (as well as the range and the semi-interquartile range) has little meaning. *S* should be considered a relative measure of the variability of the scores in the distribution.

For example, suppose a television weather reporter gives the daily temperatures in both Fahrenheit and Celsius degrees. The high temperatures for a certain week are illustrated in Figure 5-4. John Jones and

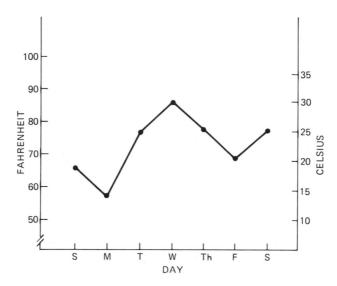

Figure 5.4   *Daily high temperatures were recorded for one week in summer using both Fahrenheit and Celsius scales.*

Sally Smith compute the standard deviation of the high temperatures; Jones reports a standard deviation of 8.78, and Smith reports a standard deviation of 4.88. Which is correct? Are both? Jones computed the standard deviation of the temperatures expressed in Fahrenheit, while Smith used the same temperature, expressed in Celsius, in computing her standard deviation.

The point of this example is to emphasize that the size of a standard deviation means little unless it is expressed with reference to other standard deviations that have been computed using the same measuring device. Comparing variabilities based on inches with variabilities based on feet simply will not tell you what you want to know, that is, which group of scores has greater variability than another? Hence, the common question, "The standard deviation of these scores is 6. What does that mean?" can be answered only when you know about the variability of other distributions. A standard deviation by itself has little meaning. (To avoid misleading you in one respect, it should be noted that if you multiply the scores in a distribution by a constant, the standard deviation of those scores will be increased by the value of that constant. Hence, you can find the Fahrenheit standard deviation if you multiply the Celsius by 1.8).

Finally, the value of the standard deviation is related to what is called *the normal distribution*. We will discuss this distribution at great length in a later chapter, but for now let us simply assume that you have encountered this distribution before—perhaps it was called *the bell-shaped curve*. In any event, the value of the standard deviation in any normal distribution is equal to the distance from the mean to the point of inflection—the point where the curve begins to turn out toward the ends of the distribution. This is shown in Figure 5-5.

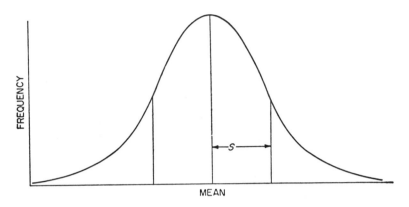

Figure 5-5 *The value of the standard deviation can be defined as the distance between the mean and a point of inflection in the normal curve.*

RELATIONSHIPS AMONG THE THREE MEASURES OF VARIABILITY

The three measures of variability discussed in this chapter—the range, the semi-interquartile range, and the standard deviation—are related to each other, and this relationship may be used as a rough check on the accuracy of your computations. We have illustrated the relationships among the three measures of variability in Figure 5-6. For the sample sizes behavioral scientists usually use, the range is roughly five times as large as the standard deviation. The semi-interquartile range is about two-thirds the size of the standard deviation. Thus, if the value of the semi-interquartile range is 2, the standard deviation would be roughly 3, and the range would be about 15. If you remember these approximate relationships, then you would be very surprised and would look for errors in your calculations if you found $S = 40$ and if the range of scores in the distribution you were considering was 30. Since you would expect that the value of the standard deviation, estimated from the range, was about $30/5 = 6$, and it was found to equal 40, you would suspect that an error had been made in your calculations. You should not expect the range to be always exactly five times the value of S. These relationships are only approximations, and the values of the range, the semi-interquartile range, and the standard deviation in any specific problem will vary somewhat from these expectations, especially if the distribution of scores departs very much from the shape of the normal curve.

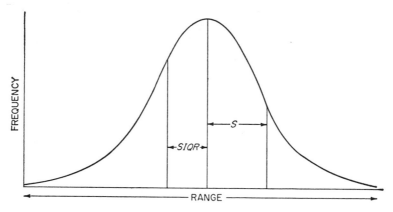

Figure 5-6 *Comparison of the range, the semi-interquartile range, and the standard deviation.*

USES OF THE THREE MEASURES OF VARIABILITY

As was the case with the various measures of central tendency, several considerations determine which measure of variability should be used, and one of these is the scale of measurement employed. When ordinal data are used, the semi-interquartile range could give a meaningful measure of variability. The standard deviation, which takes into consideration the value of every score, would not be meaningful with ordinal data because the values of the scores indicate only relative rank within the distribution. On the other hand, all three measures of variability could be used with interval data.

When speed is of importance, and a quick and easy but not necessarily accurate measure of variability is required, the range would be preferred because it can be obtained simply by looking at the two most extreme scores in the distribution—the lowest and the highest. On the other hand, when speed is not as important as accuracy, the standard deviation would be the preferred measure of variability, since it takes into consideration the value of every score in the distribution.

The standard deviation would also be used when further operations to be performed on the data require the use of a measure of variability. The reason is that the formula for the standard deviation allows subsequent algebraic manipulation, enabling a more complex statistical treatment than is the case with the range and the semi-interquartile range. Since the more complex treatments tend to be associated with statistical inference, it is reasonable to assume that the standard deviation would be used if statistical inference was the objective of the investigator. Because of this characteristic of the standard deviation, this measure of variability is used exclusively in subsequent chapters.

If a distribution is highly skewed, the best measure of variability would be the semi-interquartile range because the scores are squared in the standard deviation formula; thus, the more extreme scores are given much more weight than are the scores grouped near the mean. Then the value of the standard deviation would be markedly increased by a relatively few extreme scores. Since the values of the scores have little influence on the semi-interquartile range, it would be the preferred measure.

The semi-interquartile range is also a better measure of variability when certain scores in the distribution are missing or indeterminate. The reasoning here is much the same as in the preceding paragraph; with missing or indeterminate scores the value of the standard deviation is either distorted or impossible to compute, while the semi-interquartile range can be computed without the knowledge of the exact value of every score in the distribution.

CALCULATION OF THE VARIANCE OF LARGE SCORES

In Chapter Four we demonstrated that a constant subtracted from each score in a distribution could be added to the mean of the residual scores to yield the mean of the original scores, since $M_{X-C} = M_X - C$. This shortcut way of handling large scores is even more useful in calculating the variance, since the variance of a set of scores is not affected by subtracting a constant from every score in the distribution: $S_{X-C}^2 = S_X^2$. As an example, consider the following computation of S^2:

X	X^2	$(X - C)$	$(X - C)^2$
301	90,601	$301-300 = 1$	1
302	91,204	$302-300 = 2$	4
303	91,809	$303-300 = 3$	9
302	91,204	$302-300 = 2$	4
$\Sigma X = 1208$	$\Sigma X^2 = 364,818$	$\Sigma(X - C) = 8$	$\Sigma(X - C)^2 = 18$

$$S_X^2 = \frac{364,818 - \dfrac{(1208)^2}{4}}{3} = \frac{364,818 - \dfrac{1,459,264}{4}}{3}$$

$$= \frac{364,818 - 364,816}{3} = 0.67$$

$$S_{(X-C)}^2 = \frac{18 - \dfrac{(8)^2}{4}}{3} = \frac{18 - \dfrac{64}{4}}{3} = \frac{18 - 16}{3} = 0.67$$

Since the variance is unaffected, the standard deviation also remains the same.

CALCULATION OF THE VARIANCE FROM A FREQUENCY DISTRIBUTION

We can also compute the variance and the standard deviation from an available frequency distribution. In general, however, we would not recommend to the research worker that he routinely construct a frequency distribution and then compute the standard deviation from the distribution. This would be more laborious than simply finding the sum of the scores and the sum of the squares of the scores. Suppose we wish to compute the standard deviation from the frequency distribution of the group of scores presented in Table 5-4. The procedure we use here is similar to that used in the computation of the mean from a frequency distribution. Columns X and f indicate the frequency or number of cases with each score. To compute fX, multiply the frequency of each score by the value of the score.

For example, if $X = 8$, and $f = 4$, then $fX = 4 \times 8 = 32$. The sum of the scores in the fX column is ΣfX. To compute fX^2, we simply multiply the frequency of each score by the square of that score. For example, there are three cases equal to 4. Hence $f = 3$, $X = 4$, and $fX^2 = 3 \times 4 \times 4 = 48$. The sum of the scores in the fX^2 column is ΣfX^2.

The totals ΣfX and ΣfX^2 are exactly equal to ΣX and ΣX^2, respectively. Hence, substitution of ΣfX and ΣfX^2 into the raw-score formula will yield the standard deviation. From Table 5-4 we can see that $\Sigma fX = 144$ and $\Sigma fX^2 = 910$. Substituting these values into the raw-score formula, we find that $S^2 = 2$ and $S = 1.41$.

Table 5-4   Computation of the variance and standard deviation from a frequency distribution of 24 scores

X	f	fX	fX^2
8	4	32	256
7	5	35	245
6	7	42	252
5	4	20	100
4	3	12	48
3	1	3	9
		$\Sigma fX = 144$	$\Sigma fX^2 = 910$

$$S^2 = \frac{\Sigma fX^2 - \dfrac{(\Sigma fX)^2}{N}}{N-1} = \frac{910 - \dfrac{(144)(144)}{24}}{23} = \frac{910 - 864}{23} = \frac{46}{23} = 2$$

$$S = \sqrt{S^2} = \sqrt{2} = 1.41$$

Calculation of the standard deviation from a frequency distribution when the interval size is greater than 1.00 is done in much the same manner as for computation of the mean when the interval size was greater than 1.00. That is, find the midpoint of each interval and call the value of each midpoint the X score; then the value of f would be the number of cases within each interval.

SUMMARY

The three measures of variability may be compared roughly to the three measures of central tendency—mean, median, and mode—and also to the three measurement scales—nominal, ordinal, and interval—described previously. This comparability is represented in Table 5-5, where we have indicated the measurement scales and those measures of central tendency and variability which are similar in precision.

Table 5-5   Measurement scales and comparable measures of central tendency and variability

Scale	Central tendency	Variability
Nominal	Mode	Range
Ordinal	Median	Semi-interquartile range
Interval and ratio	Mean	Standard deviation

Since the categories of the nominal scale are not arranged in any order, computation of the mean or median when such a scale is used would have no meaning. Computation of the mode does impart some information, however. For example, if a spectator sees more football players wearing jerseys with numbers from 10 to 19 than in any other 10-unit interval, he would know that there are more quarterbacks than any other type of player on the team he is watching.

Similarly, computing the mean from data that vary along an ordinal scale would have little meaning. Although the distances between scale scores are not known, computation of the mean would tend to give the impression that they are.

Finally, the mean is the proper measure of central tendency to use when an interval scale is employed, unless other considerations force use of some other measure; if a rapid estimate is needed, the mode is used, or if the scores are highly skewed, the median would be computed.

Similar arguments can be made for the measures of variability described in this chapter. Although computation of the range for a nominal scale is impossible, since there is no ordering at all, we have placed this measure of variability at that point in the table because, like the mode, it is a fast and easy but inexact estimate of variability.

The most meaningful description of variability when an ordinal scale is employed comes from the use of the semi-interquartile range. Like the median it does not depend on the values of the scale scores, but measures the distance between the bottom and top 25 percent of the cases in the distribution of scores.

Finally, when the scale employed is an interval scale, the standard deviation is used unless other considerations force the use of some other measure of variability. For a rapid estimate of variability, the range is available; and when the distribution is highly skewed, the best estimate of variability would be the semi-interquartile range.

After this discussion as to the appropriate statistics to use with the various scales of measurement, it is well to take a little time to talk about scales of measurement. Four scales of measurement have been identified in this text. Other authors can and do identify additional scales until as many as eight or nine scales of measurement are used to cover essentially the same dimension as we have covered with four. Thus, our four scales can be thought of as fairly rough guide-posts. Many times we will be dealing with data which do not exactly meet, say, our definition of interval or ordinal data, but, instead, fall somewhere in between. When this happens, the procedure that should be selected is one that should be followed in any case; that is, if computation of a certain statistic provides meaningful information, it is appropriate to use it.

It is assumed that interval scales are used in the behavioral sciences. It is hoped, for example, that IQ scores represent an interval scale, but it is doubtful that we can show that the difference between an 80 and 90 IQ is the same as the difference between a 120 and 130 IQ. Unless these differences are the same, we are not using an interval scale. Even so, the mean and the standard deviation provide very meaningful information when computed on a distribution of IQ scores; because of this, it would seem reasonable to use these statistics in this case and other similar situations.

We have seen that a proper description of the central tendency and variability of a distribution of scores depends upon a number of considerations. Often, however, we will encounter situations in which the investigator wishes to make inferences about large populations on the basis of limited samples. When inference is the objective, we almost always limit the use of our measure of central tendency to the mean, and use the standard deviation (or the variance) to measure variability. This is because the most efficient procedures for making inferences require use of the mean and standard deviation rather than other measures of central tendency and variability. This fact will become clearer as you go through the rest of the book.

PROBLEMS

1. Compute the variance of the following scores, using the raw-score formula and then compute the variance by using the deviation-score formula: 0, 1, 7, 4, 3, 3.
2. Compute the variance of the following scores, using the raw-score formula: 3, 2, 11, 4, 8, 5, 7, 6, 8.
3. Add 2 to each score in Problem 2 and compute the variance, using the raw-score formula. Does adding a constant change the value of the variance?
4. A man measured four boards and found they were 2, 2, 3, and 5 yards long. Compute the variance of the lengths.
5. A second man measured the same four boards and found the same results, the only difference was that he measured the lengths of the boards in feet. Compute the variance of the lengths of the boards when they are expressed in feet. Does multiplying by a constant influence the value of the variance?
6. In Chapter 4 we noted that 105, 185, 175, 165, 140, 190, and 216 books were checked out of one section of the public library in one week. Find the variance of these scores.
7. A sociologist found the number of years of school attended by the adults of Centerville, a small village. These were 6, 5, 12, 3, 4, 8, 3, 4, 3, 2, 8, 2. Find the variance of these scores.

8. Compute the SIQR for these data:

X	f
6	3
5	4
4	3
3	3
2	2
1	1

9. Compute the variance of these data:

X	f
7	1
6	4
5	7
4	8
3	4
2	4
1	2

10. Compute the SIQR for these data:
    45, 22, 18, 36, 40, 32, 52, 19, 27, 33, 35, 47, 39, 42, 41, 29.
11. The range of scores in a certain distribution is 100. You compute the variance of those same scores and find a value of 4.00. What can you conclude about the distribution?
12. Find the standard deviation of the following scores—0, 8, 1, 7, 4, 1, 3, 2, 1—by both the raw-score and deviation-score methods. Do you get the same results?
13. Compute the variance for each of the following groups of scores:
    (a) 1, 0, 5, 6.
    (b) 8, 4, 5, 6, 1, 0.
    (c) 1, 2, 3, 4, 5.
    (d) 2, 5, 3, 6.
    (e) 6, 8, 2, 4, 1, 5.
14. Compute SIQR and S^2 for each of the following:

(a)

X	f
8	1
7	1
6	1
5	1
4	0
3	0
2	2
1	6

(b)

X	f
8	3
7	9
6	3
5	1
4	1
3	1

15. Compute the variance for the following group of scores; compare this with Problem 13(a):

X	f
306	1
305	1
304	0
303	0
302	0
301	1
300	1

EXERCISES

1. Compute the appropriate measure of variability from the data collected in Exercise 1 of Chapter Two. Why was this measure more appropriate than the others?
2. Select ten single-digit numbers from Appendix H and compute the standard deviation, SIQR, and range.
3. Add 10 to each number chosen in Exercise 2 and compute the standard deviation. What happened to the value of the standard deviation?
4. Multiply each of the ten numbers by 2 and compute the standard deviation. What happened to the value of the standard deviation?

1 Given two groups of four scores each:
Group I: 5, 5, 5, 5; Group II: 0, 3, 8, 9.
For Group I: $M = \sum X/N =$ _____; for Group II: $M =$ _____.

═══ STOP ═══════════════════════ STOP ═══

5; 5

---

2 I: 5, 5, 5, 5, $M = 5$; II: 0, 3, 8, 9, $M = 5$. Although the two means are _____, the scores in Group I are more uniform than the scores in Group II. That is, the scores in Group II vary more than they do in _____.

═══ STOP ═══════════════════════ STOP ═══

equal; Group I

---

3 Two groups of scores may have equal means but one group may be different from the other because the scores of one group _____ so much more than do the scores of the second group.

═══ STOP ═══════════════════════ STOP ═══

vary

---

4 Thus while two means may be equal, the scores of one group may be more _____ than the scores of the other.

═══ STOP ═══════════════════════ STOP ═══

variable

---

5 This section is devoted to measures of variability. The easiest measure of variability to understand is the range and the range is simply the difference between the largest and the _____ scores in a group of scores.

═══ STOP ═══════════════════════ STOP ═══

smallest

---

6 The range is the difference between the largest and the smallest scores. For example, with scores 4, 2, 1, 5, 3, the smallest score is _____ and the largest score is _____. The difference between the two is equal to _____ and defines the range.

═══ STOP ═══════════════════════ STOP ═══

1, 5; 4

---

7 Given scores 4, 2, 2, 7, 9, 6. The range is equal to _____.

═══════ STOP ═══════════════════════ STOP ═══════

$$9 - 2 = 7$$

---

8 Group I: 5, 5, 5, 5; Group II: 4, 3, 8, 9, For Group I the range is equal to _____ and for Group II the range is equal to _____.

═══════ STOP ═══════════════════════ STOP ═══════

zero (0), 6

---

9 Group I: 5, 5, 5, 5; Group II: 4, 3, 8, 9. The range for Group I is equal to 0 and the range for Group II is equal to 6. Since the range is one measure of variability, we then may say that Group II is more _____ than Group I.

═══════ STOP ═══════════════════════ STOP ═══════

variable

---

10 The difficulty with the range as a measure of variability is that even though there may be a large number of scores in a distribution only two are considered: the _____ and the _____.

═══════ STOP ═══════════════════════ STOP ═══════

largest, smallest

---

11 Thus the value of the range may fluctuate widely from sample to sample (all taken from the same population) because this measure of _____ is dependent only upon the largest and smallest scores in the sample.

═══════ STOP ═══════════════════════ STOP ═══════

variability

---

12 What is needed, then, is a measure of variability that considers not only the largest and smallest score but one that considers _____ scores.

═══════ STOP ═══════════════════════ STOP ═══════

all

---

13 It would appear that a measure of _____ that was dependent upon every _____ would be less likely to fluctuate widely from one sample to another.

═══════ STOP ═══════════════════════ STOP ═══════

variability, score

---

14 The standard deviation is such a measure. If the standard deviation is computed, _____ scores must be taken into consideration.

═══════ STOP ═══════════════════════════ STOP ═══════

all

---

15 The standard _____ may be defined as $S = \sqrt{\sum x^2/(N-1)}$ (where x is a deviation score).

═══════ STOP ═══════════════════════════ STOP ═══════

deviation

---

16 The square of the standard deviation (S) is called the *variance*. Thus the _____ may be defined as $S^2 = \sum x^2/(N-1)$.

═══════ STOP ═══════════════════════════ STOP ═══════

variance

---

17 We have previously defined x as the difference between the raw score and the mean. Thus _____ $= X - M$.

═══════ STOP ═══════════════════════════ STOP ═══════

x

---

18 $$x = \underline{\qquad} - M$$

═══════ STOP ═══════════════════════════ STOP ═══════

X

---

19 $$x = X - \underline{\qquad}$$

═══════ STOP ═══════════════════════════ STOP ═══════

M

---

20 and _____ $= X - M$.

═══════ STOP ═══════════════════════════ STOP ═══════

x

---

21 Thus if both sides of the equation are squared then _____ $= (X - M)^2$.

═══════ STOP ═══════════════════════════ STOP ═══════

x^2

---

22 Suppose $M = 20$ and $X = 25$, then $x =$ _____ and $x^2 =$ _____.

═══ STOP ═══ ═══ STOP ═══

5, 25

---

23 Suppose $M = 50$ and $X = 40$. Then $x =$ _____ and $x^2 =$ _____.

═══ STOP ═══ ═══ STOP ═══

-10, 100

---

24 Given scores 0, 1, 3, 4, $M =$ _____.

═══ STOP ═══ ═══ STOP ═══

2

---

25 Given scores 0, 1, 3, 4, $M = 2$. Find x for 0, 1, 3, and 4. _____, _____, _____, and _____.

═══ STOP ═══ ═══ STOP ═══

-2, -1, 1, and 2

---

26 With scores 0, 1, 3, and 4, $x_1 = -2$, $x_2 = -1$, $x_3 = 1$, $x_4 = 2$, $x_1^2 =$ _____, $x_2^2 =$ _____, $x_3^2 =$ _____, and $x_4^2 =$ _____. $\sum x^2 =$ _____.

═══ STOP ═══ ═══ STOP ═══

4, 1, 1, 4; 10

---

27 With scores 0, 1, 3, 4, $\sum x^2 = 10$, $N =$ _____.

═══ STOP ═══ ═══ STOP ═══

4

---

28 With scores 0, 1, 3, 4, $\sum x^2 = 10$, $N = 4$. The variance, which is the square of the standard deviation, has been defined $S^2 = \sum x^2/(N - 1)$. $N - 1 =$ _____, and the numerical value of $S^2 =$ _____.

═══ STOP ═══ ═══ STOP ═══

3, $10/3 = 3.33$

29
$$S^2 = \frac{\sum x^2}{(?)}$$

═══ STOP ═══ ═══ STOP ═══

$N - 1$

30

$$S^2 = \frac{(?)}{N - 1}$$

═══════ STOP ═══════════════════════ STOP ═══════

$$\sum x^2$$

31

$$\underline{\hspace{1cm}} = \sum x^2/(N - 1)$$

═══════ STOP ═══════════════════════ STOP ═══════

$$S^2$$

32

$$S^2 = \underline{\hspace{1cm}}$$

═══════ STOP ═══════════════════════ STOP ═══════

$$\frac{\sum x^2}{N - 1}$$

---

33   Given scores 1, 3, 5, 7, $M = \underline{\hspace{1cm}}$.

═══════ STOP ═══════════════════════ STOP ═══════

4

---

34   Given scores 1, 3, 5, 7, $M = 4$. $x_1^2 = \underline{\hspace{1cm}}$, $x_2^2 = \underline{\hspace{1cm}}$, $x_3^2 \underline{\hspace{1cm}}$, $x_4^2 = \underline{\hspace{1cm}}$, and $\sum x^2 = \underline{\hspace{1cm}}$.

═══════ STOP ═══════════════════════ STOP ═══════

9, 1, 1, 9, 20

---

35   With scores 1, 3, 5, 7, $\sum x^2 = 20$, $N - 1 = \underline{\hspace{1cm}}$, and the variance $S^2 = \underline{\hspace{1cm}}$.

═══════ STOP ═══════════════════════ STOP ═══════

3, 20/3 = 6.67

---

36   Given scores 0, 2, 4, 5, 4, $M = \underline{\hspace{1cm}}$ and $\sum x^2 = \underline{\hspace{1cm}}$.

═══════ STOP ═══════════════════════ STOP ═══════

3, 16

---

37   With scores 0, 2, 5, 4, 4, $\sum x^2 = 16$, and the variance $S^2 = \underline{\hspace{1cm}}$.

═══════ STOP ═══════════════════════ STOP ═══════

4

---

38 With scores 0, 2, 5, 4, 4, $S^2 = 4$ and the standard deviation (S), which is the square root of the variance, is equal to _____.

═══════ STOP ═══════════════════════════════════ STOP ═══════

2

---

39 $$X - M = \text{_____}$$

═══════ STOP ═══════════════════════════════════ STOP ═══════

x

---

40 Hence $(X - M)^2 = $ _____.

═══════ STOP ═══════════════════════════════════ STOP ═══════

x^2

---

41 Since $(X - M)^2 = x^2$, $\sum (X - M)^2 = $ _____.

═══════ STOP ═══════════════════════════════════ STOP ═══════

$\sum x^2$

---

42 $\sum x^2/(N - 1)$ is the formula for the _____.

═══════ STOP ═══════════════════════════════════ STOP ═══════

variance

---

43 And $$\frac{\sum (X - M)^2}{N - 1} = \text{_____}.$$

═══════ STOP ═══════════════════════════════════ STOP ═══════

variance (S^2)

---

44 Without proof we state that

$$\frac{\sum x^2}{N - 1} = \frac{\sum X^2 - \frac{(\sum X)^2}{N}}{N - 1}$$

and the latter formula is called the *raw-score formula for the* _____.

═══════ STOP ═══════════════════════════════════ STOP ═══════

variance (square of the standard deviation)

45 The raw-score formula for S^2 is most useful when it is inconvenient to compute deviation scores.

$$S^2 = \frac{\sum X^2 - \frac{(\sum X)^2}{N}}{(?)}$$

════════ STOP ════════════════════════ STOP ════════

$$N - 1$$

46

$$S^2 = \frac{\sum X^2 - \frac{(\sum X)^2}{(?)}}{N - 1}$$

════════ STOP ════════════════════════ STOP ════════

$$N$$

47

$$S^2 = \frac{(?) - \frac{(\sum X)^2}{N}}{N - 1}$$

════════ STOP ════════════════════════ STOP ════════

$$\sum X^2$$

48

$$S^2 = \frac{\sum X^2 - \frac{(?)}{N}}{N - 1}$$

════════ STOP ════════════════════════ STOP ════════

$$(\sum X)^2$$

49

$$S^2 = \frac{\sum X^2 - (?)}{N - 1}$$

════════ STOP ════════════════════════ STOP ════════

$$\frac{(\sum X)^2}{N}$$

50

$$S^2 = \frac{\sum X^2 \; (?) \; \frac{(\sum X)^2}{N}}{N - 1}$$

════════ STOP ════════════════════════ STOP ════════

$$- \text{ (minus)}$$

51

$$S^2 = \frac{(?)}{N - 1}$$

═══════ STOP ═══════ ═══════ STOP ═══════

$$\sum X^2 - \frac{(\sum X)^2}{N}$$

52

$$\underline{} = \frac{\sum X^2 - \frac{(\sum X)^2}{N}}{N - 1}$$

═══════ STOP ═══════ ═══════ STOP ═══════

$$S^2$$

---

53 Thus $S^2 = $ _____ .

═══════ STOP ═══════ ═══════ STOP ═══════

$$\frac{\sum X^2 - \frac{(\sum X)^2}{N}}{N - 1}$$

---

54 Given scores 0, 2, 5, 4, 4, $\sum X = $ _____ and $\sum X^2 = $ _____ .

═══════ STOP ═══════ ═══════ STOP ═══════

15, 61

---

55 With scores 0, 2, 4, 5, 4, $\sum X = 15$, $\sum X^2 = 61$,

$$S^2 = \frac{\sum X^2 - \frac{(\sum X)^2}{N}}{N - 1} \qquad \text{then} \qquad \frac{(\sum X)^2}{N} = \underline{}.$$

═══════ STOP ═══════ ═══════ STOP ═══════

$(15)^2/5 = 45$

---

56 With scores 0, 2, 5, 4, 4, $\sum X^2 = 61$, $(\sum X)^2/N = 45$, $\sum X^2 - (\sum X)^2/N = $

_____ .

═══════ STOP ═══════ ═══════ STOP ═══════

$61 - 45 = 16$

---

57 With scores 0, 2, 5, 4, 4, $\sum X^2 - (\sum X)^2/N = 16$,

$$S^2 = \frac{\sum X^2 - \dfrac{(\sum X)^2}{N}}{N-1} = \underline{\hspace{1cm}}.$$

════ STOP ════════════════════════ STOP ════

$$16/4 = 4$$

---

58 With scores 0, 2, 5, 4, 4, $S^2 = 4$, and $S = \underline{\hspace{1cm}}$.

════ STOP ════════════════════════ STOP ════

$$2$$

---

59 S^2 is called the _____ and S is called the _____ _____.

════ STOP ════════════════════════ STOP ════

variance, standard deviation

---

60 Given scores 0, 2, 1, 4, 5, 6. $\sum X^2 = \underline{\hspace{1cm}}$, $(\sum X)^2/N = \underline{\hspace{1cm}}$.

════ STOP ════════════════════════ STOP ════

82, 54

---

61 With scores 0, 2, 1, 4, 5, 6, $\sum X^2 = 82$, $(\sum X)^2/N = 54$; $\sum x^2 = \sum X^2 - (\sum X)^2/N = \underline{\hspace{1cm}}$.

════ STOP ════════════════════════ STOP ════

28

---

62 With scores 0, 2, 1, 4, 5, 6, $\sum X^2 = 82$, $(\sum X)^2/N = 54$, $\sum x^2 = 28$, $S^2 = \underline{\hspace{1cm}}$.

════ STOP ════════════════════════ STOP ════

$$28/5 = 5.60$$

---

63 $$\sum x^2 = \underline{\hspace{1cm}} - (\sum X)^2/N$$

════ STOP ════════════════════════ STOP ════

$$\sum X^2$$

---

64 $$\underline{\hspace{1cm}} = \sum X^2 - (\sum X)^2/N$$

════ STOP ════════════════════════ STOP ════

$$\sum x^2$$

65

$$\sum x^2 = \underline{\hspace{2cm}}$$

STOP ═══ STOP

$$\sum X^2 - (\sum X)^2/N$$

---

66 The raw-score formula for S^2 is \underline{\hspace{2cm}}.

STOP ═══ STOP

$$\frac{\sum X^2 - \frac{(\sum X)^2}{N}}{N - 1}$$

67

$$S^2 = \frac{\sum X^2 - \frac{(\sum X)^2}{N}}{N - 1}$$

$\sum X^2 = 30$, $\sum X = 12$, $N = 6$, $(\sum X)^2/N = \underline{\hspace{2cm}}$.

STOP ═══ STOP

$$(12)(12)/6 = 24$$

---

68 $N = 11$, $\sum X^2 = 80$, $\sum X = 22$, $(\sum X)^2/N = \underline{\hspace{2cm}}$. $\sum X^2 - (\sum X)^2/N = \underline{\hspace{2cm}}$.

STOP ═══ STOP

$$44; 36$$

---

69 $N = 20$, $\sum X^2 = 1092$, $\sum X = 130$. We wish to find S^2. $(\sum X)^2/N = \underline{\hspace{2cm}}$, $\sum x^2 = \sum X^2 - (\sum X)^2/N = \underline{\hspace{2cm}}$.

STOP ═══ STOP

$$845, 247$$

---

70 $N = 12$, $\sum X^2 = 1321$, $\sum X = 120$, $(\sum X)^2/N = \underline{\hspace{2cm}}$, $\sum x^2 = \underline{\hspace{2cm}}$, $S^2 = \underline{\hspace{2cm}}$.

STOP ═══ STOP

$$1200, 121, 11$$

---

71 $N = 10$, $\sum X^2 = 2050$, $\sum X = 130$, $\sum x^2 = \underline{\hspace{2cm}}$, $S^2 = \underline{\hspace{2cm}}$.

STOP ═══ STOP

$$360, 40$$

---

72 $N = 8, \sum X^2 = 364, \sum X = 28, S^2 = $ _____.

════════ STOP ═══════════════════════════ STOP ════════

38

---

73 $S^2 = 144, S = $ _____.

════════ STOP ═══════════════════════════ STOP ════════

12

---

74 $\sum X^2 = 3624, \sum X = 392, N = 56, S^2 = $ _____, $S = $ _____.

════════ STOP ═══════════════════════════ STOP ════════

16, 4

---

75 $\sum X^2 = 171, \sum X = 30, N = 10, S^2 = $ _____, $S = $ _____.

════════ STOP ═══════════════════════════ STOP ════════

9, 3

---

76 $\sum X^2 = 334, \sum X = 42, N = 14, S^2 = $ _____, $S = $ _____.

════════ STOP ═══════════════════════════ STOP ════════

16, 4

---

77 $\sum X^2 = 429, \sum X = 66, N = 22, S^2 = $ _____.

════════ STOP ═══════════════════════════ STOP ════════

11

---

78 With scores 1, 1, 2, 6, $\sum X^2 = $ _____, $S^2 = $ _____.

════════ STOP ═══════════════════════════ STOP ════════

42, 5.67

---

79 SIQR $= (Q_3 - Q_1)/2.$ $Q_1 = 45, Q_3 = 75.$ SIQR $= $ _____.

════════ STOP ═══════════════════════════ STOP ════════

$30/2 = 15$

---

80 $Q_1 = 40$, $Q_2 = 60$, $Q_3 = 90$. SIQR = _____.

$$\equiv\!\!\equiv\!\!\equiv \text{STOP} \equiv\!\!\equiv\!\!\equiv\!\!\equiv\!\!\equiv\!\!\equiv\!\!\equiv\!\!\equiv \text{STOP} \equiv\!\!\equiv\!\!\equiv$$

$$50/2 = 25$$

---

81   Given scores 9, 4, 6, 2, 1, 9, 4, 5, 4, 2, 1, 3. Arrange the scores in ascending order.

$$\equiv\!\!\equiv\!\!\equiv \text{STOP} \equiv\!\!\equiv\!\!\equiv\!\!\equiv\!\!\equiv\!\!\equiv\!\!\equiv\!\!\equiv \text{STOP} \equiv\!\!\equiv\!\!\equiv$$

$$1, 1, 2, 2, 3, 4, 4, 4, 5, 6, 9, 9$$

---

82   From frame 81, $Q_1 =$ _____; $Q_3 =$ _____.

$$\equiv\!\!\equiv\!\!\equiv \text{STOP} \equiv\!\!\equiv\!\!\equiv\!\!\equiv\!\!\equiv\!\!\equiv\!\!\equiv\!\!\equiv \text{STOP} \equiv\!\!\equiv\!\!\equiv$$

$$2.0, \; 5.5$$

---

83 $Q_1 = 2.0$, $Q_3 = 5.5$. SIQR = _____.

$$\equiv\!\!\equiv\!\!\equiv \text{STOP} \equiv\!\!\equiv\!\!\equiv\!\!\equiv\!\!\equiv\!\!\equiv\!\!\equiv\!\!\equiv \text{STOP} \equiv\!\!\equiv\!\!\equiv$$

$$3.5/2 = 1.75$$

---

84   Given scores 1, 3, 5, 6, 5, 7, 5, 6, 4, 1, 2, 5, 8, 1, 2, 1. $Q_1 =$ _____.

$$\equiv\!\!\equiv\!\!\equiv \text{STOP} \equiv\!\!\equiv\!\!\equiv\!\!\equiv\!\!\equiv\!\!\equiv\!\!\equiv\!\!\equiv \text{STOP} \equiv\!\!\equiv\!\!\equiv$$

$$1.50$$

---

85   From frame 84, $Q_3 =$ _____.

$$\equiv\!\!\equiv\!\!\equiv \text{STOP} \equiv\!\!\equiv\!\!\equiv\!\!\equiv\!\!\equiv\!\!\equiv\!\!\equiv\!\!\equiv \text{STOP} \equiv\!\!\equiv\!\!\equiv$$

$$5.50$$

---

86   From frame 85, SIQR = _____.

$$\equiv\!\!\equiv\!\!\equiv \text{STOP} \equiv\!\!\equiv\!\!\equiv\!\!\equiv\!\!\equiv\!\!\equiv\!\!\equiv\!\!\equiv \text{STOP} \equiv\!\!\equiv\!\!\equiv$$

$$4.00/2 = 2.00$$

Standard Scores

Measurements of objects or events can be expressed in an unlimited variety of units. Time measures, for instance, include days, hours, minutes, years, microseconds, and so on. Comparisons of sets of data are complicated by such differences in units of measurement. The purpose of this chapter is to explain methods for converting measurements from any arbitrary unit to a chosen standard.

Standardization of the unit of measurement is accomplished by equating particular distributions in terms of central tendency and variability. When the scores concern the same qualitative aspect of the observed objects or events, as in two measures of temperature, the choice of the common unit is not particularly important. The concept of standard scores, however, is particularly significant when objects and events are to be compared on qualitatively different dimensions, such as anxiety and introversion, through the use of normative data derived from large populations.

THE z-SCORE SCALE

There are many different standard units of measurement in common use. This chapter will be concerned primarily with the z-score scale and its modifications. The unit of measurement on the z-score scale is the standard deviation of the distribution of the original measurements. The mean of the z-score scale is zero. Since the z-score scale is marked in units of one standard deviation, an original score that falls one standard deviation below the mean of the original scores would be a z-score of -1.00. A raw score that corresponds to a point one-half standard deviation above the original mean would be equivalent to a z score of $+0.50$.

AN EXAMPLE Imagine yourself in the following predicament. An acquaintance of yours, who likes to pose paradoxes, discovers that you are taking a course in statistics and asks if you are heavier than you are tall. Although the question sounds silly, for some reason or other you can't quite dismiss the question as meaningless. You ask your instructor what he thinks about the question. With a grin, he says that it sounds interesting, and assigns you the task of answering it.

After some consideration, you decide that the essential difficulty lies in the fact that height and weight units are not comparable; they measure qualitatively different things. Of course, you could locate your own position on each scale separately. In fact you could also locate the position of the average person on each scale separately. This would allow you to say whether you are heavier or lighter and whether you are shorter or taller than the average person.

What you need now is some way of changing the units of each scale so that they are the same. You could then determine whether you are higher on one scale than you are on the other. The z-score scale, which has as its unit of measurement the standard deviation, provides a solution to this problem.

Suppose your height is 6 feet 1 inch, or 73 inches. Suppose further that the average adult is 70 inches tall, and that the standard deviation of adult heights is 3 inches. You know that the z-score for your height will be positive, since you are taller than the average person. The difference between your height and the mean in inches is 3. Conveniently, this is exactly equal to one standard deviation, or one z-score unit. On the z-score scale, your height is therefore 1.00.

In terms of a formula, the relationship between raw scores and z-scores for height may be expressed as follows:

$$z_H = \frac{X_H - M_H}{S_H} = \frac{73 - 70}{3} = 1.00$$

where X_H = a raw score (height in inches), M_H = mean (for height), and S_H = standard deviation (for height).

To complete the process of arriving at an answer for your instructor and your paradoxical friend, let us assume that you are also heavier at 210 pounds than the average adult. If the mean weight for all adults is 170 pounds, and the standard deviation is 20 pounds, then you can also calculate a z-score for your weight:

$$z_W = \frac{X_W - M_W}{S_W} = \frac{210 - 170}{20} = 2.00$$

After the next class period you tell your instructor that you are definitely heavier than you are tall, and demonstrate your reasoning. That evening if you corner the acquaintance who started all this and announce triumphantly, "I am heavier than I am tall!", he will probably look at you blankly for a moment, then grin, and say, "Well, no wonder you don't topple over."

OTHER STANDARD SCORES

A virtually limitless number of modifications may be derived from the basic z-score scale, simply by changing the mean of the standard-score scale from zero to some other value and/or by changing the unit of measurement to some particular multiple or fraction of one standard deviation. The most commonly used modifications are those that avoid the necessity of negative standard scores.

A convenient form for standard scores has a mean of 50 and a standard deviation of 10. We will designate such scores with a capital letter Z.[1]

Another frequently used standard score is called the *stanine*. Although originally defined as having a mean of 5.00 and a standard deviation of 1.96, the standard deviation is usually considered to be 2.00 in order to simplify hand computation.

A relatively simple formula for converting a raw score to any standard score is given below. Note that this is only an extension of the z-score formula given earlier.

$$\text{standard score} = \left(\frac{X - M_x}{S_x}\right)(S_s) + M_s$$

where X = raw score, M_x = raw-score mean, S_x = raw-score standard deviation, M_s = standard-score mean, and S_s = standard-score standard deviation.

To illustrate the use of this formula, we will use the weight we assumed for you earlier as the raw score to be converted to various standard scores. First let us define the standard score as a z score to show that the extended formula yields the same value when $M_s = 0.0$ and $S_s = 1.0$:

$$z \text{ score} = \left(\frac{210 - 170}{20}\right)(1.0) + 0.0 = 2.0$$

When $M_s = 5.0$ and $S_s = 2.0$, we obtain a stanine score:

$$\text{stanine} = \left(\frac{210 - 170}{20}\right)(2.0) + 5.0 = 9.0$$

When $M_s = 50.0$ and $S_s = 10.0$, we obtain a Z score:

$$Z \text{ score} = \left(\frac{210 - 170}{20}\right)(10.0) + 50.0 = 70.0$$

ORDINAL STANDARD SCORES (CENTILES)

In Chapter Five we discussed the points in a distribution of scores which designate the lower 25 percent (Q_1), 50 percent $(Q_2$, the median), and 75 percent (Q_3) of the cases. If we extend this notion of dividing the total number of cases into equal parts, we can establish the basis for centile scores. Instead of dividing the total number of cases into quarters (25 per-

---

[1] The so-called T score also has a mean of 50 and a standard deviation of 10, but involves special computations which force the raw scores toward a "normal" distribution (see Chapter 8). There is an infinite variety of such "nonlinear" transformations that can be applied to raw data, but the topic is beyond the scope of this book.

cent), we divide it into 100 equal parts. The various points on the score scale which separate these 1 percent parts of the distribution are called *centiles.* The median point is called the *fiftieth centile.* Similarly, the 25 percent point is called the *twenty-fifth centile.* The term *percentile* is often used interchangeably with centile.

If the total number of cases is divided into tenths, we have the basis for what are known as *deciles.* A decile score of 1 is equivalent to a centile score of 10, decile 2 equals centile 20, and so on. The derivation of still another more gross kind of ordinal standard score should be obvious from its name: *quartile.* We have already made use of them in the form of Q_1 (first quartile, 25 percent), Q_2 (second quartile, 50 percent), and Q_3 (third quartile, 75 percent). The median, which separates the upper and lower 50 percent of the cases in the distribution, is equivalent to quartile 2, decile 5, and centile 50.

It is very important to keep in mind the quite different natures of interval standard scores like z, Z, stanine, and so on, and ordinal standard scores such as quartiles, deciles, and centiles. Included in manuals for various psychological tests, one will usually find tables for converting raw scores to standard scores. Some of these tables yield percentile (centile) equivalents, while others yield interval standard scores of various kinds. Figure 6-1 illustrates that although the central points of Z-score and centile-score distributions are labeled identically (50), the ways in which the scores are distributed away from the center are quite dissimilar.

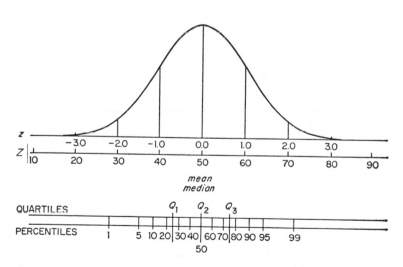

Figure 6-1 *Comparison of interval and ordinal standard-score distributions.*

The significance of the difference between the distributions of interval and ordinal standard scores will become more apparent in Chapter Eight, which deals with percentages of area under the normal curve. For the present, we need only be aware of the fact that, in terms of the number of cases involved, a centile score of 10 is much closer to the middle of the distribution than is a Z score of 10.

Computation of ordinal standard-score equivalents from a raw-score distribution is a relatively simple matter, once the raw scores have been compiled to form a *cumulative frequency distribution*. This procedure is illustrated below:

X	f	cf	C	D
5	1	10	95	9.5
4	1	9	85	8.5
3	2	8	70	7.0
2	3	6	45	4.5
1	3	3	15	1.5

$$C \text{ (centile)} = \frac{50(2cf - f)}{N}.$$

$$D \text{ (decile)} = \frac{5(2cf - f)}{N}$$

where cf = the cumulative frequency for the X score, f = the frequency for the particular X score, and N = the total number of cases.

For the first $(X = 1)$ centile:

$$C = \frac{50(2 \times 3 - 3)}{10} = \frac{50 \times 3}{10} = 15$$

For each succeeding X-score level, the corresponding f value is added to the preceding cumulative frequency to yield a new cf value. For $X = 1$, $cf = 0 + 3 = 3$. For $X = 2$, $cf = 3 + 3 = 6$, and so on.

$$C = \frac{50(2 \times 6 - 3)}{10} = \frac{50 \times 9}{10} = 45$$

For $X = 3$,

$$C = \frac{50(2 \times 8 - 2)}{10} = \frac{50 \times 14}{10} = 70$$

Note that the final cumulative frequency will always equal the total number of scores in the distribution. Note also that the decile equivalents will always be exactly one-tenth the size of the corresponding centile values.

COMPARISON OF INTERVAL AND ORDINAL STANDARD SCORES

Because of their more immediate intuitive appeal, centile (percentile) scores are widely used in reporting scores from standardized psychological instruments for the assessment of ability, achievement, interests, and so on. The statement, "He scored at a higher level than 98 percent of other students," is more easily understood than the statement, "His score was two standard deviations above the mean for all students." Unfortunately, the use of percentile scores or other ordinal standard scores imposes serious limitations on the usefulness of the data for purposes of statistical inference. The most powerful tools of statistical inference require that the data and descriptive statistics be represented on an interval scale of measurement.

On the other hand, when the raw-score distribution is highly skewed or truncated, and conversion to a standard-score scale is desirable, an ordinal measure (deciles or centiles) would more adequately represent the location of an individual's performance relative to those of others in the group.

Usually, investigators in the behavioral sciences are willing to tolerate a moderate degree of skewness in their raw-score distributions, and will compute interval standard scores such as z, Z, or stanines in order to capitalize on the power of the inferential statistical methods that require scores of this kind.

SUMMARY

Standard scores allow direct comparison of individual scores obtained from scales of measurement with quantitatively or even qualitatively different units. Two basically dissimilar types of standard scores are commonly used in the behavioral sciences. The most widely used type of standard score has the characteristics of an interval scale of measurement. The examples described in this chapter were $z(M = 0.0, S = 1.0)$, stanine $(M = 5.0, S = 2.0)$, and $Z(M = 50.0, S = 10.0)$. The general formula for computation of an interval standard score is

$$SS = \frac{X - M_x}{S_x}(S_s) + M_s$$

The other type of standard score has the properties of an ordinal scale of measurement. The examples given in this chapter were quartiles, deciles, and centiles (percentiles). Ordinal standard scores are preferable when the raw-score distribution is very skewed or truncated. These scores are not appropriate for the more powerful methods of statistical inference described later in this book, and for this reason their use is typically confined to the reporting of individual test scores with reference to some normative population. Ordinal standard-score equivalents are most economically computed through the use of a cumulative frequency distribution.

PROBLEMS

1. Compute z, Z, and stanine for each of the following raw scores:
   (a) $M = 50, S = 10, X = 55$
   (b) $M = 0, S = 5, X = 5$
   (c) $M = 30, S = 10, X = 15$
   (d) $M = 400, S = 100, X = 200$
   (e) $M = -10, S = 10, X = 0$
2. Complete the following:
   (a) Centile of 60 = decile of_____.

(b) Decile of 5 = quartile of_____.

(c) The median = decile_____, quartile_____, and centile

_____.

(d) The formula for z = _____.

(e) A z score is an_____ -scale measure.

(f) A decile is an_____ -scale measure.

3. Answer the following:

(a) Which is more unusual:

(1) stanine of 5 or a decile of 5?

(2) $z = 2$ or $Z = 2$?

(3) $Z = 45$ or a percentile of 84?

(b) If you squared every raw score, would the result be a "standard score" of some kind?

(c) If you had to discuss a child's achievement test results with his parents, which kind of standard score would you prefer to use?

(d) Figure 6-1 shows Z of 50 equivalent to a percentile of 50. Under what cir- circumstances would this equivalence not hold true?

Seven

Foundations of Statistical Inference

Nearly all male college students have had an experience like the following. Joe calls Gladys and asks her for a date on Saturday night. She replies in her most sincere tone that her grandmother just died and that she must attend the funeral, which unfortunately is to be held late Saturday afternoon. Joe calls the next week and finds that Gladys will be out of town over the weekend because her other grandmother has joined the dear departed. Joe calls the third week only to hear that Gladys' grandfather has joined his wife and Gladys must attend the funeral. And so it goes.

Joe has a problem: "Does she really want to go out with me?" Strangely enough, Joe's problem is similar to the problem faced by the statistician. Both must make a decision on the basis of incomplete information. After all, two deceased grandmothers and a departed grandfather, while being reasonable excuses for declining a date, may possibly reflect a certain disenchantment on Gladys' part. Joe must decide whether to call her again or to give up. Like Joe, the statistician is also forced to make decisions in the face of uncertainty.

Another example will be useful in explaining this decision-making process more completely. Suppose you and a friend go to the coffee shop every day for a cup of coffee. Furthermore, suppose that each day you flip a coin to see who picks up the check. Naturally, you assume that you both have an equal chance of winning, or you wouldn't be willing to flip for it.

Suppose your friend wins the first day. If you were suspicious, you might wonder if the coin was biased (that is, you wonder if your friend is cheating). Most likely you would decide that one trial (one flip) isn't enough to tell. But suppose your friend won the first four days in a row. Each time you would be faced with the same question: Is the coin biased? Table 7-1 shows the consequences of making a decision and then telling your friend about it.

Table 7-1 indicates that you could be wrong in two different ways: (1) If your friend *is not* cheating, and you accuse him of it, you run the risk of losing a friend and maybe also getting a black eye; (2) if

Table 7-1   Decision table

	He is cheating	He is not cheating
You decide he is cheating	You are correct*	Black eye (alpha error)
You decide he is not cheating	Continued support (beta error)	You are correct**

* You get a black eye, but you are ahead financially.
** He is still your friend, although an expensive one.

your friend *is* cheating, and you decide that he is not, you run the risk of supporting him all the way through school.

When the flipping started, you formed what is known as the *null hypothesis:* Events are due to chance (that is, the coin is unbiased). Statisticians also begin by stating a form of the null hypothesis, and then they collect data with which they may test their hypothesis.

In testing the null hypothesis, two kinds of errors are possible, as noted in Table 7-1:

1. *Alpha error:* the null hypothesis is true, but you reject it. In the example above, the null hypothesis was that the events were due to chance (that he was not cheating). If the events were due entirely to chance and you decided that he was cheating, you then committed an alpha error.
2. *Beta error:* the null hypothesis is false, but you decide not to reject it. Again in the above example, if the coin was biased (your friend was cheating) and you decided that there was no evidence to indicate that the coin was biased, you then would have committed a beta error.

For many years statisticians were concerned mainly with avoiding alpha errors (as you would be afraid of getting a black eye, in our example). Gradually, they began to worry about beta errors as well (in our example, the cost of supporting your friend through failure to decide that he is cheating).

You probably are wondering why we don't just reduce the probability of both alpha and beta errors so that we wouldn't have to worry. The trouble is that the two kinds of errors are related. With other things held constant, if you decrease the probability of an alpha error, you increase the probability of a beta error, and vice versa. Primarily for this reason, statisticians have generally agreed to reject the null hypothesis when the chances are equal to or less than 5 percent that the observed series of events is due to chance. Thus, when the chances are greater than 5 percent, we will say that we have no evidence to assume that anything other than chance influenced the observed series of events. In terms of the coin-flipping problem, we would reject the null hypothesis (that is, the coin is fair) if the number of heads observed in the whole series of events would have happened less than 5 percent of the time by chance.

DEFINITION OF PROBABILITY

The probability of an event may be defined as the number of *favorable* events divided by the total number of *possible* events. Suppose we roll one die, which has six sides, along the floor and observe which face ends up on top. If the die is perfectly balanced, all faces (1 to 6) have an equal chance of appearing. Thus, the total possible events would equal 6. If we were interested in a specific face, say 2 (and only 2), then the number of favorable events would be 1. Thus, the probability of rolling a 2 would be equal to 1 divided by 6, or 1/6, or 16.67 precent, or 0.17. If there were ten possible events and three favorable events, the probability would be 3/10, or 0.30. If all possible events are "favorable," as they would be if you asked the chances of rolling some number from 1 to 6 with a die, the probability would then be 1.00. On the other hand, if there would be no favorable events, as there would be if you asked the chances of rolling a 22 with a single roll of the die, then the probability would equal 0.00. It should be obvious that the probability of an event cannot be more than 1.00 (where all possible events are favorable) nor less than 0.00 (where no event is favorable).

NOTATION   If there is some event A and we wish to denote the probability (Pr) of such an event, we use the notation

$$\mathrm{Pr}(A) = \frac{N_A}{N_E}$$

and this is read in the following manner: The probability of event A is equal to the number of As divided by the number of Es (events). Thus, if $N_A = 4$, and $N_E = 12$, then $\mathrm{Pr}(A) = N_A/N_E = 4/12 = 0.33$.

THE ADDITIVE LAW OF PROBABILITY

Suppose you want to know the probability of rolling any even-numbered face in one roll of a die. To find the probability of a class (or group) of events, you simply add the probabilities of each event within the class. The class in our example is "even numbers," and there are three possible events in that class (2, 4, 6). Since the probability of each event (for any particular face) is 1/6, we add to get $3/6 \doteq 0.50$ as the probability of rolling an even number.

You must be very careful, however, to avoid counting the probability of the same event more than once. For example, the probability of drawing a king or a diamond from a deck of cards is not equal to the probability of drawing a king plus the probability of drawing a diamond, since the event "king of diamonds" is included in both "kings" and "diamonds."

The additive law of probability is used in situations where the events are related by the word *or*. In the even-numbered-face example, you wanted to find the probability of a 2, *or* a 4, *or* a 6.

Look at Table 7-2. It consists of all possible outcomes when two dice

Table 7-2 Possible outcomes with two dice

		Second die				
	1	2	3	4	5	6
1	11	12	13	14	15	16
2	21	22	23	24	25	26
3	31	32	33	34	35	36
First die 4	41	42	43	44	45	46
5	51	52	53	54	55	56
6	61	62	63	64	65	66

are rolled. The rows represent the possible outcomes of one die and the columns represent the possible outcomes of the second die. The entries in the table represent the two outcomes together. If the first die rolled were a three and the second a four, we could go down the fourth column and across the third row, and find the outcome 34 at the intersection. According to our definition of probability, the probability of some set of favorable events— say, those outcomes which are included in the circle—would be equal to the number of favorable events divided by the total number of events. There are nine outcomes in the circle and a total of 36 possible outcomes or events. Hence, the probability of rolling two dice and getting any one of the 9 events in the circle would be 9/36, or 0.25.

Returning to our earlier illustration, suppose we roll two dice and wish to determine the probability of rolling a 7 (where the two faces sum to 7) or an 11. From Table 7-3 it can be seen that the probability of rolling a 7 is 6/36, while the probability of rolling an 11 is 2/36. Hence, the probability of

Table 7-3 Possible outcomes with two dice

		Second die				
	1	2	3	4	5	6
1	11	12	13	14	15	16
2	21	22	23	24	25	26
First die 3	31	32	33	34	35	36
4	41	42	43	44	45	46
5	51	52	53	54	55	56
6	61	62	63	64	65	66

rolling a 7 or an 11 is $6/36 + 2/36 = 8/36$, or 0.22. It can be seen that the additive law is used when you wish to determine the probability of event *A or* event *B* occurring.

THE MULTIPLICATIVE LAW OF PROBABILITY

In order to understand the multiplicative law of probability, we must introduce the notion of the sequence of a series of events. You will recall that in the coin-flipping example we wanted to know the probability of a series of four heads in a row. Each coin flip is a separate event, and we want to know the probability of the series as a unit.

The multiplicative law of probability applies to those situations where the events of interest are related by the word *and*. When a specific order is also implied for a series of events, the term *and then* specifies the sequential relationship. If you wanted to know the probability of rolling a 2 *and then* a 4 *and then* a 6 with a die, you would multiply the probabilities of the separate events: $1/6 \times 1/6 \times 1/6 = 1/216$. When the sequence of the events is not specified, however, it is necessary to compute the probabilities for all possible sequences and then add them to get the probability of any "favorable" sequence.

If you flip a coin, there are two possible outcomes, H (head) or T (tail). The probability of each of these (with an unbiased coin) is 1/2, or 0.50. If you flip the coin again, there are four possible outcomes from the two flips (H-H, H-T, T-H, T-T), and each possibility is equally likely, having a probability of 1/4, or 0.25. If you flip the coin three times, there are eight possible outcomes:

H-H-H, H-H-T, H-T-H, H-T-T,

T-H-H, T-H-T, T-T-H, T-T-T

We can answer certain questions about three flips of a coin by using both the multiplicative and the additive law. For example, what is the probability of getting two heads and one tail in three flips of a coin? Three of the possible outcomes fit the requirement, each with a probability of 0.125. Thus, the answer: $0.125\,(\text{H-H-T}) + 0.125\,(\text{T-H-H}) + 0.125\,(\text{H-T-H}) = 0.375$. The probability of getting two heads and a tail in three flips of a coin is equal to 0.375.

Notice that there is a direct relationship between the number of events and the probability of any one sequence of events. The probability of H is 1/2 (0.50), of H-H is 1/4, or 0.25, and of H-H-H is 1/8, or 0.125. You might suspect (correctly) that the probability of H-H-H-H is 1/16, or 0.0625. Generally stated, the multiplicative law of probability is that the probability of a specified series of events is equal to the product of the probabilities of the separate events. Thus, the probability of H-H-T is equal to $1/2 \times 1/2 \times 1/2 = 1/8$.

The separate probabilities need not be equal. For example, what is the probability of rolling an even number and then a 6? We know already that the chances of rolling an even number are 3/6, and the chances of rolling a 6 are 1/6. In two rolls, then the chances of rolling an even number followed by a 6 are $3/6 \times 1/6 = 3/36$.

As another example of the combination of the two laws, what is the probability of rolling one and only one 6 in two rolls of a die? This means that we roll a 6 on one roll and not on the other. Initially, we should note that we can roll a 6 on the first roll and not on the second. Or we can roll a non-6 on the first-roll and then roll a 6 on the second. Either way satisfies our requirements of rolling one and only one 6 in two rolls. The probability of rolling a 6 is equal to 1/6, and the probability of not rolling a 6 would be 5/6. Thus, the probability of rolling a 6 and then not rolling a 6 by the multiplicative law would equal $1/6 \times 5/6 = 5/36$, and the probability of not rolling a 6 and then rolling a six is equal to $5/6 \times 1/6 = 5/36$. Now, by the additive law, we get $5/36 + 5/36 = 10/36$, which is the probability of rolling one and only one 6 in two rolls of a die.

Table 7-3 can also be used to illustrate the multiplicative law of probability. For example, suppose you wanted to find the probability of rolling a 6 *and then* a 5. Looking at the intersection of the sixth row (representing a 6 on the first roll) and the fifth column, we see that there is a single entry, 65, which represents a 6 and then a 5. From the table, then, we can see that the probability of a 6, and then a 5 is equal to the probability of a 6 on the first roll times the probability of a 5 on the second roll. That is, $\Pr(6,5) = \Pr(6) \times \Pr(5) = 1/6 \times 1/6 = 1/36$. Similarly, you can easily determine that the probability of a 1 or a 2 on the first roll, followed by a 3 or a 4 on the second, is equal to $2/6 \times 2/6 = 4/36$, or 1/9.

RANDOM AND INDEPENDENT SELECTION

In later chapters you will find that the statistical tests with which you become acquainted nearly always assume random and independent selection of the data. Although this process is relatively easy to understand, often its importance is overlooked. If the assumptions of random and independent selection are not met, then the results of subsequent statistical manipulations will have little if any meaning. In fact, what is worse, you may come to an erroneous conclusion.

By *random selection* we mean that all elements in a population have an equal chance of being chosen at any specified time. Thus, if every card in a deck has an equal chance of being selected on any single draw, our selection procedure is random. We should emphasize here that random selection does not mean a haphazard or an unplanned selection procedure. Instead, it is planned by the experimenter because the probabilities involved are much easier to work with mathematically than are the probabilities that occur under nonrandom selection techniques.

Tables of random numbers have been developed to help the experimenter to be sure that his selection procedure is truly random. Such a table may be found in Appendix H, where the numbers appear in random order. That is, you can go down a column, across a row, or in any other predetermined order, and numbers will appear in true random order. If you were interested in selecting two-digit numbers randomly, you could take the last two digits in any row of numbers and follow this column down. The two-digit numbers you pick will be random.

Closely related to random selection, and preferred by experimenters for the same reasons, is independent selection. By *independence*, we mean that the probability of an event does not depend on previous events. The *gambler's fallacy* is an example in which the apparently related probabilities of a series of events are actually independent. For example, suppose we have a perfectly balanced coin, which we flip five times in a row and get heads every time. The gambler's fallacy would say that it is now time for a tail to appear, and thus the probability of getting a tail on the sixth flip would be greater than 1/2, or 0.50. This is called *gambler's fallacy* because the probability of getting a tail is, and always was, 0.50; thus, the sixth flip is independent of (or uninfluenced by) the events that preceded it. An example of this reasoning occurred recently in a well-known European casino. The same number came up six times in a row on the roulette wheel. Needless to say, the probability of such a sequence of events occurring is extremely small. It is interesting to note that nobody bet on the number during its long run. Each time after the number came up, the players would say, "It has come up twice (or three times, or four times, and so on) and won't come up on the next roll." However, the events were independent and the probability of the numbers occurring on any specific roll was always the same.

Another example will help illustrate this important concept. Suppose we have a deck of 52 cards and wish to draw the ace of spades. The probability of drawing this card (or any other) is 1/52. Suppose we draw a card and find that it is the jack of clubs. Now we can either return the card to the deck, shuffle, and draw again, or we can throw the card away and draw again. If we return the card to the deck, shuffle, and draw again, the probability of drawing the ace of spades remains 1/52. No matter how many times we draw a card and return it to the deck, the probability of drawing the ace of spades is always the same (1/52). This procedure illustrates the concept of independence because the probability of the ace of spades does not depend on what happened on previous draws from the deck.

Suppose when we drew the jack of clubs we threw the card away and drew again. This time (the second draw) the probability of drawing the ace of spades is not 1/52 but 1/51 because on this draw there are only 51 cards left in the deck, one of which is the ace of spades. If on this draw we draw the two of hearts and throw this card away, then the probability of drawing the ace of spades on the third draw, when there are only 50 cards left, will be 1/50. Because the probability of drawing the ace of spades changes from draw to draw and depends upon what occurred on previous trials, we say that this is a nonindependent (or dependent) selection procedure.

It should not be surprising to find that the probabilities involved in independent selection are much simpler to work with than those involved in nonindependent selection. Apparently for this reason, nearly all the statistical techniques that have been developed have assumed random and independent selection. It should also not be surprising that the more an experimenter violates these assumptions, the greater the likelihood of making erroneous conclusions from his data.

It is not always easy for an experimenter to tell if he is meeting the assumptions of random and independent selection. Suppose an experimenter wishes to use one of several introductory English sections for an experiment. Consequently he selects a selection that meets at 11:00 A.M. MWF, and uses the members of the class as his sample. Initially, we should note that the selection was probably not independent, since it is likely that friends decided to take the course together, and it is also likely that characteristics of persons taking an 11:00 A.M. class are more similar to one another than they are to persons taking an 8:00 A.M. class.

If the experimenter were interested only in the members of the 11:00 A.M. section it would be more difficult to criticize him, but we have assumed that he wishes to generalize to the entire student population. When an experimenter wishes to generalize from a sample (in this case a class) to a population (in this case all students at a certain university), he must always randomly and independently select his sample. That is, he must use statistical techniques that assume random and independent selection.

Another important consideration is whether the sets of events are mutually exclusive. That is, we can find the probability of drawing a club or a heart by using the additive law: $\Pr(C) + \Pr(H) = 1/4 + 1/4 = 1/2$. Notice that if you applied the additive law to a situation where the events were *not* mutually exclusive, you would not arrive at the correct answer. For example, what is the probability of selecting a heart (H) or a face (F) card from a well-shuffled deck? If we were to simply add the separate probabilities, $\Pr(F) + \Pr(H) = 12/52 + 13/52 = 25/52$, we would make a mistake because we would be counting the H face cards twice. Hence, a correction procedure must be used on those occasions when the sets of events considered are not mutually exclusive. Since we counted the overlapping parts (the face cards that were hearts) twice (A,B), we can correct for this by subtracting out those parts counted twice. That is, $\Pr(A \text{ or } B) = \Pr(A) + \Pr(B) - \Pr(AB)$. This can be better understood by looking at Table 7-4, which illustrates two sets of events which are *not* mutually exclusive. Here, if we were to find the probability of rolling a 3 with the first die or a 7 with the two dice, we would first find the probability of a 3 on the first roll and then find the probability of rolling a 7 with the two dice. It can be seen from Table 7-4 that $\Pr(3 \text{ on the first die}) = 6/36$ and that $\Pr(7) = 6/36$. But simply adding these two probabilities will not give us the correct probability. Because of this, we need to find one other probability. That is, we would find the probability of rolling a 7 which included a 3 on the first die. Again, it can be seen from the table that $\Pr(3,4) = 1/36$. Hence, the probability of the two events which are not mutually exclusive is $6/36 +$

Table 7-4   Possible outcomes with two dice

		Second die					
		1	2	3	4	5	6
First die	1	11	12	13	14	15	16
	2	21	22	23	24	25	26
	3	31	32	33	34	35	36
	4	41	42	43	44	45	46
	5	51	52	53	54	55	56
	6	61	62	63	64	65	66

$6/36 - 1/36 = 11/36$. This can be easily confirmed by reference to Table 7-4. Notice now that the multiplicative law is not affected by whether or not a set of events are mutually exclusive. That is, the probability of drawing a red card and then drawing a heart (with replacement) is equal to the product of the separate probabilities; that is, $1/2 \times 1/4 = 1/8$ despite the fact that hearts and red cards are not mutually exclusive categories.

THE BINOMIAL DISTRIBUTION   The expansion of the binomial provides us with an interesting and useful way to look at certain classes of probability problems. Initially, we will limit the probabilities of the separate events such that there are only two possible events, P and Q, and that the probability of event P is equal to the probability of event Q. Such a situation would be encountered in a coin-flipping situation where $\Pr(H) = \Pr(T) = 1/2$. There are, of course, other situations where only two events are possible but where the probabilities of the two events are quite different, and these situations will be considered briefly at the end of our discussion. Briefly, the binomial expansion provides the probability of any combination of events P and Q. The answer to the question, "what is the probability of flipping five heads and three tails in eight flips of a coin" is relatively easy to find through the expansion of the binomial. The general expression is as follows:

$$(P + Q)^N$$

where P is the probability of event P, Q is the probability of event Q, and N is the number of times some event (P or Q) is observed. If we flip a coin eight times, the expression would be $(0.5 + 0.5)^8$. If we flip a coin twice, the binomial expression becomes $(H + T)^2$, where H is the probability of a head and T is the probability of a tail. The expansion of this term is easy to find, since we simply multiply $(H + T)$ by itself: $(H + T) \times (H + T)$. If we place one term over the other and multiply each top term by each bottom term, we get

$$
\begin{array}{ll}
& H + T \\
& H + T \\
\hline
H \times (H + T) = & H^2 + HT \\
T \times (H + T) = & \phantom{H^2 + {}} HT + T^2 \\
\hline
\text{summing} & H^2 + 2HT + T^2
\end{array}
$$

From expansion we see that the probability of getting two heads in a row is $H^2 = 1/2 \times 1/2 = 1/4$, that the probability of getting two tails in a row is $T^2 = 1/2 \times 1/2 = 1/4$, and that the probability of getting a head and a tail (in any order) is $2HT = 2 \times 1/2 \times 1/2 = 2/4$. The 2 indicates the number of different ways that one H and one T can be obtained: HT and TH.

If we flip a coin three times, the binomial expression becomes $(H + T)^3$, and the probabilities of various events can be found by finding the expansion of this term. That is, we can multiply $(H + T) \times (H + T) \times (H + T)$ to find $(H + T)^3$. Since we know $(H + T) \times (H + T) = H^2 + 2HT + T^2$, then if we multiply this term by $(H + T)$ we would find the expansion of $(H + T)^3$. Again placing one term over the other and multiplying each top term by each bottom term, we get

$$H^2 + 2HT + T^2$$
$$\underline{H + T}$$

$$\begin{array}{ll} \text{Row 1} \times \text{H} & H^3 + 2H^2T + HT^2 \\ \text{Row 1} \times \text{T} & \quad H^2T + 2HT^2 + T^3 \\ \text{summing} & \overline{H^3 + 3H^2T + 3HT^2 + T^3} \end{array}$$

From the expansion we see that the probability of flipping

$$\begin{array}{l} \text{Pr (3 heads)} = H^3 = 1/2 \times 1/2 \times 1/2 = 1/8 \\ \text{Pr (3 tails)} = T^3 = 1/2 \times 1/2 \times 1/2 = 1/8 \\ \text{Pr (2 tails and 1 head)} = 3HT^2 = 3 \times 1/2 \times 1/2 \times 1/2 = 3/8 \\ \text{Pr (1 tail and 2 heads)} = 3H^2T = 3 \times 1/2 \times 1/2 \times 1/2 = 3/8 \end{array}$$

Again, while we see that there is only one way to get three heads (HHH), there are three ways to get two Ts and one H: TTH, THT, and HTT.

Finally, if we flip a coin four times, the binomial becomes $(H + T)^4 = (H + T) \times (H + T) \times (H + T) \times (H + T)$, and multiplying $(H + T)^3$ by $(H + T)$ we get

$$H^3 + 3H^2T + 3HT^2 + T^3$$
$$\underline{H + T}$$
$$H^4 + 3H^3T + 3H^2T^2 + HT^3$$
$$\quad H^3T + 3H^2T^2 + 3HT^3 + T^4$$
$$\overline{H^4 + 4H^3T + 6H^2T^2 + 4HT^3 + T^4}$$

from which the probability of the appropriate events can be read. Notice in this expansion that the power to which H is raised *decreases* from left to right, while the power to which T is raised increases from left to right. Starting at the left, H^4 decreases in the next term to H^3 to H^2 to H; at the same time, T increases to T^2 to T^3 to T^4. Each of these terms indicates the number of heads and the number of tails occurring. Thus, H^4 indicates the probability of four heads occurring in four flips. On the other hand, H^2T^2 indicates the probability of two heads and two tails occurring in four flips in any specified order, while the coefficient—in this case the number 6 preceding the H^2T^2—indicates the number of ways or orders in which two heads and two tails can occur: HHTT, HTHT, HTTH,

TTHH, THTH, and THHT. Thus, the Hs and Ts indicate the number of heads and tails, and the coefficient in front indicates the number of different ways such a combination of heads and tails can occur. Thus, the general form of the expansion of the binomial will contain Hs decreasing in power, Ts increasing in power, and a coefficient. Ignoring the coefficient, the expansion of the binomial takes on this general form:

$$(H + T)^N = H^N + H^{N-1}T + H^{N-2}T^2 + \cdots + H^2T^{N-2} + HT^{N-1} + T^N$$

Notice one important thing. In the expansion of the binomial for two events, the *denominator* for each event was always 4. The probability of two heads was 1/4; two tails, was 1/4; and one head and one tail, 2/4. Similarly, in the expansion of the binomial for three events, the denominator for each event was always 8; finally, for four events, it can be shown that the denominator for each event was 16. Hence, the probability for four heads is $1/2 \times 1/2 \times 1/2 \times 1/2 = 1/16$. We can generalize these results so that we can find the denominator for *any* binomial expansion where the probabilities of the two events are equal. The denominator will always be equal to 2^N if $P = Q$. When $N = 2$, the denominator is equal to 4; when $N = 3$, $2^N = 2 \times 2 \times 2 = 8$; and when $N = 4$, $2^N = 16$. In this last example the denominator is 16 for each of the five possible events when four coins are flipped (that is, 0, 1, 2, 3, or 4 heads will occur).

What will the numerator be? The numerator will be the coefficient in the expansion of the binomial. For two flips of a coin, it was 1 (2 heads), 2 (H and T), and 1 (2 tails). For three flips, it was 1 (HHH), 3 (HHT), 3 (HTT), and 1 (TTT); and for four flips it was 1, 4, 6, 4, and 1. Although it would be difficult for us to see any relationships here, Pascal saw a relationship between these coefficients, and he developed what has come to be called Pascal's triangle, illustrated below. It can be

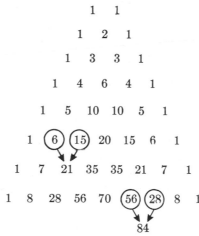

seen that the coefficients of the second row are 1, 2, 1, the numerators for the two-coin problem. In the third row the values are 1, 3, 3, 1, which are the numerators for the three-coin problem, and the values for the fourth row are the numerators for the four-coin problem. Notice that the value of each numerator is the sum of the two numbers immediately above it to its left and right. Hence, the value of 21 in the seventh row is the sum of 6 (immediately above and to the left) and 15 (immediately above and to the right).

Since each number is the sum of the two numbers immediately above (except the 1s on either end), new rows are quite easy to generate. For example, the next row after the last one given would be 1, 9, 36, 84, 126, 126, 84, 36, 9, and 1. Again each number was found by taking the sum of the two numbers above it in the row above.

With all this discussion it may not be apparent how Pascal's triangle can help in the solution of probability problems. We indicated previously that 2^N gives the value of the denominator when $P = Q = 0.5$ and that the numbers in the triangle give the value of the numerator. For an example of this, refer to the previously discussed four-coin situation. It can be seen from the expansion of the binomial by multiplication that the probability of two heads and two tails in four flips of a coin was $6H^2T^2$ or $6 \times 1/2 \times 1/2 \times 1/2 \times 1/2 = 6/16$. This same result can be obtained by recalling that the value of the denominator was $2^N = 2^4 = 16$, and the value of the numerator can be read from Pascal's triangle. To find the probability of flipping two heads in four flips, we first go down the triangle until we hit a row in which the second number (from the left or right) is 4 (that is, N). The values of this row are 1, 4, 6, 4, 1, and if we count 0, 1, 2 . . . until we come to the number of heads (or tails) we want, we can find the value of the numerator. Hence, the value of the numerator is 6, the denominator is $2^4 = 16$, and $\Pr = 6/16$. While this illustration is quite simple, it would be somewhat more difficult to find the probability of five heads in eight flips of a coin—a question we asked at the beginning of this section. To do this, we know $N = 8$, hence the denominator $= 2^8 = 256$. In addition, the numerator from the row with an 8 can be found by counting over from the left 0, 1, 2, 3, 4, 5. Hence, the appropriate numerator that corresponds to 5 is 56 and the probability of five heads in eight flips is $\Pr(5H) = 56/256$.

Other more complex probability problems can be solved through the use of Pascal's triangle. For example, what is the probability of getting three *or more* heads in six flips of a coin? Referring to the triangle, the coefficients in the row with a 6 are 1, 6, 15, 20, 15, 6, and 1. In addition, $2^N = 2^6 = 64$, which is the denominator. Counting from 0, the probability of getting three heads is 20/64, four heads is 15/64, five heads is 6/64, and six heads is 1/64. Using the additive law, the probability of getting three or more heads is

$$\frac{20 + 15 + 6 + 1}{64} = \frac{42}{64}$$

Finally, the expansion of the binomial can be used to solve probability problems, even when the probabilities of the events in question are not equal. Consider a special population that consists of only two types of members: A or B. Furthermore, 75 percent of the membership in this population is B and 25 percent is A. Hence, Pr $(B) = 0.75$ and Pr $(A) = 0.25$. What is the probability of getting two Bs in three random draws from the population? If the binomial is expanded, several things should be noted. The first is the various values of the coefficients, which can be obtained from Pascal's triangle: 1, 3, 3, 1. Next expand the binomial, at first ignoring and then combining with the coefficient: $A^3 + A^2B + AB^2 + B^3$ (ignoring); $A^3 + 3A^2B + 3AB^2 + B^3$ (combining). Remember that the probabilities of A and B are different. Hence the probability of drawing one A and two Bs is equal to $3 \times \text{Pr}(A) \times \text{Pr}(B) \times \text{Pr}(B)$, and this is taken directly from the expansion of the binomial: $3AB^2$. Thus, $3 \times 1/4 \times 3/4 \times 3/4 = 27/64$, which is the probability of getting two Bs in three random draws. Note that the probability of drawing two As and one B is considerably different:

$$3 \times \text{Pr}(A) \times \text{Pr}(A) \times \text{Pr}(B) = 3 \times 1/4 \times 1/4 \times 3/4 = 9/64$$

While it is often useful to be able to solve problems where the events are not equal, it seems that the expansion of the binomial is most useful when the probabilities of both events are equal to 0.50. Under these conditions the denominator will always equal 2^N, and the numerator can be read from the triangle.

Expanding the binomial through the use of Pascal's triangle, while relatively easy when N is small, gets much more difficult as N increases. One way of making calculations easier is to expand the binomial directly from the formula which is

$$\text{Pr}(X) = \frac{N!}{X!(N-X)!} P^X Q^{N-X}$$

While the equation looks a little threatening, you should note that you are already familiar with the term $P^X Q^{N-X}$. For example, you might flip a coin 10 times $(N = 10)$ and find 8 heads $(X = 8)$ and 2 tails $(N - X = 2)$. Here, P (the probability of a head) is $1/2$ and Q (the probability of a tail) is also $1/2$. Hence, $P^X Q^{N-X} = H^8 T^2 = (1/2)^8(1/2)^2$. The only new thing introduced here is the factorial. $N!$ (N factorial) is simply a shorthand way of indicating that all values lower than N, up to N, are multiplied together. Thus, $4! = 4 \times 3 \times 2 \times 1 = 24$. Prove to yourself that $6! = 720$. The advantage of using factorials in this case is that they can cancel out each other easily and reduce the amount of calculation. Thus, with 8 heads in 10 tosses,

$$\text{Pr}(8) = \frac{10!}{8!2!} (1/2)^8(1/2)^2$$

$$= \frac{10 \times 9 \times 8 \times 7 \times 6 \times 5 \times 4 \times 3 \times 2 \times 1}{2 \times 1 \times 8 \times 7 \times 6 \times 5 \times 4 \times 3 \times 2 \times 1} \times \frac{1}{1024}$$

$$= \frac{90}{2} \times \frac{1}{1024} = \frac{45}{1024}$$

Only a few of the numbers have to be multiplied. Instead, scores in the numerator and denominator can be canceled.

Remember, although many of our examples have used $P = Q = 1/2$, such need not always be the case. Suppose of the total population 1/4 have their birthdays during the summer months. What is the probability that 6 of 8 people at a meeting would have their birthdays during the summer?

$$\Pr(6) = \frac{8!}{6!2!} (1/4)^6 \times (3/4)^2$$

$$= \frac{8 \times 7}{2} \times \frac{1}{4096} \times \frac{9}{16}$$

$$= \frac{252}{65,536} = 0.00385$$

The probability is very low that such an unlikely set of events would happen by chance. If it happened, you would believe that these people probably gathered because of something related to their birthdays—such as an astrological society meeting.

USING PROBABILITY IN DECISION MAKING

No matter how much we have learned, Joe is better off without our advice about Gladys. We may be able to help you decide about your coin-flipping friend, however. Since he won four days in a row, you are beginning to suspect that the coin is biased. Because each day's flip is a separate event in a unique series, the multiplicative law says that if the coin is not biased, the chances are $(1/2)^4 = 1/2 \times 1/2 \times 1/2 \times 1/2 = 1/16 = 0.06$ for this particular H-H-H-H sequence. Remember that the conventional probability for rejecting the null hypothesis is 0.05, so we cannot reject the null hypothesis; in other words, we cannot conclude that the coin is biased.

It is extremely important to notice that even though we did not reject the null hypothesis, we did not prove or demonstrate that the coin is unbiased. If you keep drinking coffee with your friend, he might win for the next week. In other words, the null hypothesis can never be proved; it can only be rejected or not rejected, and the decision to reject or not to reject is based solely on whether the probability of an observed set of events is greater than or less than 0.05. If the probability of an event is less than (or equal to) 0.05, the null hypothesis is rejected, but if the probability of an event is greater than 0.05, the null hypothesis is not rejected.

We say that we *do not reject* the null hypothesis rather than *accept* it, since a variety of alternate hypotheses may be true. Suppose we flipped the coin and got four heads in four tosses. If the coin is perfectly unbiased, the probability of four heads in a row is equal to $(0.50)^4 = 0.06$. The probability of such a series of events occurring is greater than 0.05, and hence we do not reject the null hypothesis. Have we demonstrated that the null hypothesis is true? No. Notice that we assumed the probability of a head to be 0.50, and we did not find any evidence to the contrary (that is, we did not reject the null hypothesis). Suppose we assume the probability of a head to be 0.60 instead of 0.50. The hypothesis $Pr(H) = 0.60$ would not be rejected either, if we obtained four heads in a row, since $(0.60)^4 = 0.13$. The point is that failure to reject the null hypothesis certainly does not prove it to be true, since there are an infinite number of other hypotheses which would also not be rejected.

Suppose we repeat the experiment, flip the coin five times, and find that five heads appear. *If the coin is unbiased*, the probability that such a series of events would occur is equal to $(0.50)^5 = 0.03$. Since our rule is that we reject the null hypothesis if the probability of an event is less than 0.05, we would now reject the hypothesis that $Pr(H) = 0.50$. Notice that we still do not know the true probability of the occurrence of a head. It may be $Pr(H) = 0.50$; if it is, we have made a mistake. That is, we have falsely rejected the hypothesis that $Pr(H) = 0.50$, and we have committed an *alpha* error.

It might be interesting and instructive to find the probability of flipping five heads in a row when the null hypothesis is false, that is, when the probability of flipping a head is *not* equal to 0.50. Now if the null hypothesis is not true—that is, if the $Pr(H) \neq 0.50$—then $Pr(H)$ could equal anything from 0.00 to 1.00 other than 0.50. Suppose that the probability of flipping a head is 0.70; then the probability of flipping five heads in a row would be equal to $(0.70)^5 = 0.168$.

It is important to remember that in an actual research situation we would never know whether or not the coin is really biased. We employ a set of procedures which lead to correct decisions on some occasions and to erroneous decisions on others. Sometimes we will decide the coin is biased when it is not, and sometimes we will not reject the null hypothesis when we should.

From Table 7-5 it can be seen that when $Pr(H) = 0.50$ and the null hypothesis is rejected, we have committed an *alpha* error by rejecting a true null hypothesis. But when $Pr(H)$ does not equal 0.50 and we do *not* reject the null hypothesis, we have committed a *beta* error—failure to reject a false null hypothesis. Finally, the *power* of a test is defined as $1 -$ beta.

Table 7-5   Probabilities of outcomes in flipping five coins when Pr(H) = 0.50 and when Pr(H) = 0.70

	Pr(H)	
Your Decision	**0.50**	**0.70**
COIN IS UNBIASED (DO NOT REJECT NULL HYPOTHESIS)	0.969	0.832 (beta error)
COIN IS BIASED (REJECT NULL HYPOTHESIS)*	0.031 (alpha error)	0.168 (power)

* The null hypothesis is rejected only when five heads are flipped in a row.

From Table 7-5 it can be seen that when Pr(H) = 0.70, beta = 0.832, and power = 0.168. That is, when Pr(H) = 0.70, we will flip five heads in a row about 17 percent of the time. When five heads occur, we reject the null hypothesis; hence, the null hypothesis would be rejected about 17 percent of the time. This represents the power of our test. On the other hand, about 83 percent of the time we would not flip five heads in a row; therefore, 83 percent of the time we would not reject the null hypothesis. With Pr(H) = 0.70, the null hypothesis is false, and 83 percent of the time we incorrectly fail to reject the false null hypothesis. This failure to reject a false hypothesis represents the beta error for this particular illustration.

Since all the coins flipped had to be heads if we were to reject the null hypothesis, there obviously was little room for the negative case in the previous example. To allow for the negative case and yet allow the possibility of rejecting the null hypothesis, the number of observations or cases would usually be increased. For example, when ten coins are flipped and nine heads appear, we would reject the null hypothesis and conclude that the coin is biased. (Prove this to yourself by expanding the binomial and showing that the probability of nine *or more* heads in ten flips is 11/1024.) As the number of cases increases, even more negative cases can be tolerated. For example, if we wished to test the hypothesis that handedness (that is, left versus right) is distributed equally in the population, we could select 100 people at random and determine the preferred hand of each. Even if only 61 people were right-handed in the group of 100, we would still reject the null hypothesis and conclude that there is a bias in the population toward being right-handed; that is, people tend to be right-handed. (You can work this probability out by expanding the binomial, but it would take you quite a while. Other procedures are available which make the solution much easier. These are presented in Chapter Fifteen.)

SUMMARY

Basic to our understanding of inferential statistics is the understanding of simple rules of probability. Statistical inference uses probability statements such as "the probability is 0.05 that we will reject a true hypothesis." While you can make such statements without any real understanding of basic probability, it does seem that such an understanding is important. Because of the similarity of the binomial expansion to the normal distribution, we have looked at some length at the binomial. Similarly, we will learn a considerable amount about the normal distribution in Chapter Eight. The normal distribution and the basic laws of probability are central to the development of inferential statistics.

PROBLEMS

1. What is the probability of flipping three heads in three flips of a coin?
2. What is the probability of flipping two (and only two) heads in three flips of a coin?

3. If the probability of drawing a red ball from an urn is 0.40 and the probability of drawing a white ball from a second urn is 0.30, what is the probability of drawing a red ball and then a white ball?
4. What is the probability of rolling one or more 6s in two rolls of a die?
5. What is the probability of drawing two of a kind from an ordinary deck of cards?
6. What is the probability of drawing three of a kind from an ordinary deck of cards?
7. If the probability of being blonde is 0.20 and the probability of being blue-eyed is 0.30, what is the probability of being both blonde and blue-eyed? What assumption must you make here?
8. What is the probability of rolling one (and only one) 6 in two rolls of a die?
9. What is the probability of drawing a red card and a club in two draws (with replacement after each draw)?
10. Using the binomial expansion, find the probability of flipping seven or more heads in ten flips of a coin.
11. What is the probability of
    (a) Rolling two 6s in a row?
    (b) Not rolling a 6 in two rolls of a die?
    (c) Rolling none, one, or two 6s in two rolls of a die?
    (d) Rolling a 12 in two rolls of a die?
    (e) Rolling an 11 in two rolls of a die?
    (f) Rolling a 7 in two rolls of a die?
    (g) Drawing two aces in a row (with replacement)?
    (h) Drawing a spade and then a face card (with replacement)?
    (i) Drawing a diamond and then a heart (with replacement)?

EXERCISES

1. The probability is about 0.50 that a group of 22 people will have two people with the same birthday. Choose a group that is as large (or larger) and count the number of people with the same birthday.
2. If single-digit numbers are drawn at random, what is the probability that both members of a pair will be odd? Look at 20 such pairs drawn from Appendix H (randomly, of course). You should find that about 5 pairs are composed of odd numbers. How many did you get?

1   There are _____ sides to every coin.

══════ STOP ══════════════════ STOP ══════

two

---

2   There are two sides to every coin. Thus when a coin is flipped there are _____ possible events.

══════ STOP ══════════════════ STOP ══════

two

---

3   The total possible events when a coin is flipped are heads and _____.

══════ STOP ══════════════════ STOP ══════

tails

---

4   Suppose you want heads to occur (and only heads). In a single flip of the coin there is (are) _____ favorable event(s).

══════ STOP ══════════════════ STOP ══════

one

---

5   Thus when a coin is flipped and you want heads to occur there is (are) _____ total possible event(s) and _____ favorable event(s).

══════ STOP ══════════════════ STOP ══════

two, one

---

6   A die has six sides. For any roll of the die the total possible events are _____.

══════ STOP ══════════════════ STOP ══════

six

---

7   Suppose you roll a die and hope that a 2 will appear. In this case there is (are) _____ favorable event(s).

══════ STOP ══════════════════ STOP ══════

one

---

8   Suppose you roll a die again and hope that a 2 or a 4 will appear. In this case there are _____ favorable events and a total of _____ possible events.

======== STOP ======================================= STOP ========

two, six

---

9   The probability of an event is defined by the ratio of the favorable events to the _____ _____ events.

======== STOP ======================================= STOP ========

total possible

---

10   Thus the probability of a head when a coin is flipped is equal to 1 divided by _____.

======== STOP ======================================= STOP ========

2

---

11 $Pr(A) = N_A/N_E$ is one way of indicating that the probability of event A is equal to the _____ _____ _____ divided by the total number of events.

======== STOP ======================================= STOP ========

number of As (number of favorable events)

---

12 $Pr(A) = N_A/N_E$. N_A means the number of _____ events.

======== STOP ======================================= STOP ========

favorable

---

13 $Pr(A) = N_A/N_E$. N_E means the _____ _____ events.

======== STOP ======================================= STOP ========

total possible

---

14 $Pr(A) = N_A/N_E$. A verbal statement of this equation is that the probability of A is equal to _____ _____ _____ _____ _____ _____ _____ _____ _____ _____ _____.

======== STOP ======================================= STOP ========

the number of As divided by the total number of events

---

15 When the probability of an event is defined as $\Pr(A) = N_A/N_E$ an assumption implied by the definition is that all events have the same probability of occurring. That is $\Pr(E) = 1/(?)$ where E is any single event.

══════════ STOP ══════════════════════ STOP ══════════

$$N_E$$

---

16 Thus the probability of a specified card being drawn from an ordinary fifty-two-card deck = _____.

══════════ STOP ══════════════════════ STOP ══════════

$$\tfrac{1}{52} = .019$$

---

17 And the probability of drawing any face card is equal to _____.

══════════ STOP ══════════════════════ STOP ══════════

$$\tfrac{12}{52}$$

---

18 $\Pr(A) = N_A/N_E$. If there are ten As and forty possible events then $\Pr(A) =$ _____.

══════════ STOP ══════════════════════ STOP ══════════

$$\tfrac{10}{40} = .25$$

---

19 $\Pr(A) = N_A/N_E$. If there are twenty As and thirty possible events then $\Pr(A) =$ _____.

══════════ STOP ══════════════════════ STOP ══════════

$$\tfrac{20}{30} = .67$$

---

20 $\Pr(A) = N_A/N_E$. If there are zero As and twenty possible events, then $\Pr(A) =$ _____.

══════════ STOP ══════════════════════ STOP ══════════

$$\tfrac{0}{20} = .00$$

---

21 Can there ever be fewer than zero favorable events? _____.

══════════ STOP ══════════════════════ STOP ══════════

No

---

22 If there can never be fewer than zero favorable events then $\Pr(A)$ can *never* be less than _____.

══════════ STOP ══════════════════════ STOP ══════════

zero

---

23 $\Pr(A) = N_A/N_E$. Suppose there are 15 As and 15 possible events. $\Pr(A) = $ _____.

═══════ STOP ═══════════════════════ STOP ═══════

$$^{15}\!/_{15} = 1.00$$

---

24   Can there ever be more than all (that is 100 percent) favorable events? _____.

═══════ STOP ═══════════════════════ STOP ═══════

No

---

25   If the maximum number of favorable events are 100 percent, then $\Pr(A)$ can never be more than _____.

═══════ STOP ═══════════════════════ STOP ═══════

1.00

---

26   If there can never be less than zero favorable events nor more than all favorable events, then the smallest $\Pr(A) = $ _____ and the largest $\Pr(A) = $ _____.

═══════ STOP ═══════════════════════ STOP ═══════

.00, 1.00

---

27 $$\Pr(A) = \frac{N_A}{(?)}$$

═══════ STOP ═══════════════════════ STOP ═══════

N_E

---

28 $$\Pr(A) = \frac{(?)}{N_E}$$

═══════ STOP ═══════════════════════ STOP ═══════

N_A

---

29 $$\text{_____} = \frac{N_A}{N_E}$$

═══════ STOP ═══════════════════════ STOP ═══════

$\Pr(A)$

---

30 $\quad\quad\quad\quad\quad\quad$ Pr $(A) =$ _____

═══════ STOP ═══════════════════════════ STOP ═══════

$$N_A/N_E$$

---

31 $\quad$ There are fifty-two cards in an ordinary deck. What is the probability of drawing an ace of spades? _____. (State in fractional form.)

═══════ STOP ═══════════════════════════ STOP ═══════

$$\tfrac{1}{52}$$

---

32 $\quad$ There are four suits (hearts, clubs, diamonds, and spades) each with thirteen cards in an ordinary deck of cards. In fractional form what is the probability of drawing a spade? _____.

═══════ STOP ═══════════════════════════ STOP ═══════

$$\tfrac{13}{52} = \tfrac{1}{4}$$

---

33 $\quad$ There are six sides to a die. What is the probability of rolling a 2? _____; a 4? _____; a 6? _____.

═══════ STOP ═══════════════════════════ STOP ═══════

$$\tfrac{1}{6}; \tfrac{1}{6}; \tfrac{1}{6}$$

---

34 $\quad$ Pr (2) $= \tfrac{1}{6}$; Pr (4) $= \tfrac{1}{6}$; Pr (6) $= \tfrac{1}{6}$, what is the probability of rolling an even number (2, 4, or 6) when rolling a die once? _____.

═══════ STOP ═══════════════════════════ STOP ═══════

$$\tfrac{1}{6} + \tfrac{1}{6} + \tfrac{1}{6} = \tfrac{3}{6} = \tfrac{1}{2} = .50$$

---

35 $\quad$ What is the probability of rolling a 1, 2, 4, or 6 when rolling a die? _____.

═══════ STOP ═══════════════════════════ STOP ═══════

$$\tfrac{4}{6} = .67$$

---

36 $\quad$ The last few problems illustrate what may be called the *additive law of probability*. Simply stated it says that the probability of any one of a set of events is the _____ of the separate probabilities.

═══════ STOP ═══════════════════════════ STOP ═══════

$$\text{sum}$$

---

37 The probability of rolling an even number with a die is equal to Pr (2) + Pr (4) + Pr (6) = $\frac{1}{6} + \frac{1}{6} + \frac{1}{6} = \frac{3}{6} = \frac{1}{2}$. The theorem stating that the probability of any one of a set of events is equal to the sum of the separate probabilities is called the _____ *law of* _____.

═══ STOP ═══════════════════════ STOP ═══

additive, probability

---

38 Suppose Pr $(A) = \frac{1}{3}$, Pr $(B) = \frac{1}{2}$; Pr $(A$ or $B) = $ _____.

═══ STOP ═══════════════════════ STOP ═══

$$\frac{1}{3} + \frac{1}{2} = \frac{2}{6} + \frac{3}{6} = \frac{5}{6} = .83$$

---

39 Suppose Pr $(A) = \frac{1}{10}$, Pr $(B) = \frac{3}{20}$, Pr $(C) = \frac{1}{5}$; Pr $(A$ or B or $C) = $

_____.

═══ STOP ═══════════════════════ STOP ═══

$$[\frac{1}{10} = \frac{2}{20}; \frac{1}{5} = \frac{4}{20}]\frac{2}{20} + \frac{3}{20} + \frac{4}{20} = \frac{9}{20} = .45$$

---

40 Suppose Pr $(A) = \frac{1}{3}$, Pr $(B) = \frac{1}{4}$, and Pr $(C) = \frac{1}{5}$; Pr $(A$ or B or $C) = $

_____.

═══ STOP ═══════════════════════ STOP ═══

$$[\frac{1}{3} = \frac{20}{60}; \frac{1}{4} = \frac{15}{60}; \frac{1}{5} = \frac{12}{60}]\frac{20}{60} + \frac{15}{60} + \frac{12}{60} = \frac{47}{60} = .78$$

---

41 Suppose the probability of an event is equal to p. Then the probability that the event will not occur is equal to $1 - p$. For example, the probability of rolling a 6 is equal to $\frac{1}{6}$ and the probability of not rolling a 6 is equal to _____.

═══ STOP ═══════════════════════ STOP ═══

$$\frac{5}{6} = .83$$

---

42 Pr $(A) = \frac{1}{3}$, then what is the probability that event A will not occur?

_____.

═══ STOP ═══════════════════════ STOP ═══

$$1 - \text{Pr }(A) = 1 - \frac{1}{3} = \frac{2}{3} = .67$$

---

43 Pr $(A) = \frac{1}{5}$, then the probability of A not occurring is equal to _____.

═══ STOP ═══════════════════════ STOP ═══

$$\frac{4}{5} = .80$$

---

44 The probability of drawing a spade from an ordinary deck is $\frac{1}{4}$. What is the probability of *not* drawing a spade? _____.

═══════ STOP ═══════ STOP ═══════

$$\frac{3}{4} = .75$$

---

45 The multiplicative law of probability deals with sequences of events. For example, to find the probability of flipping a head and then a tail would require the use of the _____ law of probability.

═══════ STOP ═══════ STOP ═══════

multiplicative

---

46 To understand the multiplicative law of probability it is helpful to look at all possible outcomes of two flips of a coin. If you flip a coin twice you can get H-H (a head and then a second head), H-T (a head and then a tail), T-H, or _____.

═══════ STOP ═══════ STOP ═══════

T-T

---

47 The four possible outcomes of a two-flip experiment are H-H, H-T, T-H, and T-T. If the probabilities of all these events are equal, what is the probability of flipping two heads? _____.

═══════ STOP ═══════ STOP ═══════

$$\frac{1}{4} = .25$$

---

48 The possible outcomes of a two-flip experiment are H-H, H-T, T-H, and T-T. Pr (H-H) $= \frac{1}{4}$. What is Pr (H) when a coin is flipped once? _____.

═══════ STOP ═══════ STOP ═══════

$$\frac{1}{2} = .50$$

---

49 Pr (H-H) $= \frac{1}{4}$, Pr (H) $= \frac{1}{2}$, Pr (H) $\times$ Pr (H) $=$ _____.

═══════ STOP ═══════ STOP ═══════

$$\frac{1}{2} \times \frac{1}{2} = \frac{1}{4} = .25$$

---

50 Pr (H-H) $= \frac{1}{4}$ and Pr (H) $\times$ Pr (H) $= \frac{1}{4}$. The multiplicative law states that the probability of a sequence of events is equal to the product of the separate probabilities. Thus Pr (H) $= \frac{1}{2}$ and Pr (T) $= \frac{1}{2}$. Then the probability of a head and then a tail, Pr (H-T), $=$ _____.

═══════ STOP ═══════ STOP ═══════

$$\text{Pr (H)} \times \text{Pr (T)} = \frac{1}{2} \times \frac{1}{2} = \frac{1}{4} = .25$$

51 $\Pr(A) = \frac{1}{3}$, $\Pr(B) = \frac{1}{4}$, $\Pr(A\text{-}B) = $ _____.

═══════ STOP ═══════════════════════════════ STOP ═══════

$$\frac{1}{3} \times \frac{1}{4} = \frac{1}{12} = .083$$

---

52 $\Pr(A) = \frac{8}{10}$, $\Pr(B) = \frac{1}{2}$, $\Pr(A\text{-}B) = $ _____.

═══════ STOP ═══════════════════════════════ STOP ═══════

$$\frac{8}{10} \times \frac{1}{2} = \frac{8}{20} = \frac{4}{10} = .40$$

---

53 $\Pr(H) = \frac{1}{2}$. The probability of flipping three heads in a row would equal

_____.

═══════ STOP ═══════════════════════════════ STOP ═══════

$$\frac{1}{2} \times \frac{1}{2} \times \frac{1}{2} = \frac{1}{8} = .13$$

---

54 The probability of rolling a 5 with a die is equal to $\frac{1}{6}$. The probability of rolling two 5s in a row is equal to _____.

═══════ STOP ═══════════════════════════════ STOP ═══════

$$\frac{1}{6} \times \frac{1}{6} = \frac{1}{36} = .028$$

---

55 On the other hand, the probability of not rolling a 5 is equal to _____.

═══════ STOP ═══════════════════════════════ STOP ═══════

$$\frac{5}{6} = .83$$

---

56 The probability of not rolling a 5 is equal to $\frac{5}{6}$. Hence by the multiplicative law, the probability of not rolling a 5 on two rolls would equal _____.

═══════ STOP ═══════════════════════════════ STOP ═══════

$$\frac{5}{6} \times \frac{5}{6} = \frac{25}{36} = .69$$

---

57 The probability of rolling a 5 is $\frac{1}{6}$ and the probability of not rolling a 5 is $\frac{5}{6}$. Hence by the multiplicative law, the probability of rolling a 5 on the first roll and then not rolling a 5 on the second roll is equal to _____.

═══════ STOP ═══════════════════════════════ STOP ═══════

$$\frac{1}{6} \times \frac{5}{6} = \frac{5}{36} = .14$$

---

58 The probability of not rolling a 5 on the first roll and then rolling a 5 on the second is equal to _____.

═══════ STOP ═══════════════════════════════ STOP ═══════

$$\frac{5}{6} \times \frac{1}{6} = \frac{5}{36} = .14$$

---

59 The probability of rolling a 5 and then not rolling a 5 is equal to $\frac{5}{36}$. The probability of not rolling a 5 and then rolling a 5 is equal to $\frac{5}{36}$. By the additive law what is the probability of rolling one (and only one) 5 in two rolls of a die? _____.

════════ STOP ════════ STOP ════════

$$\frac{5}{36} + \frac{5}{36} = \frac{10}{36} = .28$$

60 The probability of rolling two 5s equals $\frac{1}{36}$, the probability of rolling only one 5 equals $\frac{10}{36}$ and the probability of rolling no 5s is equal to $\frac{25}{36}$. By the additive law what is the probability that either 0, 1, or 2 5s will be thrown in two rolls of a die? _____.

════════ STOP ════════ STOP ════════

$$\frac{1}{36} + \frac{10}{36} + \frac{25}{36} = \frac{36}{36} = 1.00$$

61 If a coin is flipped twice then Pr (H-H) = _____.

════════ STOP ════════ STOP ════════

$$\frac{1}{4} = .25$$

62 Pr (T-T) = _____.

════════ STOP ════════ STOP ════════

$$\frac{1}{4} = .25$$

63 Pr (T-H) = _____.

════════ STOP ════════ STOP ════════

$$\frac{1}{4} = .25$$

64 And Pr (H-T) = _____.

════════ STOP ════════ STOP ════════

$$\frac{1}{4} = .25$$

65 Pr (H-T) = $\frac{1}{4}$; Pr (T-H) = $\frac{1}{4}$. By the additive law what is the probability that one (and only one) head will appear in two flips of a coin? _____.

════════ STOP ════════ STOP ════════

$$\frac{1}{4} + \frac{1}{4} = \frac{1}{2} = .50$$

66   Pr (T-T-H) = _____ .

══════════ STOP ══════════════════════════ STOP ══════════

$$\frac{1}{2} \times \frac{1}{2} \times \frac{1}{2} = \frac{1}{8} = .13$$

---

67   Pr (H-T-T) = _____ .

══════════ STOP ══════════════════════════ STOP ══════════

$$\frac{1}{2} \times \frac{1}{2} \times \frac{1}{2} = \frac{1}{8} = .13$$

---

68   Pr (T-H-T) = _____ .

══════════ STOP ══════════════════════════ STOP ══════════

$$\frac{1}{8} = .13$$

---

69   Pr (T-T-H) = $\frac{1}{8}$, Pr (T-H-T) = $\frac{1}{8}$, Pr (H-T-T) = $\frac{1}{8}$. What is the probability of flipping one (and only one) head in three flips of a coin? _____ .

══════════ STOP ══════════════════════════ STOP ══════════

$$\frac{1}{8} + \frac{1}{8} + \frac{1}{8} = \frac{3}{8} = .38$$

---

70   The probability of flipping one (and only one) head in three flips of a coin is equal to $\frac{3}{8}$. If Pr (A) = $\frac{3}{8}$ what is the probability that A will *not* occur? _____ .

══════════ STOP ══════════════════════════ STOP ══════════

$$1 - \frac{3}{8} = \frac{5}{8} = .63$$

---

71   The probability of flipping one (and only one) head in three flips of a coin is equal to $\frac{3}{8}$. What is the probability of flipping 0, 2, or 3 heads in three flips of a coin? _____ .

══════════ STOP ══════════════════════════ STOP ══════════

$$1 - \frac{3}{8} = \frac{5}{8} = .63$$

---

72   Pr (T-T-T) = _____ .

══════════ STOP ══════════════════════════ STOP ══════════

$$\frac{1}{2} \times \frac{1}{2} \times \frac{1}{2} = \frac{1}{8} = .13$$

---

73   Pr (T-T-T) = $\frac{1}{8}$; Pr of one head = $\frac{3}{8}$; Pr (H-H-H) = _____ .

══════════ STOP ══════════════════════════ STOP ══════════

$$\frac{1}{2} \times \frac{1}{2} \times \frac{1}{2} = \frac{1}{8} = .13$$

74　What is the probability that 0, 1, 2, or 3 heads will occur in three flips of a coin? _____.

════ STOP ════════════════════════ STOP ════

$$\frac{1}{8} + \frac{3}{8} + \frac{3}{8} + \frac{1}{8} = 1.00$$

75　Pr (T-T-T) = $\frac{1}{8}$; Pr (H-H-H) = $\frac{1}{8}$; Pr of one head = $\frac{3}{8}$. By the additive law Pr of 0, 1, or 3 heads = _____.

════ STOP ════════════════════════ STOP ════

$$\frac{1}{8} + \frac{1}{8} + \frac{3}{8} = \frac{5}{8} = .63$$

76　Pr of 0, 1, or 3 heads is equal to $\frac{5}{8}$. What is the probability of flipping two (and only two) heads in three flips of a coin? _____.

════ STOP ════════════════════════ STOP ════

$$1 - \frac{5}{8} = \frac{3}{8} = .38$$

77　The probability of flipping two heads in a row is equal to $\frac{1}{2} \times \frac{1}{2} = (\frac{1}{2})^2$. The probability of flipping three heads in a row is equal to $\frac{1}{2} \times \frac{1}{2} \times \frac{1}{2} = (\frac{1}{2})^3$. The probability of flipping N heads in a row is equal to _____.

════ STOP ════════════════════════ STOP ════

$$(\tfrac{1}{2})^N$$

78　The probability of rolling two 5s in a row with a pair of dice is equal to $\frac{1}{6} \times \frac{1}{6} = (\frac{1}{6})^2$. What is the probability of rolling three 5s in a row? _____.

════ STOP ════════════════════════ STOP ════

$$\frac{1}{6} \times \frac{1}{6} \times \frac{1}{6} = (\tfrac{1}{6})^3 = \tfrac{1}{216} = .0046$$

79　The probability of rolling N 5s in a row is equal to _____.

════ STOP ════════════════════════ STOP ════

$$(\tfrac{1}{6})^N$$

80　If P is the probability of a favorable event, then what is the probability of N favorable events in a row? _____.

════ STOP ════════════════════════ STOP ════

$$P^N$$

81   The binomial distribution is derived from the expansion of the expression $(P + Q)^N$. If a coin was flipped three times $N =$ _____.

═══════ STOP ═══════════════════════ STOP ═══════

3

---

82 $(P + Q) = 1$, if $P = .75$, then $Q =$ _____.

═══════ STOP ═══════════════════════ STOP ═══════

.25

---

83   A coin is flipped four times. In the expression $(H + T)^N$, Pr (H) = _____, Pr (T) = _____, $N =$ _____.

═══════ STOP ═══════════════════════ STOP ═══════

.50, .50, 4

---

84   A coin is flipped four times. $(H + T)^N = (.5 + .5)^4$. If three heads are flipped, then _____ tails are also flipped.

═══════ STOP ═══════════════════════ STOP ═══════

one

---

85   Eight coins are flipped, five heads appear. In the expansion of the binomial five heads would be represented H^5, the tails which were flipped would be represented _____.

═══════ STOP ═══════════════════════ STOP ═══════

T^3

---

86   Thus if six coins were tossed, the expression $15H^4T^2$ would indicate that _____ heads and _____ tails were tossed.

═══════ STOP ═══════════════════════ STOP ═══════

four, two

---

87 The coefficient of the expression 15 H^4T^2 indicates there are fifteen ways of getting four heads and two tails. With $3H^2T$, there are _____ ways of getting two heads and one tail.

═══════ STOP ═══════════════════════ STOP ═══════

three

───

88 The expression $10H^3T^2$ indicates there are _____ ways to get _____ heads and _____ tails.

═══════ STOP ═══════════════════════ STOP ═══════

ten, three, two

───

89 The expression $4H^3T$ indicates there are four ways to get three heads and a tail. Flipping one coin at a time, these four ways include H–H–H–T, H–H–T–H, H–T–H–H, and _____.

═══════ STOP ═══════════════════════ STOP ═══════

T–H–H–H

───

90 The terms in the binomial expansion are arranged such that the power of P decreases while the power of Q increases. Thus with P^4Q^2 and P^3Q^3 the next term would be _____.

═══════ STOP ═══════════════════════ STOP ═══════

P^2Q^4

───

91 If four coins are flipped, H^4 would indicate four heads were flipped, H^3T would indicate that three heads and one tail were flipped. The next term would be _____ and would indicate _____ heads and _____ tails were flipped.

═══════ STOP ═══════════════════════ STOP ═══════

H^2T^2, two, two

───

92 $H^5 + H^4T + H^3T^2 \cdot \cdot \cdot$ Complete the series.

══════ STOP ══════════════════════ STOP ══════

$$H^2T^3 + HT^4 + T^5$$

---

93 Four coins are flipped. Arrange the series.

══════ STOP ══════════════════════ STOP ══════

$$H^4 + H^3T + H^2T^2 + HT^3 + T^4$$

---

94 Each of the terms in the expansion of the binomial has a coefficient. For example, the second term properly should be $4H^3T$. Three coins are flipped, arrange the series without the coefficients.

══════ STOP ══════════════════════ STOP ══════

$$H^3 + H^2T + HT^2 + T^3$$

---

95 With $3H^2T$, if Pr (H) = .50 and Pr (T) = .50, the probability of flipping two heads and one tail equals $3 \times (\frac{1}{2})^2 \times \frac{1}{2} = \frac{3}{8} = .38$. If $2HT$, then the probability of flipping one head and one tail in two flips equals_____.

══════ STOP ══════════════════════ STOP ══════

$$2 \times \frac{1}{2} \times \frac{1}{2} = \frac{2}{4} = .50$$

---

96 If $10H^3T^2$, then the probability of flipping three heads and two tails in five flips equals _____ $\times (\frac{1}{2})^3 \times (\frac{1}{2})^2$.

══════ STOP ══════════════════════ STOP ══════

10

---

97 If $20H^3T^3$ then _____ $\times (\frac{1}{2})^3 \times$ _____.

══════ STOP ══════════════════════ STOP ══════

$$20, (\frac{1}{2})^3$$

---

98 In the expansion of the binomial the coefficients can be found by generating Pascal's triangle. This is done by adding adjacent coefficients and putting 1s on both ends. Thus with 1 2 1, $1 + 2$ on the left equals 3 and $2 + 1$ on the right equals 3, putting 1s on each end, the coefficients are _____.

═══════ STOP ═══════════════════ STOP ═══════

1 3 3 1

---

99 Three coins are flipped, the coefficients are _____.

═══════ STOP ═══════════════════ STOP ═══════

1 3 3 1

---

100 Three coins are flipped. Arrange the series (without the coefficients).

═══════ STOP ═══════════════════ STOP ═══════

$$H^3 + H^2T + HT^2 + T^3$$

---

101 Three coins are flipped. Arrange the series with the proper coefficients.

═══════ STOP ═══════════════════ STOP ═══════

$$H^3 + 3H^2T + 3HT^2 + T^3$$

---

102 1 3 3 1 are the coefficients when three coins are flipped, i.e., $N = 3$. Adding adjacent coefficients, indicate the coefficients when $N = 4$.

═══════ STOP ═══════════════════ STOP ═══════

1 4 6 4 1

---

103 When $N = 4$ the coefficients are 1 4 6 4 1. Adding adjacent coefficients (with 1s on both ends) when $N = 5$, the coefficients are _____.

═══════ STOP ═══════════════════ STOP ═══════

1 5 10 10 5 1

104   When $N = 6$, the coefficients are _____.

════════ STOP ════════════════════════ STOP ════════

1 6 15 20 15 6 1

---

105   When $N = 6$, the coefficients are 1, 6, 15, 20, 15, 6, 1. Six coins are flipped. Arrange the series with the proper coefficients.

════════ STOP ════════════════════════ STOP ════════

$$H^6 + 6H^5T + 15H^4T^2 + 20H^3T^3 + 15H^2T^4 + 6HT^5 + T^6$$

---

106   Six coins are flipped. What is the probability of flipping three heads? _____.

════════ STOP ════════════════════════ STOP ════════

$$20 \times (\tfrac{1}{2})^3 \times (\tfrac{1}{2})^3 = {}^{20}\!/_{64} = .31$$

---

107   Five coins are flipped. Arrange the series with the proper coefficients.

════════ STOP ════════════════════════ STOP ════════

$$H^5 + 5H^4T + 10H^3T^2 + 10H^2T^3 + 5HT^4 + T^5$$

---

108   What is the probability of flipping three heads in five tosses of a .coin? _____.

════════ STOP ════════════════════════ STOP ════════

$$10 \times (\tfrac{1}{2})^3 \times (\tfrac{1}{2})^2 = {}^{10}\!/_{32} = .31$$

---

109   What is the probability of flipping two heads in four tosses of a coin? _____.

════════ STOP ════════════════════════ STOP ════════

$$6 \times (\tfrac{1}{2})^2 \times (\tfrac{1}{2})^2 = {}^{6}\!/_{16}$$

---

110 What is the probability of flipping five or *more* heads in six tosses? _____.

═══ STOP ═══ ═══ STOP ═══

Pr (5H) = $\frac{6}{64}$; Pr (6H) = $\frac{1}{64}$; Pr (5 or more) = $\frac{7}{64}$ = .11

---

111 Three As or Bs are randomly drawn. Arrange the series with the proper coefficients.

═══ STOP ═══ ═══ STOP ═══

$$A^3 + 3A^2B + 3AB^2 + B^3$$

---

112 If Pr (A) = $\frac{3}{4}$, Pr (B) = $\frac{1}{4}$, what is the probability of drawing two As out of three? _____.

═══ STOP ═══ ═══ STOP ═══

$3 \times (\frac{3}{4})^2 \times (\frac{1}{4})$ = $\frac{27}{64}$ = .42

---

113 Pr (A) = $\frac{3}{5}$, Pr (B) = $\frac{2}{5}$. What is the probability of drawing three or more As in four draws? _____.

═══ STOP ═══ ═══ STOP ═══

Pr (4 As) = $(\frac{3}{5})^4$, Pr (3 As) = $4(\frac{3}{5})^3(\frac{2}{5})$, Pr (3 or more As) =
$\frac{81}{625} + \frac{216}{625}$ = $\frac{297}{625}$ = .48

---

114 $Pr(X) = \dfrac{N!}{X!(N - X)!} P^X Q^{N-X}$. In the formula for the binomial, which term is the coefficient?

═══ STOP ═══ ═══ STOP ═══

$$\frac{N!}{X!(N - X)!}$$

---

115 $N!$ is a shorthand expression for multiplying all the numbers from 1 to N together. For example, $3! = 3 \times 2 \times 1$. $5! =$ _____.

═══ STOP ═══ ═══ STOP ═══

120

116 $X = 4$, $X! =$ _____.

==== STOP ==================== STOP ====

24

---

117 $N = 8$, $X = 5$, $N - X =$ _____, $(N - X)! =$ _____.

==== STOP ==================== STOP ====

3, 6

---

118 $N = 8$, $X = 5$, $N - X = 3$. The coefficient $\dfrac{N!}{X!(N - X)!} =$ _____.

==== STOP ==================== STOP ====

$$\frac{8!}{5!3!} = \frac{8 \times 7 \times 6}{3 \times 2 \times 1} = 56$$

---

119 $N = 10$, $X = 2$. The coefficient $\dfrac{N!}{X!(N - X)!} =$ _____.

==== STOP ==================== STOP ====

$$\frac{10!}{2!8!} = \frac{10 \times 9}{2} = 45$$

---

120 $N = 15$, $X = 13$. The coefficient $=$ _____.

==== STOP ==================== STOP ====

$$\frac{15!}{13!2!} = \frac{15 \times 14}{2} = 105$$

---

121 The formula for the coefficient in the binomial expansion is _____.

==== STOP ==================== STOP ====

$$\frac{N!}{X!(N - X)!}$$

---

122 A coin is flipped ten times. Eight heads are flipped. $N =$ _____,
$X =$ _____.

$=$ STOP $=$ $=$ STOP $=$

10, 8

---

123 Eight heads are flipped in ten tosses. $N - X =$ _____.

$=$ STOP $=$ $=$ STOP $=$

2

---

124 With eight heads in ten tosses, the coefficient = _____.

$=$ STOP $=$ $=$ STOP $=$

$$\frac{N!}{X!(N-X)!} = \frac{10!}{8!2!} = \frac{10 \times 9}{2} = 45$$

---

125 A coin is flipped 20 times; 15 heads appear. $\frac{N!}{X!(N-X)!} P^X Q^{N-X}$. In this expression, $Q =$ _____.

$=$ STOP $=$ $=$ STOP $=$

T, or $\frac{1}{2}$

---

126 Similarly, the value of $P =$ _____.

$=$ STOP $=$ $=$ STOP $=$

$\frac{1}{2}$

---

127 With 15 heads in 20 flips, $P^X Q^{N-X} = (\frac{1}{2})^{15}(\frac{1}{2})^5$. With 7 heads in 12 flips, $P^X Q^{N-X} =$ _____.

$=$ STOP $=$ $=$ STOP $=$

$(\frac{1}{2})^7(\frac{1}{2})^5$

---

128 With 60 heads in 100 flips, $\frac{N!}{X!(N-X)!} = \frac{100!}{60!40!}$ and $P^X Q^{N-X} =$

_____.

$=$ STOP $=$ $=$ STOP $=$

$(\frac{1}{2})^{60}(\frac{1}{2})^{40}$

---

129 With 40 heads in 60 flips, $\dfrac{N!}{X!(N-X)!}P^X Q^{N-X} = $ _____.

══════ STOP ═══════════════════════════ STOP ══════

$$\frac{60!}{40!20!}(\tfrac{1}{2})^{40}(\tfrac{1}{2})^{20}$$

---

130 With 12 heads in 18 flips, $\dfrac{N!}{X!(N-X)!}P^X Q^{N-X} = $ _____.

══════ STOP ═══════════════════════════ STOP ══════

$$\frac{18!}{12!6!}(\tfrac{1}{2})^{12}(\tfrac{1}{2})^{6}$$

---

131 Suppose $P = \tfrac{2}{3}$, $Q = \tfrac{1}{3}$. Eight Ps and 2 Qs occur in ten draws. $\dfrac{N!}{X!(N-X)!}P^N Q^{N-X} = $ _____.

══════ STOP ═══════════════════════════ STOP ══════

$$\frac{10!}{8!2!}(\tfrac{2}{3})^{8}(\tfrac{1}{3})^{2}$$

---

132 The formula for the binomial expansion is _____.

══════ STOP ═══════════════════════════ STOP ══════

$$P(X) = \frac{N!}{X!(N-X)!}P^X Q^{N-X}$$

Eight

The Normal Distribution

In previous chapters we discussed two kinds of descriptive statistics for summarizing groups of data, which were called *measures of central tendency* (the mean, the median, and the mode) and *measures of variability* (the standard deviation, the semi-interquartile range, and the range).

Description of data is not usually the final goal of research. While its importance should not be underestimated, description is usually only a preliminary step toward statistical inference. You will recall that statistical inference was defined as the process of making decisions in the absence of complete information. Because an inference is a "rational guess," the conclusions we draw must necessarily be made in terms of probability statements.

In any problem involving statistical inference, we attempt to infer or estimate the nature of a population from the characteristics of a particular sample taken from the population. A population is a group that has some common characteristic, such as height. A sample is a subgroup or simply a part of the population from which it was taken.

In order to understand techniques of statistical inference, we must be concerned with distributions of data. As we noted in previous chapters, a distribution may be considered as the relationship between two variables: (1) the *value* of the score, and (2) the *frequency* with which the scores occur. Distributions may be expressed in tabular or graphic form. Why are distributions important for an understanding of techniques of statistical inference? The reason is that statistics such as the sample mean and sample standard deviation are themselves distributed in characteristic ways. Techniques of statistical inference are based on knowledge of these distributions.

The normal distribution is of great importance in statistical inference. It occurs widely in nature, being characteristic of such diverse measures as height, intelligence, diameters of tree leaves, and sometimes even bowling scores. Suppose we collected a score from each bowler in the United States last year. The distribution would represent some 20 million scores and would probably look something like Figure 8-1.

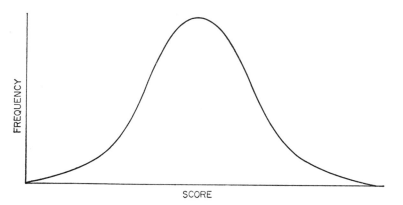

Figure 8-1   *Distribution of a population of bowling scores.*

The graph shows the relationship between the bowling score variable on the abscissa (x axis) and the frequency variable on the ordinate (y axis). By inspection of the graph we can determine roughly the proportion of scores that fall above or below a particular point, or the proportion of scores falling between two points on the baseline.

For purposes of illustration let us assume that both authors of this book have nearly the same bowling average. As is apparent in Figure 8-2, however, author A is much more erratic than author B. In statistical terms, the two distributions have the same mean, but they differ in variability. Now consider the distribution for expert bowler C, which is included in Figure 8-2 also, and compare it to the curve for bowler B. Both distributions have the same variability, but bowler C has a substantially higher average.

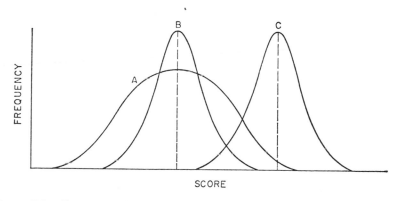

Figure 8-2   *Frequency distributions of scores of three bowlers (A, B, and C).*

The important thing about these three distributions is that they are all "normal" distributions. Thus, two points should be remembered: (1) the normality of a distribution does not depend on its mean, and (2) the normality of a distribution does not depend on its variability. What does make a distribution normal? Let us be content with the knowledge that the exact shape of a normal distribution can be specified by a certain mathematical equation.

In actual practice, research workers seldom encounter truly normal distributions. Even though we may assume a certain population to be normally distributed, we would not expect samples drawn from that population to be exactly normal. Figure 8-3 shows a normally distributed population

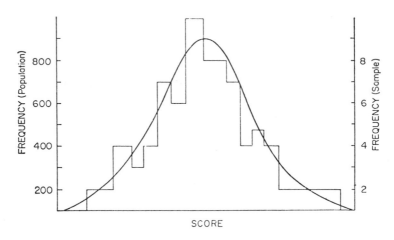

Figure 8-3 *Frequency distributions of a population and one sample.*

and a sample drawn from that population. Although this sample is close to normal, it is not exactly so. Because of our knowledge of the characteristics of normal distributions, it is usually worth the risk to apply methods to sample data that are, strictly speaking, applicable only to the theoretical normal distribution.

Now, if we were to compute the mean and standard deviation of the population, we would find that they were equal to 22.00 and 8.00, respectively. On the other hand, the mean and standard deviation of the sample are 21.66 and 7.57, respectively. Even though the sample was taken from the population, we would be quite surprised if the mean and standard deviation were equal to the population mean and standard deviation. We would expect, though, from our own experience, that the mean of a sample often comes close to that of a population.

Statisticians find it very convenient to distinguish between those measures of central tendency and variability computed from a population and those computed from a sample. When the mean and standard deviation of a sample are computed, they are called *statistics*, and these familiar symbols are used:

$$M = \text{mean of a sample}$$
$$S = \text{standard deviation of a sample}$$

But when the mean and standard deviation of a population are computed, they are called *parameters*. Then these rather unfamiliar symbols are used:

μ = mean of a population
σ = standard deviation of a population

Throughout this chapter you will be using the symbols μ (the Greek letter mu; pronounced myü) and σ (the lower-case Greek letter sigma, also pronounced sigma like the capital Σ). Don't let this bother you. Just remember that since you are dealing with populations which are normally distributed, the mean of that population is symbolized as μ and the standard deviation of that population is symbolized as σ.

CHARACTERISTICS OF NORMAL DISTRIBUTIONS

Because of its characteristic shape, the normal distribution is sometimes called a *bell-shaped curve*. Normal distributions are exactly symmetrical, and therefore exactly half of the area under the curve falls on each side of the middle. For instance, if the middle of bowler A's distribution of scores is the 150 point, then half of his scores (the area under the curve) fall below 150, and the other half above 150. We have been using the term *middle* up to now because any of the measures of central tendency that we discussed earlier could be applied here. Since the curve is symmetrical, has a single mode, and half of the observations lie on either side of the center, it follows from what we know about measures of central tendency that the mode, the median, and the mean of a normal distribution are all identical and all lie exactly in the middle of the distribution.

AREAS UNDER THE NORMAL CURVE

Suppose you take a long multiple-choice examination and get a score of 125. Knowing only your own score, you could not know how well you did on the test (especially if your instructor marks on the "curve"). Let us assume that the test scores are normally distributed. You are now told that the mean of the scores is 110, and that the standard deviation is 10. You also know that 50 students took the test. The rest of this chapter will explain the methods by which you could determine how many students scored higher or lower than you did, or how many students scored somewhere between your score and the mean. The use of the normal distribution in solving probability problems will also be discussed.

In Figure 8-4 the baseline is marked off in z-score units. Since this distribution is normal, it can be demonstrated mathematically that the shaded area between the mean ($z = 0$) and the point that is one standard deviation away ($z = 1$) will contain 34.13 percent of the total area under the curve. If 50 students took the test, then 34.13 percent, or about 17, of them would score between $z = 0$ and $z = 1$. In other words, about 34 percent of the observations fall within one standard deviation from the mean

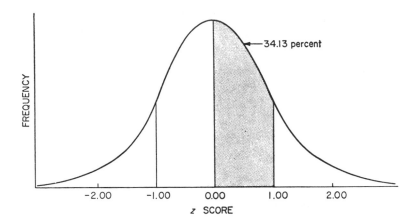

Figure 8-4   *Area under the normal curve between the mean and z = 1.00.*

in one direction. Because the normal distribution is symmetrical, an equal percentage of the area will fall between the mean and the $z = -1$ point. Therefore, 34.13 percent + 34.13 percent, or 68.26 percent, of the area will fall within one standard deviation of the mean.

   Figure 8-5 shows that 47.72 percent of the area is included between the mean and a point that is two standard deviations from the mean in one direction. The points located at $z = 2$ and $z = -2$ include 47.72 percent + 47.72 percent, or 95.44 percent, of the total area under the curve.

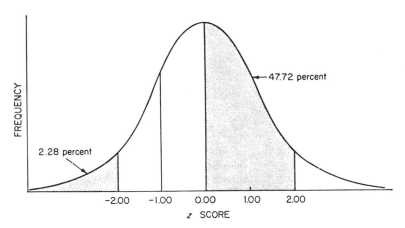

Figure 8-5   *Areas under the normal curve between the mean and z = 2.00 and beyond z = -2.00.*

   How do we find the area outside boundaries such as those indicated in Figure 8-5? The figure shows that 2.28 percent of the area lies below the $z = -2$ point. This follows from the fact that 50 percent of the total area lies below the mean, and the difference between 50 percent and 47.72

percent (the percentage of cases between the mean and $z = -2$) is indeed 2.28 percent. Since there is a "tail" of the distribution at each end of the curve, we can multiply 2.28×2 and find that 4.56 percent of the area lies outside the $z = \pm 2$ limits. Notice that this could also have been determined by subtracting 95.44 percent from 100 percent, which is the total area under the curve.

Now consider a different question. What is the percentage of area lying between the $z = 1$ and $z = 2$ points? In Figure 8-6 we see that since

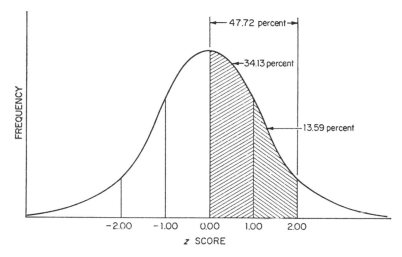

Figure 8-6   *Areas under the normal curve between* $z = 1.00$ *and* $z = 2.00$, *and* $z = 0.00$.

47.72 percent of the area lies between the mean and the $z = 2$ point, and 34.13 percent of the area lies between the mean and the $z = 1$ point, the space between the two points must be equal to the difference between the two percentages (13.59 percent).

The key to all such problems involving percentages of the area under the normal curve is the use of the mean of the distribution ($z = 0$) as a reference point for the other z scores involved in the problem.

So far, you are still unable to determine how well you did on the test mentioned earlier. The mean and your score were given in raw-score units. That is, to determine how well you did, you first must translate your test (or raw) score into a z score and then translate your z score into a percentage. The procedure for doing this is always the same, and is shown here to emphasize this fact:

$$X \rightarrow \left[z = \frac{X - \mu}{\sigma} \right] \rightarrow z \rightarrow [\text{Appendix B}] \rightarrow \%$$

That is, given X, you use the formula $z = (X - \mu)/\sigma$ to go from X to z, and then you use Appendix B to translate the z score into a percentage. Substituting your score on the multiple choice test into the z-score formula, we find

$$z = \frac{X - u}{\sigma} = \frac{125 - 110}{10} = \frac{15}{10} = 1.50$$

To find a raw score from a percentage, the process must be reversed:

$$\% \rightarrow [\text{Appendix B}] \rightarrow z \rightarrow [X = \mu + z \cdot \sigma] \rightarrow X$$

That is, the percentage can be translated into a z score by using Appendix B. Then the formula $X = \mu + z \cdot \sigma$ is used to translate a z score into a raw score. This latter formula may look new, but it is only a rearrangement of the first z-score formula. Appendix B is a highly useful table because it enables us to find areas above, below, and between various points on any normal curve as long as we know the mean and standard deviation.

READING APPENDIX B   The first column on the left-hand side of Appendix B contains z scores representing various distances from the mean of a normal distribution. Column B contains percentages of the total area under the normal curve corresponding to the sections between the mean and the various z-score points in column A. Similarly, columns C and D contain the percentages in the larger portion of the curve or the smaller portion of the curve at the various designated z-score distances from the mean.

To find the percentage of the area under the normal curve for $z = 1.54$, first go down the z-score list, column A, until you reach 1.54. At that point, column B indicates that 43.82 percent of the area falls between the mean and $z = 1.54$. Similarly, column C indicates that 93.82 percent of the area falls below $z = 1.54$ and 6.18 percent of the area falls above $z = 1.54$. Similarly, $z = -0.59$ contains 22.24 percent of the area under the normal curve between that point and the mean. Also, column C indicates that 72.24 percent falls *above* $z = -0.59$, and column D indicates that 27.76 percent falls *below* $z = -0.59$. Notice that when you are looking at a point *above* the mean, a positive z score, the larger portion (column C) represents the area *below* that point. But when you are looking at a point *below* the mean, then column C represents the area *above* that point. You should always keep this in mind as you use this table.

PROBLEMS INVOLVING USE OF THE NORMAL CURVE

We now have the basic tools for answering a number of different questions raised earlier in the test-score example. In solving these problems, we will illustrate the primary types of questions that can be answered by use of the normal distribution. You will recall that we have the following information at this point: (1) Fifty students took the test, (2) the average score was 110, (3) the standard deviation of the scores was 10, and (4) your own score was 125. We also transformed your score into a z score so that it could be used with the z-score table in Appendix B. We found that your raw score of 125 was equivalent to a z score of 1.50.

(1) *How many scores were higher than yours?* Your score was 1.50 standard deviations above the mean, so anybody who scored higher must fall in that portion of the area under the curve above a z score of 1.50. We know that 50 percent of the total area lies above the mean, and by looking up a z score of 1.50 in Appendix B, we learn (from column D) that 6.68 percent of the area (and therefore of the students who took the test) lies above $z = 1.50$. Since 50 students took the test, we know that approximately three of them scored higher than you did. Figure 8-7 illustrates this problem.

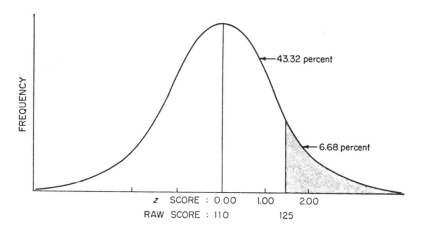

Figure 8-7   *Area under the normal curve beyond a raw score of 125.*

(2) *How many students got scores between your score and the mean?* We find the answer to this question simply by looking under Column B in the row where $z = 1.50$. Here the entry is 43.32 percent, which represents the area between $z = 1.50$ and the mean. Multiplying the total number of students (50) by the percentage (43.32 percent) yields an answer of approximately 22 students. Notice that the number of students with scores between yours and the mean (22) plus those scoring higher than you (3) equals 25, or exactly half of the total class (the half that scored above the mean).

You may be wondering which of these two subgroups contains *your* score. This question cannot be answered, since the score of 125 is treated as a point falling exactly between the percentages yielded by the normal curve table.

(3) *What is the raw score that separates the upper 25 percent of the class from the rest?* To answer this question we have to reverse the process we used before. Instead of transforming a raw score to a z score and then looking up the percentage of area in Appendix B, we begin here

with a percentage. Since we are interested in the score above which **25** percent of the area falls, we would use Column D.

So, by searching among the percentages in Column D, we locate the figure closest to 25, which is 25.14. By looking at the labels for the row and column corresponding to this entry, we determine that 25 percent of the area falls between the mean and $z = 0.67$. This is the dividing line for the upper 25 percent of the class. We still have to convert this z score into a raw score, however. By changing our previous formula around algebraically, we obtain the formula we need: $z = (X - \mu)/\sigma$ becomes $X = (z)(\sigma) + \mu$. Substituting our present scores in this formula gives us the raw score we need: $X = (0.67)(10) + 110 = 116.7$. Approximately **25** percent of the students scored higher than **117** on the test. Since your score was **125**, you are well into this high-scoring group.

(4) *How many of the students made scores between 85 and 125?* Since we already know the percentage of students scoring between the mean and 125, this would be an easy question to answer if 85 was just as far below the mean as 125 is above it. We would only have to double the percentage to get the answer. Since this is not the case, we will have to go at the problem differently.

Remember always to use the mean as a starting point in finding areas. With this in mind, it is clear that we must add together two areas—one from each side of the mean—after determining them separately. This technique is illustrated in Figure 8-8.

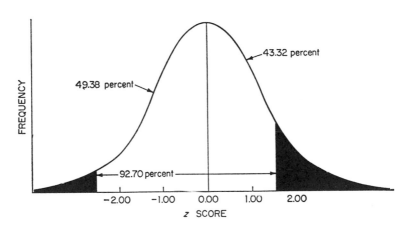

Figure 8-8   *Area under the normal curve between raw scores of 85 and 125.*

We already know that the area we want above the mean is **43.32** percent; we found that in studying your test score. To find the other part of the area—between the mean and a raw score of 85—we must first convert 85 to a z score. Using the formula in the usual way, we find that a raw score of 85 is equivalent to $z = -2.50$. A negative sign on a z score *always* indicates that it lies below the mean.

The next step is to determine the percentage of area below this point. The fact that the z score is negative can be ignored in using the table, since the normal curve is symmetrical. Column B shows 49.38 percent for $z = -2.50$. That is, 49.38 percent of the area falls between 85 and 110. We now add the two areas because they lie on opposite sides of the mean. The total area between $z = -2.50$ and $z = 1.50$ is 49.38 percent $+$ 43.32 percent, or 92.70 percent. So, approximately 93 percent, or 46, students scored between 85 and 125.

PROBABILITY AND THE NORMAL CURVE

In Chapter Seven we defined the probability of an event as the ratio of favorable to total possible events. The equation stating the probability of event A was given as $\Pr(A) = N_A/N_E$, where N_A is the number of As and N_E is the total number of possible events.

It may already have occurred to you that areas under the normal curve offer the possibility of answering questions dealing with probability. In terms of our test-score example, for instance, we could think of all the test scores as all possible events. Using the probability formula and what we have determined concerning the number of scores above 125, we could substitute in the formula in the following manner:

$$\Pr(A) = N_A/N_E = 3/50 = 0.06$$

The probability that a student in the class picked at random would have a score greater than 125 is about 3 in 50, or 0.06, or 6 percent. Notice that this is the same (except for rounding error) as the percentage of area under the normal curve that we previously calculated for scores above $z = 1.50$.

Many questions stated as problems in probability determination may be successfully approached through the use of the normal curve. When used in this way, the total area under the curve corresponds to the probability of all possible events, and some portion of the total area is identified as the set of favorable events. If the events can be measured on a scale so that the mean and standard deviation can be found, and if the total of all possible events can be assumed to be normally distributed, computations of probability can be achieved by the use of the methods described in this chapter.

SUMMARY

This chapter has presented an extended discussion of the normal distribution, a distribution which is central to any discussion of inferential statistics. It should be apparent by now that any normal distribution can be "translated" or transformed into a z distribution, a distribution with $\mu = 0$, $\sigma = 1$. Through the use of Appendix B, areas under any normal distribution can be readily found when only two scores are known: the mean and the standard deviation. This gives us powerful tools with which to deal with large numbers of populations and to make inferences about these populations.

PROBLEMS

The following problems can be most efficiently solved if you will draw a normal curve to aid you in keeping the relationships given in the problem in mind. For example, for the third question you might draw the curve shown below.

1. $\mu = 50$, $\sigma = 10$, $X = 65$, $z = $ _____.
2. $\mu = 120$, $\sigma = 20$, $z = -1.50$, $X = $ _____.
3. $\mu = 75$, $\sigma = 5$. What percent of the scores fall between the mean and 83?
4. $\mu = 90$, $\sigma = 2$. What percent of the area under the normal curve falls above 93?
5. $\mu = 18$, $\sigma = 3$. What percent of the area falls between 13 and 20?
6. $\mu = 200$, $\sigma = 25$. What percent of the area falls between 220 and 245?
7. Two hundred students took a test which had $\mu = 60$, $\sigma = 20$. Twenty students made an "A." What was the minimum score a student could make and still make an "A?"
8. IQ scores were found for each of the inmates of an army Disciplinary Barracks (that is, prison). The mean was 85 and the standard deviation was 15. What percent of the prison population had average (100) IQs or higher?
9. From Question 8: If someone with an IQ of 75 or less can be classified as mentally impaired, what percent of the prison population could be classified as mentally impaired?
10. At Drizzly Tech the mean IQ is 120 with a standard deviation of 10. At Flotsam University the mean IQ is 110 with a standard deviation of 15. What percent of the Flotsam University students are below 60 percent of the Drizzly Tech students?
11. $\mu = 50$, $\sigma = 10$. What percent of the area falls above a score of 60?
12. $\mu = 20$, $\sigma = 2$. What percent of the area falls above a score of 19?
13. $\mu = 80$, $\sigma = 5$. What percent of the area falls between 78 and 86?
14. $\mu = 150$, $\sigma = 20$; 95 percent of the scores fall between what two points?
15. $\mu = 500$, $\sigma = 50$; 99 percent of the scores fall between what two points?
16. $\mu = 200$, $\sigma = 10$; 25 percent of the scores fall above what point?
17. A test was given with the following results: $\mu = 40$, $\sigma = 3$; 10 percent of the class got a grade of "A." What score was the lowest "A"?
18. $\mu = 60$, $\sigma = 4$; 5 percent of the class failed the test. What was the highest failing score?

EXERCISE

1. Using the data collected from Exercise 1 of Chapter Two, estimate the number of cases falling within certain intervals. Use your knowledge of the mean and standard deviation, and Appendix B. How close were the estimates when made from the normal distribution of the numbers of cases you actually had?

1 The area under the normal curve may be expressed in percentages. If all the area is under the normal curve what percent of the area is under the normal curve? _____.

========= STOP ================================= STOP =========

100 percent

---

2 The normal curve is symmetrical and the median and mode fall in the exact middle of the distribution. The mean is also found in the _____ of the distribution.

========= STOP ================================= STOP =========

middle

---

3 Since the curve is symmetrical and the mean falls exactly in the middle we may assume that _____ percent of the area falls above the mean.

========= STOP ================================= STOP =========

50

---

4 So too must _____ percent fall below the mean.

========= STOP ================================= STOP =========

50

---

5 In the function that specifies the normal curve there are two unknowns: the mean, and the standard deviation. Once we know the _____ and the _____ we know (with the help of tables) everything about the area under the normal curve.

========= STOP ================================= STOP =========

mean, standard deviation

---

6 Approximately 34 percent of the area under the normal curve falls between the mean and a point that is 1 standard deviation distant. The percentage of the area between the mean and the point 1 standard deviation above the mean is _____.

========= STOP ================================= STOP =========

34 percent

---

7  The percentage of the area under the normal curve between the mean and 1 standard deviation is about 34 percent. What is the percentage below a point 1 standard deviation below the the mean? _____.

═══ STOP ═══════════════════════ STOP ═══

16 percent

---

8 To get this answer we subtract 34 percent (the area between the mean and a point 1 standard deviation below the mean) from _____.

═══ STOP ═══════════════════════ STOP ═══

50 percent

---

9 This 50 percent is the area falling _____ ____ _____.

═══ STOP ═══════════════════════ STOP ═══

below the mean

---

10 The area falling above a point 1 standard deviation above the mean is equal to _____.

═══ STOP ═══════════════════════ STOP ═══

16 percent

---

11 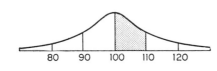 Suppose the mean is equal to 100 and the standard deviation is equal to 10. Approximately what percent of the area falls between 100 and 110? _____.

═══ STOP ═══════════════════════ STOP ═══

34 percent

---

12 Suppose $\mu = 100$ and $\sigma = 10$; 34 percent of the area under the normal curve falls between 90 and _____.

═══ STOP ═══════════════════════ STOP ═══

100

---

13   Another 34 percent falls between 100 and _____.

═══════ STOP ═══════════════════════════ STOP ═══════

110

14

Again suppose $\mu = 100$ and $\sigma = 10$. What percent of the area under the normal curve falls below 90? _____.

═══════ STOP ═══════════════════════════ STOP ═══════

16 percent

---

15   To find this we subtract the percentage of the area between 90 and 100 from 50 percent. What percent of the area falls between 90 and 100? _____.

═══════ STOP ═══════════════════════════ STOP ═══════

34 percent

---

16   Suppose $\mu = 10$ and $\sigma = 1$; 34 percent of the area falls between 11 and _____.

═══════ STOP ═══════════════════════════ STOP ═══════

10

---

17   Again $\mu = 10$ and $\sigma = 1$; 16 percent of the area falls below what point? _____.

═══════ STOP ═══════════════════════════ STOP ═══════

9

---

18   If $\mu = 25$ and $\sigma = 5$; then 16 percent of the area falls above _____.

═══════ STOP ═══════════════════════════ STOP ═══════

30

---

19   If $\mu = 500$ and $\sigma = 100$; then 16 percent of the cases fall below _____.

═══════ STOP ═══════════════════════════ STOP ═══════

400

---

20   Let $\mu = 100$ and $\sigma = 10$. Suppose we wished to find the area between 90 and 110. The area between 90 and 100 is equal to 34 percent. So, too, is the area between 100 and 110 equal to 34 percent. What is the area between 90 and 110 equal to? _____.

══════ STOP ══════════════════════════════ STOP ══════

68 percent

---

21   This last problem illustrates a principle: If we wish to find an area under the curve, which falls on both sides of the mean, we must find the area between the lowest point and the mean and add this to the area between the highest point and the _____.

══════ STOP ══════════════════════════════ STOP ══════

mean

---

22   If $\mu = 500$ and $\sigma = 100$, then 68 percent of the area falls between 400 and _____.

══════ STOP ══════════════════════════════ STOP ══════

600

---

23   If $\mu = 1000$ and $\sigma = 200$, what percent of the area falls between 800 and 1200? _____.

══════ STOP ══════════════════════════════ STOP ══════

68 percent

---

24   If $\mu = 1000$ and $\sigma = 200$, what percent of the area falls below 800? _____.

══════ STOP ══════════════════════════════ STOP ══════

16 percent

---

25   If $\mu = 1000$ and $\sigma = 200$, what percent of the area falls above 1200? _____.

══════ STOP ══════════════════════════════ STOP ══════

16 percent

26

Therefore, if $\mu = 1000$ and $\sigma = 200$, what percent of the area falls below 800 and above 1200? _____.

═══════ **STOP** ═══════════════════════ **STOP** ═══════

32 percent

---

27 This problem illustrates another principle: To find an area falling outside two points that are under the normal curve, one must first find the area below the lowest point. Next, one must find the area above the highest point, and then _____ the two areas together.

═══════ **STOP** ═══════════════════════ **STOP** ═══════

add

---

28 Suppose 30 percent of the area falls below 70 and 15 percent of the area falls above 90. What percent of the area falls below 70 and above 90?

═══════ **STOP** ═══════════════════════ **STOP** ═══════

45 percent

---

29 If $\mu = 100$ and $\sigma = 15$, what percent of the area falls below 85 and above 115? _____.

═══════ **STOP** ═══════════════════════ **STOP** ═══════

32 percent

---

30 Let us review: Given $\mu = 100$ and $\sigma = 10$, what percent of the area falls between 90 and 100? _____.

═══════ **STOP** ═══════════════════════ **STOP** ═══════

34 percent

---

31 $\mu = 100$, $\sigma = 10$, what percent of the area falls between 100 and 110? _____.

═══════ **STOP** ═══════════════════════ **STOP** ═══════

34 percent

---

32 $\mu = 100$ and $\sigma = 10$, what percent of the area falls between 90 and 110?
_____.

══════ STOP ══════════════════ STOP ══════

68 percent

33 $\mu = 100$, $\sigma = 10$, what percent of the area falls below a score of 90?
_____.

══════ STOP ══════════════════ STOP ══════

16 percent

34 $\mu = 100$, $\sigma = 10$, what percent falls above a score of 110? _____

══════ STOP ══════════════════ STOP ══════

16 percent

35 $\mu = 100$, $\sigma = 10$, what percent of the area falls below 90 and above 110? _____.

══════ STOP ══════════════════ STOP ══════

32 percent

36 Approximately 47.5 percent of the area falls between the mean and a point 2 standard deviations away from the mean. If $\mu = 100$ and $\sigma = 10$, then what percent of the area would fall between 100 and 120? _____.

══════ STOP ══════════════════ STOP ══════

47.5 percent

37 Again $\mu = 100$ and $\sigma = 10$, what percent of the area under the normal curve falls between 80 and 100? _____.

══════ STOP ══════════════════ STOP ══════

47.5 percent

38 Therefore, if 47.5 percent of the area falls between 80 and 100 and 47.5 percent falls between 100 and 120, approximately how much of the area falls between 80 and 120? _____.

══════ STOP ══════════════════ STOP ══════

95 percent

39   Again if $\mu = 100$ and $\sigma = 10$, what percent of the area falls outside of the 80–120 range? _____.

═══════ STOP ═══════════════════════ STOP ═══════

5 percent

---

40 $\mu = 500$, $\sigma = 100$, 95 percent of the area falls between 300 and _____.

═══════ STOP ═══════════════════════ STOP ═══════

700

---

41 $\mu = 50$, $\sigma = 2$, 95 percent of the area under the normal curve falls between _____ and _____.

═══════ STOP ═══════════════════════ STOP ═══════

46 and 54

---

42 $\mu = 1000$ and $\sigma = 100$, 16 percent of the scores fall below what point? _____.

═══════ STOP ═══════════════════════ STOP ═══════

900

---

43 $\mu = 1000$ and $\sigma = 100$, 2.5 percent of the scores fall above what point? _____.

═══════ STOP ═══════════════════════ STOP ═══════

1200

---

44 $\mu = 1000$ and $\sigma = 100$, what percent of the scores fall below 900 and above 1200? _____.

═══════ STOP ═══════════════════════ STOP ═══════

18.5 percent

---

45   To get this answer we find the area below _____, the area above _____, and then we _____.

═══════ STOP ═══════════════════════ STOP ═══════

900, 1200, add

---

46 $\mu = 80$, $\sigma = 3$, what percent of the area falls between 77 and 86? _____.

══════ STOP ═══════════════════════════ STOP ══════

81.5 percent

---

47   To find this last answer we first find the area between 77 and _____ (remember $\mu = 80$ and $\sigma = 3$).

══════ STOP ═══════════════════════════ STOP ══════

80 (the mean)

---

48   This area is equal to _____.

══════ STOP ═══════════════════════════ STOP ══════

34 percent

---

49   Next we find the area between 86 and _____ (remember $\mu = 80$ and $\sigma = 3$).

══════ STOP ═══════════════════════════ STOP ══════

80 (the mean)

---

50   This second area is equal to what percent ? _____.

══════ STOP ═══════════════════════════ STOP ══════

47.5 percent

---

51   Since 34 percent of the area falls between 77 and 80 and since 47.5 percent of the area falls between 80 and 86, then _____ percent of the area falls between 77 and 86.

══════ STOP ═══════════════════════════ STOP ══════

81.5

---

52 $\mu = 10$, $\sigma = 1$, what percent of the area falls below 8 and above 11? _____.

══════ STOP ═══════════════════════════ STOP ══════

18.5 percent

---

53 $\mu = 10$, $\sigma = 1$, what percent of the area falls between 9 and 12? _____.

══════ STOP ═══════════════════════════ STOP ══════

81.5 percent

---

54 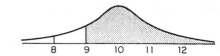 $\mu = 10$, $\sigma = 1$, what percent of the area falls above 9? _____.

════ STOP ════ ════ STOP ════

84 percent

---

55 To find this we first find the area between 9 and 10 (the mean) and then we add this to the area falling above 10. What percent of the area falls above 10? _____.

════ STOP ════ ════ STOP ════

50 percent

---

56 $\mu = 20$, $\sigma = 4$, what percent of the area falls below 24? _____.

════ STOP ════ ════ STOP ════

84 percent

---

57 $\mu = 20$, $\sigma = 4$, what percent of the area falls above 12? _____.

════ STOP ════ ════ STOP ════

97.5 percent

---

58 $\mu = 100$ and $\sigma = 10$, what percent of the area falls between 100 and 110? _____.

════ STOP ════ ════ STOP ════

34 percent

---

59 $\mu = 100$, $\sigma = 10$, what percent of the area falls between 100 and 120? _____.

════ STOP ════ ════ STOP ════

47.5 percent

---

60 $\mu = 100$ and $\sigma = 10$, if 34 percent of the area falls between 100 and 110, and if 47.5 percent of the area falls between 100 and 120, then what percent of the area falls between 110 and 120? _____.

════ STOP ════ ════ STOP ════

13.5 percent

---

61 We get this by _____ the area between 100 and 110 from the area between 100 and 120.

===== STOP ===== STOP =====

subtracting

---

62 Similarly, if $\mu = 100$ and $\sigma = 10$, then the area between 80 and 90 is equal to _____.

===== STOP ===== STOP =====

13.5 percent

---

63 While we can give rough approximations of the area under the normal curve by memory, we must use a z-score table to be accurate. A z score is a distance stated in terms of standard deviations. For example, if $\mu = 100$ and $\sigma = 10$, 110 would be 1 standard deviation above the mean and z would equal 1.00. If $\mu = 100$ and $\sigma = 10$, then if $X = 120$, $z = $ _____.

===== STOP ===== STOP =====

2.00

---

64 To find a z score, simply subtract the mean from the (raw) score and divide by σ, or $z = (X - \mu)/\sigma$. If $\mu = 1000$, $\sigma = 50$, and $X = 1100$, then X expressed in z-score units would equal _____.

===== STOP ===== STOP =====

2.00

---

65 Be careful to note that since the mean is subtracted from the score then a z score representing a score above the mean is positive while a z score representing a score below the mean is _____.

===== STOP ===== STOP =====

negative

---

66 Thus if we had $z = 2.3$ we would know this represents a score above the mean and that $z = -.42$ represents a score _____ _____ _____.

===== STOP ===== STOP =====

below the mean

---

67 $z = (X - \mu)/\sigma$. If $\mu = 100$, $\sigma = 15$, and $X = 130$, then $z = $ _____.

===== STOP ===== STOP =====

2.00

68 To get this we substituted into the formula $z = (X - \mu)/\sigma$. Since $\mu = 100$ and $\sigma = 15$, then $z = (X - 100)/15$; and if $X = 130$, then $z = (130 - 100)/15 = 30/15 = 2.00$. With the same μ and σ the z corresponding to $X = 115$ is _____.

══════ STOP ════════════════════════ STOP ══════

1.00

69
$$z = \frac{X - \mu}{(?)}$$

══════ STOP ════════════════════════ STOP ══════

σ

70
$$z = \frac{X - (?)}{\sigma}$$

══════ STOP ════════════════════════ STOP ══════

μ

71
$$z = \frac{(?) - \mu}{\sigma}$$

══════ STOP ════════════════════════ STOP ══════

X

72
$$\underline{\hspace{2cm}} = \frac{X - \mu}{\sigma}$$

══════ STOP ════════════════════════ STOP ══════

z

73
$$z = \underline{\hspace{2cm}}$$

══════ STOP ════════════════════════ STOP ══════

$(X - \mu)/\sigma$

---

74 $\mu = 100$, $\sigma = 15$, if $X = 85$, then $z = $ _____.

══════ STOP ════════════════════════ STOP ══════

-1.00

---

75 The sign of the z score was negative because X was _____ the mean.

══════ STOP ════════════════════════ STOP ══════

below

---

76   Again $\mu = 100$, $\sigma = 15$, if $X = 120$, then $z =$ _____ .

══════ STOP ══════════════════════════ STOP ══════

1.33

---

77   Again $\mu = 100$, $\sigma = 15$, if $X = 145$, then $z =$ _____ .

══════ STOP ══════════════════════════ STOP ══════

3.00

---

78 $\mu = 100$, $\sigma = 15$, if $X = 90$, $z =$ _____ .

══════ STOP ══════════════════════════ STOP ══════

−0.67

---

79   Again $\mu = 100$, $\sigma = 15$, if $X = 80$, then $z =$ _____ .

══════ STOP ══════════════════════════ STOP ══════

−1.33

---

80 $\mu = 50$, $\sigma = 2$, if $X = 53$, then $z =$ _____ .

══════ STOP ══════════════════════════ STOP ══════

1.50

---

81 $\mu = 10$, $\sigma = 3$, if $X = 6$, then $z =$ _____ .

══════ STOP ══════════════════════════ STOP ══════

−1.33

---

82 $\mu = 1000$, $\sigma = 100$, if $X = 966$, then $z =$ _____ .

══════ STOP ══════════════════════════ STOP ══════

−0.34

---

83   The process can be reversed. That is, we can find a score if the z score is given. Since $z = (X - \mu)/\sigma$, then $X = z \times \sigma + \mu$. If $\mu = 100$, $\sigma = 15$, $z = 1.00$; then $X =$ _____ .

══════ STOP ══════════════════════════ STOP ══════

115

84 $X = z \times \sigma + \mu$. If $\mu = 100$, $\sigma = 15$, $z = -0.67$, then $X =$ _____.
════════ STOP ═══════════════════════════════ STOP ═══════

90

85 $X = z \times \sigma +$ _____
════════ STOP ═══════════════════════════════ STOP ═══════

μ

86 $X =$ _____ $\times \sigma + \mu$
════════ STOP ═══════════════════════════════ STOP ═══════

z

87 $X = z \times$ _____ $+ \mu$
════════ STOP ═══════════════════════════════ STOP ═══════

σ

88   _____ $= z \times \sigma + \mu$
════════ STOP ═══════════════════════════════ STOP ═══════

X

89 $X =$ _____
════════ STOP ═══════════════════════════════ STOP ═══════

$z \times \sigma + \mu$

90   Given $\mu = 50$, $\sigma = 5$, $z = 2.00$, then $X =$ _____.
════════ STOP ═══════════════════════════════ STOP ═══════

60

91 $\mu = 500$, $\sigma = 100$, $z = -1.50$, $X =$ _____.
════════ STOP ═══════════════════════════════ STOP ═══════

350

92 $\mu = 80$, $\sigma = 3$, $z = -3.00$, $X =$ _____.
════════ STOP ═══════════════════════════════ STOP ═══════

71

93 $\mu = 80$, $\sigma = 3$, $z = 0.00$, then $X =$ _____.

═══════ STOP ═══════════════════════════ STOP ═══════

80

───

94 From our previous discussion (questions 6–35) we know that approximately 34 percent of the area falls between the mean and a point 1 standard deviation from the mean. Thus the area under the normal curve between the mean and $z = 1.00$ is equal to _____.

═══════ STOP ═══════════════════════════ STOP ═══════

34 percent

───

95 The area above $z = 1.00$ is equal to _____.

═══════ STOP ═══════════════════════════ STOP ═══════

16 percent

───

96 The area below $z = -1.00$ is equal to _____.

═══════ STOP ═══════════════════════════ STOP ═══════

16 percent

───

97   Hence the area above $z = -1.00$ is equal to _____.

═══════ STOP ═══════════════════════════ STOP ═══════

84 percent

───

98 The area between $z = -1.00$ and $z = 1.00$ is equal to _____.

═══════ STOP ═══════════════════════════ STOP ═══════

68 percent

───

99 The area between a point 2 standard deviations below the mean and a point 2 standard deviations above the mean is equal to 95 percent. Hence the area between $z = -2.00$ and $z = 2.00$ is equal to _____.

═══════ STOP ═══════════════════════════ STOP ═══════

95 percent

───

100   The area between $z = -2.00$ and $z = 0.00$ is equal to what percent? _____.

════════ STOP ════════════════════════ STOP ════════

47.5 percent

---

101   The area between $z = -2.00$ and $z = 1.00$ is equal to _____.

════════ STOP ════════════════════════ STOP ════════

47.5 percent + 34 percent = 81.5 percent

---

102   The area between $z = 1.00$ and $z = 2.00$ is equal to _____.

════════ STOP ════════════════════════ STOP ════════

13.5 percent

---

103   The reason for this is that 34 percent of the area falls between $z = 0.00$ and $z = 1.00$ and that _____ percent falls between $z = 0.00$ and $z = 2.00$.

════════ STOP ════════════════════════ STOP ════════

47.5

---

104   If 34 percent falls between $z = 0.00$ and $z = 1.00$ and if 47.50 percent falls between $z = 0.00$ and $z = 2.00$ then _____ percent falls between $z = 1.00$ and $z = 2.00$.

════════ STOP ════════════════════════ STOP ════════

13.5

---

105   Similarly, the area between $z = -2.00$ and $z = -1.00$ is equal to _____.

════════ STOP ════════════════════════ STOP ════════

13.5 percent

---

106   Look at Appendix B. The column on the left is labeled _____.

════════ STOP ════════════════════════ STOP ════════

"A" or "z"

---

107   Go down the "A" column to a number reading 0.25. Follow the 0.25 row over one column until you are under the column labeled "B" or "area between mean and z." The table entry is _____.

════════ STOP ════════════════════════ STOP ════════

9.87

---

108 Go on to the next column in the same row. The column is labeled "C," or "area in larger portion," and the tabled entry is _____.

═══════ STOP ═══════════════════════ STOP ═══════

59.87

---

109 This percentage is the column B value plus 50% and represents the area in the larger portion under the normal curve. Similarly, column D was found by subtracting the tabled entry (9.87) from 50%. This value is

_____.

═══════ STOP ═══════════════════════ STOP ═══════

40.13

---

110 The numbers in the body of the table are areas (in percentages) under the normal curve between the mean and _____ or the areas above or below _____.

═══════ STOP ═══════════════════════ STOP ═══════

z, z

---

111 Hence, if you wish to find the area between the mean and $z = 2.22$, you go down the z column to $z = 2.22$ and go across the row to the _____ column. The tabled entry here is _____.

═══════ STOP ═══════════════════════ STOP ═══════

B, or "area between the mean and z"; 48.68

---

112 If you wished to find the area below $z = 2.05$, you would go down the z column to $z =$ _____ and then go over to the _____ column. Here the tabled entry is _____.

═══════ STOP ═══════════════════════ STOP ═══════

2.05; C, or "area in larger portion"; 97.98

---

113 If you wished to find the area above $z = 1.55$, you would go down the z column to $z = 1.55$ and then go over to the _____ column. Here the tabled entry is _____.

═══════ STOP ═══════════════════════ STOP ═══════

D, or "area in smaller portion"; 6.06

---

114 If $z = 0.85$, then the area between the mean and that point equals _____; the area below that point equals _____; and the area above that point equals _____.

═══════ STOP ═══════════════════════ STOP ═══════

30.23, 80.23, 19.77

---

115 According to the table what percent of the area falls between the mean and $z = 1.55$? _____

═══════ STOP ═══════════════════════════ STOP ═══════

44 percent (43.94)

---

116 If $z = 2.00$, then according to the table, the percentage of the area falling below $z = 2.00$ is equal to _____.

═══════ STOP ═══════════════════════════ STOP ═══════

97.72 percent

---

117 The mean can be expressed as a z score. Since the mean does not differ from the reference point (that is, the mean), what is the z score corresponding to the mean? _____.

═══════ STOP ═══════════════════════════ STOP ═══════

$z = 0$

---

118 Then instead of asking for the area between the mean and $z = 2.00$ we could ask: $z_1 = 0.00$; $z_2 = 2.00$; what is the area between z_1 and z_2? _____.

═══════ STOP ═══════════════════════════ STOP ═══════

48 percent (47.72)

---

119 $z_1 = 0.00$; $z_2 = 0.50$, what percent of the area falls between z_1 and z_2? _____.

═══════ STOP ═══════════════════════════ STOP ═══════

19 percent (19.15)

---

120 The area between $z_1 = 0.00$ and $z_2 = 1.50$ is equal to _____.

═══════ STOP ═══════════════════════════ STOP ═══════

43 percent (43.32)

---

121 If the area between $z = 0.00$ and $z = 1.50$ is equal to 43 percent and if the area between $z = 0.00$ and $z = 0.50$ is equal to 19 percent then the area between $z = 0.50$ and $z = 1.50$ is _____.

═══════ STOP ═══════════════════════════ STOP ═══════

24 percent

---

122 $z_1 = 1.00$; $z_2 = 2.00$, what percent of the area under the normal curve falls between z_1 and z_2? _____.

══════ STOP ══════════════════════════ STOP ══════

14 percent (13.59)

---

123 The reason for this answer is that 34 percent of the area falls between the mean ($z = 0.00$) and $z = 1.00$ and 48 percent falls between $z = 0.00$ and $z = 2.00$. To find the area between $z = 1.00$ and $z = 2.00$ you _____ the smaller from the larger.

══════ STOP ══════════════════════════ STOP ══════

subtract

---

124 The area between $z = -0.50$ and $z = 0.00$ is equal to _____.

══════ STOP ══════════════════════════ STOP ══════

19 percent (19.15)

---

125 The area between $z = 0.00$ and $z = 0.75$ would be equal to _____.

══════ STOP ══════════════════════════ STOP ══════

27 percent (27.34)

---

126 If the area between $z = -0.50$ and $z = 0.00$ is equal to 19 percent and the area between $z = 0.00$ and $z = 0.75$ is equal to 27 percent then the area between $z = -0.50$ and $z = 0.75$ is equal to _____.

══════ STOP ══════════════════════════ STOP ══════

19 percent + 27 percent = 46 percent

---

127 The area between $z_1 = -1.00$ and $z_2 = 1.00$ is equal to _____.

══════ STOP ══════════════════════════ STOP ══════

68 percent (68.26)

---

128 To get this we find the area between $z = -1.00$ and $z = 0.00$. Then we find the area between $z = 0.00$ and $z = 1.00$. Finally we _____ the two areas.

══════ STOP ══════════════════════════ STOP ══════

add

129 If we wish to find the area between two z scores on opposite sides of the mean we _____ the areas between each of the z scores and the mean.

═══════ STOP ═══════════════════════ STOP ═══════

add

---

130 But if both z scores are on the same side of the mean and we wish to find the area between the two, then we _____ the smaller from the larger.

═══════ STOP ═══════════════════════ STOP ═══════

subtract

---

131 For example, $z_1 = -0.50$ and $z_2 = 1.50$. We wish to find the area between z_1 and z_2. We find the area between $z = -0.50$ and $z = 0.00$, and we find the area between $z = 0.00$ and $z = 1.50$. What do we do now? _____ _____ _____ _____.

═══════ STOP ═══════════════════════ STOP ═══════

Add the two areas

---

132 If $z_1 = 0.50$ and $z_2 = 1.50$ and we wish to find the area between the two points, we find the area between $z = 0.50$ and $z =$ _____. Then we find the area between $z = 1.50$ and $z =$ _____, and then we _____.

═══════ STOP ═══════════════════════ STOP ═══════

zero; zero, subtract

---

133 The area between $z_1 = 0.75$ and $z_2 = 1.75$ is _____.

═══════ STOP ═══════════════════════ STOP ═══════

19 percent (18.65)

---

134 Find the area between $z_1 = -0.75$ and $z_2 = 1.75$. _____.

═══════ STOP ═══════════════════════ STOP ═══════

73 percent (73.33)

---

135 Suppose $\mu = 100$ and $\sigma = 10$. We wish to find the area between 100 and 120. Since the percentage of the area is dependent on μ and σ, the scores must be converted to z scores. $z_1 =$ _____ and $z_2 =$ _____.

═══════ STOP ═══════════════════════ STOP ═══════

0.00, 2.00

---

136   The area between $z_1 = 0.00$ and $z_2 = 2.00$, according to the table, is equal to _____.

═══════ STOP ═══════════════════════ STOP ═══════

47.72 percent

---

137   Hence, if $\mu = 100$ and $\sigma = 10$, then _____ percent of the area falls between 100 and 120.

═══════ STOP ═══════════════════════ STOP ═══════

47.72

---

138   Suppose $\mu = 50$, $\sigma = 5$, and we wish to find the area between 45 and 50. $z_1 =$ _____ and $z_2 =$ _____.

═══════ STOP ═══════════════════════ STOP ═══════

-1.00, 0.00

---

139   The area between $z_1 = 0.00$ and $z_2 = -1.00$ is equal to _____.

═══════ STOP ═══════════════════════ STOP ═══════

34 percent (34.13)

---

140   Hence the area between 45 and 50 (with $\mu = 50$ and $\sigma = 5$) is equal to _____.

═══════ STOP ═══════════════════════ STOP ═══════

34 percent (34.13)

---

141 $\mu = 200$, $\sigma = 15$, find the area between 185 and 200.

170   185   200   215   230

═══════ STOP ═══════════════════════ STOP ═══════

34 percent (34.13)

---

142   We wish to find the area between 9 and 12. $\mu = 10$, $\sigma = 2$, what percent of the area falls between 9 and 10? _____.

═══════ STOP ═══════════════════════ STOP ═══════

19 percent (19.15)

---

143 $\mu = 10$, $\sigma = 2$. We wish to find the area between 9 and 12. The area between 9 and 10 is equal to 19 percent. What is the area between 10 and 12 equal to? _____.

════════ STOP ══════════════════════════ STOP ════════

34 percent (34.13)

---

144 $\mu = 10$, $\sigma = 2$. We wish to find the area between 9 and 12. The area between 9 and 10 is equal to 19 percent and the area between 10 and 12 is equal to 34 percent. To find the area between 9 and 12 we must _____ _____ _____ _____.

════════ STOP ══════════════════════════ STOP ════════

add the two areas

---

145 $\mu = 10$, $\sigma = 2$. The area between 9 and 10 is equal to 19 percent and the area between 10 and 12 is 34 percent. Therefore the area between 9 and 12 is equal to _____.

════════ STOP ══════════════════════════ STOP ════════

53 percent

---

146 $\mu = 30$, $\sigma = 5$. We wish to find the area between 26 and 32. The area between 26 and 30 is equal to _____.

════════ STOP ══════════════════════════ STOP ════════

29 percent (28.81)

---

147 $\mu = 30$, $\sigma = 5$. We wish to find the area between 26 and 32. The area between 26 and 30 is equal to 29 percent and the area between 30 and 32 is equal to _____.

════════ STOP ══════════════════════════ STOP ════════

16 percent (15.54)

---

148 $\mu = 30$, $\sigma = 5$. The area between 26 and 30 is equal to 29 percent. The area between 30 and 32 is equal to 16 percent. What is the area between 26 and 32 equal to? _____.

════════ STOP ══════════════════════════ STOP ════════

45 percent

---

149 $\mu = 30$, $\sigma = 5$. We wish now to find the area between 21 and 25. The area between 25 and 30 is equal to _____.

34 percent (34.13)

---

150 $\mu = 30$, $\sigma = 5$. We wish to find the area between 21 and 25. The area between 25 and 30 is equal to 34 percent. What is the area between 21 and 30 equal to? _____.

46 percent (46.41)

---

151 $\mu = 30$, $\sigma = 5$. The area between 25 and 30 is equal to 34 percent and the area between 21 and 30 is equal to 46 percent. Thus, the area between 21 and 25 is equal to _____.

12 percent

---

152 $\mu = 60$, $\sigma = 4$, what is the area between 57 and 65 equal to? _____.

$27.34 + 39.44 = 66.78$ percent (67)

---

153 $\mu = 72$, $\sigma = 2$, what is the area between 69 and 71? _____.

24 percent (24.17)

---

154   If we wished to go from a raw score to a z score we would use the formula: $z = (X - \mu)/\sigma$. But if we wished to go from a z score to a raw score we would use what formula? _____.

$$X = z \times \sigma + \mu$$

---

155   For example, $\mu = 50$, $\sigma = 4$. If $z = 1.50$, then $X =$ _____.

56

156 Look at Appendix B. What z score is associated with an area of 43.32 percent between itself and the mean? _____.

═══════ STOP ═══════════════════════════ STOP ═══════

$$z = 1.50$$

---

157 16.64 percent of the area under the normal curve falls between the mean and $z =$ _____.

═══════ STOP ═══════════════════════════ STOP ═══════

$$z = 0.43$$

---

158 The tabled z score that most nearly contains 25 percent between itself and the mean is $z =$ _____.

═══════ STOP ═══════════════════════════ STOP ═══════

0.67

---

159 Suppose we wanted to find the score which between itself and the mean a certain percentage of the area fell. To do this we would first translate the percentage into a _____.

═══════ STOP ═══════════════════════════ STOP ═══════

z score

---

160 Then we would translate the z score into a _____ score.

═══════ STOP ═══════════════════════════ STOP ═══════

raw

---

161 $\mu = 100$, $\sigma = 10$. We wish to find the point that includes 20 percent of the area under the normal curve between itself and the mean. The z most nearly corresponding to 20 percent is _____.

═══════ STOP ═══════════════════════════ STOP ═══════

$$z = 0.52$$

---

162 $\mu = 100$, $\sigma = 10$, 20 percent of the area falls between the mean and a z score equal to 0.52. A z score of 0.52 corresponds to a raw score of

_____.

═══════ STOP ═══════════════════════════ STOP ═══════

105.2

---

163 Thus 20 percent of the area falls between 100 and 105.2 when $\mu = 100$ and $\sigma = 10$. The top 30 percent of the area falls above what point? _____.

═══════ STOP ═══════════════════════ STOP ═══════

105.2

---

164 $\mu = 40$, $\sigma = 4$. We wish to find the point above which 20 percent of the area falls. To do this we must go from a z score to a raw score. To find the top 20 percent, we must use the column labeled "area in smaller portion." The z score corresponding to the top 20 percent is

═══════ STOP ═══════════════════════ STOP ═══════

0.84

---

165 We wish to find the point below which 60 percent of the area falls. In this case we use the column labeled "area in larger portion." That z score below which 60 percent of the area falls is _____.

═══════ STOP ═══════════════════════ STOP ═══════

$z = 0.25$

---

166 Be careful to notice how the table works. For example, which column in Appendix B would you use to find the z score below which 40 percent of the cases fall?

═══════ STOP ═══════════════════════ STOP ═══════

D, or "area in smaller portion"

---

167 Suppose you wish to find the point below which 10 percent of the area falls. The z score corresponding to that point equals _____.

═══════ STOP ═══════════════════════ STOP ═══════

-1.28

---

168 Similarly, to find the area above which 80 percent of the area falls, we would use the column labeled _____ in Appendix B.

═══════ STOP ═══════════════════════ STOP ═══════

C, or "area in larger portion"

---

169 The z score corresponding to the point above which 80 percent of the area falls is _____.

============ STOP ============================== STOP ============

−0.84

---

170 Suppose a very large class was given a test in which $\mu = 60$ and $\sigma = 5$; 20 percent of the class was given a grade of "A." To find the lowest "A" scores, we must first find the z score corresponding to that 20 percent. We look in Appendix B in column _____.

============ STOP ============================== STOP ============

D, or "area in smaller portion"

---

171 The z score corresponding to that 20 percent is _____.

============ STOP ============================== STOP ============

0.84

---

172 $\mu = 60$, $\sigma = 5$, the raw score corresponding to a z score of 0.84 is equal to _____.

============ STOP ============================== STOP ============

64.20

---

173 Hence everyone scoring above 64.20 would receive a grade of _____.

============ STOP ============================== STOP ============

"A"

---

174 With reference to the same test ($\mu = 60$, $\sigma = 5$) 20 percent of the class received a "B." We know 64 was the highest "B" and we wish to find the lowest "B." What percent of the area falls above the lowest "B?" _____.

============ STOP ============================== STOP ============

40 percent

---

175 This is because the "A"s that comprise 20 percent of the area fell above the lowest "B" as well as the _____ percent of the students who received a grade of "B."

============ STOP ============================== STOP ============

20

---

176 We wish to find the point above which 40 percent of the scores fall. This corresponds to a z of _____.

================ STOP ================ STOP ================

0.25

---

177 If $\mu = 60$, $\sigma = 5$, and $z = 0.25$, then $X =$ _____.

================ STOP ================ STOP ================

61.25

---

178 Therefore students who received scores between _____ and 64 received a grade of "B."

================ STOP ================ STOP ================

61 (61.25)

---

179 In a second test $\mu = 70$ and $\sigma = 8$. This time only 10 percent of the class got an "A" while 20 percent of the class got "B." We wish to find the highest and lowest "B" scores. The area above the highest "B" equals _____ percent. The area above the lowest "B" equals _____ percent.

================ STOP ================ STOP ================

10; 30

---

180 The two z scores corresponding to the points above which 10 percent and 30 percent of the area fall are _____ and _____.

================ STOP ================ STOP ================

1.28 and 0.52

---

181 If $\mu = 70$, $\sigma = 8$, $z_1 = 0.52$, and $z_2 = 1.28$, then the raw scores (and the highest and lowest "B"s) are equal to _____ and _____.

================ STOP ================ STOP ================

74.16 and 80.24

---

182 Suppose on the same test 55 percent of the class was given a grade of "C." (Remember $\mu = 70$, $\sigma = 8$ and 30 percent of the class got "A" or "B.") What percent of the class members got "C" and were above the mean? _____.

================ STOP ================ STOP ================

20 percent

---

183 $\mu = 70$, $\sigma = 8$, what is the z score corresponding to the point which between itself and the mean contains 20 percent of the area? _____.

━━━━ STOP ━━━━━━━━━━━━━━━━━━━━━━━━ STOP ━━━━

0.52

---

184 $\mu = 70$, $\sigma = 8$, what is the raw score corresponding to a z of 0.52? _____.

━━━━ STOP ━━━━━━━━━━━━━━━━━━━━━━━━ STOP ━━━━

74.16

---

185 $\mu = 70$, $\sigma = 8$. Hence the 20 percent who scored between 74.16 and the mean received a grade of "C." But since 55 percent of the class received "C"s, then _____ percent of the class received "C"s and yet were below the mean.

━━━━ STOP ━━━━━━━━━━━━━━━━━━━━━━━━ STOP ━━━━

35

---

186   The z score corresponding to the 35 percent of the class who fell below the mean and still got a "C" would be _____.

━━━━ STOP ━━━━━━━━━━━━━━━━━━━━━━━━ STOP ━━━━

-1.04

---

187   The reason why the z is negative is that the score of the lowest "C"

_____ _____ _____ _____.

━━━━ STOP ━━━━━━━━━━━━━━━━━━━━━━━━ STOP ━━━━

falls below the mean

---

188   With $\mu = 70$, $\sigma = 8$, $z = -1.04$, then $X =$ _____.

━━━━ STOP ━━━━━━━━━━━━━━━━━━━━━━━━ STOP ━━━━

61.68

---

189   Again the score 61.68 is less than the mean because the sign of the z score was _____.

━━━━ STOP ━━━━━━━━━━━━━━━━━━━━━━━━ STOP ━━━━

negative

190 In another test $\mu = 100$, $\sigma = 10$. 10 percent of the class got "A," 25 percent of the class got "B" and 55 percent of the class got "C." We wish to find the highest and lowest "C." What percent of the class got "C" and were above the mean? _____.

═══════ STOP ═══════════════════════════ STOP ═══════

15 percent

---

191 That 15 percent corresponds to a z of _____.

═══════ STOP ═══════════════════════════ STOP ═══════

0.39

---

192 If $\mu = 100$, $\sigma = 10$, and $z = 0.39$, then $X =$ _____.

═══════ STOP ═══════════════════════════ STOP ═══════

103.9

---

193 $\mu = 100$, $\sigma = 10$, $X = 103$, if someone got a score of 103 he would then get a grade of _____.

═══════ STOP ═══════════════════════════ STOP ═══════

"C"

---

194 $\mu = 100$, $\sigma = 10$. Of the 55 percent who got "C," 15 percent were above the mean. We wish to find the lowest "C" score. The lowest "C" score would correspond to a z of _____.

═══════ STOP ═══════════════════════════ STOP ═══════

-1.28

---

195 $\mu = 100$, $\sigma = 10$, $z = -1.28$; then the lowest "C" has a score of _____.

═══════ STOP ═══════════════════════════ STOP ═══════

87.20

The Standard Error
of the Mean

Suppose that your campus newspaper editor wants to decide whether or not to drop the crossword puzzle that appears in every edition. She wants to know if the students would object, and asks you for some help in answering her question.

You decide that the best method would be to ask all students in school whether they would mind if the puzzle did not appear in the paper. You also decide that this would be impossible because of the time and effort involved. After a little thought you realize that you don't have to know exactly how many students would be for or against the idea, and that a limited poll of student opinions would give you a rather good indication of the general attitude.

Suppose for the sake of discussion that 76 percent of all students in the population are against dropping the puzzle. Also suppose that you asked the first 25 people you met on campus how they felt about it, and 20 (80 percent) said they were against dropping it. You would now have quantitative measures to deal with.

From Chapter Eight we know that a *parameter* is a quantitative characteristic of a population. The proportion of students in the whole student body who are against dropping the puzzle (76 percent) is a parameter of that population. *The mean and standard deviation of a population* have the symbols μ and σ, respectively.

We also know from Chapter Eight that a *statistic* is a quantitative characteristic of a sample. The proportion of students in your group of 25 who are against dropping the puzzle (80 percent) is a statistic of that sample. *The mean and standard deviation of a sample* are represented by the symbols M and S, respectively.

In summary, the statistician makes inferences about population parameters on the basis of sample statistics.

SELECTING THE SAMPLE

Suppose you had been standing in the hall outside an English classroom as classes were dismissed when you decided to pick your 25 people to ask about the crossword puzzle. Suppose your sample included 24 English majors and a bewildered freshman looking for a chemistry class. Even though English majors might not have any particular loyalty to crossword puzzles compared to students majoring in other fields, such a sampling procedure would make a statistician quite uncomfortable.

The point is that the safest inferences about populations are made from samples that are representative of those populations. Although there are a number of special techniques available to achieve this goal of accurate representation, we will confine our discussion to random selection, since this is the basis of all methods of sampling.

When we use the term *random sample,* we mean that the individuals that make up the sample were *selected* randomly. As you will recall, this means that every individual in the population had an equal chance of being selected. Note that this definition of random selection concerns the probabilities on any one selection of a sample member. In the population includes 100 people, the probability is 1/4 that a given person will be selected eventually for a sample of 25 people.

Now you can see why your original sampling procedure was not random. You picked 25 people from those who happened to be in a particular part of a particular building at a particular time on a particular day of the week. For all those students who were not there to be chosen the probability of selection was zero.

One method of random sampling that you could use would be to write the name of every student in the school on a separate slip of paper and put all the papers in a very large hat, mix them up thoroughly, and pull out the names of 25 students for your sample.

A more practical method would be to use the table of random numbers in the back of this book (Appendix H) to select pages from the student directory, and then use the table again to select one student from each selected page.

Now, the most important thing to remember about samples and populations is that an extremely large number of different samples could be drawn from the same population (except under the unusual condition where a sample includes almost all of the population). Since the statistics derived from these different samples would be based on different subgroups, we would not expect the value of the statistic from one random sample to equal the population parameter any more than we would expect to get exactly five heads if we flipped a coin ten times.

SAMPLE STATISTICS AS INDIVIDUAL SCORES

Suppose each of five students agreed to corner a separate random sample of students from the entire student population and ask each student to rate the desirability of having a crossword puzzle in the campus newspaper. The rating scale has nine points, 1 ("Who needs it?") through 9 ("I work it out every day!").

Your five researchers venture forth and return with a set of ten ratings in each sample. Each of them then computes a mean for his sample of ten ratings:

$$\left.\begin{array}{l} M_1 = 4.0 \\ M_2 = 7.0 \\ M_3 = 6.0 \\ M_4 = 4.0 \\ M_5 = 4.0 \end{array}\right\} \text{mean of the sample means} = 25/5 = 5.0$$

Why not treat these five means as a sample of five means from a population of all possible sample means? You decide to compute the mean of these means, and find it to be 5.0. Notice that this is the same value that would be obtained by treating the 50 students as one sample and finding the mean of this large sample. You also decide that you can estimate the variability of these five means by computing their standard deviation:

$$\text{standard deviation of the sample means} = \sqrt{\frac{133 - \dfrac{625}{5}}{5 - 1}} = \sqrt{2} = 1.41$$

You decide that a standard deviation of 1.41 is too large and that you need a better estimate than you can get with samples of ten students. You send the researchers forth again to enlarge their samples to 25 students each. This time you find that the variability of the means has been reduced, compared with that of the means of samples of 10.

$$\left.\begin{array}{l} M_1 = 6.5 \\ M_2 = 5.5 \\ M_3 = 6.5 \\ M_4 = 4.5 \\ M_5 = 4.5 \end{array}\right\} \text{mean of the sample means} = 5.50$$

$$\text{standard deviation of the sample means} = \sqrt{\frac{155.25 - \dfrac{756.25}{5}}{4}}$$

$$= \sqrt{\frac{155.25 - 151.25}{4}} = \sqrt{\frac{4}{4}} = 1.00$$

You are pleased with the results of your sampling and are now becoming less interested in the poll for the campus newspaper and more interested in the way sample means distribute themselves. While the decrease in the variability of the means may have satisfied you enough to decide the crossword puzzle should be kept, you are now curious to see if the variability of the means will continue to get smaller as the size of the samples increases. That is, as you increased the size of the sample from 10 to 25, the variability of the sample means decreased by about one-third.

This may mean that the decrease in variability of the means is related to the proportional increase or to the absolute increase in sample size. To check on these possibilities, you decide to double the sample size from 25 to 50 and see what happens to the variance of the means. Unfortunately, when you decide to conduct the experiment again, your co-workers tell you that they are tired of collecting data and that if you want the sampling done you will have to do it yourself. Somewhat less zealous than you were before you lost your helpers, you go out and collect four samples of 50 students each. The results were tabulated and the means computed; the results are presented below:

$$\left.\begin{array}{l} M_1 = 5.32 \\ M_2 = 4.34 \\ M_3 = 5.78 \\ M_4 = 5.92 \end{array}\right\} \text{ mean of the sample means } = 5.34$$

$$\text{standard deviation of the sample means} = \sqrt{\dfrac{115.59 - \dfrac{456.25}{4}}{3}} = \sqrt{\dfrac{115.59 - 114.06}{3}}$$

$$= \sqrt{\dfrac{1.53}{3}} = \sqrt{0.51} = 0.71$$

Again the variability of the means has decreased by about one-third. You conclude that you could increase your sample size still more, and find that the variability of the means would be reduced still further. (The exact relationship between the size of the sample and the variability of the means will be shown later in this chapter.) In any event, you can state a general rule: *As the sample size increases, the variability of the sample means decreases.* In other words, you can make inferences about population parameters with more precision if you are willing to expend the extra time and effort to increase your sample size. However, the law of diminishing returns operates in research strategy as it does in economics.

In the example about the campus newspaper poll we have been trying to illustrate the general idea of a sampling distribution of means. The rest of this chapter will be devoted entirely to this concept, which is difficult to grasp and which many students tend to have a hard time understanding. One possible reason for this is that they cannot easily differentiate between sample statistics and population parameters. To keep this distinction clear, you should remember that M and S represent the mean and standard deviation *of a sample*, while μ and σ represent the mean and standard deviation *of a population*.

SAMPLING DISTRIBUTIONS

One of the important and interesting reasons for studying the normal distribution is that so many variables are distributed in this fashion. For example, the expansion of the binomial $(P + Q)^N$ approaches the normal distribution when $P = Q$ and when N is large. Many phenomena in the behavioral sciences seem to be the composite result of the interaction of many specific influences. In such cases the composite will tend to have a normal distribution. Height, for example, is the result of the many variables that influence heredity and environment. All these variables together produce a normal distribution of individual heights.

There is another important reason for studying the normal distribution: The distributions of *means* of samples taken from normal *and* non-normal distributions tend to be normal. This phenomenon is so reliable that it has been formalized as the *Central Limit theorem:* As the number of cases in each sample increases, and as the number of samples increases, the distribution of sample means approaches the normal distribution.

This theorem can be illustrated by examining sampling distributions of means obtained by rolling perfectly balanced dice. Each die has six sides, and the probability of rolling any side is 1/6. If we roll a single die many times, the distribution of the means would be rectangular—obviously not a normal distribution. The reason for this is that we have a large number of samples (as the theorem says), but we have only one case in each sample. In fact, with $N = 1$ per sample, the distribution in Figure 9-1 is the population distribution from which we are sampling.

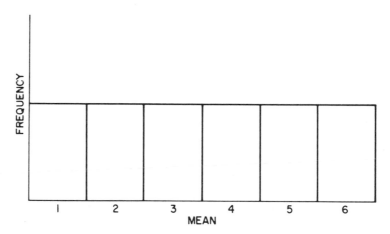

Figure 9-1   *Sampling distribution of the mean for one die.*

Now suppose we increase our sample size to 2. That is, we will roll two dice and find the mean of the number of dots, or pips, that appear each time. If we roll the dice many times, we would get a distribution similar to that in Figure 9-2. Again we see that the distribution is not normal, although it is considerably closer than that in Figure 9-1.

Finally, if we use three dice, roll them many times, and find the mean of the number of dots for each roll, the resulting distribution is an even closer approximation of the normal distribution.

Actually, our example makes some big assumptions. The three distributions pictured in Figures 9-1, 9-2, and 9-3 would only occur if we rolled our one, two, or three dice an infinite number of times. Instead, we obtained the distributions by computing the probability of each outcome and graph-

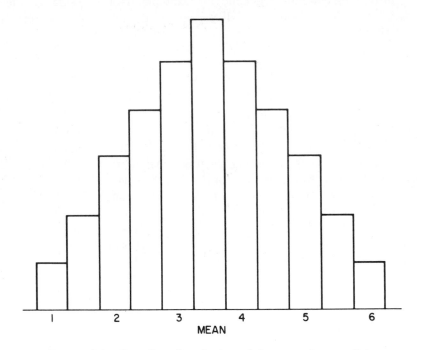

Figure 9-2   *Sampling distribution of the mean for two dice.*

Figure 9-3   *Sampling distribution of the mean for three dice.*

ing those. So, while we are sure that the Central Limit theorem is true, we have not empirically demonstrated it.

For empirical evidence we resorted to use of a computer, and to illustrate just how powerful the Central Limit theorem is, we drew samples from a non-normal population. (You can be sure that if we had samples from a normal distribution, essentially the same results would have occurred.) Such a population is composed of 3s and 9s, and nothing else. This J-shaped distribution is illustrated in Figure 9-4. It can be seen that 9s comprise 75 percent of the population and that 3s comprise 25 percent.

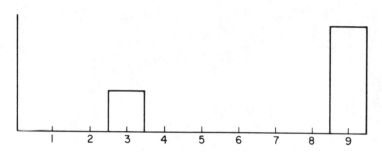

Figure 9-4   *A population distribution consisting of 3s and 9s.*

From this population we drew samples of size 5 ($N = 5$ per sample) and found the mean of each sample. We did this with a computer 10,000 times. The results are shown in Figure 9-5.

Next, we randomly selected samples of ten scores each ($N = 10$ per sample) and again found the mean for each of the 10,000 samples that were drawn. These data are also shown in Figure 9-5. Finally, we randomly selected samples of 25 cases each ($N = 25$ per sample) and again found the mean for each of the samples. This distribution of sample means is also shown in Figure 9-5.

Consider what we have done. We have sampled from a J-shaped non-normal distribution. As we increased the sample size (from 5 to 10 to 25), the shape of the distribution of sample means—the sampling distribution of the mean—became more and more like the normal distribution. While actual distributions rarely depart from the normal distribution as much as the one pictured in Figure 9-4, we still find that sample means from such a distribution tend to be normally distributed. This holds true for a wide range of distributions, both normal and non-normal, since the Central Limit theorem does not specify the shape of the population from which samples are taken.

Although we have assured you that the Central Limit theorem applies equally well to normal as to non-normal distributions, this has not yet been demonstrated. We programmed the computer to use a normal population with a mean of 50 and a standard deviation of 12 (Figure 9-6). From this population we took random samples of four scores each (that is, $N = 4$), and found the mean for each of 10,000 samples. These 10,000 means are

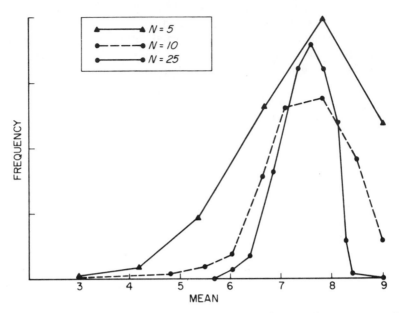

Figure 9-5 *Distributions of sample means when the sample size is equal to 5, 10, or 25.*

plotted in Figure 9-6. Next, we took random samples of 16 each, found the mean for each of these samples, and repeated this operation 10,000 times also. The sampling distribution of these 10,000 means is also plotted in Figure 9-6. Finally, we took samples of 144 scores each, found the mean for each sample, and repeated the process 10,000 times. These data are also plotted in Figure 9-6.

Notice two very important things in Figure 9-6. First, the means for samples of $N = 4$, $N = 16$, and $N = 144$ are all very nearly normally distributed. Second, the variability of the means decreases as the sample size increases. You can estimate the standard deviation of each of the distributions pictured in Figure 9-6 by estimating the range of the distribution and dividing by 6 (instead of 5, since a large number of means were used). These estimates for the distributions of means based on samples of 4, 16, and 144 cases are about 6, 3, and 1, respectively.

It is, of course, a very laborious process to generate samples, find the means of those samples, and find the standard deviation of those means. A much easier procedure—one which would save both time and effort—would be to estimate the variability of means from knowledge of variability in the population. The next formula does exactly this, for it can be easily shown that the variability of means is related to variability in the population. Hence the formula

$$\sigma_M = \frac{\sigma}{\sqrt{N}}$$

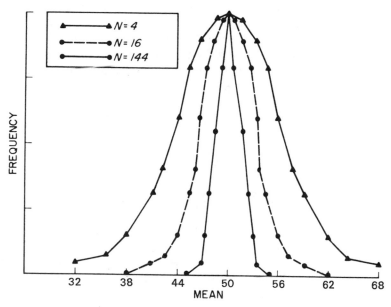

Figure 9-6   *Distributions of sample means when the sample size is equal to 4, 16, or 144, and when the population is normal.*

which relates the variability of sample means to that of the individual scores in the samples. The formula indicates that the standard deviation, of the means (σ_M) is equal to the standard deviation of the population (σ) divided by the square root of the sample size $(\sqrt{N})$. Thus, from the population pictured in Figure 9-6—which has a standard deviation of 12 and samples of size 4, 16, and 144—we would expect from the formula that

$$\sigma_M = \frac{12}{\sqrt{4}} = \frac{12}{2} = 6$$

when samples of four are taken, and that

$$\sigma_M = \frac{12}{\sqrt{16}} = \frac{12}{4} = 3$$

when $N = 16$, and that

$$\sigma_M = \frac{12}{\sqrt{144}} = \frac{12}{12} = 1$$

when $N = 144$. As can be seen from Figure 9-6, this closely approximates the standard deviations estimated previously from the ranges of the distributions.

It is interesting to note that σ_M is *not* properly called the "standard deviation of the means"; the correct term is *standard error of the mean*. No error is actually involved; the name is a holdover from previous times when variability was thought of only as error of measurement.

AN EXAMPLE   Suppose that we know the mean and standard deviation of the heights of all male students in a certain college to be $\mu = 70$ and $\sigma = 2$. Suppose further that we want to know the variability of the means of samples of 16 heights drawn from this population. We could compute M for each possible sample of 16 students and then find the standard deviation of these means, but we wouldn't live long enough to finish the task. We could also use the formula for the standard error of the mean and get exactly the same answer:

$$\sigma_M = \frac{\sigma}{\sqrt{N}} \qquad (\sigma = 2, N = 16)$$

$$\sigma_M = \frac{2}{\sqrt{16}} = \frac{2}{4} = 0.50$$

The standard deviation of the means of all possible samples of 16 scores each is equal to 0.50.

As the size of the sample increases, the size of the standard error of the mean decreases. Time and cost will limit the size of the sample, and so therefore the experimenter must decide at some point if the increase in accuracy is worth the increase in cost. Table 9-1 illustrates the relationship

Table 9-1   Relationship between N and σ_M

σ	N	σ_M
24	4	12
24	9	8
24	16	6
24	36	4
24	64	3
24	144	2
24	576	1

between sample size and the standard error of the mean when the standard deviation of the population is equal to 24.

THE STANDARD ERROR OF THE MEAN AND z-SCORES

The fact that means of reasonably large samples are normally distributed provides the necessary link to the powerful techniques we discussed previously in relation to the area under the normal curve. The z-score distribution is normally distributed with a mean of zero and a standard deviation of 1.00. Any normal distribution can be converted to a z distribution by subtracting the mean from each score and then dividing by the standard deviation. That is, if X is a normally distributed variable, then

$$z = \frac{X - \mu}{\sigma}$$

where μ and σ are the mean and standard deviation of the X distribution, respectively.

Similarly, if we have a distribution of means, then from the Central Limit theorem we know that this distribution of means will tend to be normal. If we subtract the mean of *this* distribution and divide by *this* standard deviation, then another z-score distribution is produced. Notice, however, that the appropriate mean and standard deviation must be used. In this case, μ would still be the mean, but σ_M would be the standard deviation. Thus,

$$z = \frac{M - \mu}{\sigma_M}$$

where M is a score from the distribution of means, μ is the mean of the means, and σ_M is the standard deviation of the means. Again, z is normally distributed with a mean of zero and standard deviation of 1.

In Chapter Eight techniques were developed to answer questions about areas under the normal curve involving individual scores. In precisely the same manner we can answer questions about areas under the normal curve dealing with the proportion or number of means above or below a certain point in a theoretical distribution of means.

AN EXAMPLE   Many intelligence tests are constructed to have a mean (μ) IQ of 100 and a standard deviation (σ) of 15. These parameters are based on extremely large random samples of the most widely used tests.

If we were to go into a certain community and give one of these tests to a large number of randomly selected samples of 25 people each, what might we expect to find with regard to the means of these samples? From our knowledge of the characteristics of the population we can say:

(1) The true (population) mean is known to be $\mu = 100$.
(2) The true standard deviation is known to be $\sigma = 15$.
(3) The true standard error of the mean (the true standard deviation of sample means) can be found by using the formula introduced in this chapter:

$$\sigma_M = \frac{\sigma}{\sqrt{N}} = \frac{15}{\sqrt{25}} = \frac{15}{5} = 3$$

On the basis of this information we have drawn Figure 9-7, which shows the distribution of IQ scores for individuals in the population and the distribution of sample means based on samples of 25 cases.

We can answer questions about either the proportion of individual scores above or below a certain point, or the proportion of means above or below a point. For example, what percentage of the individual IQ scores fall above 105? For a score of 105, $z = (105 - 100)/15 = 0.33$, and the entry in Appendix B for $z = 0.33$ is 37.07, rounded to 37 percent. That is, we find that 37 percent of the individual IQ scores fall above 105.

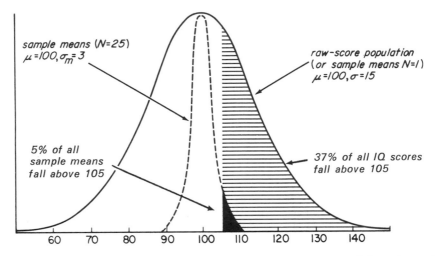

Figure 9-7 *Relationship between population variability and variability of sample means.*

Similarly, what percentages of the sample means fall above 105? Since this question concerns the variability of the means, the appropriate standard deviation to use is σ_M. Hence,

$$z = \frac{M - \mu}{\sigma_M} = \frac{105 - 100}{3} = 1.67$$

and the tabled value for $z = 1.67$ is 4.75, rounded to 5 percent. That is, we find that 5 percent of the sample means fall above 105. This may be compared to 37 percent of the individual IQ scores that fall above 105. This is also illustrated in Figure 9-7.

A second example may be helpful. Suppose that we want to find the points equidistant from the mean within which 50 percent of the *scores* fall and also the points within which 50 percent of the *means* fall. To do this, we want to find the z score corresponding to 25 percent above (and below) the mean. Using Appendix B, we find that $z = 0.67$ most closely approximates 25 percent. Since $X = \mu \pm z\sigma$, the point above the mean which includes 25 percent of the area between itself and the mean is $X = 100 + (0.67)(15) = 100 + 10 = 110$. Similarly, the corresponding point below the mean is $X = 100 - (0.67)(15) = 100 - 10 = 90$. Hence, 50 percent of the scores fall between 90 and 110.

A similar procedure is employed to find the points within which 50 percent of the means fall. In this case the question is concerned with the variability of means, and σ_M is the appropriate standard deviation to use. Substituting σ_M for σ in the formula just used, the point above the mean which includes 25 percent of the sample means between itself and the grand mean is $\mu + z\sigma_M$, or $100 + (0.67)(3) = 100 + 2 = 102$. The corresponding point below the mean is $\mu - z\sigma_M$, or $100 - (0.67)(3) = 100 - 2 = 98$. Hence, 50 percent of the sample means would fall between 98 and 102.

TESTING HYPOTHESES ABOUT SAMPLE MEANS

For several years a school administrator has given a questionnaire to students in sixth-grade classes to assess their attitudes toward their teachers. In prior years the questionnaire has been given in October, and a composite score for each teacher found. A mean of 178 and a standard deviation of 24 have been found to be representative of the population of classes sampled. This year the questionnaire was administered in May. A sample of 36 classes was selected, and the mean of the 36 composite teacher scores was found to be 170. Did the pupils' attitudes change?

The process of statistical inference used in this case may be outlined as follows:

When the questionnaires are given in October, the average teacher score is 178. Since this score is representative of a large number of teacher scores taken over the years, it seems reasonable to assume that 178 is the mean of this population, that is, $\mu = 178$. By a similar reasoning, $\sigma = 24$.

From the May questionnaire, a sample of 36 classes are selected, and 36 teacher-opinion scores are obtained. The mean of these 36 scores will vary from one sample to another. However, since $\sigma = 24$, then the variability of the mean of a group of 36 scores can be found with the formula $\sigma_M = \sigma/\sqrt{N}$. Hence

$$\sigma_M = 24/\sqrt{36} = 24/6 = 4.00.$$

A sample mean was obtained in the Spring; the null hypothesis would state that the true means for Spring and Fall would be the same, that is, 178.

You will recall that statisticians have generally agreed that if an event occurs by chance less than five times out of 100—($\Pr \leq 0.05$)—then (when such an event occurs) the null hypothesis is rejected (that is, the hypothesis of chance occurrence is rejected). The sample mean is 170, and if 170 is too far from 178, then the hypothesis that 178 is the true mean of the population is rejected. How far is too far? This, of course, would depend on the variability of the scores, and we know that 1.96 standard deviations above and below the mean include 95% of the area under the normal curve.

Since we are talking about means, then the distance is measured in σ_M units because σ_M is the unit of variability among means. Hence, a distance greater than $1.96\sigma_M$ would be considered "too far." The distance between two points, expressed in σ_M units, can be found by the formula $z = (M - \mu)/\sigma_M$. In our example,

$$z = \frac{170 - 178}{4} = \frac{8}{4} = 2.00$$

That is, 178 is 2.00 standard deviations (of the mean) away from 170. By our definition, that is "too far." In other words, the probability is less than 0.05 that a sample mean of 170 could have come from a population which has a mean of 178. Because this is not a likely event, we reject the hypothesis that $\mu = 178$.

In summary, $N = 36$, $\sigma = 24$, $M = 170$. We wish to use the z test to test the hypothesis (H) that the true mean of the May questionnaire is equal to 178. Then

$$\sigma_M = \frac{\sigma}{\sqrt{N}} = \frac{24}{\sqrt{36}} = \frac{24}{6} = 4$$

and

$$z = \frac{M - \mu}{\sigma_M} = \frac{170 - 178}{4} = 2$$

(In this case the sign is ignored.) Since 2 is larger than 1.96, H_0: $\mu = 178$ is rejected.

Let's give another example. Suppose over the years that student nurses have been taking the temperature of "Archie," a manikin heated to a constant temperature. These measures have been found to have a mean of 100°F and a standard deviation of 2. Suppose also that a group of 25 nursing students have been put into a new pilot program that emphasizes the proper use of various types of equipment. The director of the project wishes to see if the new program has any influence on thermometer readings. Each of the 25 student nurses takes a temperature and records it. The mean temperature reading is found to equal 99°F. Does this reading differ from that expected by previous classes?

Here, we wish to test the hypothesis, that is, H_0: $\mu = 100$. To do this, we express the difference between M and our hypothetical μ in σ_M units. The value of $\sigma_M = \sigma/\sqrt{N} = 2/\sqrt{25} = 2/5 = 0.40$. Substituting into the formula for the z test:

$$z = \frac{M - \mu}{\sigma_M} = \frac{99 - 100}{0.40} = 2.50$$

With $z = 2.50$, the difference between M and μ is too great, and we reject the hypothesis that the true mean is 100. That is, the mean of the temperatures taken under the new program is less than the mean of the temperatures found under the old program.

The director of the pilot project can conclude that the new program does result in a change in temperature readings. Notice that we did not say what temperature Archie, the manikin, was heated to. It is not necessary to know what Archie's temperature is to conclude that a change occurred. However, if Archie's temperature is 97°F, then the project has resulted in a more accurate reading by the student nurses.

Certain assumptions have to be made before the z test can be used. These are listed below.

ASSUMPTIONS OF THE z TEST

    (1) Sample comes from a normal distribution.
    (2) The standard deviation of the population (that is, σ) is known.
    (3) Independence of sampling.
    (4) Random sampling.
    (5) Measurement scale is at least an interval scale.

ESTABLISHING THE CONFIDENCE INTERVAL FOR THE MEAN

In the preceding example the investigator was interested in determining if his sample could have been drawn from a population that had a true mean of 100. Since his sample mean was of a size that would have occurred less than five times out of one hundred by chance (if $\mu = 100$), the hypothesis that the true mean was 100 was rejected. In this section we will give procedures that will enable the investigator to estimate the limits within which the true mean falls. The investigator will never be able to say that the true mean is equal to a certain score; at best, all he can say is that the true mean falls between two specified points. These two points are chosen so that, on the average, the probability is equal to $1 - \alpha$ (usually 0.95) that the true mean falls between them. The true mean is, of course, not a variable, and once a confidence interval has been set up, the true mean is either included or not included within this interval. However, if 100 confidence intervals are set up, and if $\alpha = 0.05$, then we would be right 95 times out of 100 when we say that the true mean falls between these two points.

    Suppose the true standard deviation (σ) of some population is equal to 40, but that the true mean is unknown. Furthermore, suppose an investigator wishes to estimate the value of the true mean and randomly selects a sample of 16 cases from this population. Since $\sigma_M = \sigma/\sqrt{N}$, then $\sigma_M = 10$. Although the reason may not be immediately apparent, we can set up limits for the true mean in a manner very similar to the procedure used to determine the limits within which sample means will fall. Thus, in the absence of the true mean, we can use the same formula as before, *substituting* the sample mean for the true mean:

$$X_{U(\text{upper limit})} = (z)(\sigma_M) + M = (1.96)(10) + M$$
$$X_{L(\text{lower limit})} = (-z)(\sigma_M) + M = (-1.96)(10) + M$$

or

$$\mu = M \pm (z)(\sigma_M)$$
$$\mu = M \pm (1.96)(10)$$

    Suppose in this example that we found the mean of our 16 cases to be equal to 120. We would then substitute 120 into the equation so that $\mu = 120 \pm (1.96)(10)$. Thus, $X_U = 120 + 19.6 = 139.6$ and $X_L = 120 - 19.6 = 100.4$, and we estimate that μ falls between 100.4 and 139.6.

    The logic of this procedure is illustrated in Figure 9-8. Suppose $\mu = 500$ and $\sigma_M = 10$; then, using Appendix B, we would know that 95 percent of

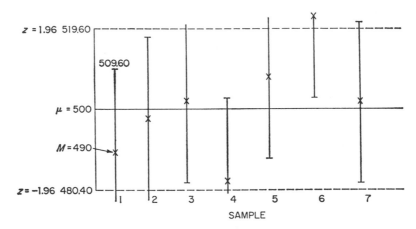

Figure 9-8 *Several samples are drawn. The two dashed lines represent 480.40 and 519.60. The Xs represent the values of the sample means. The lines extending from the Xs represent the 95 percent confidence interval established for each sample. Since 95 percent of the sample means fall between 480.40 and 519.60, then 95 percent of the confidence intervals will contain μ, the true mean.*

all sample means fall between 480.40 and 519.60 (that is, $\mu \pm 1.96 \times \sigma_M$). However, an investigator would not know the true mean and would set up a confidence interval for the true mean as it is illustrated in Figure 9-8.

For example, the mean of sample 1 was found to equal 490. Using the preceding formula $\mu = M \pm (z)(\sigma_M)$, $\mu = 490 \pm (1.96)(10) = 490 \pm 19.60$. Thus, the investigator sets the upper limit for the true mean at 509.60 and the lower limit at 470.40. The investigator would then say that the 95 percent confidence interval for the true mean (μ) is from 470.40 to 509.60. Only when a sample mean falls outside the 480.40 to 519.60 interval, as is the case for sample 6 in Figure 9-8, does the confidence interval for the true mean not include the true mean. But since a sample mean will fall 95 percent of the time from 480.40 to 519.60, then the confidence interval for the mean will include the true mean 95 percent of the time. On the other hand, 5 percent of the time (or five times in every one hundred attempts) the investigator will be wrong when he says that the true mean falls within the confidence interval.

If the investigator does not want to be wrong 5 percent of the time, he can set up wider confidence intervals. Another common interval is the 99 percent confidence interval in which the investigator sets his limits sufficiently wide so that he will be wrong only one time out of one hundred. From Appendix B we see that 99 percent of all sample means would fall between $\pm z = 2.58$. Therefore, if we substitute 2.58 into the formula for establishing the confidence interval for the mean, we have the values for the 99 percent confidence interval: $\mu = M \pm (2.58)(\sigma_M)$. For example the mean of sample 1 was equal to 490, and the 99 percent confidence interval would be equal to $490 \pm (2.58)(10)$, or 464.20 to 515.80. Even wider confidence intervals could be used if they were desired. However, as the probability of not including the true mean in the confidence interval decreases, the width of the interval increases. For example, to decrease the chances of making an error by 4 percent (from 5 percent to 1 percent) you had to increase the width of the interval about 25 percent (from $z = 1.96$ to $z = 2.58$). Because the increase in the width of the interval is so much greater than the decrease in the percentage of errors, the 95 percent confidence interval tends to be used more often than the 99 percent confidence interval.

SUMMARY

We learned that the variability of sample means is related to the variability of individual scores by the formula $\sigma_M = \sigma/\sqrt{N}$. Since sample means are normally distributed, we can use Appendix B to determine the probability that a specific sample mean will fall between two points. If these two points are chosen to include 95 percent of the area under the normal curve, then we can reject the null hypothesis whenever a sample mean falls beyond either of the limits. The null hypothesis we test in such cases is that the true mean of the sample is equal to a theoretical true mean. We can also establish the confidence interval for the true mean. Using the formula presented in this chapter, we can use a sample mean to find the interval within which the true mean will fall. To make such inferences, however, we must know the true standard deviation (σ) of the population, and this is not usually the case. In Chapter Ten we will describe procedures for making inferences of this kind when σ is not known and when the only estimate of population variability we have is S.

PROBLEMS

1. An experimenter selects 4 samples of 16 scores each from a population which has a mean of 10 and a standard deviation of 4. The means of the samples are $M_1 = 9$, $M_2 = 17$, $M_3 = 11$, and $M_4 = 11$. What is the standard deviation of these means?

2. From Problem 1, what is the standard deviation of the means estimated from the standard deviation of the population?

3. For some population, $\mu = 40$, $\sigma = 5$, samples of 25 scores each are selected and the mean is computed from each. Two-thirds of the scores fall between what two points (equidistant from the mean)?

4. From Problem 3, two-thirds of the sample means fall between what two points?

5. $\mu = 50$, $\sigma = 24$. If $N = 4$, two-thirds of the sample means fall between what two points? If $N = 9$, two-thirds of the sample means would fall between what two points? If $N = 16$? If $N = 36$? If $N = 64$?

6. From some population, $\mu = 200$, $\sigma = 30$, a sample of 100 scores is randomly selected. The probability is 0.50 that the sample mean falls between what two scores? On the other hand, 50 percent of the raw scores would fall between what two points?

7. From some population, $\mu = 75$, $\sigma = 15$, a sample of 25 scores is randomly selected. About 95 percent of the scores would fall between what two points? The probability is 0.95 that the mean would fall between what two points?

8. In a new parole program, the mean age of the 25 parolees is found to be 30. In the past, the mean age at the time of parole was 32, with a standard deviation of 4.00. Is this group of parolees younger than might otherwise be expected?

9. The mean date of the latest killing frost in a southern city is March 3, with a standard deviation of six days. A climatologist has looked at the dates of the last killing frost for each of the past nine years and has found that the last frost occurs, on the average, on March 7. Is there any evidence to suggest that the date of the last killing frost is changing?

10. It is known that the mean number of sit-ups that can be done by draftees upon entering the army is 25 (with a standard deviation of 5). Captain Snarf, ex-statistics professor at Jetsam U., is in charge of basic training at Ft. Maelstrom and wishes to determine if the physical training during basic training actually increases a soldier's performance. To test this, he takes a sample of 100 soldiers just finishing basic training and has them do sit-ups. Their mean is 29. What does Captain Snarf conclude?

11. (a) $\sigma = 20$, $N = 16$, $\sigma_M = $ _____.

    (b) $\sigma = 10$, $N = 25$, $\sigma_M = $ _____.

    (c) $\sigma = 48$,

        (1) $N = 4$, $\sigma_M = $ _____.

        (2) $N = 9$, $\sigma_M = $ _____.

        (3) $N = 16$, $\sigma_M = $ _____.

        (4) $N = 64$, $\sigma_M = $ _____.

    (d) $\mu = 80$, $\sigma = 8$, $N = 64$:

        (1) 68 percent of the scores would fall between what two points?

        (2) 68 percent of the sample means would fall between what two points?

    (e) $M = 90$, $\sigma = 12$, $N = 36$:

        (1) Establish the 95 percent confidence interval for the true mean.

        (2) Establish the 99 percent confidence interval for the true mean.

EXERCISES

1. There are equal numbers of 0s, 1s, 2s, and so on in Appendix H, as a histogram constructed from these numbers would illustrate. Take a random sample of five single-digit numbers from Appendix H. Find the mean. Do this for 20 such samples of five cases each. How are these means distributed?

2. The σ of the single-digit numbers in Appendix H is equal to 2.87. Take ten single-digit numbers at random from Appendix H. Establish the 95 percent confidence interval for the true mean. How does this interval compare with other intervals established by other members of the class?

1   A population is defined as all individuals with a common characteristic. For example, all people who live in Nebraska constitute a _____.

══════ STOP ══════════════════════ STOP ══════

population

---

2   All people who are 6 feet or taller also constitute a _____.

══════ STOP ══════════════════════ STOP ══════

population

---

3   A sample, on the other hand, is a group of individuals that forms a part of a population. Considering all people who live in Ohio, those people who live in Columbus would be a _____.

══════ STOP ══════════════════════ STOP ══════

sample

---

4   The statistician will usually select his sample randomly from some defined _____.

══════ STOP ══════════════════════ STOP ══════

population

---

5   Thus if we consider all the members of the freshman class at some university and randomly select ten people, we have selected a _____.

══════ STOP ══════════════════════ STOP ══════

sample

---

6   A quantitative characteristic of a population is called a *parameter*. Thus if we find that the mean height of all males at some college is 5 feet 10 inches, this would be a _____.

══════ STOP ══════════════════════ STOP ══════

parameter

---

7   On the other hand a quantitative characteristic of a sample is known as a *statistic*. Thus, if we take a sample of ten freshmen men and find their mean height, we have computed a _____.

══════ STOP ══════════════════════ STOP ══════

statistic

---

8   While the variance of a population would be a parameter, the variance of a sample would be a _____.

═══════ STOP ═══════════════════════════ STOP ═══════

statistic

---

9   A statistic may be used to make an inference (or an estimate) about a parameter. Thus, if we compute a mean from a randomly selected sample we are able to make an inference about a parameter. This parameter is the _____ _____.

═══════ STOP ═══════════════════════════ STOP ═══════

population mean (true mean)

---

10   Commonly we use Greek letters to represent parameters and Roman letters to represent statistics. Thus μ represents the true mean of the population while M represents the mean of the _____.

═══════ STOP ═══════════════════════════ STOP ═══════

sample

---

11   Similarly, σ is used to represent the (true) standard deviation of a population while S is used to represent _____ _____ _____ _____ _____ _____.

═══════ STOP ═══════════════════════════ STOP ═══════

the standard deviation of a sample

---

12   If we use μ and σ we are talking about a (population/sample) _____.

═══════ STOP ═══════════════════════════ STOP ═══════

population

---

13   But if we were talking about the mean and standard deviation of a sample we would use the symbols _____ and _____.

═══════ STOP ═══════════════════════════ STOP ═══════

M and S

---

14   If we compute M we can make inferences about μ and if we compute S we can make inferences about _____.

═══════ STOP ═══════════════════════════ STOP ═══════

σ

---

15 For any specific population, the value of μ and σ are constant from sample to sample. Will the value of M and S be the same from sample to sample? (Yes/No) _____.

═══════ STOP ═══════════════════════════════ STOP ═══════

No

---

16 For a certain population the value of M will vary from sample to sample while the value of μ will remain _____.

═══════ STOP ═══════════════════════════════ STOP ═══════

constant (the same)

---

17 For example, if the mean of a certain population equals 100, we would expect M to _____ around 100 but we rarely would expect M to equal 100.

═══════ STOP ═══════════════════════════════ STOP ═══════

vary

---

18 Similarly if $\sigma = 10$, we would expect S to _____ around 10 but rarely would we expect S to equal _____.

═══════ STOP ═══════════════════════════════ STOP ═══════

vary, 10

---

19 If a large number of samples are selected and the Ms are computed, these Ms will be normally distributed with _____ as the mean.

═══════ STOP ═══════════════════════════════ STOP ═══════

μ

---

20 A distribution of Ms is normally distributed with a mean of μ. What percent of the sample means will fall above μ? _____.

═══════ STOP ═══════════════════════════════ STOP ═══════

50 percent

---

21 Stated in another way, the probability that M will be larger than μ is equal to _____.

═══════ STOP ═══════════════════════════════ STOP ═══════

.50

---

22 The standard deviation of all possible sample means is symbolized as σ_M (the standard error of the mean) and is analogous to S or σ. Thus, 34 percent of the Ms would fall between a point 1 σ_M above the true mean (μ) and _____.

══════ STOP ══════════════════════════════ STOP ══════

μ

───

23 Sample means are normally distributed. What percent of the sample means would fall between μ and a point 2 σ_Ms from μ? _____.

══════ STOP ══════════════════════════════ STOP ══════

47.7 percent (47.72)

───

24 As with S or σ, z scores can be computed for σ_M with the formula $z = (M - \mu)/\sigma_M$. If $M = 105$, $\mu = 100$, and $\sigma_M = 10$, then $z = $ _____.

══════ STOP ══════════════════════════════ STOP ══════

$5/10 = 0.50$

───

25 $z = (M - \mu)/\sigma_M$. $M = 160$, $\mu = 130$, $\sigma_M = 20$, $z = $ _____.

══════ STOP ══════════════════════════════ STOP ══════

$30/20 = 1.50$

───

26 $M = 40$, $\mu = 50$, $\sigma_M = 5$, then $z = $ _____.

══════ STOP ══════════════════════════════ STOP ══════

$-10/5 = -2$

───

27 $M = 60$, $\mu = 50$, $\sigma_M = 15$, then $z = $ _____.

══════ STOP ══════════════════════════════ STOP ══════

$10/15 = 0.67$

───

28 Since Ms are normally distributed we can use Appendix B. For example, if $M = 105$, $\mu = 100$, and $\sigma_M = 10$, then $z = 0.50$. Appendix B tells us that 19 percent of the sample means would fall between 100 and 105. What percent of the sample means fall between 100 and 106.7? _____.

══════ STOP ══════════════════════════════ STOP ══════

25 percent (24.86)

───

29   Thus if $\mu = 100$ and $\sigma_M = 10$, then 47.7 percent of all sample means would fall between 80 and _____.

═══════ STOP ═══════════════════════ STOP ═══════

100

---

30   Similarly if $\mu = 100$ and $\sigma_M = 10$, then 47.7 percent of all sample means would fall between 120 and _____.

═══════ STOP ═══════════════════════ STOP ═══════

100

---

31   If 47.7 percent of the sample means fall between 80 and 100 and an additional 47.7 percent between 100 and 120, then 95.4 percent of the sample means fall between _____ and _____.

═══════ STOP ═══════════════════════ STOP ═══════

80 and 120

---

32 $\mu = 60$, $\sigma_M = 10$. We wish to find the percent of Ms between 65 and 70. What percent of Ms fall between 60 and 65? _____.

═══════ STOP ═══════════════════════ STOP ═══════

19 percent (19.15)

---

33 $\mu = 60$, $\sigma_M = 10$, and we wish to find the percentage of Ms between 65 and 70. 19 percent of the sample means fall between 60 and 65. What percent fall between 60 and 70? _____.

═══════ STOP ═══════════════════════ STOP ═══════

34 percent (34.13)

---

34 $\mu = 60$, $\sigma_M = 10$, and we wish to find the percentage of Ms falling between 65 and 70. 19 percent fall between 60 and 65, 34 percent fall between 60 and 70. What percent of the sample means fall between 65 and 70?

_____.

═══════ STOP ═══════════════════════ STOP ═══════

15 percent

---

35 σ_M is related to σ by the equation $\sigma_M = \sigma/\sqrt{N}$, where N is the number of individuals or scores in the sample. If $N = 16$, then $\sqrt{N} =$ _____.

═══════ STOP ═══════════════════════ STOP ═══════

4

36 $\sigma_M = \sigma/\sqrt{N}$. If the number of scores in each sample were equal to 9, then σ would be divided by _____ to get σ_M.

══════ STOP ══════════════════════ STOP ══════

3

---

37   The standard deviation of all possible sample means, called the *standard error of the mean*, is symbolized _____.

══════ STOP ══════════════════════ STOP ══════

σ_M

---

38 $\sigma_M = \sigma/\sqrt{N}$; $\sigma = 12$. The standard deviation of all possible sample means each based on sixteen scores is equal to _____.

══════ STOP ══════════════════════ STOP ══════

$\sigma_M = 3$

39

$$\sigma_M = \frac{\sigma}{(?)}$$

══════ STOP ══════════════════════ STOP ══════

$\sqrt{N}$

40

$$\sigma_M = \frac{(?)}{\sqrt{N}}$$

══════ STOP ══════════════════════ STOP ══════

σ

41

$$\underline{\hspace{2cm}} = \frac{\sigma}{\sqrt{N}}$$

══════ STOP ══════════════════════ STOP ══════

σ_M

---

42 $\sigma_M = $ _____

══════ STOP ══════════════════════ STOP ══════

$$\frac{\sigma}{\sqrt{N}}$$

---

43 $\mu = 100$, $\sigma = 12$, and samples of nine scores each are randomly selected. We wish to find the percent of sample means falling between 100 and 104. $\sigma_M = $ _____.

===== STOP ===== ===== STOP =====

$$12/3 = 4$$

---

44 $\mu = 100$, $\sigma = 12$, $\sigma_M = 4$, what percent of the sample means fall between 100 and 104? _____.

===== STOP ===== ===== STOP =====

34 percent (34.13)

---

45   Note that the percentage of scores and percentage of sample means falling between two points will not necessarily be the same. For example if $\sigma = 8$, $N = 4$, and $\mu = 140$, then $\sigma_M = $ _____.

===== STOP ===== ===== STOP =====

$$8/2 = 4$$

---

46 $\mu = 140$, $\sigma = 8$, $\sigma_M = 4$, what percent of the sample means fall between 136 and 140? _____.

===== STOP ===== ===== STOP =====

34 percent (34.13)

---

47 $\mu = 140$, $\sigma = 8$, $\sigma_M = 4$, on the other hand, what percent of the individual scores fall between 136 and 140? _____.

===== STOP ===== ===== STOP =====

19 percent (19.15)

---

48 $\sigma = 8$, $N = 4$, $\mu = 140$, the reason why 34 percent of the sample means fall between 136 and 140 while only 19 percent of the scores fall between the same points is that the standard deviation of the scores is equal to _____ while the standard deviation of the means equals _____.

===== STOP ===== ===== STOP =====

8, 4

---

49 σ, the variability of scores in the population and σ_M, the variability of sample means, each based on N scores, are related by the formula $\sigma_M = $ _____.

===== STOP ===== ===== STOP =====

$$\sigma/\sqrt{N}$$

50   Thus while the variability of individual scores is dependent only upon σ, the variability of sample means is dependent upon _____ as well as the _____ of scores in each sample.

════════ STOP ════════════════════ STOP ════════

σ, number

---

51   For example, suppose $\sigma = 24$, find σ_M if $N = 9$. _____ ; $N = 16$. _____ ; $N = 64$. _____

════════ STOP ════════════════════ STOP ════════

$8; 6; 3$

---

52 $\sigma = 10$. A very large number of samples of twenty-five are taken at random and the mean is computed for each. Find the value of the standard deviation of these sample means. _____ .

════════ STOP ════════════════════ STOP ════════

$\sigma_M = 2$

---

53   The name commonly used for the standard deviation of a group of means would be the _____ _____ _____ _____ _____ .

════════ STOP ════════════════════ STOP ════════

standard error of the mean

---

54   If we wish to convert from a z score to a raw score we use the formula $X = \mu + z\sigma$. Similarly if we wish to convert from a z score to a sample mean, we would use the formula $M = \mu + z$ _____ .

════════ STOP ════════════════════ STOP ════════

σ_M

---

55   95 percent of the sample means fall between $z = -1.96$ and $z = 1.96$ (or $z = \pm 1.96$) using the formula $M = \mu + z\,\sigma_M$ and if $\mu = 100$ and $\sigma_M = 10$, then 95 percent of the sample means would fall between 80.40 and _____ .

════════ STOP ════════════════════ STOP ════════

119.60

---

56   If $\mu = 100$, $\sigma_M = 10$, what percent of the sample means fall above 119.6? _____ .

════════ STOP ════════════════════ STOP ════════

2.5 percent

---

57 $\mu = 50$, $\sigma_M = 5$, 95 percent of the sample means would fall between $z = $ _____ and _____.

======= STOP ==================================== STOP =======

-1.96 and $+1.96$

---

58 $\mu = 50$, $\sigma_M = 5$. 95 percent of the sample means fall between $M = \mu \pm 1.96\sigma_M$. 95 percent of the sample means fall between raw scores of _____ and

_____.

======= STOP ==================================== STOP =======

40.20 and 59.80

---

59   Suppose $\mu = 200$, $\sigma = 30$, and we wish to find the scores between which 95 percent of the sample means would fall if each sample had thirty-six scores. $\sigma_M = $ _____.

======= STOP ==================================== STOP =======

$30/6 = 5$

---

60 $\mu = 200$, $\sigma = 30$, $\sigma_M = 5$, 95 percent of the sample means would fall between _____ and _____.

======= STOP ==================================== STOP =======

190.20 and 209.80

---

61 $\mu = 70$, $\sigma = 20$, successive samples of twenty-five cases each were selected and then the mean was computed for each; 95 percent of the sample means would fall between _____ and _____.

======= STOP ==================================== STOP =======

62.16 and 77.84

---

62   Note that 95 percent of the scores and 95 percent of the sample means will not necessarily fall between the same two points. For example, if $\mu = 140$, $\sigma = 8$, $N = 4$, then $\sigma_M = $ _____.

======= STOP ==================================== STOP =======

4

---

63 $\mu = 140$, $\sigma = 8$, $\sigma_M = 4$, 95 percent of the sample means will fall between _____ and _____.

======= STOP ==================================== STOP =======

132.16 and 147.84

---

64 $\mu = 140$, $\sigma = 8$, $\sigma_M = 4$, 95 percent of the sample means will fall between 132.16 and 147.84. On the other hand 95 percent of the scores will fall between _____ and _____.

═══════ STOP ═══════════════════════════ STOP ═══════

124.32 and 155.68

---

65 $\mu = 180$, $\sigma = 27$, $N = 81$, 95 percent of the scores fall between _____ and _____; 95 percent of the sample means fall between _____ and _____.

═══════ STOP ═══════════════════════════ STOP ═══════

127.08 and 232.92; 174.12 and 185.88

---

66 In our previous discussion we have said that when the probability of an event is equal to or less than _____ we would reject the hypothesis that our results are due to chance (that is, sampling error).

═══════ STOP ═══════════════════════════ STOP ═══════

.05

---

67 We do this even though we will be wrong _____ percent of the time if our results are due to chance.

═══════ STOP ═══════════════════════════ STOP ═══════

5

---

68 $\sigma_M = 5$, if $\mu = 250$, the probability of an M falling outside the 240.20–259.80 interval is equal to .05. Hence if we select a single sample and M falls outside of the 240.20–259.80 interval, we reject the hypothesis that $\mu = $ _____.

═══════ STOP ═══════════════════════════ STOP ═══════

250

---

69 $\mu = 250$, $\sigma_M = 5$, again if we select one sample and $M = 240$, then we must (reject/not reject) _____ the hypothesis that $\mu = 250$.

═══════ STOP ═══════════════════════════ STOP ═══════

reject

70 We reject the hypothesis that $\mu = 250$ when we have a single sample $M = 240$ because the probability that such an event would happen is less than _____.

STOP ——————————— STOP

.05

71 $\mu = 100$, $\sigma_M = 20$. We will reject the hypothesis that $\mu = 100$ anytime we observe an M that falls outside the interval _____ to _____.

STOP ——————————— STOP

60.80 to 139.20

72 $\mu = 60$, $\sigma = 15$, $N = 25$, $\sigma_M =$ _____.

STOP ——————————— STOP

$15/5 = 3$

73 $\mu = 60$, $\sigma = 15$, $N = 25$, $\sigma_M = 3$, and we will reject the hypothesis that $\mu = 60$ anytime we observe an M that falls outside the interval _____ to _____.

STOP ——————————— STOP

54.12 to 65.88

74 $\mu = 500$, $\sigma = 100$, $N = 100$, we will reject the hypothesis that $\mu = 500$ whenever we take a single sample that has a mean outside the interval _____ to _____.

STOP ——————————— STOP

480.40 to 519.60

75 We know that the mean IQ of the population is equal to 100 and that $\sigma = 15$. An experimenter feels that he can raise IQ by injection of Vitamin Q. He randomly selects twenty-five subjects and injects Vitamin Q. After a time he tests their IQ and finds $M = 106$. $\sigma_M =$ _____.

STOP ——————————— STOP

3

76 $\mu = 100$ in the normal population, $\sigma_M = 3$, $M = 106$. If the same experiment were repeated many times (and if Vitamin Q *did not* improve IQ), 95 percent of the Ms would fall between _____ and _____.

STOP ——————————— STOP

94.12 and 105.88

77 If Vitamin Q does not improve IQ the probability would be .95 that M would fall between 94.12 and 105.88. $M = 106$. We (reject/do not reject) _____ the hypothesis that $\mu = 100$.

═══════ STOP ═══════ STOP ═══════

reject

---

78 We reject the hypothesis that $\mu = 100$. We conclude that Vitamin Q (does/does not) _____ raise IQ.

═══════ STOP ═══════ STOP ═══════

does

---

79 The formula $z = (M - \mu)/\sigma_M$ is used to test the hypothesis that the true mean is equal to some specific value. If $\sigma = 30$, $N = 9$, $\sigma_M = $ _____?

═══════ STOP ═══════ STOP ═══════

10

---

80 If you wished to test the hypothesis that the true mean was equal to 100, then in the equation above $\mu = $ _____.

═══════ STOP ═══════ STOP ═══════

100

---

81 $\sigma_M = 10$, $\mu = 100$. Suppose a sample mean was found to equal 110; then $z = $ _____.

═══════ STOP ═══════ STOP ═══════

$(110 - 100)/10 = 1.00$

---

82 $\sigma_M = 20$, $M = 150$, and we wish to test the hypothesis that $\mu = 120$. $z = $ _____.

═══════ STOP ═══════ STOP ═══════

$(150 - 120)/20 = 1.50$

---

83 $\sigma_M = 2$. We wish to test the hypothesis that $\mu = 20$. We find a sample mean of 24. $z = $ _____.

═══════ STOP ═══════ STOP ═══════

$(24 - 20)/2 = 2.00$

---

84   The mean for the past several years is 10. This year's mean is 13, if we substitute into the equation $M =$ _____, $\mu =$ _____.

════════ STOP ═══════════════════════════ STOP ════════

$$M = 13,\ \mu = 10$$

---

85   The mean number of bushels of wheat per acre which can be grown in a certain area is 40. In a pilot project using irrigation, the mean number of bushels is 45. In the z-score equation, $M =$ _____, $\mu =$ _____.

════════ STOP ═══════════════════════════ STOP ════════

$$M = 45,\ \mu = 40$$

---

86   The mean number of dental cavities per year per person in a certain city is 3.25. Fluoride is added to the water. A sample of 100 people averaged 2.20 cavities in a year. $M =$ _____, $\mu =$ _____.

════════ STOP ═══════════════════════════ STOP ════════

$$M = 2.20,\ \mu = 3.25$$

---

87 $\mu = 40,\ \sigma = 4,\ N = 4.\ \sigma_M =$ _____.

════════ STOP ═══════════════════════════ STOP ════════

$$2.00$$

---

88 $\mu = 40,\ \sigma = 4,\ N = 4,\ \sigma_M = 2.00$. The mean of a sample of four scores has been found to equal 46. $z =$ _____.

════════ STOP ═══════════════════════════ STOP ════════

$$(46 - 40)/2 = 3.00$$

---

89 $$z = \frac{M - \mu}{(?)}$$

════════ STOP ═══════════════════════════ STOP ════════

$$\sigma_M$$

---

90 $$z = \frac{(?) - \mu}{\sigma_M}$$

════════ STOP ═══════════════════════════ STOP ════════

$$M$$

91

$$z = \frac{(?)}{(?)}$$

━━━━ STOP ━━━━━━━━━━━━━━━━━━━━━ STOP ━━━━

$$\frac{M - \mu}{\sigma_M}$$

---

92 The statistician often wishes to establish a confidence interval for the mean. That is, he wishes to estimate the interval within which the true _____ falls.

━━━━ STOP ━━━━━━━━━━━━━━━━━━━━━ STOP ━━━━

mean

---

93 That is, the statistician knows the value of the sample mean and wishes to estimate the value of the _____.

━━━━ STOP ━━━━━━━━━━━━━━━━━━━━━ STOP ━━━━

true mean

---

94 To estimate the value of the true mean, the statistician finds the confidence limits, which are the two points bounding the confidence _____ for the true mean.

━━━━ STOP ━━━━━━━━━━━━━━━━━━━━━ STOP ━━━━

interval

---

95 Since we do not know the value of μ we cannot use the formula $M = \mu \pm z\,\sigma_M$. Instead we must use the formula $\mu = M \pm z\,\sigma_M$ to establish the _____ interval for the mean.

━━━━ STOP ━━━━━━━━━━━━━━━━━━━━━ STOP ━━━━

confidence

---

96 If we wish to find the confidence limits for the true mean we must use the formula $\mu = M \pm z\,\sigma_M$, that is, the lower limit is found by using $\mu = M - z\,\sigma_M$ while the _____ limit is found by using $\mu = M + z\,\sigma_M$.

━━━━ STOP ━━━━━━━━━━━━━━━━━━━━━ STOP ━━━━

upper

---

97 $\mu = M \pm z\, \sigma_M$. If $M = 105$, $\sigma_M = 5$, and $z = 1.96$, then the lower limit equals 95.20 and the upper limit equals _____.

═══════ STOP ═══════════════════════════ STOP ═══════

114.80

---

98 $\mu = M\ (?)\ z\, \sigma_M$

═══════ STOP ═══════════════════════════ STOP ═══════

$\pm$ [plus or minus]

---

99 $\mu = \underline{\hspace{1cm}} \pm z\, \sigma_M$

═══════ STOP ═══════════════════════════ STOP ═══════

M

---

100 $\mu = M \pm z\, \underline{\hspace{1.5cm}}$

═══════ STOP ═══════════════════════════ STOP ═══════

σ_M

---

101 $\mu = \underline{\hspace{1.5cm}}$

═══════ STOP ═══════════════════════════ STOP ═══════

$M \pm z\, \sigma_M$

---

102 When we wish to establish the 99 percent confidence interval we would use $z = 2.58$, but when we wish to establish the 95 percent confidence interval we would use $z = \underline{\hspace{1.5cm}}$.

═══════ STOP ═══════════════════════════ STOP ═══════

1.96

---

103 $M = 40$, $\sigma_M = 4$, $z = 1.96$, find the confidence interval for the true mean. _____.

═══════ STOP ═══════════════════════════ STOP ═══════

32.16 to 47.84

---

104 $M = 80$, $\sigma_M = 3$, $z = 2.58$, find the confidence interval for the true mean. _____.

═══════ STOP ═══════════════════════════ STOP ═══════

72.26 to 87.74

---

105 $M = 30$, $\sigma_M = 5$, find the 95 percent confidence interval for the true mean. _____.

═══════ STOP ═══════════════════════ STOP ═══════

20.20 to 39.80

---

106 $M = 80$, $\sigma = 20$, $N = 25$, $\sigma_M =$ _____, and the 95 percent confidence limits are _____ and _____.

═══════ STOP ═══════════════════════ STOP ═══════

4, 72.16 and 87.84

---

107 We will be wrong only 1 percent of the time when we assert the true mean falls within the 99 percent confidence interval. In this case $z =$ _____.

═══════ STOP ═══════════════════════ STOP ═══════

2.58

---

108 $M = 120$, $\sigma = 24$, $N = 16$, the 99 percent confidence limits for the true mean are _____ and _____.

═══════ STOP ═══════════════════════ STOP ═══════

104.52 and 135.48

---

109 We will be wrong 5 percent of the time when we assert that the true mean falls within the 95 percent confidence interval; $M = 200$, $\sigma = 49$, $N = 49$. The 95 percent confidence limits for the true mean are _____ and _____.

═══════ STOP ═══════════════════════ STOP ═══════

186.28 and 213.72

---

110 $M = 50$, $\sigma_M = 4$, a score of 60 falls (inside/outside) _____ the 95 percent confidence interval.

═══════ STOP ═══════════════════════ STOP ═══════

outside

---

111 $M = 50$, $\sigma_M = 4$, 60 falls outside the 95 percent confidence interval. Is H: $\mu = 60$ tenable, that is, is the hypothesis $\mu = 60$ tenable? (Yes/No) _____.

═══════ STOP ═══════════════════════ STOP ═══════

No

---

112 $M = 5$, $\sigma_M = 1$, you (reject/do not reject) _____ the hypothesis that $\mu = 3$.

═════════ STOP ═══════════════════════ STOP ═══════

reject

---

113 $M = 40$, $\sigma = 10$, $N = 16$, H: $\mu = 45$. You (reject/not reject) _____.

═════════ STOP ═══════════════════════ STOP ═══════

reject

---

114 $M = 1000$, $\sigma = 200$, $N = 100$, find the 95 percent confidence limits for the true mean. _____.

═════════ STOP ═══════════════════════ STOP ═══════

960.80 to 1039.20

---

115 $M = 200$, $\sigma = 50$, $N = 25$, find the 95 percent confidence limits for the true mean. _____.

═════════ STOP ═══════════════════════ STOP ═══════

180.40 to 219.60

---

116 $M = 40$, $\sigma = 24$, $N = 36$, find the 95 percent confidence limits for the true mean. _____.

═════════ STOP ═══════════════════════ STOP ═══════

32.16 to 47.84

Ten

The *t* Distribution

We have studied the z distribution and have found that z is normal with $\mu = 0$ and $\sigma = 1$. We will now study a second type of distribution, which is called t. An Englishman named Gosset developed the theory of the t distribution, and he did so for a very practical reason. He worked in a brewery, measuring, let us assume, the water content of newly produced stout. He had to sample many, many barrels of stout to be sure that his sample standard deviation (S) was a stable estimate of the true standard deviation (σ). For some unknown reason, he became interested in developing a technique for using smaller samples than he had before. Perhaps he just did not like the smell of green stout. Whatever his motives, he developed the theory of the extremely important t distribution and prepared the way for the subsequent development of small-sample statistical theory. Gosset published his paper under the pseudonym "Student," and for this reason the t distribution as presented in this chapter is often called *Student's t distribution*. More than one student taking introductory statistics has completed the course thinking that Student's t distribution was a simplified version of the t-distribution.

Like the z, the t distribution is important in problems in which the true mean (μ) is estimated from the sample mean (M). Unlike the z, the t distribution does not require knowledge of the true standard deviation (σ) and no attempt is made to estimate the true standard deviation (σ) from the sample standard deviation (S). All that is required by the t is that a sample be drawn from a normally distributed population and that this sample have a mean (M) and a standard deviation (S).

In Chapter Nine you learned how to establish a confidence interval around a sample mean. The basic procedure was to set an upper and a lower limit, and to say that the probability of the true means falling inside the limits was $1 - \alpha$. We usually set α at 0.05, although in cases where we must be sure that the confidence interval includes the true mean, we might set α at 0.01. In order to do this, we used our knowledge of the z distribution and the formula $z = (M - \mu)/\sigma_M$, and then changed the formula around to solve for the true mean (μ); that is, $\mu = M + (z) \cdot (\sigma_M)$. If α was set at 0.05, we would substitute 1.96 for z into the equation (see Appendix B). In this equation the value of σ_M was found by use of the formula $\sigma_M = \sigma/\sqrt{N}$. But note that we are required to know a parameter of the population (σ); in order to establish the confidence interval for the mean, the true standard deviation (σ) must be known. There is, obviously, one gigantic difficulty involved in using this method: It is a rare situation indeed where we know the standard deviation of a population (that is, σ), but do not know the mean of the population (μ). Far more common is the situation in which we do not know any more about the population in question than what our sample tells us, namely, the sample mean (M) and the sample standard deviation (S).

Prior to the development of the t-distribution theory by Gosset, the only possible procedure for establishing confidence intervals was to take a very large sample and hope that the standard deviation (S) of such a large sample would closely approximate σ. If it did, then the probability statements made were exact; but if it did not, then a great deal of error could be made in setting up the confidence interval for the true mean.

THE STANDARD ERROR OF THE MEAN

In Chapter Nine we developed the use of the standard error of the mean (σ_M). We noted that the standard deviation of the means of all possible samples of size N is related to the standard deviation by the formula

$$\sigma_M = \frac{\sigma}{\sqrt{N}}$$

where σ is the standard deviation of the population, and σ_M, computed from the formula, is used to determine confidence intervals for the mean. Unfortunately, σ—as are other parameters—is rarely known, and the use of S requires that σ be estimated from S, the standard deviation of the sample. As was noted above, the extent to which S approximates σ determines the accuracy of the estimate of σ_M. This in turn determines the accuracy of the confidence interval for the mean. Clearly, this is an unfortunate situation.

Gosset's work eliminated this necessity of estimating parameters when he developed the theory of the t-distribution. Initially, let us present the formula for the statistic called the *standard error of the mean* (S_M). Again we must note that no error is implied by use of the word *error* and it is equally proper to say that S_M is the standard deviation of a group of sample means. This statistic S_M should be distinguished from the parameter σ_M, which represents the standard deviation of all possible sample means. S_M is an estimate of σ_M and S_M can be computed directly by finding the standard deviation of a group of means, or, more conveniently, from a single sample by the use of a formula. The formula for S_M parallels that for σ_M:

$$\sigma_M = \frac{\sigma}{\sqrt{N}} \quad \text{and} \quad S_M = \frac{S}{\sqrt{N}}$$

where N is the size of the sample.

THE t AND z DISTRIBUTIONS

We have noted that the distribution of sample means (M) will tend to be normal even when the population from which a sample is drawn is not normal. This was demonstrated in Chapter Nine by the dice-rolling experiment and, with the help of the computer, by the experiment sampling means from a highly skewed distribution. We have seen that subtracting the true mean from the sample mean $(M - \mu)$ will not change the shape of the distribution. That is, the distribution will still be normal. In fact, if we divide by σ_M, the distribution is normal (and becomes the z distribution) because, for every possible sample mean, we are dividing by a constant (σ_M). However, if we take a sample mean (M), subtract the true mean $(M - \mu)$, and divide by the standard error of the sample mean (S_M), the resulting distribution will not be normal. The reason for this is that the size of S (and hence S_M) will vary from sample to sample, while σ, the parameter, will not.

We have said that the ratio $(M - \mu)/S_M$ is not normally distributed. Rather than being normally distributed, this ratio yields the t distribution. We should first compare the formulas for z and t:

$$z = \frac{(M - \mu)}{\sigma_M} \quad \text{and} \quad t = \frac{(M - \mu)}{S_M}$$

Note that the only difference between the two formulas is that the denominator for z is σ_M and that the denominator for t is S_M. The result of this apparently slight difference is that z is normally distributed and t is not. Actually, the t distribution differs from the z distribution in another important way: The t distribution is really several distributions. The shape of the t distribution changes with the size of the sample (N). As shown in Figure 10-1, as the sample size becomes smaller, the t distribution becomes more spread out or variable when compared to the z distribution. As we might expect, the larger the sample size, the more closely the distribution approaches the shape of the z distribution, but when the sample becomes infinitely large, the z and t distributions are exactly the same.

Figure 10-1 illustrates what happens to the two points that include 95 percent of the distribution as the sample size decreases—they are displaced farther and farther from the mean of the distributions. For example, the distance ± 1.96 standard deviations from the mean $(z = \pm 1.96)$ will include 95 percent of the area under the distribution (df $= \infty$). But if $N = 20$, then ± 2.09 standard deviations $(t = 2.09)$ will include 95 percent of the area under the t distribution (df $= 19$); and if $N = 3$, then $t = 4.30$ will include 95 percent of the area (df $= 2$).

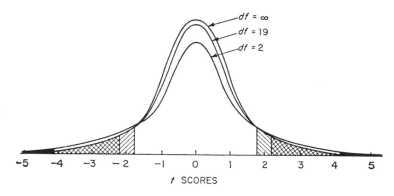

Figure 10.1   *Comparison of three t distributions.*

The importance of the t distribution is that we can use it to establish a confidence interval for the mean without knowing the parameter σ. That is, instead of using the formula $z = (M - \mu)/\sigma_M$ to determine the confidence interval, we can now use the formula $t = (M - \mu)/S_M$ and work with t rather than with z.

DEGREES OF FREEDOM

You will notice that the three distributions in Figure 10-1 are labeled df = 2, 19, and ∞, rather than $N = 3$, 20, and ∞; df stands for *degrees of freedom*, and it is something we will be working with continuously throughout the remainder of the book. Degrees of freedom in a set of scores are related to the number of restrictions placed upon a set of scores. For example, suppose the mean of three scores is equal to 4.00. This would mean that the sum of the three scores has to equal 12. Any two of these scores are free to take on any value, but once these two scores are determined, the third has to have a value such that the sum of the three is equal to 12. Thus, if the first two scores were 2 and 3, then $2 + 3 + X = 12$ and X would have to equal 7. The restriction placed on the scores in this example was, of course, that the mean was equal to 4.00. In general, a single df is "lost" every time a parameter is estimated. In the present chapter you will find that df will always be equal to $N - 1$, since by our procedures we are estimating only one parameter, the true mean μ.

In any event, the distribution labeled df = 2 is what the t distribution (or the sampling distribution of t) will look like when ts are computed from a large number of samples of three scores each. The mean (M) and the standard deviation (S) are computed for each sample of three scores. Using the formula $S_M = S/\sqrt{N}$, S_M is then computed from S for each sample. Finally, t is computed for each sample from the formula $t = (M - \mu)/S_M$. The distribution of these t values, presented in the df = 2 distribution of Figure 10-1, is such that 2.5 percent of the t values will be larger than 4.30 (and 2.5 percent will be smaller than -4.30). This area is represented by the shaded portions of the df = 2 distribution. Similarly, we can see from the distribution labeled df = 19 that 5 percent of the sample *t*s will fall outside the area ±2.09. The distribution labeled df = ∞ is exactly the same as the *z*-score distribution with 5 percent of the area falling beyond ±1.96.

READING APPENDIX C   In our previous discussion we noted that the *t* distribution is a group of distributions rather than a single distribution. We also noted that the shape of the distribution depends upon the number of df. When the number of df is small, we must go farther out on the *t* distribution to include 95 percent (or, for that matter, 99 percent) of the area than we do when the number of df is larger. Appendix C may be used to tell us how far out we must go to include a certain percentage of the area when we have a specified number of df. Appendix C is divided into three columns. The column on the left is labeled "df," the middle column is labeled ".05," and the right-hand column is labeled ".01." This table tells you how far out on the *t* distribution you must go to include 95 percent (that is, all but 5 percent) of the area or 99 percent (that is, all but 1 percent) of the area. For example, if $N = 11$, then df = 10, and a *t* value of ±2.23 will include 95 percent of the area under the *t* distribution. Or if $N = 6$, then df = 5, and $t = 4.03$ will include 99 percent of the area under the curve.

TESTING HYPOTHESES ABOUT SAMPLE MEANS

The hypothesis that the true mean is equal to some specific value can be tested with the *t* test as easily as with the *z* test. Since the formula for the *t* does not require knowledge of any population parameter, we might find its use even more convenient than the *z*, which does require knowledge of the standard deviation of the population. In contrast to the *z* test, with the *t* test the standard error of the mean is computed from the sample standard deviation. As in the *z* test, the value of μ does not necessarily represent the value of the mean of the population, which is what we are trying to determine. We are testing to see if the population mean can be equal to the value we have given μ in the equation. If the difference between M and μ is too great, then we reject the hypothesis that the population mean is equal to the value given μ in the equation. If the difference is too great, then we conclude that the value of the population mean is equal to some other value.

AN EXAMPLE   The recommended daily intake of calcium for adolescent boys is 1400 milligrams. A nutritionist studies the diets of a random sample of 64 impoverished boys in an urban ghetto and finds that their average calcium intake is 1200 milligrams a day (the standard deviation is found to equal 200 milligrams). We wish to answer the question, "Does this sample represent a population with adequate calcium nutrition?"

The process of statistical inference used in this case may be outlined as follows:

The recommended calcium intake is 1400 milligrams. Since a population with adequate calcium nutrition would have (at least) a mean of 1400 milligrams, it seems reasonable to test the hypothesis that $\mu = 1400$ in the specific population studied. With the mean of the sample found to equal 1200, then if the difference between M and μ (the sample mean and the hypothetical population mean) is too great, we reject the hypothesis that the population mean of the particular population studied is 1400.

Remember that if the probability of an event is less than 0.05, the hypothesis is rejected. It may be that the probability of getting a sample mean of 1200 from a population with a mean of 1400 is less than 0.05. If so, we would then reject the hypothesis that the mean of the ghetto population studied is equal to 1400. That is, 1400 is too far from 1200 to be thought of as the population mean.

How far is too far? That depends upon the unit of distance, and the unit of distance here is the variability of sample means. This can be found from the formula for the standard error of the mean; that is, if $S = 200$ and $N = 64$, then

$$S_M = \frac{S}{\sqrt{N}} = \frac{200}{\sqrt{64}} = \frac{200}{8} = 25$$

With the three needed values determined, we can now substitute into the t formula:

$$t = \frac{M - \mu}{S_M} = \frac{1200 - 1400}{25} = \frac{200}{25} = 8.00$$

The distance between M and the hypothesized true mean is 8.00 standard deviation units. With $N = 64$, df $= 63$, reference to Appendix C indicates that with 63 df, a t of 2.00 would include 95 percent of the area. Hence, a t of 8.00 is much larger than is needed to reject the hypothesis that the true mean is 1400. Under these conditions we would conclude that the true mean is less than 1400 milligrams of calcium per day. This would indicate that the boys in the group studied do not get enough calcium in their diets.

On the other hand, if we had not rejected the hypothesis that $\mu = 1400$, then μ could have equaled 1400. (We would not have rejected the hypothesis if the value of t had been less than 2.00.) While the average calcium intake of the sample was less than that recommended, we would have concluded that we had no evidence that the population studied had an average intake which was less than adequate.

ASSUMPTIONS OF THE t TEST

1. Random sampling.
2. Scores in sample are independent of one another.
3. Sample comes from a normal distribution.
4. The population standard deviation is unknown and the sample standard deviation is used.
5. Measurement scale is at least an interval scale.

ESTABLISHING THE CONFIDENCE INTERVAL FOR THE TRUE MEAN

If we take the formula previously presented for t and solve for μ, we have then the formula for the confidence interval for the true mean when the true standard deviation (σ) is not known. The formula is

$$\mu = M \pm (t)(S_M)$$

where M = sample mean, S_M = standard error of the mean computed from S, and t = value obtained from the t table.

Suppose a sample of nine scores has been selected. $M = 55$ and $S = 12$, and we wish to establish the 95 percent confidence interval for the mean. To solve the problem, we must know M, S_M, and t. We know $M = 55$, and we can find S_M, since $S_M = S/\sqrt{N} = 12/3 = 4$. To find the value of t, we must look in the t table. We have $N - 1 = 8$ df, and the t required for the 95 percent confidence interval is found to equal 2.31. Since the 95 percent confidence interval for the mean is $\mu = M \pm (t)(S_M)$, then $\mu = 55 \pm (2.31)(4) = 55 \pm 9.24 = 45.76$ to 64.24. The 95 percent confidence interval for the true mean is from 45.76 to 64.24.

AN EXAMPLE Suppose an anthropologist studies an obscure tribe living in a rain forest of the Philippines. Casual observation on his part convinces him that the adult members of the tribe are considerably shorter than other inhabitants of the same island. Because of the obvious differences in height, he is not interested in testing the hypothesis that they are the same mean height as other people, although this could easily be done. Instead he wishes to make a statement about how tall they are. To do this, he would need to establish a confidence interval for the mean. This is essentially the same confidence interval you learned about in Chapter Nine. The only difference is that here you do not know the standard deviation of the population. In this case you must use the standard deviation of the sample in setting up your confidence interval.

After being assured that other members of the tribe are just like the people he was talking to, the anthropologist measured their heights. (He was concerned whether the few people he had contacted were representative of the population as a whole. Once he had decided this was the case, he was then satisfied he had met the assumptions of random and independent selection.) He then established confidence intervals for the nine men he had contacted. Here are their heights in inches: 52, 45, 47, 50, 53, 46, 45, 48, 46. From these scores he computes the following:

$$\Sigma X = 432 \qquad M = \frac{432}{9} = 48.00$$

$$\Sigma X^2 = 20,808$$

$$S^2 = \frac{\Sigma X^2 - \dfrac{(\Sigma X)^2}{N}}{N - 1} = \frac{20,808 - \dfrac{432 \times 432}{9}}{8} = \frac{72}{8} = 9$$

$$S = 3$$

$$S_m = \frac{S}{\sqrt{N}} = \frac{3}{\sqrt{9}} = \frac{3}{3} = 1.00$$

With $N = 9$, df $= 8$, and the entry in Appendix C under the 0.05 column is 2.31. With $M = 48.00$, $S_m = 1.00$, and $t = 2.31$, we can substitute into the formula

$$\mu = M \pm t \cdot S_m = 48 \pm 2.31 \times 1.00$$

$$\mu \text{ (upper limit)} = 48 + 2.31 \times 1.00 = 50.31$$

$$\mu \text{ (lower limit)} = 48 - 2.31 \times 1.00 = 45.69$$

Hence, the 95 percent confidence interval for the true mean ranges from 45.69 to 50.31. Although the standard deviation will vary from one sample to the next, the t distribution, by being somewhat more variable than the z distribution, takes this into account and adjusts for it.

Figure 10-2 illustrates the process of establishing the confidence interval for the mean when the sample rather than the population standard

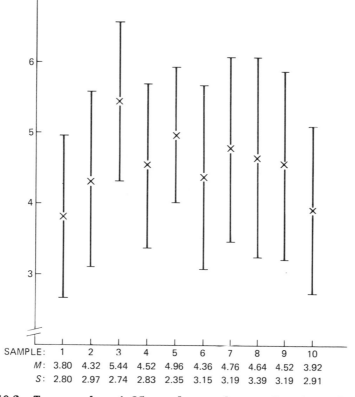

SAMPLE:	1	2	3	4	5	6	7	8	9	10
M:	3.80	4.32	5.44	4.52	4.96	4.36	4.76	4.64	4.52	3.92
S:	2.80	2.97	2.74	2.83	2.35	3.15	3.19	3.39	3.19	2.91

Figure 10-2   *Ten samples of 25 numbers each are taken from the first 10 columns of Appendix H. M and S are computed for each, and the 95 percent confidence interval is also established for each sample.*

deviation is known. This was done by taking ten samples of 25 numbers each from Appendix H. The first five rows—consisting of 25 single-digit numbers—of each of the ten columns of numbers on the first page of Appendix H were used. For each sample, M and S_m were computed, and with 24 df a t of 2.06 is found in Appendix C. Next, these values were substituted into the formula $\mu = M \pm t \cdot S_M$. After the limits were found for each sample, they were graphed.

Suppose someone asks you, "Which set of limits is the correct one?" About the only answer you can give is that any set of limits may or may not include the true mean. But if you did the same experiment over and over again, 95 percent of the intervals would include the true mean. Thus, when the sample is being selected, the probability is 0.95 that the true mean will fall within the 95 percent confidence interval for the mean.

TESTING THE DIFFERENCE BETWEEN TWO MEANS

The t test can be used in other ways. Suppose, for example, you conducted an experiment in which you wished to determine if the performance of the experimental group was different from that of the control group. A t test could be used to answer that question. Rather than give you the formula for the t test for two independent groups, we refer you to the F test in Chapter Thirteen. In that chapter the analysis of variance is introduced, and this statistical test can be used to test the difference between two independent groups. It has the additional advantage of being able to test the difference between more than two groups. That is, it is more general than the t test. Nevertheless, the t test is a widely used test, and you will often encounter use of the t test between two means in your reading. Just as z and t are related, so too are t and F. Hence, instead of giving you a new formula, just remember that $t^2 = F$ with two groups (and that $t = \sqrt{F}$). When you learn to compute the F in the analysis of variance, take the square root of the F value, and that will be exactly equal to t.

The t can also be used to test changes in performance. That is, if we get a "before and after" score on each person in a group, we can determine if the change is greater than you would expect on the basis of chance. To do this, you can use the t formula you have available to you. But again, since the analysis of variance is a much more general test, we have postponed discussion of this until Chapter Fourteen—the second chapter dealing with the analysis of variance. Again, the square root of the F you get is exactly equal to the t you would obtain if you used the appropriate t formula.

SUMMARY

The t test was a major breakthrough in the development of inferential statistics. In contrast to the z test, the t test does not require the knowledge of any parameter, since it relies only on statistics. That is, the z test requires knowledge of σ—the standard deviation of the population. In contrast, the t test does not require the population standard deviation; instead it uses only the sample standard deviation, S. Since only rarely is the population standard deviation known, development of the *t* early in the twentieth century allowed statisticians to use S without worrying about the consequences of getting a poor estimate of σ from it. In earlier research using *z*, "large" samples—generally more than 30 cases—were required to be sure that an adequate estimate of σ could be made from S. With the development of the *t* test, smaller samples could be, and were, used by research workers without fear of distorted results.

PROBLEMS

1. *E* wishes to test the hypothesis that the true mean of some sample is equal to 100. He has the sample mean and knows the standard deviation of the population. What is the appropriate statistic (that is, give the formula) to use in this case?

2. On the other hand, suppose *E* wanted to test the same hypothesis, but had available only the information from a single sample. In this case, what would be the appropriate statistical test to use? Again give the appropriate formula.

3. *E* wishes to test the hypothesis that $\mu = 50$. Her sample of 25 scores has a standard deviation of 10. The standard error of the mean is equal to what value? What is the number of degrees of freedom? What *t* value must be obtained before *E* can reject the hypothesis that $\mu = 50$?

4. *E* selects a sample of 16 scores and finds that the mean of the sample is 40 and the standard deviation of the sample is 12. *E* wishes to test the hypothesis that $\mu = 46$. Do everything necessary. What are your conclusions?

5. Recruits going into the army at a certain camp are given a physical fitness test which includes doing pushups. In one group, 100 soldiers had a mean of 15 pushups and a standard deviation of 3. Establish the 95 percent confidence interval for the true mean.

6. Since time is money, a pickle packer was interested in a machine that its maker claimed would pack a gross of pickle jars faster than his old machine. The old machine packed a gross of jars in 6 minutes (360 seconds). As part of the demonstration, the maker packed 9 gross of pickles. Unobstrusively, the pickle packer timed each gross as it was packed and then gave the data to a friendly statistician, who found the mean time to pack a gross was 330 seconds and the standard deviation 30.00 seconds. Since the pickle packer wanted to be very sure that the new machine was faster than the old machine, the statistician used the 99 percent level. Did he recommend that the pickle packer buy the machine?

7. The traffic safety director of a five-country area decided to put on an advertising campaign stressing safe driving. After a month of the campaign she noted that the number of accidents involving fatalities in each of the counties was 0, 6, 2, 4, and 3. The usual number of such accidents per county per month is 6.00. Has the advertising campaign done any good?

8. Prior to a small public relations campaign to read history, the mean number of history books checked out of the library was 40 per week. In the four weeks

since the campaign, this mean has increased to 50 with a standard deviation of 8. Has the number of history books taken out increased? What, if any, assumptions may have been violated in answering this question?

9. One species of fish, the Bigeye Shiner, has a maximum of 40 rows of scales. A zoologist found a sample of fish which looks like the Bigeye Shiner but which had a mean of 41.00 rows of scales. There were 25 fish in the sample, and their row standard deviation was 2.00. Test the hypothesis that the true mean is 40.00. Interpret the results of your analysis.

10. A light-bulb manufacturer claims in his advertising that his bulbs will burn for 1000 hours. An independent testing firm randomly took 16 light bulbs from the assembly line and burned them until each went out. The number of hours burned for each was recorded. The mean number of hours the lights burned was 950 hours and the standard deviation was found to equal 100 hours. Can we reject the manufacturer's claims as inaccurate?

11. $M = 32$, $S_M = 5$, $N = 7$. Establish the 95 percent confidence interval for the mean.

12. $M = 65$, $S = 10$, $N = 25$. Establish the 95 percent confidence interval for the mean.

13. $M = 120$, $S = 18$, $N = 9$. Establish the 95 percent confidence interval for the mean.

14. Establish the 99 percent confidence interval for the mean in Problem 12.

15. $M = 82$, $S = 12$, $N = 4$. Test the hypothesis that $\mu = 100$. Compute the value of t and tell what you do with the hypothesis.

16. $M = 150$, $S = 28$, $N = 16$. Test the hypothesis that $\mu = 135$. Compute the value of t and tell what you do with the hypothesis.

EXERCISES

1. Try to guess the correct suit of eight cards before they are drawn randomly from an ordinary bridge deck. Count the number of times you are correct and write it down. Do this experiment several times. Now you have several numbers with a possible range of 0–8. Find M, S, S_M, and establish the 95% confidence interval for the true mean. If you were guessing randomly, the true mean would be 2.0. Why? Test the hypothesis $\mu = 2.0$.

2. Find the height of each man in your group. Find M, S, S_M, and establish the 95 percent confidence interval for the mean. Next find the height of each woman in your group. Again find M, S, and S_M for the women. Using the male mean as μ, test the hypothesis that the mean height of the women is the same as that of the men.

Comparison of z and t distributions

Value of z or t	z	Percent of the area between mean and t (df = 20)	t (df = 2)
.20	08	08	07
.40	16	15	14
.60	23	22	19
.80	29	27	24
1.00	34	32	28
1.20	38	37	32
1.40	42	41	35
1.60	45	44	37
1.80	46	45	39
1.96	47.5	47	40
2.00	48	47	40
2.09	48	47.5	41
2.20	49	48	41
2.40	49	48	42
2.58	49.5	49	43
2.60	49.5	49	43
2.80	49.7	49.5	44
4.30			47.5
9.93			49.5

1   It will be recalled that a parameter is a quantitative characteristic of a _____.

══════ STOP ═══════════════════ STOP ══════

population

---

2   A quantitative characteristic of a sample is called a _____.

══════ STOP ═══════════════════ STOP ══════

statistic

---

3   In the previous chapter we learned how to establish the confidence interval for the mean using the formula $\mu = M \pm (z)(\sigma_M)$. σ_M is a (parameter/statistic) _____ and M is a (parameter/statistic) _____.

══════ STOP ═══════════════════ STOP ══════

parameter, statistic

---

4   We do not usually know the value of σ since it is a parameter. However, if N is large, the value of S will closely approximate the value of _____.

══════ STOP ═══════════════════ STOP ══════

σ

---

5   Unless N is infinitely large we would not always expect S to equal _____.

══════ STOP ═══════════════════ STOP ══════

σ

---

6   Since S is a statistic, its value will vary from sample to sample. However, S would be expected to vary around _____.

══════ STOP ═══════════════════ STOP ══════

σ

---

7   Similarly we do not usually know the value of σ_M. However, we can use S_M to estimate _____.

══════ STOP ═══════════════════ STOP ══════

σ_M

---

8 If M is normally distributed, both $(M - \mu)/\sigma_M$ and $(M - \mu)/S_M$ will have means equal to zero. However while the ratio $(M - \mu)/\sigma_M$ is normally distributed, $(M - \mu)/S_M$ is not _____ _____.

===== STOP ================================= STOP =====

normally distributed

---

9 $(M - \mu)/S_M$ is t distributed while $(M - \mu)/\sigma_M$ is _____ distributed.

===== STOP ================================= STOP =====

z (normally)

---

10 While the t and z distributions are quite similar, the exact shape of the t distribution depends on the number of degrees of freedom (df). When the number of df is infinite, $S_M = \sigma_M$ and the t distribution is exactly the same as the _____.

===== STOP ================================= STOP =====

z distribution

---

11 The abbreviation for degrees of freedom is _____.

===== STOP ================================= STOP =====

df

---

12 The number of df is inversely related to the number of restrictions placed on a set of scores. When a set of N scores must add to a certain sum then one restriction is placed on that set of scores. In general there are $N - 1$ _____ in a set of scores.

===== STOP ================================= STOP =====

df

---

13 There are $N - 1$ df in a set of numbers. If there are ten scores then there would be _____ df.

===== STOP ================================= STOP =====

9

---

14 If there were twenty-five scores then there would be _____ df.

===== STOP ================================= STOP =====

24

15   But if there were 15 df there would then be _____ scores.

---

16   As the number of df decrease the variability of the t distribution increases with more area in the tails. Distribution (A/B) _____ is a t distribution.

---

17   One consequence of an increased variability in the t distribution is that the width of the 95 percent confidence interval for the mean would be _____ for the t than for the z distribution.

---

18   Thus $\mu \pm 1.96\sigma_M$ would include 95 percent of the sample means. But if df were small, $\mu \pm 1.96 S_M$ would include _____ 95 percent of the sample means.

---

19   Hence if we were to use S_M rather than σ_M (and if df were small) then we would have to go _____ out on the t than the z distribution to include the same percent of the area.

---

20   Similarly, the values of $M \pm (S_M)(t)$ would have to be _____ than the values of $M \pm (\sigma_M)(z)$ to have the same probability that the true mean was included.

---

21 Since the shape of the t distribution is dependent on df, then the width of the 95 percent confidence interval would also be dependent upon _____ _____ _____ _____ .

$\equiv$ **STOP** $\equiv$ $\equiv$ **STOP** $\equiv$

the number of df

---

22 Consider the table entitled *Comparison of z and t distributions* facing the first page of the program. The column on the left is labeled _____ _____ _____ _____ _____ .

$\equiv$ **STOP** $\equiv$ $\equiv$ **STOP** $\equiv$

"Value of z or t"

---

23 The column on the extreme right is labeled "Percent of the area between mean and _____."

$\equiv$ **STOP** $\equiv$ $\equiv$ **STOP** $\equiv$

$t(\mathrm{df} = 2)$

---

24 Thus for a given value of z, for example, we can find the area between the mean and z. What percent of the area under the normal curve falls between the mean and $z = 1.40$? _____ .

$\equiv$ **STOP** $\equiv$ $\equiv$ **STOP** $\equiv$

42 percent

---

25 What percent of the area under the t distribution (df = 20) falls between the mean and $t = 1.40$? _____ .

$\equiv$ **STOP** $\equiv$ $\equiv$ **STOP** $\equiv$

41 percent

---

26 What percent of the area under the t distribution (df = 2) falls between the mean and $t = 1.40$? _____ .

$\equiv$ **STOP** $\equiv$ $\equiv$ **STOP** $\equiv$

35 percent

---

27 Although 34 percent of the area under the normal curve falls between the mean and $z = 1$, what percent falls between the mean and $t = 1$ when df = 20? _____ ; when df = 2? _____ .

$\equiv$ **STOP** $\equiv$ $\equiv$ **STOP** $\equiv$

32 percent; 28 percent

---

28   47.5 percent of the area under the normal distribution falls between the mean and $z =$ _____.

$\equiv\equiv\equiv$ STOP $\equiv\equiv\equiv\equiv\equiv\equiv\equiv\equiv\equiv\equiv$ STOP $\equiv\equiv\equiv$

1.96

---

29   47.5 percent of the area under the t distribution (df $= 20$) falls between the mean and $t =$ _____

$\equiv\equiv\equiv$ STOP $\equiv\equiv\equiv\equiv\equiv\equiv\equiv\equiv\equiv\equiv$ STOP $\equiv\equiv\equiv$

2.09

---

30   47.5 percent of the area under the t distribution (df $= 2$) falls between the mean and $t =$ _____.

$\equiv\equiv\equiv$ STOP $\equiv\equiv\equiv\equiv\equiv\equiv\equiv\equiv\equiv\equiv$ STOP $\equiv\equiv\equiv$

4.30

---

31   95 percent of the area under the normal curve is included between $z_1 =$ _____ and $z_2 =$ _____.

$\equiv\equiv\equiv$ STOP $\equiv\equiv\equiv\equiv\equiv\equiv\equiv\equiv\equiv\equiv$ STOP $\equiv\equiv\equiv$

$-1.96, 1.96$

---

32   And 95 percent of the area under the t distribution is included between $t_1 =$ _____ and $t_2 =$ _____ (df $= 20$); $t_1 =$ _____ and $t_2 =$ _____ (df $= 2$).

$\equiv\equiv\equiv$ STOP $\equiv\equiv\equiv\equiv\equiv\equiv\equiv\equiv\equiv\equiv$ STOP $\equiv\equiv\equiv$

$-2.09, 2.09; -4.30, 4.30$

---

33   Finally, 99 percent of the area under the normal curve is included between $z = \pm2.58$. 99 percent of the area when df $= 20$ is included between $t = \pm$_____ and 99 percent of the area when df $= 2$ is included between $t = \pm$_____.

$\equiv\equiv\equiv$ STOP $\equiv\equiv\equiv\equiv\equiv\equiv\equiv\equiv\equiv\equiv$ STOP $\equiv\equiv\equiv$

2.80, 9.93

---

34   If σ_M is known, then 95 percent of the sample means would fall between $z_1 =$ _____ and $z_2 =$ _____.

$\equiv\equiv\equiv$ STOP $\equiv\equiv\equiv\equiv\equiv\equiv\equiv\equiv\equiv\equiv$ STOP $\equiv\equiv\equiv$

$-1.96, 1.96$

---

35 If σ_M is unknown but S_M is known, and df $= 20$, then 95 percent of the sample means would fall between $t =$ _____ and $t =$ _____.

═══════ **STOP** ═══════════════════════ **STOP** ═══════

$$-2.09,\ 2.09$$

---

36 Given M and σ_M, the confidence interval for the mean is given by the formula _____.

═══════ **STOP** ═══════════════════════ **STOP** ═══════

$$\mu = M \pm (z)(\sigma_M)$$

---

37 On the other hand if σ_M was unknown and if S_M was available, the confidence interval for the mean could be found by the formula $\mu =$ _____.

═══════ **STOP** ═══════════════════════ **STOP** ═══════

$$M \pm (t)(S_M)$$

---

38 If $\mu = M \pm (t)(S_M)$ with $M = 100$, $S_M = 10$ and df $= 20$, then the 95 percent confidence limits for the mean would be _____ and _____.

═══════ **STOP** ═══════════════════════ **STOP** ═══════

79.10 and 120.90

---

39 $\mu = M \pm (t)(S_M)$ with $M = 100$, $S_M = 10$, and df $= 2$, then the 95 percent confidence limits for the mean would be _____ and _____.

═══════ **STOP** ═══════════════════════ **STOP** ═══════

57.0 and 143.0

---

40 $\mu = M \pm (t)(S_M)$ with $M = 50$, $S_M = 3$, and $N = 3$, then the 95 percent confidence interval falls between _____ and _____.

═══════ **STOP** ═══════════════════════ **STOP** ═══════

37.10 and 62.90

---

41 $M = 70$, $S_M = 5$, $N = 21$, then the 95 percent confidence interval for the mean would fall between _____ and _____.

═══════ **STOP** ═══════════════════════ **STOP** ═══════

59.55 and 80.45

---

42 A sample of twenty-one persons is randomly selected from a population in which the mean IQ = 100. Vitamin Q is injected and IQ is measured later. We find $M = 108$, $S_M = 3$. If Vitamin Q did not raise IQ, using the t distribution to approximate z, 95 percent of the sample means would fall between _____ and _____.

════════ STOP ════════════════════════ STOP ════════

93.73 and 106.27 $[100 \pm 2.09 \times 3]$

───

43 If Vitamin Q *does not* influence IQ, then the probability that we would find a sample with $M = 108$ is less than _____.

════════ STOP ════════════════════════ STOP ════════

.05 (5 percent)

───

44 Since the probability that we would select a sample by chance with $M = 108$, is less than .05, we must _____ the hypothesis that Vitamin Q does not influence IQ.

════════ STOP ════════════════════════ STOP ════════

reject

───

45 If we solve for t in the formula $\mu = M \pm (t)(S_M)$ then $t = (M - \mu)/S_M$. In the last example $M = 108$, $S_M = 3$, and we tested the hypothesis that $\mu = 100$, $t =$ _____.

════════ STOP ════════════════════════ STOP ════════

$(108 - 100)/3 = 2.67$

───

46 $t = 2.67$ is larger than tabled value of $t =$ _____ (df $= 20$), which includes 95 percent of the area.

════════ STOP ════════════════════════ STOP ════════

2.09

───

47 In other words, any time $t = (M - \mu)/S_M$ is equal to or larger than 2.09 (df $= 20$) then we _____ the hypothesis that the true mean is equal to μ.

════════ STOP ════════════════════════ STOP ════════

reject

───

48 $t = (M - \mu)/S_M$, we wish to test the hypothesis that $\mu = 200$. $M = 180$, $S_M = 4$, $N = 3$, $t =$ _____.

════════ STOP ════════════════════════ STOP ════════

5.00

───

49 We wish to test the hypothesis H: $\mu = 200$. $M = 180$, $S_M = 4$, $N = 3$, and $t = 5.00$. How large must the value of t be with df $= 2$ before we reject H: $\mu = 200$? _____ .

═══ STOP ═══════════════════════ STOP ═══

$$t = 4.30$$

---

50 $M = 180$, $S_M = 4$, $N = 3$, $t = 5.00$, we wish to test H: $\mu = 200$. If $\mu = 200$ (and $S_M = 4$, df $= 2$) 95 percent of the sample means would fall within the area $\mu \pm (S_M)(4.30)$. Since obtained value of t is _____ than _____ we must _____ H: $\mu = 200$.

═══ STOP ═══════════════════════ STOP ═══

larger, 4.30, reject

---

51 We have noted many times that any time the probability of an event is less than _____ percent we _____ the hypothesis and anytime the probability is greater than _____ percent we ____ ____ ____ the hypothesis.

═══ STOP ═══════════════════════ STOP ═══

5, reject, 5, do not reject

---

52 If we are interested in testing hypotheses we will not be interested in areas under the t distribution except those which indicate the points within which _____ percent of the area falls.

═══ STOP ═══════════════════════ STOP ═══

95

---

53 In other words if we know these limits we know that only _____ percent falls outside by chance.

═══ STOP ═══════════════════════ STOP ═══

5

---

54 But when the probability is .05 or less we _____ the hypothesis of chance occurrence.

═══ STOP ═══════════════════════ STOP ═══

reject

---

55 If we are only interested in rejecting or not rejecting hypotheses and since this decision is dependent upon the number of df, then if we had a table showing only those points beyond which 5 percent of the area fell, we could decide to (reject/not reject) _____ for all possible df.

═══════ STOP ═══════════════════════ STOP ═══════

reject

---

56 Look at Appendix C. The column on the left is labeled _____ and the column in the middle is labeled _____.

═══════ STOP ═══════════════════════ STOP ═══════

"df," ".05"

---

57 Therefore, if df = 20, we would go down the left column to df = 20, and then go across the row to the middle column score of _____. Thus a t of _____ or larger is required to reject the hypothesis of chance occurrence.

═══════ STOP ═══════════════════════ STOP ═══════

2.09; 2.09

---

58 If $N = 30$, a t of _____ or larger is required to reject the null hypothesis.

═══════ STOP ═══════════════════════ STOP ═══════

2.05

---

59 If $N = 8$, a t of _____ or larger is required to reject.

═══════ STOP ═══════════════════════ STOP ═══════

2.37

---

60 Suppose we wished to reject the null hypothesis incorrectly only 1 percent of the time. If $N = 15$, then a t of _____ would be required to reject.

═══════ STOP ═══════════════════════ STOP ═══════

2.98

---

61 Suppose $S_M = 3$, $M = 80$, $N = 16$, and we wish to test H: $\mu = 72$. $t = (M - \mu)/S_M =$ _____

═══════ STOP ═══════════════════════ STOP ═══════

$(80 - 72)/3 = 2.67$

---

62 $M = 80$, $S_M = 3$, $N = 16$, $t = 2.67$, and we are testing H: $\mu = 72$. A t of _____ or larger is required to reject the hypothesis.

═══════ STOP ══════════════════════════ STOP ═══════

2.13

63 $M = 80$, $S_M = 3$, $N = 16$, df $= 15$, $t = 2.67$ and we are testing H: $\mu = 72$. A t of 2.13 or larger is required to reject the hypothesis. Therefore we _____ H: $\mu = 72$.

═══════ STOP ══════════════════════════ STOP ═══════

reject

64   With df $= 15$, we reject H: $\mu = 72$ because our obtained t [the t obtained from the fraction $(M - \mu)/S_M$] is larger than _____.

═══════ STOP ══════════════════════════ STOP ═══════

2.13 (the tabled t)

65   Suppose we take a sample of nine cases and find $M = 300$, $S = 45$; we wish to test H: $\mu = 335$, $S_M =$ _____.

═══════ STOP ══════════════════════════ STOP ═══════

15

66 $M = 300$, $S = 45$, $N = 9$, $S_M = 15$, and we wish to test H: $\mu = 335$. $t = (M - \mu)/S_M =$ _____.

═══════ STOP ══════════════════════════ STOP ═══════

2.33 (ignoring the sign)

67   With $N = 9$, a t of _____ or larger will occur 5 percent of the time by chance.

═══════ STOP ══════════════════════════ STOP ═══════

2.31

68   Our obtained $t = 2.33$ and a t of 2.31 or larger will occur 5 percent of the time. What do we do with the hypothesis that $\mu = 335$? _____.

═══════ STOP ══════════════════════════ STOP ═══════

Reject

69

$$t = \frac{M - \mu}{(?)}$$

════ STOP ════════════════════════════ STOP ════

$$S_M$$

70

$$t = \frac{(?) - \mu}{S_M}$$

════ STOP ════════════════════════════ STOP ════

$$M$$

71

$$t = \underline{\hspace{1.5cm}}$$

════ STOP ════════════════════════════ STOP ════

$$\frac{M - \mu}{S_M}$$

72

$$z = \frac{M - \mu}{(?)}$$

════ STOP ════════════════════════════ STOP ════

$$\sigma_M$$

---

73 The only difference between the equations for t and z is that _____ is the denominator for t while _____ is the denominator for z.

════ STOP ════════════════════════════ STOP ════

$$S_M, \sigma_M$$

---

74 $M = 40$, $S = 5$, $N = 16$, we wish to test H: $\mu = 37$. $t = $ _____.

════ STOP ════════════════════════════ STOP ════

$$(40 - 37)/1.25 = 2.40$$

---

75 $M = 40$, $S = 5$, $N = 16$, $t = 2.40$, what do we do with the hypothesis: $\mu = 37$? _____.

════ STOP ════════════════════════════ STOP ════

Reject

---

76 $N = 20$, $t = 2.13$, what do we do with the hypothesis? _____.

════ STOP ════════════════════════════ STOP ════

Reject

77 $N = 10$, $t = 2.13$, what do we do with the hypothesis? _____.

═══════ STOP ═══════════════════════════ STOP ═══════

Do not reject

---

78 We have the following scores: 1, 1, 2, 3, 3. We wish to test H: $\mu = 0$. $\sum X =$ _____, $\sum X^2 =$ _____.

═══════ STOP ═══════════════════════════ STOP ═══════

10, 24

---

79 With scores 1, 1, 2, 3, 3, $\sum X = 10$, $\sum X^2 = 24$, $M =$ _____.

═══════ STOP ═══════════════════════════ STOP ═══════

2

---

80 With scores 1, 1, 2, 3, 3, $\sum X = 10$, $\sum X^2 = 24$, $M = 2.00$, $\sum x^2 = \sum X^2 - (\sum X)^2/N =$ _____.

═══════ STOP ═══════════════════════════ STOP ═══════

4

---

81 With scores 1, 1, 2, 3, 3, $\sum x^2 = 4$, $S^2 =$ _____, $S =$ _____.

═══════ STOP ═══════════════════════════ STOP ═══════

1, 1

---

82 $M = 2.00$, $S = 1.00$, $N = 5$, and we wish to test H: $\mu = 0$. S_M _____.

═══════ STOP ═══════════════════════════ STOP ═══════

$1/\sqrt{5} = 0.45$

---

83 $M = 2.00$, $S = 1.00$, $S_M = 0.45$, $N = 5$, and we wish to test H: $\mu = 0$. $t =$ _____.

═══════ STOP ═══════════════════════════ STOP ═══════

4.44

---

84 $M = 2.00$, $S = 1.00$, $S_M = 0.45$, $N = 5$, and $t = 4.44$, can we reasonably assume $\mu = 0$? _____.

═══════ STOP ═══════════════════════════ STOP ═══════

No

---

85   Stated in statistical terms we _____ the hypothesis.

===== STOP ===================================== STOP =====

reject

---

86   If $M = 2.00$, $S_M = 0.45$, $N = 5$, and we wished to test H: $\mu = 1.00$, then $t =$ _____ and we _____ the hypothesis.

===== STOP ===================================== STOP =====

2.22, do not reject

---

87   Suppose we had scores 1, 5, 1, 1, and we wished to test H: $\mu = 0$. $\sum X =$ _____, $M =$ _____, $\sum X^2 =$ _____.

===== STOP ===================================== STOP =====

8, 2, 28

---

88   With scores of 1, 5, 1, 1, we wish to test H: $\mu = 0$, $\sum X = 8$, $M = 2.00$, $\sum X^2 = 28$. $S =$ _____, $S_M =$ _____.

===== STOP ===================================== STOP =====

2.00, 1.00

---

89   With scores of 1, 5, 1, 1, we wish to test H: $\mu = 0$, $M = 2.00$, $S_M = 1.00$. $t =$ _____ and we _____ the hypothesis.

===== STOP ===================================== STOP =====

2.00, do not reject

---

90   Given scores 0, 2, 5, 5. We wish to test H: $\mu = 0$. $M =$ _____, $S_M =$ _____, $t =$ _____.

===== STOP ===================================== STOP =====

3.00, 1.23, 2.44

---

91   With scores 0, 2, 5, 5, the value of t required to reject H: $\mu = 0$ is equal to _____. We _____ the hypothesis.

===== STOP ===================================== STOP =====

3.18; do not reject

---

92   Four students were given a test and got scores of 2, 2, 4, 8. S_M = _____ and the 95 percent confidence interval for the mean is established from _____ _____ _____.

════════ STOP ═══════════════════════ STOP ════════

1.41, −0.48 to 8.48

───

93   The 95 percent confidence interval for the mean is from −0.48 to 8.48. H: μ = 0.00. You (reject/do not reject) _____.

════════ STOP ═══════════════════════ STOP ════════

do not reject

───

94   Four other students were given a test and they had scores of 3, 3, 9, 9. S_M = _____.

════════ STOP ═══════════════════════ STOP ════════

1.73

───

95   With scores 3, 3, 9, 9, S_M = 1.73. Test H: μ = 0. t = _____ and you (reject/do not reject) _____.

════════ STOP ═══════════════════════ STOP ════════

3.47, reject

───

Eleven

The F Test

Comparison of Variances

A newly appointed judge once complained to one of his more experienced colleagues about the difficulty of impartiality in the administration of the law. The older judge smiled and said that did not bother him a bit. The defendants he did not like were jailed for a long time, while those he did like were given suspended sentences. "I figure it all evens out in the end," he said.

Just as the "Hanging Judge" gives out longer sentences than do his colleagues, the older judge mentioned in the anecdote above is more variable in his assignment of sentences. In giving some defendants many years in jail, and others few if any time in jail for the same crimes, the variance (S^2) of his sentences is greater than those of his younger colleagues.

In the present chapter we will look at those research situations where the variance of one group of scores is compared with the variance of another group of scores. If one sample variance is much larger than the other, then we will reject the hypothesis that the population variances from which the sample variances were drawn are equal. How much is "much larger?" Most of the remainder of this chapter is devoted to deciding that. Before we go on, let's consider another example.

A sociologist interested in the effects of diet upon growth measured the heights of two samples of adult males. One group was randomly selected from a population which had grown up in poverty-level surroundings, while the second group, also randomly selected, had grown up in a middle-class environment. Her hypothesis was that the variance of the heights of the lower-class group would be greater because the growth of some members of this group would be impaired by poor diet. For the 25 members of the lower-class group, $M = 69$, $S^2 = 6$; for the 66 members of the middle-class group, $M = 70$, $S^2 = 4$. Do the variances of the two groups differ?

How might these data be evaluated? Let us outline briefly the theory of the chi-square and F distributions. The test able to evaluate differences between variances is derived from these distributions.

CHI SQUARE

The distribution of sample variances is closely related to a distribution called the *chi square*. The shape of the chi-square distribution, like the *t* distribution, changes with changes in the number of degrees of freedom (df), but, unlike the *t* distribution, the chi-square distribution tends to be positively skewed (at least when *N* is small). Suppose you take a sample of 20 scores from some normally distributed population and find the variance of this sample; take another 20 scores, find the variance of this sample; take another 20 and find the variance, and so on.

Now, if you divide each sample variance by the true variance of that population (that is, S^2/σ^2), then multiply that fraction by $N - 1$ (one less than the sample size; in this case, $20 - 1 = 19$), and then plot the distribution of these scores, this will yield the chi-square distribution. This same procedure is expressed as

$$\chi^2 = \frac{(N - 1)S^2}{\sigma^2}$$

An interesting characteristic of the chi-square distribution is that the mean of the distribution is equal to the number of df associated with it. In our example, df $= N - 1 = 19$, and the mean of the distribution is 19. A graph of the chi-square distribution for three different df is presented in Figure 11-1. The three chi-square distributions have df of 3, 5, and

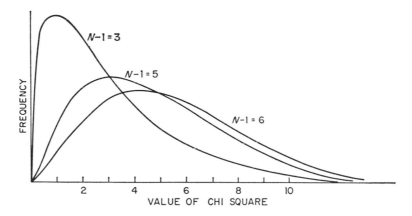

Figure 11-1   *The chi-square distribution for df = 3, 5, 6. The means of the three distributions are equal to 3, 5, and 6, respectively.*

6, and they also have means of 3, 5, and 6. We will go into the application of the chi-square distribution to statistical questions in much greater detail in Chapter Fifteen. For now, it is enough to remember that the chi-square distribution is closely related to the distribution of sample variances.

If we divide a chi-square distribution by the number of df associated with it, we would have, as for the distribution of S^2, another distribution closely related to the chi-square distribution:

$$\frac{\chi^2}{df} = \frac{(N-1)S^2}{\sigma^2} \times \frac{1}{N-1} = \frac{S^2}{\sigma^2}$$

In words, if we divide a chi-square distribution by its df (that is, $N-1$), we have a distribution of sample variances, each divided by the true variance. The mean of the S^2/σ^2 distribution is equal to 1.00. This should, of course, not be too surprising, since the mean of the sample variances is equal to the true variance. That is, the mean $S^2 = \sigma^2$. In actual practice we would not know what the true variance is, but the fact that we can divide by σ^2 will come in handy later on.

Suppose that we have two populations, X and Y. For each population we take samples and find their variances. For each of the variances taken from the X population we divide by the true variance of X and for each of the Y variances we divide by the true variance of Y. Now we would have two S^2/σ^2 distributions, each with a mean of 1.00. The F distribution is derived from the ratio of these two distributions.

THE *F* DISTRIBUTION

Suppose you take a sample from the X population, find the variance, and divide by the true X variance; then you take a sample from the Y population, find the variance, and divide this variance by the true Y variance. Finally, you divide one into the other. This fraction is shown as

$$\frac{\dfrac{S_x^2}{\sigma_x^2}}{\dfrac{S_y^2}{\sigma_y^2}}$$

Since both the X and Y sample variances will vary from sample to sample, we can be sure that the ratio above will itself vary from sample to sample. However, the average value of the ratio will also equal 1.00.[1] The sampling distribution of these ratios is the F distribution, and its shape depends on the number of df in both the numerator and denominator.

As you may have guessed, the chi-square and F distributions are closely related. In fact, we can define one in terms of the other. That is,

$$F = \frac{\dfrac{\chi_x^2}{df_x}}{\dfrac{\chi_y^2}{df_y}}$$

---

[1] We say *average* because with the F distribution the various measures of central tendency all cluster around 1.00, but only rarely will any exactly equal 1.00. For example, the median = 1.00 only when the df of the numerator and denominator are equal. The mean of the F distribution is equal to $df_2/(df_2-2)$, where df_2 refers to the df in the denominator. Even so, the central tendency of the F distribution is sufficiently close to 1.00 for us to say for purposes of exposition that the average is equal to 1.00.

The *F* distribution is defined as the ratio of two independent chi-square distributions, each divided by their respective numbers of df. This definition is much more general than would ever be needed for our purposes. All we would need to remember is that an *F* distribution is generated when we divide one S^2/σ^2 ratio into another, repeating this process until we have a frequency distribution of such ratios.

Now suppose the true variance of $X(\sigma_x{}^2)$ equals the true variance of $Y(\sigma_y{}^2)$. If the two true variances are equal, then the true variances cancel in the ratio and the equation can be simplified:

$$\frac{\dfrac{S_x{}^2}{\sigma_x{}^2}}{\dfrac{S_y{}^2}{\sigma_y{}^2}} = \frac{S_x{}^2}{S_y{}^2}$$

Thus, if the true variance of X is equal to the true variance of Y, then $S_x{}^2/S_y{}^2$ is F distributed with an average of 1.00. If the true variances are not equal, the distribution of the ratios of $S_x{}^2/S_y{}^2$ will still be F distributed, but the average will not equal 1.00. In fact, the average F will be equal to the ratio of the two true variances. For example, if $\sigma_x{}^2 = 100$ and $\sigma_y{}^2 = 25$, then the average F will equal 4.00.

When $\sigma_x{}^2 = \sigma_y{}^2$, then the ratio $S_x{}^2/S_y{}^2$ generates an F distribution with an average of 1.00. We will call this distribution the *tabled F distribution* because it is this F distribution which is represented in Appendix D. The tabled F distribution should be distinguished from all other F distributions that would be encountered when $\sigma_x{}^2 \neq \sigma_y{}^2$. These other distributions of $S_x/S_y{}^2$, which occur when $\sigma_x{}^2 \neq \sigma_y{}^2$ are still F distributions but are not tabled in Appendix D. The distinction between the tabled F and other Fs, then, is whether or not the average value of $S_x{}^2/S_y{}^2 = 1.00$. If the average is 1.00 then the F distribution is tabled in Appendix D, but if the average value of $S_x{}^2/S_y{}^2$ does not equal 1.00, then the generated F distribution is not presented in Appendix D.

For practical reasons, interest has been focused primarily on F distributions that have an average of 1.00 (that is, the tabled F distribution). Such a distribution is shown in Figure 11-2, and is based on 9 and 9 df (9 df in the numerator and 9 df in the denominator). From the figure it is apparent that the average F equals 1.00 and that 5 percent of the area falls to the right of 3.18. Thus, if we could select an F ratio at random from the distribution, the probability is equal to 0.05 that an F with a value greater than 3.18 will be selected.

Once more let us look at the theory of the F distribution. We have said that the distribution of sample variances divided by the population variance $(S_x{}^2/\sigma_x{}^2)$ or $(S_y{}^2/\sigma_y{}^2)$ will have a mean (μ) of 1.00. Furthermore, we noted that the ratio of these two ratios,

$$F = \frac{\dfrac{S_x{}^2}{\sigma_x{}^2}}{\dfrac{S_y{}^2}{\sigma_y{}^2}}$$

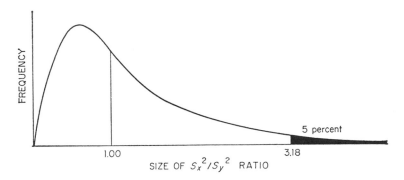

Figure 11-2   *Frequency distribution of S_x^2/S_y^2 with 9 df and 9 df, and when $\sigma_x^2 = \sigma_y^2$.*

will be F distributed with an average of 1.00. Now if the variances of the two populations are equal $(\sigma_x^2 = \sigma_y^2)$, then the F ratio may be simplified by dropping the parameters from the equation. The equation then becomes

$$F = \frac{S_x^2}{S_y^2}$$

This new equation results in exactly the same distribution with the same average F (1.00) as before if $\sigma_x^2 = \sigma_y^2$, and the probability of randomly selecting an F larger than 3.18 (with 9 and 9 df) is still equal to 0.05.

Suppose we selected two samples of ten scores each, computed the variances, and found that the ratio of the two variances was larger than 3.18. If the true variances of the two populations from which the samples were drawn were equal, then the probability of selecting at random an F larger than 3.18 would be less than 0.05.

You should remember from the discussion in Chapter Nine that we reject the hypothesis of chance occurrence if the probability of an event is less than 0.05. Thus, if we get an F larger than 3.18 (with 9 and 9 df), we would reject the hypothesis. What do we mean when we reject the hypothesis of chance occurrence, and what can we conclude when we do reject that hypothesis? The idea that only 5 percent of the F values will be larger than 3.18 is based on the assumption that the true variances are equal, and this assumption is called the *null hypothesis*. If we get an F larger than 3.18, the chances that such an event will occur are so small $(Pr < 0.05)$ that we reject the null hypothesis and conclude that the true variances are not equal.

In all our previous discussions we have indicated that rarely do we know the value of a parameter. We can test the hypothesis that the parameters $(\sigma_x^2$ and $\sigma_y^2)$ are equal (null hypothesis) simply by following the procedure outlined previously; that is, find the ratio of two sample variances (S_x^2/S_y^2). Next, find the number of df in the numerator and the number of df in the denominator. Since the number of df in both the numerator and denominator determine the shape of the F distribution, we must next refer to an F table. The values in the F table are taken from F distributions with an average of 1.00, and will tell you the size of the F ratio that is exceeded in value by exactly 5 percent (or 1 percent) of the F values. When the F you compute is larger than the tabled value of F, you reject the null hypothesis and conclude that the true variances of the two samples are not equal.

Now, getting back to the effects of diet on the variability of heights, recall that the variance of the 25 members of the poverty-level sample was 6.00, while the variance of the 66 members of the middle-class group was 4.00.

In this case,

$$F = \frac{S_x^2}{S_y^2} = \frac{6}{4} = 1.50$$

With 24 and 65 df it can be seen from Appendix D that an F of 1.68 is required to reject the null hypothesis at the 0.05 level. With an obtained F of 1.50 the null hypothesis cannot be rejected, and the hypothesis of equivalent variances of heights of the two groups cannot be rejected.

Consider another example dealing with a different kind of height. You may not have noticed, but the heights of adult books seem to be more standardized than those of children's books—or at least that is the way it appeared to a librarian. To check on his hunch, he selected a sample of nine adult books and ten children's books and measured their respective heights. These data (in inches) are tabulated below.

Children's books (X)	Adult books (Y)
4	8
10	7
5	10
9	8
9	9
5	8
7	7
7	7
6	8
8	
M_x 7.00	M_y 8.00
S_x^2 4.00	S_y^2 1.00
N_x 10	N_y 9

The two samples were selected randomly. One sample has 9 df and the other has 8 df. The two variances were found to equal 4.00 and 1.00, and the F ratio between the two was found to equal 4.00.

We do not know the true variances of the adults' or children's books. However, we start with the assumption that the two true variances are equal (null hypothesis). If the null hypothesis is true, then the ratios of sample variances result in an F distribution with an average of 1.00, and given the 9 and 8 df, the probability is 0.05 that an F greater than 3.39 will be randomly selected from that distribution. The obtained value of F in our experiment is 4.00, which is larger than 3.39. Since the probability is less than 0.05 that we would have randomly selected an F this large, we reject the null hypothesis: the hypothesis that $\sigma_x{}^2 = \sigma_y{}^2$. By rejecting it, we conclude that the average of the F distribution is not equal to 1.00; therefore the two true variances are not equal, since only when they are equal can the F distribution $(S_x{}^2/S_y{}^2)$ have an average equal to 1.00. In short, we always reject the hypothesis that the two true variances are equal (the null hypothesis) if the value of the obtained F is greater than the value obtained from the table in Appendix D.

READING APPENDIX D

The preceding discussion may have convinced you that 5 percent of the Fs in all distributions fall above 3.39. This, of course, is not the case. The shape of each F distribution is determined by the number of df in the numerator and the number of df in the denominator. Since the F distribution is a family of curves (like the t distribution), it is obvious that a table is necessary to find the point beyond which 5 percent of the Fs will fall (when the null hypothesis is true) for each combination of dfs. In Appendix D (the F table), the columns are headed by the number of df in the numerator, and the rows are headed by the number of df in the denominator. For example, if you wished to find the point beyond which 5 percent of the F values would fall with 10 and 20 df (10 df in the numerator and 20 df in the denominator), you would go to the column labeled "10" and then down that column until you came to the row labeled "20." The tabled value at that point is 2.35, which is the value exceeded by 5 percent of the Fs. To be sure you understand how the table works, prove to yourself that the tabled value for 2 and 10 df is 4.10.

ASSUMPTIONS OF F TEST FOR VARIANCE $(F = Sx^2/Sy^2)$

1. Random sampling.
2. All scores in each sample are independent of one another.
3. Sample comes from a normal distribution.
4. Measurement scale is at least an interval scale.

ALPHA AND BETA ERRORS

Let us return once again to the experiment on the effects of diet upon the variability of heights. We found that the variance of the heights of the lower-class group, while larger than the variance of the middle-class group, was not significantly larger. That is, the ratio of the two variances was not large enough to permit us to reject the null hypothesis, the hypothesis that the two true variances are equal. Notice that these results fail to confirm the hypothesis of the experimenter who conducted the study.

You will recall our discussion in Chapter Seven concerning statistical decision making. No matter what decision we make, there is some probability that we will be wrong. Table 11-1 is another decision table based on the diet experiment.

Table 11-1   Decision table

	Heights are more variable	Heights are not more variable
You decide heights are more variable.	You are correct.	Alpha error
You decide heights are not more variable.	Beta error	You are correct.

As was the case in Chapter Seven, there are two kinds of errors we can make:

(1) *Alpha error:* The null hypothesis is true, but you reject it. In Table 11-1 the null hypothesis is that the true variance of the lower-class sample was the same as the true variance of the middle-class sample and that the observed difference between the two was due to chance. An alpha error will occur when we compute an F larger than the tabled value of F, even though the null hypothesis is true. It should be apparent that this type of error will be made 5 percent of the time when the null hypothesis is true.

(2) *Beta error:* The null hypothesis is false, but you decide not to reject it. In the present experiment, the lower-class population may be more variable in their heights than the middle-class population. If we get an F smaller than the tabled value of F, we would not reject the null hypothesis. If we failed to reject the null hypothesis when it was in fact false, we would have committed a beta error.

Let us note once again that these errors are related in that as we reduce the probability of one type of error, we increase the probability of the second. At the present time the usual procedure is to set alpha equal to 0.05, as we have done. This procedure results, statisticians feel, in a reasonable balance between the probability of committing alpha and beta errors.

POWER

We noted that a beta error is committed if the null hypothesis is false and it is not rejected. A concept closely related to beta is that of *power*. The power of a test is the probability that a false null hypothesis will be rejected. The concept of power as well as of alpha and beta can be best illustrated by an example.

Suppose you randomly selected a large number of samples from each of two populations where $\sigma_x^2 = \sigma_y^2$ (that is, $\sigma_x^2/\sigma_y^2 = 1.00$). If you computed the F ratios between successive pairs of samples, you would find that 5 percent of your F values would be larger than the tabled value of F, and you would reject the null hypothesis 5 percent of the time. Next suppose you selected a large number of samples from each of two populations where the ratio $\sigma_x^2/\sigma_y^2 = 2.00$. If you computed the F ratios between successive pairs of samples, you would find that you would reject the (false) null hypothesis more than 5 percent of the time. If you sampled from populations where $\sigma_x^2/\sigma_y^2 = 3.00$, you would reject the false null hypothesis an even greater proportion of the time.

This procedure is illustrated in Figure 11-3. The x axis is the value of the ratio σ_x^2/σ_y^2 and the y axis is the probability that the null hypothesis will be rejected. You can see that when $\sigma_x^2/\sigma_y^2 = 1.00$, the probability of rejecting the (true) null hypothesis is 0.05. This probability increases as the value of the σ_x^2/σ_y^2 ratio increases. When the value of the σ_x^2/σ_y^2 ratio equals 2.00, the probability in this case of rejecting the null hypothesis is 0.20. That is, only 20 percent of the sample S_x^2/S_y^2 values will be larger than the tabled value, and 80 percent of the time a beta error will be made; that is, the false null hypothesis is not rejected. As the value of σ_x^2/σ_y^2 increases, the probability of a beta error decreases and power increases until it is great at large values of σ_x^2/σ_y^2, and there is very little chance that the null hypothesis will not be rejected.

The power function illustrated in Figure 11-3 depends upon a large number of variables. For example, we noted before that alpha and beta

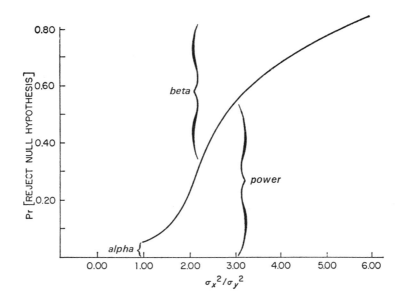

Figure 11-3   *The probability of rejecting the null hypothesis is shown as a function of the value of the ratio of the true variances. When $\sigma_x^2/\sigma_y^2 = 1$, the null hypothesis is true and alpha $= .05$, but when $\sigma_x^2/\sigma_y^2 \neq 1$ beta and power are as shown above.*

errors are related, so that as one increases the other decreases. If alpha were set at 0.01, then beta would increase for all other values of $\sigma_x{}^2/\sigma_y{}^2$.

The ideal statistical test would be one where alpha could be set very low and yet the power of the test would be very high. Even though the F test is extremely powerful compared with other statistical tests, you can see from Figure 11-3 that errors, both alpha and beta, are often made. Thus, the decisions that are made will always have some probability of being incorrect because, in the application of statistical methods, we will never know the real value of the ratio $\sigma_x{}^2/\sigma_y{}^2$. All we can do is attempt to adopt a research strategy that will reduce the effects of making a wrong decision. One common practice here is to set the alpha level at 0.05. This is something of a compromise between the statistician's fears of falsely rejecting a true hypothesis and of failing to reject a false one. The former error is an alpha error and the latter is a beta error.

TESTING VARIANCE RATIOS WITHOUT A PRIOR HYPOTHESIS

The F test may be used to test the null hypothesis even when you have no particular reason to expect one or the other of the two sample variances to be larger. In the preceding example we put the lower-class variance in the numerator, and would have done so no matter what the sizes of the two variances had been. We did this because we had a particular hypothesis in mind.

When we wish simply to test whether or not two variances have been drawn from the same population, and have no expectations about the result, we put the larger of the two sample variances in the numerator and use the F table as usual. Since the value of F under these circumstances can never be less than 1.00, we are precluding the use of the part of the F distribution below 1.00. The effect of this is to double the probability of getting an F larger than the tabled value. Thus, if we had had no reason to look for a greater variability in the children's books in the previous example, and had put the children's variance in the numerator simply because it was larger, the F value tabled for the 5 percent level of significance would actually represent the 10 percent level of significance and, similarly, the probability for the F value at the 1 percent level would really be 0.02.

Despite this change in the probability levels when no prior hypotheses are being entertained, the F table is generally used in the same manner to make decisions. It is important, however, to be aware of the increased risk of making an alpha error whenever the larger variance is put in the numerator simply because it is larger.

SUMMARY

The F test is an extremely useful test to the statistician. The wide range of uses for the F test has not been fully illustrated in the present chapter. Instead, we have generally restricted our discussion to a development of the F statistic. While a comparison of variances is occasionally of interest to the research worker, it is not often used. The F test would be a relatively unused test if its use were restricted to the comparison of variances. Where the F test is of great importance and finds enormous use is in the comparison of means. The comparison of means, called the *analysis of variance*, is another use to which the F test can be put, and we will devote Chapters Thirteen and Fourteen to the development of this extremely important test.

PROBLEMS

1. $\chi^2 = \dfrac{(N-1)S^2}{\sigma^2}$; then $\chi^2/df = $ _____.

2. Given the information in Problem 1, $F = \dfrac{\chi_1^2/df_1}{\chi_2^2/df_2} = $ _____.

3. The ratio S_1^2/S_2^2 is distributed according to the tabled F distribution only if _____.

4. Similarly an F distribution which is not tabled in Appendix D is generated if _____.

5. Two samples are taken from two different populations and the variance is found for each. For population 1, the sample variance is 20 and the true variance is 40; for population 2, $S^2 = 10$ and $\sigma^2 = 5$. If the experiment were repeated over and over, the average value of the F ratio would be equal to what?

6. E predicts that the true variance of X is greater than the true variance of Y. However, $\sigma_x^2 = \sigma_y^2$. E takes a random sample of 15 cases from population X and a similar size sample from population Y. In this case, $\Pr = 0.50$ that the obtained F is larger than what value? Also, $\Pr = 0.05$ that the obtained F is larger than what other value?

7. $\sigma_x^2 = 3.18$, $\sigma_y^2 = 1.00$. E predicts $\sigma_x^2 > \sigma_y^2$. E selects two random samples of ten scores each. What is the probability that the null hypothesis will be rejected?

8. E predicts $\sigma_x^2 > \sigma_y^2$, $\Sigma X = 30$, $\Sigma X^2 = 198$, $N_x = 10$, $\Sigma Y = 24$, $\Sigma Y^2 = 116$, $N_y = 6$. Is E's prediction confirmed?

9. E predicts $\sigma_x^2 > \sigma_y^2$. Can you help him make a decision?

X	Y
3	2
3	2
5	3
9	4
	4

10. Bubbly Cola, angered by the claims of Bouncy Cola, says that Bubbly Cola is not only as good as Bouncy, but also that its quality is much more consistent. An eager experimenter set out to test the consistency claim of Bubbly and randomly sampled 25 bottles of Bouncy and another 25 of Bubbly. Drink-

ing at a rate to control for possible order bias and lack of independence, he rates each bottle on a 1–10 (bad–good) scale. The results are shown below. Are Bubbly Cola's claims justified?

Bubbly Cola	Bouncy Cola
$M = 6.50$	$M = 7.00$
$S = 2.00$	$S = 3.00$

11. In each of the following problems compute $F(S_x^2/S_y^2)$ and indicate whether the null hypothesis should be rejected or not rejected.
    (a) $S_x^2 = 50$, $S_y^2 = 15$, $N_x = 11$, $N_y = 13$.
    (b) $S_x^2 = 40$, $S_y^2 = 15$, $N_x = 7$, $N_y = 15$.
    (c) $S_x^2 = 24$, $S_y^2 = 8$, $N_x = 7$, $N_y = 13$.
12. A medical research worker has a group of 10 interns and a group of 20 nurses who estimate the number of heart beats per minute for a certain patient (each is allowed to hold the patient's wrist for 10 seconds). For the interns, the results were: mean = 75, standard deviation = 4.00. For the nurses, the results were: mean = 72, standard deviation = 2.00. What sensible hypothesis can you test in this experiment? Test it. What are your results and conclusions?
13. Let's look at those two judges mentioned at the beginning of the chapter. For a certain type of crime the mean number of years sentenced by the older judge ($N = 200$) was 10 with a standard deviation of 5.00; the younger judge has a mean of 12 ($N = 50$) and a standard deviation of 4.00. Is there any evidence to suggest that the older judge is more variable in his sentencing than the younger judge?

EXERCISES

1. Rap your fist on a desk or a wall. Count 10 seconds and rap again. Ask the people who heard you to write down their estimate of the time between knocks. For another group of people do the same thing. This time let 50 seconds elapse between knocks. Again ask those who heard you to write down their estimate of the time. Test the hypothesis that the variability of the 10-second guesses is equal to the variability of the 50-second guesses.
2. Select two random samples of five single-digit numbers from Appendix H. Multiply the numbers in *one* sample by 3. Test the hypothesis that the variability of the two samples is the same.

Special *F* table

With df = 1 and 20, the tabled *F* distribution will be approximately that listed below and will look approximately like Figure 11-4

F VALUE	PERCENT TO RIGHT
0.2	66
0.4	55
0.6	45
0.8	37
1.0	32
1.2	28
1.4	24
1.6	22
1.8	20
2.0	17
2.5	12
3.0	10
3.5	7
4.0	6
4.35	5
4.5	4
5.0	3
6.0	2
8.10	1

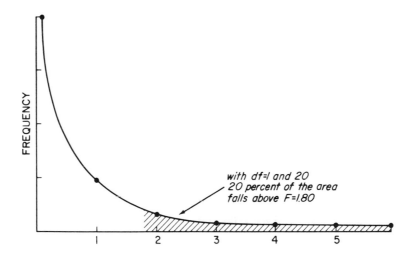

Figure 11-4

1 If we select sample variances (S^2) over and over from a population which has a true mean equal to μ and a true variance equal to σ^2, we would not expect every sample variance (S^2) to equal _____.

========= STOP ============================ STOP =========

$$\sigma^2$$

2 That is, S^2 will _____ around the true variance σ^2

========= STOP ============================ STOP =========

vary

3 If we select sample variances and plot them (with size of S^2 on the x axis and frequency on the y axis) we would have a _____ of sample variances.

========= STOP ============================ STOP =========

distribution

4 While sample means (Ms) are _____ distributed, sample variances have a differently shaped distribution.

========= STOP ============================ STOP =========

normally

5 The shape of the S^2 distribution tends to be positively skewed. This means that the sample variances tend to pile up at the _____ of the distribution.

========= STOP ============================ STOP =========

left

6 The distribution of S^2 is related to the chi-square distribution. χ^2 is the symbol for _____.

========= STOP ============================ STOP =========

chi square

7 When each S^2 in the distribution is multiplied by $(N-1)$ and divided by σ^2, the resulting distribution of $(N-1)S^2/\sigma^2$ has a mean equal to $(N-1)$ and is _____ distributed.

========= STOP ============================ STOP =========

chi-square

8 Like the t distribution, the χ^2 distribution is really a family or group of curves. And like the t distribution the shape of the χ^2 distribution depends on the _____ of df.

═══ STOP ═══════════════════════ STOP ═══

number

---

9 $\chi^2 = (N - 1)S^2/\sigma^2.$ If $N = 25$, the mean of the χ^2 distribution is equal to _____ and df = _____.

═══ STOP ═══════════════════════ STOP ═══

24, 24

10
$$\chi^2 = \frac{(?)S^2}{\sigma^2}$$

═══ STOP ═══════════════════════ STOP ═══

$(N - 1)$

11
$$\chi^2 = \frac{(N - 1)S^2}{(?)}$$

═══ STOP ═══════════════════════ STOP ═══

σ^2

12
$$\chi^2 = \text{_____}$$

═══ STOP ═══════════════════════ STOP ═══

$(N - 1)S^2/\sigma^2$

---

13 Suppose samples of size ten were randomly selected from some population and their variances computed. A χ^2 distribution based on these samples would have a mean equal to _____.

═══ STOP ═══════════════════════ STOP ═══

$N - 1 = 9$

---

14 The ratio S^2/σ^2 is related to the χ^2 distribution and has a mean of 1.00. If the ratios S^2/σ^2 are each multiplied by $(N-1)$, this results in a _____ _____ with a mean of _____.

═══ STOP ═══════════════════════ STOP ═══

χ^2 distribution, $(N - 1)$

15   We said that the distribution S^2/σ^2 is related to the χ^2 distribution and has a mean of _____ .

═══ STOP ═══════════════════════════════════ STOP ═══

1.00

---

16   The ratio of two independent χ^2 distributions, each divided by their respective df yields still another distribution: F. Like χ^2 and t, the F distribution is a family (or group) of _____ .

═══ STOP ═══════════════════════════════════ STOP ═══

curves (distributions)

---

17   And as is true with the t and χ^2 distributions, the shape of the F distribution is dependent upon the number of _____ .

═══ STOP ═══════════════════════════════════ STOP ═══

df

---

18 $F = \dfrac{\chi_1^2/df_1}{\chi_2^2/df_2}$ Note however that in the F distribution there are df in the numerator as well as in the _____ .

═══ STOP ═══════════════════════════════════ STOP ═══

denominator

---

19 $F = \dfrac{\chi_1^2/df_1}{\chi_2^2/df_2}$ and $\chi^2 = $ _____ .

═══ STOP ═══════════════════════════════════ STOP ═══

$(N-1)S^2/\sigma^2$

---

20   Thus

$$\chi^2/df = \frac{(N-1)S^2}{\sigma^2} \times \frac{1}{N-1}$$

which reduces to _____ .

═══ STOP ═══════════════════════════════════ STOP ═══

S^2/σ^2

21 $\chi^2/df = S^2/\sigma^2$. Thus

$$\frac{S_x^2}{\sigma_x^2} \Big/ \frac{S_y^2}{\sigma_y^2} = \frac{\chi_x^2/df_x}{\chi_y^2/df_y}$$

and is _____ distributed.

══════ STOP ═══════════════════════════ STOP ══════

F

22

$$F = \frac{S_x^2}{\sigma_x^2} \Big/ \frac{S_y^2}{\sigma_y^2}$$

If $\sigma_x^2 = \sigma_y^2$ the fraction reduces to $F =$ _____.

══════ STOP ═══════════════════════════ STOP ══════

S_x^2/S_y^2

---

23   Samples of size ten are randomly selected:

$$S^2/\sigma^2 = \chi^2/df. \qquad df = \underline{\hspace{1.5cm}}.$$

══════ STOP ═══════════════════════════ STOP ══════

$N - 1 = 9$

---

24   With $N = 25$ df = _____.

══════ STOP ═══════════════════════════ STOP ══════

24

---

25   Suppose we select ten scores and compute the variance (S_x^2) and then we select twenty scores and compute the variance (S_y^2), then the ratio S_x^2/S_y^2 is F distributed with _____ df in the numerator and _____ df in the denominator.

══════ STOP ═══════════════════════════ STOP ══════

9,19

---

26   Suppose $S_x^2 = 36$, $N = 12$; $S_y^2 = 9$, $N = 23$. Then the value of $F =$ _____ with _____ and _____ df. (For convenience we will always put S_x^2 in the numerator and S_y^2 in the denominator.)

══════ STOP ═══════════════════════════ STOP ══════

4.00, 11, 22

---

27 Suppose we have two sets of scores: 0, 1, 5, and 2, 5, 7, 2. We wish to compute F. For set 1 $\sum_{1}^{N} X = $ _____ and $\sum_{1}^{N} X^2 = $ _____.

═══ STOP ═══ ═══ STOP ═══

6, 26

---

28 Suppose we have scores 0, 1, 5 and 2, 5, 7, 2 and we wish to find F. For the first set $\sum_{1}^{N} X = 6$ and $\sum_{1}^{N} X^2 = 26$; for the second set $\sum_{1}^{N} X = $ _____ and $\sum_{1}^{N} X^2 = $ _____.

═══ STOP ═══ ═══ STOP ═══

16, 82

---

29 We have said $\sum_{1}^{N} x^2 = \sum_{1}^{N} X^2 - (\sum_{1}^{N} X)^2/N$. For set 1, with scores 0, 1, 5, $\sum_{1}^{N} X = 6$ and $\sum_{1}^{N} X^2 = 26$, find $\sum_{1}^{N} x^2$ _____.

═══ STOP ═══ ═══ STOP ═══

14

---

30 $\sum_{1}^{N} x^2 = \sum_{1}^{N} X^2 - (\sum_{1}^{N} X)^2/N$. For set 2 the scores are 2, 5, 7, 2; $\sum_{1}^{N} X = 16$, $\sum_{1}^{N} X^2 = 82$, find $\sum_{1}^{N} x^2$. _____.

═══ STOP ═══ ═══ STOP ═══

18

31
$$\sum_{1}^{N} x^2 = \sum_{1}^{N} X^2 - \frac{\left(\sum_{1}^{N} X\right)^2}{(?)}$$

═══ STOP ═══ ═══ STOP ═══

N

32
$$\sum_{1}^{N} x^2 = \sum_{1}^{N} X^2 - \frac{(?)}{N}$$

═══ STOP ═══ ═══ STOP ═══

$$\left(\sum_{1}^{N} X\right)^2$$

33

$$\sum_{1}^{N} x^2 = (?) - \frac{\left(\sum_{1}^{N} X \right)^2}{N}$$

══════ STOP ══════════════════════════ STOP ══════

$$\sum_{1}^{N} X^2$$

---

34   Thus $\sum_{1}^{N} x^2 = $ —————.

══════ STOP ══════════════════════════ STOP ══════

$$\sum_{1}^{N} X^2 - \left(\sum_{1}^{N} X \right)^2 / N$$

35

$$S^2 = \frac{\sum_{1}^{N} x^2}{N - 1}.$$

Thus for set 1 with scores 0, 1, 5, $\sum_{1}^{N} x^2 = 14$ and $S^2 = $ —————.

══════ STOP ══════════════════════════ STOP ══════

7

36

$$S^2 = \frac{\sum_{1}^{N} x^2}{N - 1}.$$

For set 2 with scores 2, 5, 7, 2, $\sum_{1}^{N} x^2 = 18$ and $S^2 = $ —————.

══════ STOP ══════════════════════════ STOP ══════

$18/3 = 6$

37

$$S^2 = \frac{(?)}{N - 1}$$

══════ STOP ══════════════════════════ STOP ══════

$$\sum_{1}^{N} x^2$$

---

38 With two different groups of scores we use S_x^2 and S_y^2 to identify these groups. With scores 0, 1, 5, $S_x^2 = 7$ and df = _____ ; with scores 2, 5, 7, 2, $S_y^2 = 6$ and df = _____ .

══════════ STOP ══════════ ══════════ STOP ══════════

2; 3

---

39 $S_x^2 = 7$, df = 2; $S_y^2 = 6$, df = 3; $F =$ _____ with _____ and _____ df.

══════════ STOP ══════════ ══════════ STOP ══════════

1.17, 2 and 3

---

40 Set 1 (X) scores: 0, 3, 7, 6; set 2 (Y) scores: 0, 1, 1. $S_x^2 =$ _____, $S_y^2 =$ _____.

══════════ STOP ══════════ ══════════ STOP ══════════

10, 0.33

---

41 $S_x^2 = 10$ ($N_x = 4$); $S_y^2 = 0.33$ ($N_y = 3$). $F =$ _____ with _____ and _____ df.

══════════ STOP ══════════ ══════════ STOP ══════════

30, 3, 2

---

42 $F = 30.00$ where $S_x^2 = 10$ and $S_y^2 = 0.33$. Do we know the true variance of S_x^2 or S_y^2? _____ .

══════════ STOP ══════════ ══════════ STOP ══════════

No

---

43 We have seen that if $\sigma_x^2 = \sigma_y^2$ then the ratio S_x^2/S_y^2 is F distributed. Similarly if σ_x^2 does not equal σ_y^2, the ratio S_x^2/S_y^2 is still _____ distributed.

══════════ STOP ══════════ ══════════ STOP ══════════

F

---

44 However, if $\sigma_X^2 \neq \sigma_Y^2$ then the F distribution obtained from the ratio S_X^2/S_Y^2 is not presented in Appendix D. The F distribution obtain when $\sigma_X^2 = \sigma_Y^2$ is called the _____ F distribution or simply the F distribution.

══════════ STOP ══════════ ══════════ STOP ══════════

tabled

---

45   Finally we said that the mean of S_x^2/σ_x^2 was equal to _____.

═══════ STOP ═══════════════════════════════════ STOP ═══════

$$1.00$$

---

46   Thus if we would divide the sample variances S_x^2 and S_y^2 by their respective true variances (σ_x^2 and σ_y^2) each ratio (fraction) would have a distribution with a mean of _____.

═══════ STOP ═══════════════════════════════════ STOP ═══════

$$1.00$$

---

47   Thus the ratio $\dfrac{S_x^2/\sigma_x^2}{S_y^2/\sigma_y^2}$ would be the ratio of two distributions each with a mean equal to 1.00. The resulting ratio (fraction) would be _____ distributed with an average of _____.

═══════ STOP ═══════════════════════════════════ STOP ═══════

$$F,\ 1.00$$

---

48   Normally σ_x^2 and σ_y^2 will be unknown. Hence it is obvious that the ratio S_x^2/σ_x^2 cannot be _____.

═══════ STOP ═══════════════════════════════════ STOP ═══════

computed (determined)

---

49   We have said before that the average (mean) value of S_x^2 would equal

_____.

═══════ STOP ═══════════════════════════════════ STOP ═══════

$$\sigma_x^2$$

---

50   If the average value of S_x^2 is σ_x^2 and the average value of S_y^2 is σ_y^2, then we would suspect that the average value of the ratio S_x^2/S_y^2 would be equal to the ratio _____.

═══════ STOP ═══════════════════════════════════ STOP ═══════

$$\sigma_x^2/\sigma_y^2$$

---

51 Thus if we sample from a population with a true variance of 1000 and find $S_x^2 = 800$ and then sample from a second population with a true variance of 100 and find $S_y^2 = 200$, the average value of the ratio S_x^2/S_y^2 would equal _____.

=== STOP === STOP ===

$$\sigma_x^2/\sigma_y^2 = 1000/100 = 10$$

52 And thus if we sample from two populations where $\sigma_x^2 = 500$ and $\sigma_y^2 = 100$ and find S_x^2 and S_y^2, the average value of the F ratio would be _____.

=== STOP === STOP ===

5

53 More generally the average value of any F ratio computed from populations with true variances of σ_x^2 and σ_y^2 would be equal to _____.

=== STOP === STOP ===

$$\sigma_x^2/\sigma_y^2$$

54 But if (and only if) the true variance of population X were equal to the true variance of the population Y then the average value of the F ratio would equal _____. This F is the _____ F distribution.

=== STOP === STOP ===

1.00; tabled

55 For example $\sigma_x^2 = 100$, $S_x^2 = 150$ and $\sigma_y^2 = 100$, $S_y^2 = 50$. The value of F computed for this experiment would equal _____ while the average value of F' would equal _____.

=== STOP === STOP ===

3.00, 1.00

56 $\sigma_x^2 = 50$, $S_x^2 = 100$; $\sigma_y^2 = 50$, $S_y^2 = 25$. $F =$ _____ and average $F =$ _____.

=== STOP === STOP ===

4.00, 1.00

57 $\sigma_x^2 = 40$, $S_x^2 = 50$; $\sigma_y^2 = 20$, $S_y^2 = 10$, $F = $ _____ and average $F = $ _____.

═══════ STOP ═══════ ═══════ STOP ═══════

5.00, 2.00

---

58   We have seen that the *shape* of the F distribution depends upon the degrees of freedom in the numerator and denominator. The average F depends upon _____ and _____.

═══════ STOP ═══════ ═══════ STOP ═══════

σ_x^2 and σ_y^2

---

59   We have said that the ratio of two independent χ^2 distributions divided by their respective df yields what kind of distribution? _____.

═══════ STOP ═══════ ═══════ STOP ═══════

F

---

60   Furthermore, we have seen the ratio S_x^2/σ_x^2 yields a distribution with a mean of _____.

═══════ STOP ═══════ ═══════ STOP ═══════

1.00

---

61   Finally, we have seen that $\dfrac{S_x^2/\sigma_x^2}{S_y^2/\sigma_y^2}$ yields an _____ distribution with an average of _____.

═══════ STOP ═══════ ═══════ STOP ═══════

F, 1.00

---

62   Consider the ratio $\dfrac{S_x^2/\sigma_x^2}{S_y^2/\sigma_y^2}$ and suppose $\sigma_x^2 = \sigma_y^2$. The ratio can be simplified merely by cancelling σ_x^2 and σ_y^2. The new ratio would be _____.

═══════ STOP ═══════ ═══════ STOP ═══════

S_x^2/S_y^2

---

63   Now, if $\sigma_x^2 = \sigma_y^2$ then S_x^2/S_y^2 would also be _____ distributed with average equal to _____.

═══════ STOP ═══════ ═══════ STOP ═══════

F, 1.00

---

64 Thus even if we do not know the values of σ_x^2 and σ_y^2, the F ratio S_x^2/S_y^2 will be F distributed with an average of 1.00 if (and only if) $\sigma_x^2 = $ _____.

═══ STOP ═══════════════════════ STOP ═══

$$\sigma_y^2$$

---

65 No matter how many df in the numerator or denominator, as long as $\sigma_x^2 = \sigma_y^2$ then S_x^2/S_y^2 is _____ distributed with average equal to _____.

═══ STOP ═══════════════════════ STOP ═══

$$F, \ 1.00$$

---

66 Remember, however, that even though S_x^2/S_y^2 is F distributed with average equal to 1 (if $\sigma_x^2 = \sigma_y^2$) the shape of the F distribution depends upon ____ ____ ____ _____ ____ _____.

═══ STOP ═══════════════════════ STOP ═══

df in the numerator and demoninator

---

67 Even if the average of the F ratios is equal to 1.00, since the shape of the F distribution is dependent upon the df in the numerator and the df in the denominator, there are an _____ number of F distributions.

═══ STOP ═══════════════════════ STOP ═══

infinite

---

68 One of these F distributions has an average $= 1.00$ with df $= 1$ and 20. The values for this distribution are presented in the *Special F table* facing the first page of the program. The table differs from the z table in that the percent values are the areas to the right (larger) of some point. For example _____ percent of the area lies to the right of an F of 1.20.

═══ STOP ═══════════════════════ STOP ═══

28

---

69 This means that when df $= 1$ and 20 and if the average of the F distribution is equal to 1.00 then 32 percent of the F values will be larger than _____.

═══ STOP ═══════════════════════ STOP ═══

1.00

---

70 Suppose you take two samples from the same population. $S_x^2 = 40.00$, with 1 df and $S_y^2 = 8.00$, with 20 df $F =$ _____.

≡≡≡≡≡≡≡ STOP ≡≡≡≡≡≡≡≡≡≡≡≡≡≡≡≡≡≡≡≡≡≡≡ STOP ≡≡≡≡≡≡≡

5.00

---

71 S_x^2 and S_y^2 from the same population are 40 (df $= 1$) and 8 (df $= 20$) respectively and $F = 5.00$. Approximately what percent of the Fs would be larger than this if we repeated the experiment again and again? _____.

≡≡≡≡≡≡≡ STOP ≡≡≡≡≡≡≡≡≡≡≡≡≡≡≡≡≡≡≡≡≡≡≡ STOP ≡≡≡≡≡≡≡

3 percent

---

72 If we draw a large number of pairs of samples of 2 and 21 scores (df $= 1$ and 20) from the same population, approximately 5 percent of the Fs would be larger than _____.

≡≡≡≡≡≡≡ STOP ≡≡≡≡≡≡≡≡≡≡≡≡≡≡≡≡≡≡≡≡≡≡≡ STOP ≡≡≡≡≡≡≡

4.35

---

73 And about 1 percent of the Fs would be larger than what value? _____.

≡≡≡≡≡≡≡ STOP ≡≡≡≡≡≡≡≡≡≡≡≡≡≡≡≡≡≡≡≡≡≡≡ STOP ≡≡≡≡≡≡≡

8.10

---

74 Finally 20 percent of the Fs would be larger than what value? _____.

≡≡≡≡≡≡≡ STOP ≡≡≡≡≡≡≡≡≡≡≡≡≡≡≡≡≡≡≡≡≡≡≡ STOP ≡≡≡≡≡≡≡

1.80

---

75 Suppose we sampled from two populations with $\sigma_x^2 = \sigma_y^2$. The samples from population X had 1 df and the samples from population Y had df $= 20$. If we randomly paired the variances from the two populations what percent of the F values would be larger than $F = 3.00$? _____.

≡≡≡≡≡≡≡ STOP ≡≡≡≡≡≡≡≡≡≡≡≡≡≡≡≡≡≡≡≡≡≡≡ STOP ≡≡≡≡≡≡≡

10 percent

---

76 If we sampled from two populations where $\sigma_x^2 = 1000$ and $\sigma_y^2 = 10$, the average F would equal _____.

≡≡≡≡≡≡≡ STOP ≡≡≡≡≡≡≡≡≡≡≡≡≡≡≡≡≡≡≡≡≡≡≡ STOP ≡≡≡≡≡≡≡

100

77 Thus when $\sigma_x^2 = 1000$ and $\sigma_y^2 = 10$ the average $F = 100$; however when $\sigma_x^2 = 100$ and $\sigma_y^2 = 100$ the average F would equal _____.

═══════ STOP ═══════════════════════════ STOP ═══════

1.00

---

78 According to the table (average $= 1.00$, df $= 1, 20$) 55 percent of the Fs would be larger than what value of F? _____.

═══════ STOP ═══════════════════════════ STOP ═══════

0.40

---

79 We have said before that we will reject the hypothesis of chance occurrence if the probability of an event is equal to _____ or less.

═══════ STOP ═══════════════════════════ STOP ═══════

.05 (5 percent)

---

80 Suppose we compute an F and we do not know the values of σ_x^2 and σ_y^2. With df $= 1$ and 20 if $\sigma_x^2 = \sigma_y^2$ then the average S_x^2/S_y^2 will be equal to _____ and 5 percent of the Fs will be larger than _____.

═══════ STOP ═══════════════════════════ STOP ═══════

1.00, 4.35

---

81 Suppose we get an F larger than 4.35. We would reject the hypothesis that $\sigma_x^2 =$ _____.

═══════ STOP ═══════════════════════════ STOP ═══════

σ_y^2

---

82 If σ_x^2 does in fact equal σ_y^2 then _____ percent of the time we will be wrong.

═══════ STOP ═══════════════════════════ STOP ═══════

5

---

83 $S_x^2 = 24$; $S_y^2 = 4$; $F =$ _____.

═══════ STOP ═══════════════════════════ STOP ═══════

6

84   We do not know the values of $\sigma_z{}^2$ and $\sigma_y{}^2$ but we state the hypothesis $\sigma_z{}^2 = \sigma_y{}^2$. We find $F = 6.00$ (df $= 1$ and 20). If $\sigma_z{}^2 = \sigma_y{}^2$ approximately what percent of the values of sample Fs would be larger than 6.00? _____.

═══ STOP ═══════════════════════════════════ STOP ═══

2 percent

---

85   Since the probability that an F will be 6.00 or larger is less than $.05$, then we must _____ the hypothesis that $\sigma_z{}^2 =$ _____.

═══ STOP ═══════════════════════════════════ STOP ═══

reject, $\sigma_y{}^2$

---

86   Suppose we sample from two populations (df $= 1$ and 20) $S_z{}^2 = 15$ and $S_y{}^2 = 3$. The hypothesis we wish to test is _____.

═══ STOP ═══════════════════════════════════ STOP ═══

$\sigma_z{}^2 = \sigma_y{}^2$

---

87   With df $= 1$ and 20, $S_z{}^2 = 15$ and $S_y{}^2 = 3$, $F =$ _____.

═══ STOP ═══════════════════════════════════ STOP ═══

5

---

88   df $= 1$ and 20, $F = 5.00$. The hypothesis is $\sigma_z{}^2 = \sigma_y{}^2$. If $\sigma_z{}^2 = \sigma_y{}^2$ the approximate percentage of F values larger than 5.00 would be _____ and for this reason we would _____ the hypothesis.

═══ STOP ═══════════════════════════════════ STOP ═══

3 percent, reject

---

89   If we reject the hypothesis then we would conclude that $\sigma_z{}^2$ _____ _____ _____.

═══ STOP ═══════════════════════════════════ STOP ═══

does not equal $\sigma_y{}^2$

---

90   In the preceding problems we decided to reject the hypothesis $\sigma_z{}^2 = \sigma_y{}^2$ when F was equal to or greater than _____.

═══ STOP ═══════════════════════════════════ STOP ═══

4.35

---

91   If our only interest is in the decision to reject or not to reject the hypothesis, then the only information we need is that point on each of the many F distributions beyond which _____ percent of the cases (sample Fs) would fall.

══════ STOP ══════════════════════ STOP ══════

5

---

92   The shape of the F distribution (even when the average equals 1.00) differs from one set of df to another. For this reason we would expect that the point beyond which 5 percent of the area falls would _____ from one set of df to another.

══════ STOP ══════════════════════ STOP ══════

differ

---

93   Look at Appendix D. The columns are headed by the number of df in the numerator of the F ratio and the rows are headed by the number of df in the _____ of the F ratio.

══════ STOP ══════════════════════ STOP ══════

denominator

---

94   Suppose, as in your previous examples, you had an F with 1 df in the numerator and 20 df in the denominator. The 5 percent entry in the F table for 1 and 20 df is _____.

══════ STOP ══════════════════════ STOP ══════

4.35

---

95   This means that _____ percent of the sample Fs would be larger than _____ if $\sigma_x^2 = \sigma_y^2$ (that is, the average value of $F = 1.00$).

══════ STOP ══════════════════════ STOP ══════

5, 4.35

---

96   Suppose the numerator had 10 df while the denominator had 30 df. If $\sigma_x^2 = \sigma_y^2$, then 5 percent of the sample Fs would fall beyond what point?

_____.

══════ STOP ══════════════════════ STOP ══════

$F = 2.16$

---

97 With 10 and 10 df, if the average of the F distribution is equal to 1.00 then 5 percent of the sample Fs will fall beyond what point? _____.

═══════ STOP ═══════════════════════ STOP ═══════

2.97

---

98 If with 20 and 20 df we get an F of 2.50, we would reject the hypothesis that $\sigma_x{}^2 = $ _____.

═══════ STOP ═══════════════════════ STOP ═══════

$\sigma_y{}^2$

---

99 With 6 df in the numerator, 25 in the denominator and if $\sigma_x{}^2 = \sigma_y{}^2$, 5 percent of the sample Fs would be greater than what value ?_____.

═══════ STOP ═══════════════════════ STOP ═══════

2.49

---

100 With 1 and 60 df, if $\sigma_x{}^2 = \sigma_y{}^2 = 50$, you would expect what percent of your sample Fs to be larger than 4.00? _____.

═══════ STOP ═══════════════════════ STOP ═══════

5 percent

---

101 On the other hand with 1 and 60 df, if $\sigma_x{}^2 = 200$ and $\sigma_y{}^2 = 50$, you would expect much more than _____ percent of your sample Fs to be larger than 4.00.

═══════ STOP ═══════════════════════ STOP ═══════

5

---

102 We wish to determine if the true variance $(\sigma_x{}^2)$ of one sample differs from the true variance of a second $(\sigma_y{}^2)$ sample. We first state the hypothesis _____.

═══════ STOP ═══════════════════════ STOP ═══════

$\sigma_x{}^2 = \sigma_y{}^2$

---

103 We wish to determine if the true variances of two samples differ. We state H: $\sigma_x{}^2 = \sigma_y{}^2$. From our two sample variances $S_x{}^2$ and $S_y{}^2$ we next compute _____.

═══════ STOP ═══════════════════════ STOP ═══════

F

104 We wish to determine if the true variances of two samples differ. $H: \sigma_z^2 = \sigma_y^2$ and F has been computed with $N_z - 1$ and $N_y - 1$ df. Next the value of the sample F is compared with the _____ ____ ____ listed in the F table.

═══════════ STOP ═══════════════════════ STOP ═══════

value of F

---

105 If the value of our sample F is greater than the tabled value of F we then _____ the hypothesis that $\sigma_z^2 = \sigma_y^2$.

═══════════ STOP ═══════════════════════ STOP ═══════

reject

---

106 We would then conclude that the true variance of the first population _____ from the true variance of the second population.

═══════════ STOP ═══════════════════════ STOP ═══════

differs

---

107 Note that at no time in the computation and evaluation of F do we know what the respective true variances σ_z^2 and σ_y^2 are. Rather we either reject or do not reject the hypothesis that _____.

═══════════ STOP ═══════════════════════ STOP ═══════

$\sigma_z^2 = \sigma_y^2$

---

108 Suppose $S_z^2 = 20$, $S_y^2 = 10$, then $F =$ _____.

═══════════ STOP ═══════════════════════ STOP ═══════

2.00

---

109 Suppose $S_z^2 = 36$ with 12 df and $S_y^2 = 12$ with 10 df, then $F =$ _____.

═══════════ STOP ═══════════════════════ STOP ═══════

3.00

---

110 $S_z^2 = 36$ with 12 df; $S_y^2 = 12$ with 10 df; $F = 3.00$. What is the hypothesis? _____.

═══════════ STOP ═══════════════════════ STOP ═══════

$H: \sigma_z^2 = \sigma_y^2$

---

111 $S_x^2 = 36$ with 12 df; $S_y^2 = 12$ with 10 df; $F = 3.00$. What do we do with the hypothesis? _____.

═══════ STOP ═══════════════════════ STOP ═══════

Reject

---

112 We reject the hypothesis that $\sigma_x^2 = \sigma_y^2$ because if the hypothesis were true we would expect an F of 3.00 to occur less than _____ percent _____ _____ _____ by chance.

═══════ STOP ═══════════════════════ STOP ═══════

5, of the time

---

113 $S_x^2 = 40$ with 20 df and $S_y^2 = 20$ with 25 df, $F = $ _____ and we _____ the hypothesis.

═══════ STOP ═══════════════════════ STOP ═══════

2, reject

---

114 $S_x^2 = 20$ with 1 df and $S_y^2 = 5$ with 25 df. $F = $ _____ and what do we do with the hypothesis? _____.

═══════ STOP ═══════════════════════ STOP ═══════

4, Do not reject

---

115 $S_x^2 = 10$ with 5 df and $S_y^2 = 20$ with 8 df, $F = $ _____ and what do we do with the hypothesis? _____.

═══════ STOP ═══════════════════════ STOP ═══════

0.50, Do not reject

Twelve

Notation

Multiple Summation

Up to the present time we have been looking at one group of scores at a time. This group has been composed of N scores, and we have indicated summation either by ΣX or by $\sum_{1}^{N} X$. One of the most important aspects of statistical inference is in determining if the means of two or more groups differ from one another. For this reason we now must develop additional notation to enable us to add (or sum) the scores in several groups. To do this, we simply add another Σ sign to the ΣX we already have. This means to add the sum of the scores in each group. How many groups do we have? Tradition tells us we have K groups. Thus, if we have two groups of 12 scores each, then $N = 12$ and $K = 2$, and if we have five groups of three scores each, then $K = 5$ and $N = 3$. Thus, $K \times N$ (or KN) equals the total number of scores.

Sometimes it is necessary to indicate the sum of all squared scores in all groups

$$\sum_{1}^{K} \sum_{1}^{N} X^2$$

or the sum of the squared sums of the scores in each group

$$\sum_{1}^{K} \left(\sum_{1}^{N} X \right)^2$$

or perhaps the square of the sum of all scores in all groups

$$\left(\sum_{1}^{K} \sum_{1}^{N} X \right)^2$$

Before going on, let us consider an example or two. Suppose we give three groups of students the same list of words to memorize. The first group is told to learn the list by repeating over and over the words in the list. The second group is told to make up a story connecting and using the words in the list, while the third group is told to string the words in the list together by making a mental image of each word and having each image interact in some bizarre way with the next word in the list. Below are the data which represent the number of words remembered after a week.

I	II	III
0	5	8
1	7	8
4	9	10
2	8	9
1	10	9
0	9	10
1	7	7
2	9	9
11	64	70

Initially, we wish to find the sum of all the scores in all the groups. To do this, we must first sum over the N scores in each group and then sum over the group of K sums. That is,

$$\sum_1^K \sum_1^N X = 11 + 64 + 70 = 145$$

Similarly, if we wish to find the sum of all the squared scores, we would first square each score, sum all N scores for each group, and then sum over the group of K sums. That is,

$$\sum_1^K \sum_1^N X^2 = 27 + 530 + 620 = 1177$$

(Note that for Group I, $0^2 + 1^2 + 4^2 + 2^2 + 1^2 + 0^2 + 1^2 + 2^2 = 27$.)

There are only two other operations which are usually done when two summation signs are used. The first is to find the sum of the squared sums for each group. (Trying to say *that* is another good reason to have a notational system.) That is, notice that the operations to be done,

$$\sum_1^K \left(\sum_1^N X \right)^2 = 11^2 + 64^2 + 70^2 = 9117$$

must be done in the same sequence. The operations inside the parentheses (if any) are done first, followed by squaring if indicated, and then summing from right to left, in contrast to our usual left-to-right procedure. Thus, to find

$$\sum_1^K \left(\sum_1^N X \right)^2$$

we first found the sums for each group (11, 64, and 70), then each sum was squared (121, 4096, 4900) and then finally summed (121 + 4096 + 4900 = 9117).

The last operation usually performed in this type of experiment is to find the square of the sum of all scores. That is,

$$\left(\sum_1^K \sum_1^N X \right)^2 = 145^2 = 21{,}025$$

Notice again that the operations inside the parentheses were first performed, followed by squaring.

Here are some more scores from a different experiment. Following a public service campaign that emphasized the self-improvement which comes from reading nonfiction books, an experimenter at the checkout counter in the local library counted the number of nonfiction books taken out by each of five people for each of 0, 1, 2, and 3 weeks after the end of the campaign.

Time

0 week	1 week	2 weeks	3 weeks
6	4	1	0
4	2	2	1
5	1	0	0
2	2	1	1
3	2	1	1

Demonstrate to yourself that

$$\sum_1^K \sum_1^N X = 39$$

$$\sum_1^K \sum_1^N X^2 = 129$$

$$\left(\sum_1^K \sum_1^N X \right)^2 = 1521$$

$$\sum_1^K \left(\sum_1^N X \right)^2 = 555$$

An additional variable makes things somewhat more complex, but the same general rules apply: *Do everything inside the parentheses first and, after squaring, work from right to left.*

The data tabulated below present day salaries of social workers who had graduated in 1955, 1965, or 1975. The data comprise salaries to the nearest thousand dollars. For each group or cell, $\sum_1^N X$ has been found. There are several sums which must be found when an analysis of such a complex experiment is to be done.

	1955 (Y_1)	1965 (Y_2)	1975 (Y_3)	
Male (S_1)	27	17	10	
	25	15	11	
	20 /72	19 /51	12 /33	S = sex
Female (S_2)	19	14	9	
	17	13	10	
	18 /54	12 /39	8 /27	Y = year

Initially, we wish to find the sum of all the scores:

$$\sum_1^S \sum_1^Y \sum_1^N X = 72 + 51 + 33 + 54 + 39 + 27 = 276$$

Next find the sum of all the squared scores:

$$\sum_1^S \sum_1^Y \sum_1^N X^2 = 27^2 + 25^2 + 20^2 \cdots + 9^2 + 10^2 + 8^2 = 4722$$

Next find the square of the sum of all the scores:

$$\left(\sum_1^S \sum_1^Y \sum_1^N X\right)^2 = 276^2 = 76{,}176$$

Next find the sum of the squared cell sums:

$$\sum_1^S \sum_1^Y \left(\sum_1^N X\right)^2 = 72^2 + 51^2 + 33^2 + 54^2 + 39^2 + 27^2$$
$$= 5184 + 2601 + 1089 + 2916 + 1521 + 729$$
$$= 14{,}040$$

The next two summations are easily confused, and the operations to be performed are not intuitively obvious. The first of these is to find the sum of the squared row sums:

$$\sum_1^S \left(\sum_1^Y \sum_1^N X\right)^2 = (72 + 51 + 33)^2 + (54 + 39 + 27)^2$$
$$= 156^2 + 120^2$$
$$= 24{,}336 + 14{,}400$$
$$= 38{,}736$$

The second summation is the sum of the squared column sums:

$$\sum_1^Y \left(\sum_1^S \sum_1^N X\right)^2 = (72 + 54)^2 + (51 + 39)^2 + (33 + 27)^2$$
$$= 126^2 + 90^2 + 60^2$$
$$= 15{,}876 + 8{,}100 + 3{,}600$$
$$= 27{,}576$$

The apparent reason for the confusion between these last two summations is that it is hard to remember, for example, that $\sum_1^Y$ means to sum from year 1 (1955) to year Y (1975). Thus:

$$\sum_1^S \left(\sum_1^Y \sum_1^N X\right)^2$$

means that the sum of each cell is found first; then the sum of the three years, under S_1 (male), is found; and finally the sum of the three years, under S_2 (female), is found. After both sums are squared, they are then summed.

Note that when the column variable is inside the parentheses and the row variable is outside, the sum of each row is found, squared, and then summed over rows. For example:

	C_1	C_2
R_1	0	4
	1	3
R_2	1	5
	0	4

(a) $\displaystyle\sum_1^R \left(\sum_1^C \sum_1^N X \right)^2$ The sums of each row are found ($R_1 = 8$; $R_2 = 10$).

(b) The sums of each row are squared ($R_1 = 64$; $R_2 = 100$).

(c) The squares of each row are summed ($64 + 100 = 164$).

(d) $\displaystyle\sum_1^C \left(\sum_1^R \sum_1^N X \right)^2$ The sums of each column are found ($C_1 = 2$; $C_2 = 16$).

(e) The sums of each column are squared ($C_1 = 4$; $C_2 = 256$).

(f) The squares of each column are summed ($4 + 256 = 260$).

Remember: $\displaystyle\sum_1^R$ does *not* mean that you sum across rows. Rather, it means that you sum from where $R = 1$ (in Row 1) to $R = R_x$ (in this example, R_2). If you keep this in mind, this complex notation will be much easier for you.

PROBLEMS

1.

	B_1	B_2	B_3
A_1	0	3	2
	1	1	3
A_2	2	4	0
	2	2	0

$N = \underline{\hspace{2cm}}$

$AN = \underline{\hspace{2cm}}$

$BAN = \underline{\hspace{2cm}}$

2. From Problem 1 $\displaystyle\sum_1^A \sum_1^B \sum_1^N X = \underline{\hspace{2cm}}$

3. From Problem 1 $\displaystyle\sum_1^A \sum_1^B \sum_1^N X^2 = \underline{\hspace{2cm}}$

4. From Problem 1 $\displaystyle\sum_1^A \sum_1^B \left(\sum_1^N X \right)^2 = \underline{\hspace{2cm}}$

5. From Problem 1 $\displaystyle\sum_1^A \left(\sum_1^B \sum_1^N X \right)^2 = \underline{\hspace{2cm}}$

6. From Problem 1 $\displaystyle\sum_1^B \left(\sum_1^A \sum_1^N X \right)^2 = \underline{\hspace{2cm}}$

7. From Problem 1 $\displaystyle\left(\sum_1^B \sum_1^A \sum_1^N X \right)^2 = \underline{\hspace{2cm}}$

8.

	C_1	C_2
	0	3
R_1	1	1
	1	1
	2	0
R_2	0	0
	0	0
	1	2
R_3	0	1
	0	2

$$\sum_1^R \sum_1^C \left(\sum_1^N X \right)^2 = \underline{\hspace{2cm}}$$

$$\sum_1^R \left(\sum_1^C \sum_1^N X \right)^2 = \underline{\hspace{2cm}}$$

9.

	Q_1	Q_2	Q_3
	1	2	1
P_1	2	0	2
	3	1	1
	0	1	1
P_2	0	1	1
	0	1	0

$$\sum_1^P \sum_1^Q \sum_1^N X^2 = \underline{\hspace{2cm}}$$

$$\sum_1^P \left(\sum_1^Q \sum_1^N X \right)^2 = \underline{\hspace{2cm}}$$

10.

	C_1	C_2
	0	2
R_1	0	0
	1	1
	1	1
	4	1
R_2	1	2
	0	2
	0	1

$$\sum_1^R \sum_1^C \sum_1^N X^2 = \underline{\hspace{2cm}}$$

$$\sum_1^R \sum_1^C \left(\sum_1^N X \right)^2 = \underline{\hspace{2cm}}$$

$$\sum_1^R \left(\sum_1^C \sum_1^N X \right)^2 = \underline{\hspace{2cm}}$$

$$\sum_1^C \left(\sum_1^R \sum_1^N X \right)^2 = \underline{\hspace{2cm}}$$

1 Suppose we have two groups of scores: group I: 0, 1, 3; group II: 1, 3, 2. To indicate that we wish to add all the scores in both groups we could use the notation $\sum_1^N X_\mathrm{I}$ (referring to the scores in group I) $+ \sum_1^N X_\mathrm{II}$ (referring to the scores in group II). Thus $\sum_1^N X_\mathrm{I} + \sum_1^N X_\mathrm{II} = $ _____.

═══ STOP ═══ STOP ═══

10

---

2 Group I: 0, 1, 3; group II: 1, 3, 2. Since we have used a summation sign for adding numbers 1 through N ($\sum_1^N$), we could use a summation sign to indicate adding groups. Thus $\sum_1^2 \sum_1^N X$ would mean $\sum_1^N X_\mathrm{I} + \sum_1^N X_\mathrm{II}$. $\sum_1^2 \sum_1^N X = $ _____.

═══ STOP ═══ STOP ═══

10

---

3 Group I: 0, 1, 3; group II: 1, 3, 2. If we have K groups of scores we could use the more general notation $\sum_1^K \sum_1^N X$. In this case $K = $ _____.

═══ STOP ═══ STOP ═══

2

---

4 Group I: 0, 1, 3; group II: 1, 3, 2. $\sum_1^K \sum_1^N X = $ _____.

═══ STOP ═══ STOP ═══

10

---

5 Group I: 0, 1, 3, 4; group II: 0, 2, 5, 3; group III: 4, 1, 7, 6. If we wished to find the sum of the scores in two groups rather than in all K groups, we would use, instead of the $\sum_1^K \sum_1^N X$ notation $\sum_1^2 \sum_1^N X$. Thus $\sum_1^2 \sum_1^N X = $

_____.

═══ STOP ═══ STOP ═══

18

---

6   Group I: 0, 1, 3, 4; group II: 0, 2, 5, 3; group III: 4, 1, 7, 6. If we wished to find the sum of scores in the last two groups we would use the notation $\sum\limits_{2}^{3}$. Hence $\sum\limits_{2}^{3}\sum\limits_{1}^{N} X =$ _____.

═══════ STOP ═══════════════════════════ STOP ═══════

28

───

7   Group I: 0, 1, 3, 4; group II: 0, 2, 5, 3; group III: 4, 1, 7, 6. Finally if we wished to add only the first three numbers in each group we would use the notation $\sum\limits_{1}^{K}\sum\limits_{1}^{3} X$. $\sum\limits_{1}^{K}\sum\limits_{1}^{3} X =$ _____.

═══════ STOP ═══════════════════════════ STOP ═══════

$4 + 7 + 12 = 23$

───

8   Group I: 0, 2, 5; group II: 1, 2, 0; group III: 1, 0, 7. $\sum\limits_{1}^{2}\sum\limits_{1}^{N} X =$ _____.

═══════ STOP ═══════════════════════════ STOP ═══════

$7 + 3 = 10$

───

9   I: 0, 2, 5; II: 1, 2, 0; III: 1, 0, 7. $\sum\limits_{1}^{K}\sum\limits_{1}^{2} X =$ _____.

═══════ STOP ═══════════════════════════ STOP ═══════

$2 + 3 + 1 = 6$

───

10   I: 0, 2, 5; II: 1, 2, 0; III: 1, 0, 7; IV: 0, 2, 9. What notation would we use if we wished to add the first two scores in the first three groups? _____.

═══════ STOP ═══════════════════════════ STOP ═══════

$\sum\limits_{1}^{3}\sum\limits_{1}^{2} X$

───

11   If we have five groups with six scores in each group, we could indicate that we wished to sum the last three numbers in the first three groups by the notation _____.

═══════ STOP ═══════════════════════════ STOP ═══════

$\sum\limits_{1}^{3}\sum\limits_{4}^{6} X$

───

12   Usually however we simply want to add all scores in all groups and this is indicated by the notation _____.

═════════ STOP ═══════════════════════ STOP ═════════

$$\sum_{1}^{K} \sum_{1}^{N} X$$

---

13   We have five scores 0, 1, 3, 2, 4. $\sum_{1}^{N} X =$ _____, $(\sum_{1}^{N} X)^2 =$ _____, $\sum_{1}^{N} X^2 =$ _____.

═════════ STOP ═══════════════════════ STOP ═════════

10, 100, 30

---

14   Note that when we wish to indicate the sum of a group of squared scores we used the notation $\sum_{1}^{N} X^2$ but when we wish to square a sum we use the notation _____.

═════════ STOP ═══════════════════════ STOP ═════════

$$\left(\sum_{1}^{N} X\right)^2$$

---

15   Suppose a statistician has K groups of N scores. If he wished to square each score and add all the scores he would use the notation $\sum_{1}^{K} \sum_{1}^{N} X^2$ but if he would add all the scores together and then square the sum of all the scores he would use the notation _____.

═════════ STOP ═══════════════════════ STOP ═════════

$$\left(\sum_{1}^{K} \sum_{1}^{N} X\right)^2$$

---

16   He may want to do something else. He may want to find the sum of all the scores in a group, square each of these sums, and then add. The notation used for indicating that the sum of a single group of scores is squared is _____.

═════════ STOP ═══════════════════════ STOP ═════════

$$\left(\sum_{1}^{N} X\right)^2$$

---

17 If $(\sum_{1}^{N} X)^2$ means that the scores in a group have been summed and then the sum is squared, then $\sum_{1}^{K} (\sum_{1}^{N} X)^2$ means _____ for all K groups.

═══════ STOP ═══════════════════════ STOP ═══════

$$\left(\sum_{1}^{N} X\right)^2 + \left(\sum_{1}^{N} X\right)^2 \quad \text{etc.}$$

---

18 Group I: 0, 1, 3; group II: 2, 3, 1. $\sum_{1}^{K} \sum_{1}^{N} X^2 =$ _____.

═══════ STOP ═══════════════════════ STOP ═══════

24

---

19 I: 0, 1, 3; II: 2, 3, 1, $\sum_{1}^{K} \sum_{1}^{N} X^2 = 24$. $(\sum_{1}^{K} \sum_{1}^{N} X)^2 =$ _____.

═══════ STOP ═══════════════════════ STOP ═══════

100

---

20 I: 0, 1, 3; II: 2, 3, 1. $\sum_{1}^{K} \sum_{1}^{N} X^2 = 24$, $(\sum_{1}^{K} \sum_{1}^{N} X)^2 = 100$. For group I $(\sum_{1}^{N} X)^2 =$ _____, for group II $(\sum_{1}^{N} X)^2 =$ _____. Hence $\sum_{1}^{K} (\sum_{1}^{N} X)^2 =$ _____.

═══════ STOP ═══════════════════════ STOP ═══════

16, 36; 52

---

21 I: 0, 1, 1; II: 3, 1, 4; III: 2, 5, 3. $\sum_{1}^{K} \sum_{1}^{N} X^2 =$ _____, $(\sum_{1}^{K} \sum_{1}^{N} X)^2 =$ _____, $\sum_{1}^{K} (\sum_{1}^{N} X)^2 =$ _____.

═══════ STOP ═══════════════════════ STOP ═══════

66, 400, 168

---

22 I: 0, 1, 1; II: 3, 1, 4; III: 2, 5, 3; IV: 0, 1, 3. $N =$ _____, $K =$ _____, hence KN (read K times N) = _____.

═══════ STOP ═══════════════════════ STOP ═══════

3, 4, 12

---

23   I: 0, 1, 1; II: 3, 1, 4; III: 2, 5, 3; IV: 0, 1, 3. $N = 3$; $K = 4$; $KN = 12$.

$\sum_{1}^{K} \sum_{1}^{N} X = $ _____, $\sum_{1}^{K} \sum_{1}^{N} X/KN = $ _____.

═══════ STOP ═══════════════════════════ STOP ═══════

24, 24/12 = 2.00

---

24   I: 0, 1, 1; II: 3, 1, 4; III: 2, 5, 3; IV: 0, 1, 3. $(\sum_{1}^{K} \sum_{1}^{N} X)^2 = $ _____,

$(\sum_{1}^{K} \sum_{1}^{N} X)^2/KN = $ _____.

═══════ STOP ═══════════════════════════ STOP ═══════

576, 48

---

25   I: 0, 1, 1; II: 3, 1, 4; III: 2, 5, 3; IV: 0, 1, 3. $\sum_{1}^{K} (\sum_{1}^{N} X)^2 = $ _____

$\sum_{1}^{K} (\sum_{1}^{N} X)^2/KN = $ _____.

═══════ STOP ═══════════════════════════ STOP ═══════

184, 184/12 = 15.33

---

26   Given scores 0, 2, 5, 5, $\sum_{1}^{N} X^2 = $ _____, $\sum_{1}^{N} X = $ _____, $(\sum_{1}^{N} X)^2/N = $

_____.

═══════ STOP ═══════════════════════════ STOP ═══════

54, 12, 36

---

27   With scores 0, 2, 5, 5, $\sum_{1}^{N} X^2 = 54$, $(\sum_{1}^{N} X)^2/N = 36$. $\sum_{1}^{N} X^2 - (\sum_{1}^{N} X)^2/N = $

_____.

═══════ STOP ═══════════════════════════ STOP ═══════

18

---

28   Given two groups of scores: I: 0, 2, 5, 5; II: 4, 5, 2, 5. For I $\sum_{1}^{N} X^2 -$

$(\sum_{1}^{N} X)^2/N = 18$, for II $\sum_{1}^{N} X^2 - (\sum_{1}^{N} X)^2/N = $ _____.

═══════ STOP ═══════════════════════════ STOP ═══════

70 − 64 = 6

29 Given two groups of scores: For group I $\sum_1^N X^2 - (\sum_1^N X)^2/N = 18$, for group II $\sum_1^N X^2 - (\sum_1^N X)^2/N = 6$. Hence $\sum_1^K \left[\sum_1^N X^2 - (\sum_1^N X)^2/N \right] =$ _____.

═══ STOP ═══   ═══ STOP ═══

24

---

30 Three groups of four scores each: I: 0, 1, 1, 0; II: 3, 4, 5, 2; III: 2, 3, 2, 1.

$\sum_1^K \sum_1^N X^2 =$ _____, $\sum_1^K (\sum_1^N X)^2 =$ _____, $(\sum_1^K \sum_1^N X)^2 =$ _____.

═══ STOP ═══   ═══ STOP ═══

74, 264, 576

---

31 Suppose we have eight scores. These scores are divided into two groups: R_1 and R_2 with four scores in each. Each of these two groups are in turn divided in half so that there are now four groups: R_1C_1, R_1C_2, R_2C_1, R_2C_2. How many scores are in each group? _____.

═══ STOP ═══   ═══ STOP ═══

2

32

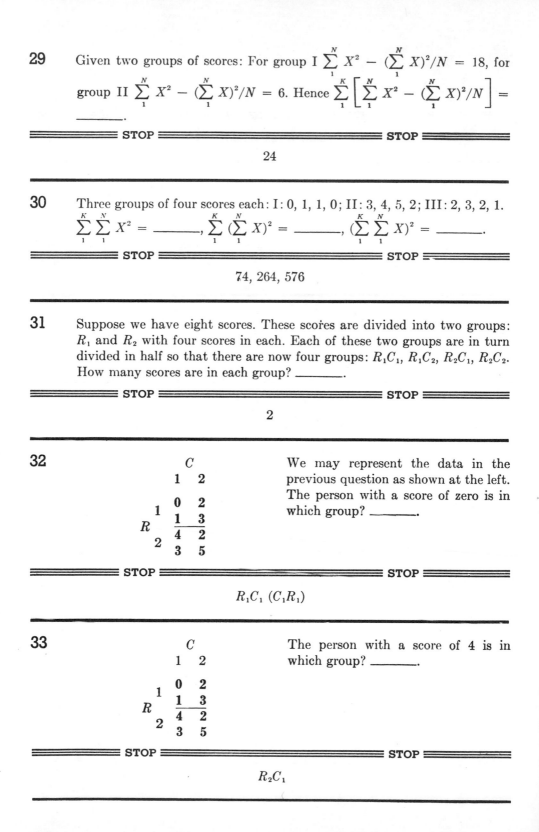

		C	
		1	2
R	1	0	2
		1	3
	2	4	2
		3	5

We may represent the data in the previous question as shown at the left. The person with a score of zero is in which group? _____.

═══ STOP ═══   ═══ STOP ═══

R_1C_1 (C_1R_1)

33

		C	
		1	2
R	1	0	2
		1	3
	2	4	2
		3	5

The person with a score of 4 is in which group? _____.

═══ STOP ═══   ═══ STOP ═══

R_2C_1

34

$$
\begin{array}{c|cc}
 & \multicolumn{2}{c}{C} \\
 & 1 & 2 \\
\hline
 & 0 & 2 \\
1 & 1 & 3 \\
R & \multicolumn{2}{c}{} \\
 & 4 & 2 \\
2 & 3 & 5 \\
\end{array}
$$

For group R_1C_2 $\displaystyle\sum_1^N X = $ _____.

═══ STOP ═══════════════════════════════ STOP ═══

5

35

$$
\begin{array}{c|cc}
 & \multicolumn{2}{c}{C} \\
 & 1 & 2 \\
\hline
 & 0 & 2 \\
1 & 1 & 3 \\
R & \multicolumn{2}{c}{} \\
 & 4 & 2 \\
2 & 3 & 5 \\
\end{array}
$$

For group R_2C_2 $\displaystyle\sum_1^N X^2 = $ _____.

═══ STOP ═══════════════════════════════ STOP ═══

29

36

$$
\begin{array}{c|cc}
 & \multicolumn{2}{c}{C} \\
 & 1 & 2 \\
\hline
 & 0 & 2 \\
1 & 1 & 3 \\
R & \multicolumn{2}{c}{} \\
 & 4 & 2 \\
2 & 3 & 5 \\
\end{array}
$$

For group R_2C_1 $\left(\displaystyle\sum_1^N X\right)^2 = $ _____.

═══ STOP ═══════════════════════════════ STOP ═══

49

37

$$
\begin{array}{c|cc}
 & \multicolumn{2}{c}{C} \\
 & 1 & 2 \\
\hline
 & 0 & 2 \\
1 & 1 & 3 \\
R & \multicolumn{2}{c}{} \\
 & 4 & 2 \\
2 & 3 & 5 \\
\end{array}
$$

For group R_1C_1 $\displaystyle\sum_1^N X = $ _____, for group R_2C_1 $\displaystyle\sum_1^N X = $ _____. Hence for group C_1 $\displaystyle\sum_1^R \sum_1^N X = $ _____.

═══ STOP ═══════════════════════════════ STOP ═══

1, 7; 8

38

$$
\begin{array}{c}
\ \ C\\
\ 1\ \ 2
\end{array}
$$

	C	
	1	2
R 1	0	2
	1	3
	4	2
R 2	3	5

For group $R_1 C_2 \ \sum\limits_1^N X = $ _____, for

group $R_2 C_2 \ \sum\limits_1^N X = $ _____, for group

$C_2 \ \sum\limits_1^R \sum\limits_1^N X = $ _____.

═══ STOP ═══════════════════ STOP ═══

5, 7, 12

39

	C	
	1	2
R 1	0	2
	1	3
	4	2
R 2	3	5

For group $C_2 \ \sum\limits_1^R \sum\limits_1^N X = 12$, for group

$C_1 \ \sum\limits_1^R \sum\limits_1^N X = 8$. Hence $\sum\limits_1^C \sum\limits_1^R \sum\limits_1^N X = $

_____.

═══ STOP ═══════════════════ STOP ═══

20

40

	C	
	1	2
R 1	0	2
	1	3
	4	2
R 2	3	5

For group $R_1 C_1 \ \sum\limits_1^N X = $ _____, for

group $R_1 C_2 \ \sum\limits_1^N X = $ _____. Hence

for group $R_1 \ \sum\limits_1^C \sum\limits_1^N X = $ _____.

═══ STOP ═══════════════════ STOP ═══

1, 5; 6

41

	C	
	1	2
R 1	0	2
	1	3
	4	2
R 2	3	5

For group $R_2 C_1 \ \sum\limits_1^N X = $ _____, for

group $R_2 C_2 \ \sum\limits_1^N X = $ _____. Hence

for group $R_2 \ \sum\limits_1^C \sum\limits_1^N X = $ _____.

═══ STOP ═══════════════════ STOP ═══

7, 7; 14

42

	C	
	1	2
R 1	0	2
	1	3
R 2	4	2
	3	5

For group $R_1 \sum_1^C \sum_1^N X = 6$, for group $R_2 \sum_1^C \sum_1^N X = 14$. Hence $\sum_1^R \sum_1^C \sum_1^N X = \underline{\quad\quad}$.

═══ STOP ═══ ═══ STOP ═══

20

43

	C	
	1	2
R 1	0	2
	1	3
R 2	4	2
	3	5

$\sum_1^C \sum_1^R \sum_1^N X = \sum_1^R \sum_1^C \sum_1^N X = \underline{\quad\quad}$.

═══ STOP ═══ ═══ STOP ═══

20

44

	C	
	1	2
R 1	0	2
	1	3
R 2	4	2
	3	5

For group $R_1C_1 \sum_1^N X^2 = \underline{\quad\quad}$, for group $R_2C_1 \sum_1^N X^2 = \underline{\quad\quad}$. Hence for group $C_1 \sum_1^R \sum_1^N X^2 = \underline{\quad\quad}$

═══ STOP ═══ ═══ STOP ═══

1, 25; 26

45

	C	
	1	2
R 1	0	2
	1	3
R 2	4	2
	3	5

For group $R_1C_2 \sum_1^N X^2 = \underline{\quad\quad}$, for group $R_2C_2 \sum_1^N X^2 = \underline{\quad\quad}$. Hence for group $C_2 \sum_1^R \sum_1^N X^2 = \underline{\quad\quad}$.

═══ STOP ═══ ═══ STOP ═══

13, 29; 42

46

$$C$$
$$1 \quad 2$$

$$R \begin{array}{c} 1 \\ \\ 2 \end{array} \begin{array}{cc} 0 & 2 \\ 1 & 3 \\ 4 & 2 \\ 3 & 5 \end{array}$$

For group $C_1 \sum_1^R \sum_1^N X^2 = 26$, for group $C_2 \sum_1^R \sum_1^N X^2 = 42$. Hence $\sum_1^C \sum_1^R \sum_1^N X^2 = $ _____.

═══ STOP ═══════════════════ STOP ═══

68

47

$$C$$
$$1 \quad 2$$

$$R \begin{array}{c} 1 \\ \\ 2 \end{array} \begin{array}{cc} 0 & 2 \\ 1 & 3 \\ 4 & 2 \\ 3 & 5 \end{array}$$

For group $R_1 \sum_1^C \sum_1^N X^2 = $ _____.

═══ STOP ═══════════════════ STOP ═══

14

48

$$C$$
$$1 \quad 2$$

$$R \begin{array}{c} 1 \\ \\ 2 \end{array} \begin{array}{cc} 0 & 2 \\ 1 & 3 \\ 4 & 2 \\ 3 & 5 \end{array}$$

For group $R_1 \sum_1^C \sum_1^N X^2 = 14$, for group $R_2 \sum_1^C \sum_1^N X^2 = $ _____.

═══ STOP ═══════════════════ STOP ═══

54

---

49 For group $R_1 \sum_1^C \sum_1^N X^2 = 14$, for group $R_2 \sum_1^C \sum_1^N X^2 = 54$. Hence $\sum_1^R \sum_1^C \sum_1^N X^2 = $ _____.

═══ STOP ═══════════════════ STOP ═══

68

---

50 $\sum_1^R \sum_1^C \sum_1^N X^2 = 68$, $\sum_1^C \sum_1^R \sum_1^N X^2 = 68$. Hence $\sum_1^R \sum_1^C \sum_1^N X^2 = $ _____.

═══ STOP ═══════════════════ STOP ═══

$$\sum_1^C \sum_1^R \sum_1^N X^2$$

51

$$\begin{array}{cc} & C \\ & 1 \quad 2 \end{array}$$

$$R \begin{array}{c} 1 \\ \\ 2 \end{array} \begin{array}{cc} 0 & 2 \\ 1 & 3 \\ 4 & 2 \\ 3 & 5 \end{array}$$

For group R_1C_1 $(\sum_1^N X)^2 =$ _____,

for group R_1C_2 $(\sum_1^N X)^2 =$ _____.

Hence for group R_1 $\sum_1^C (\sum_1^N X)^2 =$ _____.

════ STOP ════════════ STOP ════

1, 25; 26

52

$$\begin{array}{cc} & C \\ & 1 \quad 2 \end{array}$$

$$R \begin{array}{c} 1 \\ \\ 2 \end{array} \begin{array}{cc} 0 & 2 \\ 1 & 3 \\ 4 & 2 \\ 3 & 5 \end{array}$$

For group R_2C_1 $(\sum_1^N X)^2 =$ _____,

for group R_2C_2 $(\sum_1^N X)^2 =$ _____.

Hence for group R_2 $\sum_1^C (\sum_1^N X)^2 =$ _____.

════ STOP ════════════ STOP ════

49, 49; 98

---

53 For group R_1 $\sum_1^C (\sum_1^N X)^2 = 26$, for group R_2 $\sum_1^C (\sum_1^N X)^2 = 98$. Hence $\sum_1^R \sum_1^C (\sum_1^N X)^2 =$ _____.

════ STOP ════════════ STOP ════

124

54

$$\begin{array}{ccc} & & C \\ & 1 \quad 2 \quad 3 \end{array}$$

$$R \begin{array}{c} 1 \\ \\ 2 \end{array} \begin{array}{ccc} 1 & 2 & 4 \\ 1 & 3 & 2 \\ 5 & 2 & 0 \\ 4 & 3 & 1 \end{array}$$

For group R_1 $\sum_1^C \sum_1^N X =$ _____,

for R_2 $\sum_1^C \sum_1^N X =$ _____. Hence

$\sum_1^R \sum_1^C \sum_1^N X =$ _____.

════ STOP ════════════ STOP ════

13, 15; 28

55

$$
\begin{array}{cc}
 & C \\
 & 1\ \ 2\ \ 3 \\
R \begin{array}{c} 1 \\ \\ 2 \end{array} & \begin{array}{ccc} 1 & 2 & 4 \\ 1 & 3 & 2 \\ \hline 5 & 2 & 0 \\ 4 & 3 & 1 \end{array}
\end{array}
$$

For group R_1 $\displaystyle\sum_1^C \sum_1^N X^2 =$ _____,

for group R_2 $\displaystyle\sum_1^C \sum_1^N X^2 =$ _____.

Hence $\displaystyle\sum_1^R \sum_1^C \sum_1^N X^2 =$ _____.

═══ STOP ═══════════════════════════ STOP ═══

35, 55; 90

56

$$
\begin{array}{cc}
 & C \\
 & 1\ \ 2\ \ 3 \\
R \begin{array}{c} 1 \\ \\ 2 \end{array} & \begin{array}{ccc} 1 & 2 & 4 \\ 1 & 3 & 2 \\ \hline 5 & 2 & 0 \\ 4 & 3 & 1 \end{array}
\end{array}
$$

$\displaystyle\sum_1^R \sum_1^C \sum_1^N X = 28,\ (\sum_1^R \sum_1^C \sum_1^N X)^2 =$

_____.

═══ STOP ═══════════════════════════ STOP ═══

784

57

$$
\begin{array}{cc}
 & C \\
 & 1\ \ 2\ \ 3 \\
R \begin{array}{c} 1 \\ \\ 2 \end{array} & \begin{array}{ccc} 1 & 2 & 4 \\ 1 & 3 & 2 \\ \hline 5 & 2 & 0 \\ 4 & 3 & 1 \end{array}
\end{array}
$$

For C_1 $(\displaystyle\sum_1^R \sum_1^N X)^2 =$ _____;

$\displaystyle\sum_1^C (\sum_1^R \sum_1^N X)^2 =$ _____.

═══ STOP ═══════════════════════════ STOP ═══

121; 270

58

$$
\begin{array}{cc}
 & C \\
 & 1\ \ 2\ \ 3 \\
R \begin{array}{c} 1 \\ \\ 2 \end{array} & \begin{array}{ccc} 1 & 2 & 4 \\ 1 & 3 & 2 \\ \hline 5 & 2 & 0 \\ 4 & 3 & 1 \end{array}
\end{array}
$$

$\displaystyle\sum_1^R \sum_1^C (\sum_1^N X)^2 =$ _____.

═══ STOP ═══════════════════════════ STOP ═══

$2^2 + 5^2 + 6^2 + 9^2 + 5^2 + 1^2 = 172$

59

$$\begin{array}{ccc} & C & \\ 1 & 2 & 3 \end{array}$$

$$\sum_{1}^{R} (\sum_{1}^{C} \sum_{1}^{N} X)^2 = \text{_____}.$$

$$R \begin{array}{c} 1 \\ \\ 2 \end{array} \begin{array}{ccc} 1 & 2 & 4 \\ 1 & 3 & 2 \\ \hline 5 & 2 & 0 \\ 4 & 3 & 1 \end{array}$$

═══ STOP ═══════════════════════ STOP ═══

$$13^2 + 15^2 = 394$$

───────────────────────────────

60

$$\begin{array}{cc} & C \\ 1 & 2 \end{array}$$

For $R_1 \sum_{1}^{C} (\sum_{1}^{N} X)^2 = \text{_____},$

$$\sum_{1}^{R} \sum_{1}^{C} (\sum_{1}^{N} X)^2 = \text{_____}.$$

$$R \begin{array}{c} 1 \\ \\ 2 \end{array} \begin{array}{cc} 0 & 2 \\ 0 & 3 \\ \hline 1 & 1 \\ 4 & 1 \end{array}$$

═══ STOP ═══════════════════════ STOP ═══

25, 54

───────────────────────────────

61

$$\begin{array}{cc} & C \\ 1 & 2 \end{array}$$

For $C_1 (\sum_{1}^{R} \sum_{1}^{N} X)^2 = \text{_____},$

$$\sum_{1}^{C} (\sum_{1}^{R} \sum_{1}^{N} X)^2 = \text{_____}.$$

$$R \begin{array}{c} 1 \\ \\ 2 \end{array} \begin{array}{cc} 1 & 5 \\ 2 & 0 \\ \hline 2 & 1 \\ 2 & 3 \end{array}$$

═══ STOP ═══════════════════════ STOP ═══

49, 130

───────────────────────────────

62

$$\begin{array}{ccc} & C & \\ 1 & 2 & 3 \end{array}$$

$$\sum_{1}^{R} (\sum_{1}^{C} \sum_{1}^{N} X)^2 = \text{_____}.$$

$$R \begin{array}{c} 1 \\ \\ 2 \end{array} \begin{array}{ccc} 0 & 2 & 1 \\ 0 & 3 & 2 \\ \hline 5 & 1 & 2 \\ 2 & 2 & 4 \end{array}$$

═══ STOP ═══════════════════════ STOP ═══

$$64 + 256 = 320$$

───────────────────────────────

$$\sum_{1}^{R} \sum_{1}^{C} (\sum_{1}^{N} X)^2 = \underline{\qquad}.$$

		C		
		1	2	3
R	1	0	2	1
		0	3	2
	2	5	1	2
		2	2	4

═══════ STOP ═══════════════════════ STOP ═══════

$$0^2 + 5^2 + 3^2 + 7^2 + 3^2 + 6^2 = 128$$

Thirteen

The F Test

Analysis of Variance and the Comparison of Means

Nearly all of us at one time or another have made some generalizations concerning different populations of individuals. Men are taller than women, girls are better students than boys, people like one television program better than another, Englishmen drink more tea than Americans, and so on. There are literally thousands of comparisons that can be made, and as many guesses seem to be made each day. In each of these cases we are asserting something about the characteristics of the populations being compared—the mean of one population is different (larger, stronger, smarter, and so on) from the mean of the other population. The purpose of this chapter is to develop techniques by which such assertions can be tested statistically.

Suppose an experimenter is interested in evaluating a new method of teaching arithmetic. From the material presented in Chapter Ten on establishing the confidence interval for the mean, he might decide to find the mean of the new teaching technique and compare this with the true mean for the traditional teaching method. However, this procedure assumes that the true mean for the traditional teaching technique is known and that the experimenter is able to use it in his calculations. As we have mentioned before, it is an extremely rare case in which the experimenter has knowledge of the parameter. Even if norms did exist, they would be based on performance measures taken some time in the past. If the mean performance of students had changed over a period of time, then such norms would be worse than useless; they would be misleading.

In the absence of knowledge of the true mean for the traditional technique, the experimenter must compare the two teaching methods. Suppose he selects a group of pupils from a certain population and randomly assigns half of them to learn arithmetic by the traditional method, and assigns the second half of the pupils to learn by the new method of teaching. Now the experimenter can compare the performances of the two groups. Since both groups of pupils have been randomly selected from the same population, the experimenter can be reasonably sure that any differences (beyond those expected on the basis of chance) between the performances of the two groups are due to differences in teaching method. (This assumes that there are no differences introduced by a teacher variable. In an actual experiment, such a possibility must be taken into consideration, and the experiment must be designed to account or control for possible teacher differences.)

To make a comparison between the new teaching technique and the traditional teaching method, the experimenter will randomly select two groups from the same population, give one group the new method, give the second group the old method, and compare the performances of the two groups. Once the experiment has been conducted and the data collected, how might the data be analyzed? In this chapter you will learn the statistical techniques that will do just this. The technique we are introducing is the *analysis of variance* and the theory of this technique is based upon the *F* distribution.

REVIEW: VARIABILITY OF MEANS

Before we go on, let us first recall some relationships described in previous chapters. The basic relationship, upon which the entire theory of the analysis of variance is based, is that which exists between the variability of means and the variability of individual scores:

$$\sigma_M = \frac{\sigma}{\sqrt{N}}$$

Although we call the term on the left-hand side of the equation (σ_M) the *standard error of the mean*, no error is implied. The term on the left is the standard deviation of an extremely large number of sample means with N scores in each sample. If we square both sides of the equation, the relationship between the variability of sample means and individual scores remains. And if we multiply both sides of the equation by N, the equation becomes

$$N\sigma_M{}^2 = \sigma^2$$

This new equation means that the variance of all possible sample means based on samples of size N (that is, $\sigma_M{}^2$) multiplied by N is equal to σ^2, the true variance. Note also that the ratio of the left- and right-hand sides of the equation ($N\sigma_M{}^2/\sigma^2$) is equal to 1.00 if $N\sigma_M{}^2 = \sigma^2$. As has been noted many times before, we rarely know the values of parameters, and we must, instead, depend upon statistics. Thus, the above equation can be expressed in terms of statistics rather than in terms of parameters. This equation is

$$NS_M{}^2 = S^2$$

and, as in the parametric equation, this indicates that N times the variance of the sample means will (*on the average*) equal the sample variance. This equation also means that there are two different ways of computing an estimate of the true variance. The first way is simply by computing the variance of the scores (and using S^2 as an estimate of σ^2), and the second way is through the use of the relationship $NS_M{}^2 = S^2$. This can be illustrated by an example taken from Chapter Nine. You will recall that five samples of ten cases each were selected with the following results:

$M_1 = 4.0$

$M_2 = 7.0$

$M_M = 25/5 = 5.00$ (the mean of the means)

$M_3 = 6.0$

$S_M^2 = 2.00$ (the variance of the means)

$M_4 = 4.0$

$N = 10$

$M_5 = 4.0$

$NS_M^2 = 20$

Since $S_M^2 = 2.00$, then $NS_M^2 = 20$; and since NS_M^2 on the average is equal to S^2, then we can also use NS_M^2 as an estimate of σ^2. In this case we estimate that $\sigma^2 = 20$. Of course we do not know what S^2 would equal if it were computed directly from the variability of individual scores. Since both NS_M^2 and S^2 are estimates of σ^2, it becomes apparent that *on the average* the two terms will equal one another. Although the ratio $N\sigma_M^2/\sigma^2 = 1.00$, the ratio NS_M^2/S^2 will rarely be exactly equal to 1.00, but will on the average equal 1.00.

The fact that we are looking at the ratio of two variances means, of course, that we are discussing variables that are F distributed. Let us place the discussion in that light. Initially, let us note that

$$\frac{\dfrac{NS_M^2}{N\sigma_M^2}}{\dfrac{S^2}{\sigma^2}}$$

is F distributed with an average equal to 1.00. Furthermore, if $N\sigma_M^2 = \sigma^2$, then the equation can be simplified:

$$F = \frac{\dfrac{NS_M^2}{N\sigma_M^2}}{\dfrac{S^2}{\sigma^2}} = \frac{\dfrac{NS_M^2}{\sigma^2}}{\dfrac{S^2}{\sigma^2}} = \frac{NS_M^2}{S^2}$$

and hence the ratio of NS_M^2/S^2 will be F distributed with an average equal to 1.00.

Under what conditions will $N\sigma_M^2 = \sigma^2$, and under what conditions will the equation not be true? The equation will always be true if samples are taken from the same population. That is, if successive samples are taken from the same population, then $N\sigma_M^2 = \sigma^2$. But, if successive samples are taken from populations with different means, then the equation will not be true (that is, $N\sigma_M^2 \neq \sigma^2$).

What effect will the taking of samples from populations with different means have upon the value of $N\sigma_M^2$? It can be shown that the value of $N\sigma_M^2$ will be increased relative to the value of σ^2. How much larger σ_M^2 will get will depend upon the difference between the true means $(\mu_x - \mu_y)$. If the difference is large, then the increase in the value of $N\sigma_M{}^2$ will also be large.

As we have stated elsewhere, the average value of the F obtained from the ratio $NS_M{}^2/S^2$ will be equal to the ratio of the parameters $N\sigma_M{}^2/\sigma^2$. Hence, if the value of the ratio $N\sigma_M{}^2/\sigma^2$ increases because of a difference between population means, then the average value of the ratio of $NS_M{}^2/S^2$ will also increase. And as the size of the F ratio increases, the probability that the null hypothesis will be rejected will also increase.

AN EXAMPLE   Before going on, let us look at an example to illustrate what we have been talking about in the previous paragraphs. A sociologist interested in historical factors in delinquency selected a group of five teenage boys who had been convicted of some serious crime. Next, he selected a comparison group of five teenage boys who were as similar as possible to the first five in terms of socioeconomic level, age, and other background factors. He was interested in determining if early life differences could be detected even before any pattern of misbehavior had been found. To do this, he found the number of days absent from school during a three-month period in the first grade. The data are tabulated below.

Delinquent	Nondelinquent
13	14
18	13
14	9
16	11
14	13
$\Sigma X = 75$	$\Sigma Y = 60$
$M_x = 15$	$M_y = 12$
$\Sigma X^2 = 1141$	$\Sigma Y^2 = 736$

$$S_x{}^2 = \frac{1141 - \dfrac{(75)^2}{5}}{4} = \frac{16}{4} = 4.00$$

$$S_y{}^2 = \frac{736 - \dfrac{(60)^2}{5}}{4} = \frac{16}{4} = 4.00$$

The variances of the delinquent group and the nondelinquent group both happen to be 4.00. Next we must compute $NS_M{}^2$. The procedure for computing $S_M{}^2$, the variance of the means, is exactly the same as that for the computation of the variance; the only difference is that the two means are treated as if they were raw scores. $M_{\text{del}} = 15$ and $M_{\text{nondel}} = 12$.

$$\Sigma M = 15 + 12 = 27$$
$$\Sigma M^2 = 12^2 + 15^2 = 144 + 225 = 369$$

$$S_M{}^2 = \frac{\Sigma M^2 - \dfrac{(\Sigma M)^2}{K}}{K - 1} = \frac{369 - \dfrac{27 \times 27}{2}}{1} = \frac{369 - 364.5}{1} = 4.5$$

Notice that in the computation of the variance of the means, K was used rather than N. In the computation of $S_M{}^2$, we treated the means as if they were raw scores. Since there were K scores (two means), K was used rather than N. Since $S_M{}^2 = 4.5$, then $NS_M{}^2 = 5 \times 4.5 = 22.5$ (N refers to the number of scores within each group).

In our previous discussion we said that if $N\sigma_M{}^2 = \sigma^2$, then the ratio $NS_M{}^2/S^2$ would be F distributed and would, on the average, equal 1.00. In the present experiment, $N\sigma_M{}^2 = \sigma^2$ would be true only if the null hypothesis were true. That is, if the mean days absent for the delinquent and non-delinquent groups were equal, then the ratio $NS_M{}^2/S^2$ would on the average equal 1.00, and $N\sigma_M{}^2$ would equal σ^2. But if the mean days absent for the two groups were not equal, then the average of the ratio $NS_M{}^2/S^2$ would be greater than 1.00. An increase in the magnitude of the difference between the two population means results in larger F ratios (again, on the average).

The F distribution is in reality a family of curves (or distributions) with a different distribution for each combination of degrees of freedom in the numerator and in the denominator. Because of the large number of F distributions, only those distributions that have an average score of 1.00 have been tabled. These distributions are of great importance because we can use them to test the null hypothesis. If we conclude that the value of our obtained F ratio is larger than would be expected on the basis of sampling error, we then reject the null hypothesis, and in this case we conclude that there is a difference between our means.

What is the value of the F ratio from our experiment? We found that $NS_M{}^2 = 5 \times 4.5 = 22.5$. We also found that each of the S^2 values we found was equal to 4.00 (and thus the mean S^2 is also equal to 4.00). Hence, the F ratio $= NS_M{}^2/S^2 = 22.5/4.00 = 5.63$. If we evaluate $F = 5.63$ with the proper degrees of freedom, we can then decide whether or not to reject the null hypothesis. We have 4 df associated with the variance of each group; hence, we have a total of 8 df in the denominator. We had two scores (means) with which to compute $S_M{}^2$; hence, we have 1 df in the numerator. Thus, we evaluate our F ratio of 5.63 with 1 and 8 df. An F of 5.32 is required for significance at the 0.05 level. That is, if $N\sigma_M{}^2 = \sigma^2$, then 5 percent of the area under the F distribution with 1 and 8 df falls beyond the point equal to 5.32. If we are randomly taking F values from such a distribution, then the probability is equal to 0.05 that we will select a value larger than 5.32. On the other hand, if $N\sigma_M{}^2 > \sigma^2$, then the probability is greater than 0.05 that we will find an F value beyond 5.32, and hence there is a greater probability that the null hypothesis will be rejected.

The obtained F of 5.63 is larger than the tabled F of 5.32, and for this reason we reject the null hypothesis. In rejecting the null hypothesis, we conclude that even early in life before delinquent behavior is apparent, certain behavioral differences can be found between our two groups; thus, the mean number of days absent for the delinquent group is greater than the mean days absent for the nondelinquent group.

In summary, to test the hypothesis that the true mean of one group is equal to the true mean of another (or the hypothesis that the means of several groups are equal), we find the variance of the means $(S_M{}^2)$, the variance of the individual scores (S^2), and the value of the ratio $NS_M{}^2/S^2$. If the true means are equal, then the ratio $NS_M{}^2/S^2$ will be F distributed and will have an average F ratio of 1.00. Only 5 percent of the time (again if the true means are equal) will an F be obtained that has a value larger than that given in the F table. If the F is larger than that given in the table, the null hypothesis is rejected and we conclude that the true means are not equal. If the F value is smaller than that given by the table we do not reject the null hypothesis and we thus cannot conclude that the means are different from one another. (Note, however, that we cannot conclude that the means are equal to one another.) In our example, the F we computed was larger than the tabled value of F; we then rejected the null hypothesis and concluded that the true means did in fact differ.

COMPUTER-GENERATED F DISTRIBUTIONS

We have used the computer again to illustrate the changes in the F distribution when the difference between the means of two populations increases from no difference to a relatively large difference. To do this, we started with population X, which was normal and had a mean of 50 and a standard deviation of 12. A second population, Y, was also developed with the same mean and standard deviation. Then we selected random samples of ten scores each from populations X and Y. An analysis of variance was next computed on these 20 scores to test the difference between the sample means, and an F value was obtained. Next, two more samples of ten each were taken and again an F value was obtained. This was done again and again for 10,000 times. The value of each F was noted, and the distribution of these F values is presented in Figure 13-1. Since the null hypothesis was true ($\mu_1 = \mu_2 = 50$), the tabled F distribution for 1 and 18 df was generated. From Appendix D with 1 and 18 df, 5 percent of the F values should be larger than 4.41 and 1 percent should be larger than 8.28. From Figure 13-1 it can be seen that about 5 percent of the F values in the graph (5 percent of the area under the curve) are larger than 4.41 and that about 1 percent of the F values exceed 8.28. Thus, the results of the experiment closely approximate what would be expected from knowledge of the F distribution tabled in Appendix D.

There are also two other distributions illustrated in Figure 13-1. These were generated in much the same way as was the first distribution. These other two graphs illustrate the types of F distributions which occur when the null hypothesis is false. In the second experiment, population X remained the same, with $\mu = 50$, $\sigma = 12$. However, the mean of population Y was changed so that $\mu = 60$, $\sigma = 12$. Again successive random samples of ten each were selected from the two populations and F values were computed. This was done again and again until 10,000 F values had been computed. Again the value of each F was noted, and again the distribution of these scores was plotted in Figure 13-1. Finally, a third experiment

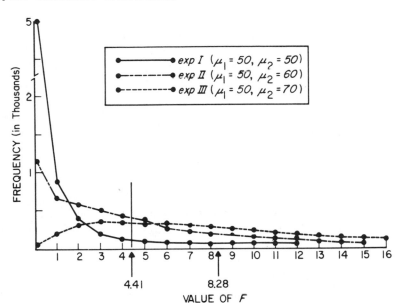

Figure 13-1   *Three different F distributions are generated when the population mean differences are 0, 10, and 20.*

was conducted. In this case, population X had $\mu = 50$, $\sigma = 12$ and population Y had $\mu = 70$, $\sigma = 12$. Again samples of ten scores each were taken from the two populations and an analysis of variance was computed to test the difference between means. The sampling was continued until 10,000 F values were obtained; these data are also presented in Figure 13-1.

Note that when $\mu_1 = \mu_2$, the tabled F distribution with 1 and 18 df was generated; when $\mu_1 \neq \mu_2$, two other F distributions were generated. The shape of these other F distributions depended upon the difference between the true means of the two populations. Casual observation of Figure 13-1 will show you that the area to the right of $F = 4.41$ *increases* from one distribution to another as the difference between the true means increases. That is, when $\mu_1 = \mu_2$, only 5 percent of the area falls to the right of (that is, is larger than) 4.41. From Figure 13-1, when $\mu_1 = 50$ and $\mu_2 = 60$, about half the area falls to the right of 4.41, and when $\mu_1 = 50$ and $\mu_2 = 70$, it appears that about three-fourths of the area falls to the right of 4.41.

The fact that the percent of the area to the right of 4.41 increases simply reflects the fact that as the difference between the two true means increases, the average value of F also increases. The average F value increases because the $\mu_1 - \mu_2$ difference contributes to the value of S_M^2, and as the $\mu_1 - \mu_2$ difference increases, so too does the value of S_M^2.

Finally, notice that if you had sampled from population X and population Y, you would not know if $\mu_1 = \mu_2$. Consequently, if your F value with 1 and 18 df exceeded 4.41, you would have rejected the null hypothesis and concluded that the two true means were different. If $\mu_1 = \mu_2 = 50$, you would have made an incorrect conclusion and you would have committed an alpha error. However, if you had gotten an F value of less than 4.41, you would not have rejected the null hypothesis. But if the null hypothesis were not true (say, $\mu_1 = 50$, $\mu_2 = 60$), you would also have made an error, a beta error. Since you will never know if the null hypothesis is true, the appropriate procedure is to compare the obtained F value with the tabled F value, and if the obtained F is larger than the tabled value, the null hypothesis is rejected. If the null hypothesis is false, the percent of the time you will be right will depend upon the difference between the population means.

COMPUTING FORMULAS

It would be a laborious process indeed to compute $S_M{}^2$ and then compute S^2 for each of two or more groups. For this reason and because an interesting relationship holds between what are known as *sums of squares*, we can use a computing formula that makes computation of our F ratios relatively easy. The computational formulas are based on the fact that the difference between the grand mean (the mean of all the scores and not just the mean of one group) and any raw score can be divided into two parts: (1) the difference between the grand mean and the sample mean, and (2) the difference between the sample mean and the individual score.

In using the computing formulas, we initially compute sums of squares. As you will see, a sum of squares is very similar to a variance. Because we can partition the difference between a raw score and the grand mean into two parts, we can also find a similar relation between the sums of squares associated with these different parts: total (*tot*) sum of squares (*ss*) = between-group (*bg*) sum of squares + within-group (*wg*) sum of squares, or

$$tot_{ss} = bg_{ss} + wg_{ss}$$

The great advantage of this relationship is that once two of the three sums of squares are computed, the third can be determined by addition or subtraction.

Let us look at our study of teenage delinquents once again. You will notice that the data are exactly the same as we used previously ($K = 2$ and $N = 5$)

Delinquent group	Nondelinquent group
13	14
18	13
14	9
16	11
14	13

Summing over all scores:

$$\sum_{1}^{K} \sum_{1}^{N} X = 13 + 18 + \cdots + 11 + 13 = 135$$

Squaring and summing:

$$\sum_{1}^{K} \sum_{1}^{N} X^2 = 13^2 + 18^2 + \cdots + 11^2 + 13^2 = 1877$$

Next find the *correction factor (CF)*:

$$CF = \frac{\left(\sum_{1}^{K} \sum_{1}^{N} X\right)^2}{KN} = \frac{(135)^2}{10} = \frac{18,225}{10} = 1822.5$$

Use of the correction factor (CF) enables us to change from raw scores to deviation scores. If, for example, a score of 10 were added to each of the scores in both groups, the values of

$$\sum_{1}^{K} \sum_{1}^{N} X \quad \text{and} \quad \sum_{1}^{K} \sum_{1}^{N} X^2$$

would change, but the values of the variances we are computing would remain the same because the CF corrects for the size of the score when variances are computed.

We now find the total sum of squares (tot_{ss}):

$$tot_{ss} = \sum_{1}^{K} \sum_{1}^{N} X^2 - \frac{\left(\sum_{1}^{K} \sum_{1}^{N} X\right)^2}{KN} = \sum_{1}^{K} \sum_{1}^{N} X^2 - CF$$

Notice that if we were to divide the tot_{ss} by $KN - 1$ (all the scores minus one), we would have computed the variance of all our scores, ignoring the groups they were in.

$$tot_{ss} = \sum_{1}^{K} \sum_{1}^{N} X^2 - CF = 1877 - 1822.5 = 54.5$$

BETWEEN-GROUP SUM OF SQUARES The next sum of squares we will compute is the between-group sum of squares, or bg_{ss}. The bg_{ss} is analogous to the S_M^2 which we computed previously. The formula for bg_{ss} is

$$bg_{ss} = \frac{\sum_{1}^{K} \left(\sum_{1}^{N} X\right)^2}{N} - CF$$

$$= \frac{(13 + 18 + 14 + 16 + 14)^2}{5} + \frac{(14 + 13 + 9 + 11 + 13)^2}{5} - CF$$

$$= \frac{75^2 + 60^2}{5} - CF$$

$$= \frac{5625 + 3600}{5} - 1822.5$$

$$= 1845 - 1822.5$$

$$= 22.5$$

WITHIN-GROUP SUM OF SQUARES   The final sum of squares we will compute is the within-group sum of squares, or wg_{ss}. The wg_{ss} is analogous to the S^2 computed in the previous example. The formula for wg_{ss} is

$$wg_{ss} = \sum_1^K \sum_1^N X^2 - \frac{\sum_1^K \left(\sum_1^N X \right)^2}{N}$$

But we have previously noted that $tot_{ss} = bg_{ss} + wg_{ss}$. Since we already have computed tot_{ss} and bg_{ss}, we can compute wg_{ss} simply by subtracting bg_{ss} from tot_{ss}. That is, $wg_{ss} = tot_{ss} - bg_{ss}$. Since $tot_{ss} = 54.5$ and $bg_{ss} = 22.5$, then $wg_{ss} = 54.5 - 22.5 = 32.00$.

The next step, after we have computed tot_{ss}, bg_{ss}, and wg_{ss}, is the construction of a source table. The source table enables us to present our calculations in a concise manner and helps us to remember some additional calculations that must be made. Rather than include these additional calculations in a formula, the additional necessary operations are indicated by the source table itself.

Source	ss	df	ms	F
bg	22.5	1	22.5	5.63
wg	32.0	8	4.0	
tot	54.5	9		

The first column in the table shows the source or origin of the various sums of squares shown in the column labeled "ss." Thus we can see that bg, for example, has a sum of squares of 22.5 and that the tot_{ss} is equal to 54.5. It can be seen from the table that the sum of the values for bg_{ss} and wg_{ss} is equal to the tot_{ss}.

The third column is labeled "df," or degrees of freedom. Determination of the proper number of degrees of freedom associated with each sum of squares is a most important aspect of the computation of the analysis of variance. As we have noted before, a single df is lost every time a parameter is estimated. This is also true in the analysis of variance where the total number of df is always one less than the total number of scores. In this case a df is lost because the *true grand mean* is estimated. This parameter is based on the grand mean of all the scores regardless of groups. Because 1 df is always lost, the total number of df in any analysis of variance is always one less than the total number of scores. In the delinquent versus nondelinquent experiment, there were two groups $(K = 2)$ with five scores in each group $(N = 5)$. Thus there were a total of $2 \times 5 = 10$ scores $(KN = 10)$. Since 1 df is lost, the total df $= 9$ (tot df $= KN - 1$). These 9 df are then divided between the bg and wg sums of squares. The number of df associated with bg_{ss} is always one less than the number of groups (bg df $= K - 1$). In the delinquency experiment there were two groups and hence there are $2 - 1 = 1$ df associated with the bg_{ss}. The remainder of the df are associated with the wg_{ss} and may be determined by subtraction in exactly the same manner that the wg_{ss} was computed.

The wg df may be found directly. The df in the wg are always equal to the sum of the df associated with each group, and the number of df in each group is $N - 1$. Hence, the wg df are $K(N - 1)$. In the delinquency experiment, since $K = 2$ and $N = 5$, the wg df $= 2(5 - 1) = 8$. Notice finally that bg df $+ wg$ df $=$ tot df; in the delinquency experiment, $1 + 8 = 9$. The formula for df in the general case is $(K - 1) + K(N - 1) = KN - 1$. The wg df $+ bg$ df will always equal the total df.

The reason why computation of the df is so important is that the df associated with each sum of squares are divided into that sum of squares, resulting in the *mean square* in the fourth column. For example, bg_{ss}/bg df $= 22.5/1 = 22.5$, and wg_{ss}/wg df $= 32/8 = 4.00$. As you can see from the table, $bg_{ms} = 22.5$ and $wg_{ms} = 4.00$. These scores may be familiar to you, since $bg_{ms} = NS_M^2$ and $wg_{ms} = S^2$. Thus the computational formulas have done exactly what was done previously: NS_M^2 and S^2 have been computed.

Once bg_{ms} and wg_{ms} have been computed, we do exactly what we did before when we divided NS_M^2 by S^2. If we divide bg_{ms} by wg_{ms}, we compute the F ratio of 5.63. Notice that this F ratio is exactly the same as we computed previously from the ratio NS_M^2/S^2, and it is interpreted in exactly the same manner.

A word should be said about the computation of wg_{ms}. In our example we found wg_{ms} equal to S^2. In actual practice, wg_{ms} is the mean of all the S^2 computed from each of the several groups. For example, suppose we were to compute an analysis of variance in which there were three groups. If $S_x^2 = 3$, $S_y^2 = 4$, and $S_z^2 = 8$, then the mean of all the S^2 would equal 5.00 and hence wg_{ms} would also equal 5.00.

ASSUMPTIONS OF THE ANALYSIS OF VARIANCE

1. Random sampling.
2. Independent sampling.
3. Samples come for a normal distribution.
4. True variances of each group are equal (that is, $\sigma_1^2 = \sigma_2^2 = \cdots = \sigma_k^2$).
5. At least an interval scale of measurement is used.
6. Independence of means and variances.

AN EXAMPLE   The preceding discussion and explanation have tended to obscure the ease with which an analysis of variance may be computed and interpreted. By way of illustration we have chosen the following example:

Suppose a medical research worker believes that a physician's attitude toward his nurse is related to the number of years he has spent practicing medicine. Hence, she selects three groups of physicians who have spent a short time (median of 3 years), medium time (median of 12 years), or a long time (median of 24 years) in practice. Next she has each physician rate on a 0–12 scale (from poor to good) his nurse (or head nurse). (The average age of the nurses in each group have been equated.) The results of the experiment are presented below, with the data being the ratings of the nurses by the physicians.

Group 3	Group 12	Group 24
0	5	5
7	2	11
2	3	9
1	9	9
1	6	4
7	5	10
$\Sigma X = 18$	30	48

$$\sum_1^K \sum_1^N X = 0 + 7 + 2 + \cdots + 9 + 4 + 10 = 96$$

$$\sum_1^K \sum_1^N X^2 = 0^2 + 7^2 + 2^2 + \cdots + 9^2 + 4^2 + 10^2 = 708$$

$$CF = \frac{\left(\sum_1^K \sum_1^N X\right)^2}{KN} = \frac{96 \times 96}{18} = 512$$

$$tot_{ss} = \sum_1^K \sum_1^N X^2 - CF$$

$$= 708 - 512 = 196$$

$$bg_{ss} = \frac{\sum_1^K \left(\sum_1^N X\right)^2}{N} - CF$$

$$= \frac{18^2 + 30^2 + 48^2}{6} - CF$$

$$= 588 - 512 = 76$$

$$wg_{ss} = tot_{ss} - bg_{ss}$$

$$= 196 - 76 = 120$$

We next set up a source table.

Source	ss	df	ms	F
bg	76.0	2	38.0	4.75
wg	120.0	15	8.0	
tot	196.0	17		

There are 18 scores, and hence *tot* df $= 17$. There are three groups, and hence *bg* df $= K - 1 = 2$. Subtracting 2 from 17 gives us the number of *wg* df $= 15$. Dividing df into the sums of squares associated with them results in mean squares of 38.0 and 8.0 for *bg* and *wg*, respectively. Finally, the ratio between bg_{ms} and wg_{ms} is found to equal $F = 4.75$. An F of 3.68 is required to reject the null hypothesis at the 0.05 level with 2 and 15 df. The obtained F is larger than the tabled F, and the probability that an F this large would occur by chance is less than 0.05. Hence, the null hypothesis is rejected. The medical research worker concludes that the ratings of the nurses are dependent upon the length of time the physician has been in practice. The experimenter's own personal hypothesis is thus confirmed. Note, however, that procedures involved in statistical inference have little to do with the experimenter's beliefs. Statistically speaking, what is being tested is not the opinion of the experimenter but rather the null hypothesis—the hypothesis that there are no differences between true means and that differences between the sample means are simply due to chance. The null hypothesis was tested by the experiment, and the results indicated that a difference between the means as large as those observed would occur less than 5 percent of the time if the null hypothesis were true. Since we have arbitrarily chosen 5 percent as the cutoff point, we then say that we reject the null hypothesis and that the results must represent differences between the true means.

Note that the conclusions are stated in terms of inferences about the populations from which the samples were drawn. We do *not* simply conclude that the sample means are different; we can determine that fact by inspection. Our inferences are always about the populations from which the samples were drawn.

ANALYSIS OF VARIANCE WITH UNEQUAL *N*

One final example will allow us to demonstrate the analysis of variance with groups of unequal sizes. No new formulas are necessary, and the theory is exactly the same. There are, however, slight changes in the computational procedures.

The data presented below were taken from an experiment in which attitudes toward children were measured in student teachers majoring in

Elementary	Secondary
5	3
4	4
5	3
5	4
4	5
5	3
4	2
5	4
5	3
4	4
	3
	4
$\Sigma X = \overline{46}$	$\Sigma Y = \overline{42}$

summing over all scores: $5 + 4 + \cdots + 3 + 4 = 88$
squaring and summing: $5^2 + 4^2 + \cdots + 3^2 + 4^2 = 368$
next find the correction factor (CF)

elementary and secondary education. The null hypothesis to be tested here is that the population means of the secondary and elementary education majors are equal. Two samples of student teachers were randomly selected, and each student was given a questionnaire in which a high score indicated a positive attitude toward children. Included in the survey were 12 secondary and 10 elementary education majors.

$$CF = \frac{\left(\sum\limits_{1}^{K}\sum\limits_{1}^{N} X\right)^2}{N_x + N_y} = \frac{(46 + 42)^2}{10 + 12} = \frac{(88)^2}{22} = 352$$

$$tot_{ss} = \sum\limits_{1}^{K}\sum\limits_{1}^{N} X^2 - CF$$

$$= 368 - 352 = 16.00$$

$$bg_{ss} = \frac{\left(\sum\limits_{1}^{N} X\right)^2}{N_x} + \frac{\left(\sum\limits_{1}^{N} Y\right)^2}{N_y} - CF$$

$$= \frac{(46)^2}{10} + \frac{(42)^2}{12} - 352$$

$$= 211.60 + 147.00 - 352 = 6.60$$

$$wg_{ss} = tot_{ss} - bg_{ss}$$

$$= 16.00 - 6.60 = 9.40$$

Source	ss	df	ms	F
bg	6.60	1	6.60	14.04
wg	9.40	20	0.47	
tot	16.00	21		

Since the F ratio exceeds the value tabled for 1 and 20 df at the 0.01 level, we conclude that the two samples do not come from populations with the same mean, and we reject the null hypothesis. We conclude that elementary education majors ($M = 4.60$) have a more positive attitude toward children than do secondary education majors ($M = 3.50$).

This analysis is nearly identical to previous analyses which have been done. Only a few minor changes have been introduced. Since N is not constant from group to group, KN will not give us the total number of scores. For this reason, when computing CF we found the total number of scores by adding. Thus, $N_x + N_y = 10 + 12 = 22$, which was then divided into the sums of all the scores. For similar reasons we found the bg_{ss} by summing and squaring the scores in each of the two groups and then dividing by the number of scores in each group—10 in one and 12 in the other. The wg_{ss} was found by subtraction: $wg_{ss} = tot_{ss} - bg_{ss}$.

Finally, df was computed for tot_{ss} by finding the sum of all scores minus 1: $N_x + N_y - 1 = 10 + 12 - 1 = 21$. The df for bg_{ss}, as before, is equal to $K - 1 = 2 - 1 = 1$, and the df for wg_{ss} was found by subtraction: wg df $= tot$ df $- bg$ df $= 21 - 1 = 20$.

t TEST BETWEEN MEANS

We noted in Chapter Ten that use of the *t* was not restricted to the single-sample situation and that it could be used in comparing two sample means. The *t* and *F* distributions are related such that $t^2 = F$ when *F* has a single degree of freedom in the numerator (that is, in the two-group case). Hence, if you were to compute *F* in the usual way and then find the square root of the *F* value, this would be equal to *t*. For example, in the delinquent versus nondelinquent experiment, $F = 5.63$. The square root of this is 2.37. Hence, $t = 2.37$ with 8 df, and this can be evaluated in Appendix C. Similarly, $t^2 = F$ in the two-group case when the number of cases in one group differs from that in the other. For example, in the experiment comparing elementary and secondary education students, $F = 14.04$. Taking the square root of that, we find $t = 3.75$, which would be evaluated in Appendix C with 20 df.

SUMMARY OF USES OF *z*, *t*, AND *F*

As described in Chapter Nine, the *z* distribution may be used to establish a confidence interval for the true mean. Similarly, the *z* distribution, by use of the formula $z = (M - \mu)/\sigma_M$, may be used to determine if the true mean for a sample is μ. Knowledge of the parameter σ is required with use of the *z* distribution.

The *t* distribution was discussed in Chapter Ten. The *t* was developed for situations where the parameter σ is unknown but where the statistic *S* is known. Such a situation would be encountered when the number of cases in a sample were too small to make an adequate estimate of σ. When σ is not known, it is possible, by use of the *t* distribution, to answer the same questions as were answered through use of the *z* distribution. That is, the *t* distribution may be used to establish a confidence interval for the true mean, and the formula $t = (M - \mu)/S_M$ may be used to determine if the true mean for a sample is μ. Since the shape of the *t* distribution depends upon the number of degrees of freedom, the df must be found, and the appropriate value of *t* can then be determined with the help of Appendix C.

In Chapter Eleven the *F* distribution was introduced. Like the *t*, the shape of the *F* distribution is dependent upon the number of df. In this case, df in both the numerator and denominator determine the shape of the *F* distribution. A ratio of sample variances S_x^2/S_y^2 is reduced to an *F* value, which is compared to the appropriate value of *F* in Appendix D. If the tabled value is exceeded, the hypothesis that the variances have the same true variance is rejected. It should be noted that this use of the *F* distribution is concerned only with differences between variances, and is independent of differences between the means of samples or populations.

Finally, the present chapter (as well as the next) describes the use of the *F* distribution in answering questions about differences between the means of populations from which two or more samples have been drawn.

The comparison of sample means by use of the analysis of variance is probably the most widely used statistical technique employed by research workers in the behavioral sciences. This technique depends upon the relationship between the variability of individual scores and the variability of sample means: $S_M{}^2 = S^2/N$ or $NS_M{}^2 = S^2$. The formula $F = NS_M{}^2/S^2$ was first used to demonstrate that the ratio between these two sources of variance should equal 1.00 if the null hypothesis were true. Subsequently, an equivalent formula, $F = bg_{ms}/wg_{ms}$, was introduced for ease in computation.

It is crucially important that the two uses of the F distribution be carefully differentiated. The comparison of sample variances and the comparison of sample means are entirely different operations because the means and variances of samples taken from a normally distributed population are independent of one another. A comparison of sample variances tells us nothing about the values of the means, and vice versa.

The following table may help you recall this summary.

Distribution	Number of samples	Basic formula	Purpose of test
z	1	$z = (M - \mu)/\sigma_M$	Test $H: \mu = X$
		$\mu = M \pm z\sigma_M$	Location of μ
t	1	$t = (M - \mu)/S_M$	Test $H: \mu = X$
		$\mu = M \pm tS_M$	Location of μ
F	2	$F = S_x{}^2/S_y{}^2$	Comparison of variances
F	2 or more	$F = bg_{ms}/wg_{ms}$	Comparison of means

PROBLEMS

1. In an experiment which has four groups of ten scores each, an analysis of variance is computed. What are the df associated with each source of variance?

2. Complete the following table. Four groups of three subjects each served in the experiment. What is the null hypothesis? Can it be rejected?

Source	ss	df	ms	F
bg	24			
wg				
tot	40			

3. Three groups of four subjects each were run in an experiment. Compute CF, bg_{ss}.

I	II	III
1	4	6
2	0	0
3	1	2
0	1	4

4. In an experiment with six groups of ten scores each, the total sum of squares is equal to 990 and the between-groups sum of squares is equal to 450. What is the value of the standard error of the mean (that is, the square root of the variance of the means) and the average of each group?

5. Compute an analysis of variance for the following data. Do the means differ?

I	II	III	IV
0	4	5	6
0	2	3	4

6. Compute an analysis of variance for the following data. Do the means differ?

A_1	A_2	A_3
0	1	8
0	4	5
4	5	4
0	2	3

7. This is a classic question taken from an old exam. A graduate student had collected the data for his master's thesis and had stored it in his car. He went for a ride, had an accident, and drove off a bridge. He lost his data, but was able to remember the summary statistics, which are presented in the following table ($N = 9$ per group). Do the means differ?

I	II	III	IV
$M = 6.00$	$M = 10.00$	$M = 9.00$	$M = 15.00$
$S = 5.00$	$S = 1.00$	$S = 7.00$	$S = 3.00$

8. Do the means of the following groups differ? Do an analysis of variance.

Ex	C
7	3
5	2
6	4
6	1
	5
	3
	2
	4

9. Given the information in the following table, answer the following questions:
(a) How many groups were there in the experiment?
(b) How many scores per group?
(c) What was the average variance of the scores in each groups?
(d) $S_M^2 = $ _____.

Source	ss	df	ms	F
bg		3		
wg	112			
tot	160	31		

10. To test her physical fitness program, one research worker randomly assigned students to two groups of 20 each. The first, after preliminary training, jogged 5 miles a day; the second served as a control. After six months on this program, the respiration rate (in liters per minute) after no exertion was checked for both groups. In the no-special-exercise group the mean rate was 7.00 ($S = 2.00$), while in the exercise group $M = 5.50$, $S = 1.50$. Test the hypothesis that the rates of respiration are the same for both groups.

11. Given the following information, set up a source table showing ss, df, ms, and F. Do the groups differ?

$$\sum_1^K \sum_1^N X = 45; \quad \sum_1^K \sum_1^N X^2 = 175; \quad \sum_1^K \left(\sum_1^N X\right)^2 = 755; \quad K = 3; N = 5$$

12. Given the following information, set up a source table. Find ss, df, ms, and F.

$$\sum_1^K \sum_1^N X = 40; \quad \sum_1^K \sum_1^N X^2 = 172; \quad \frac{\sum_1^K \left(\sum_1^N X\right)^2}{N} = 136; \quad N = 4; K = 4$$

13. You conduct an experiment with the following results:

Group A	Group B	Group C
0	4	6
1	8	9
2	2	5
5	2	4

Do the analysis, set up a source table, and find F. Do you reject the null hypothesis?

14. You conduct an experiment with the following results:

I	II	III	IV
0	1	2	4
1	2	4	5
2	4	4	7

Do the analysis, set up a source table, and find F. Do you reject the null hypothesis?

15. Given the following scores from an experiment, compute S_M^2 and S^2 directly from the data.

I	II	III	IV
1	4	4	4
3	7	8	8
3	6	5	8
1	7	7	4

EXERCISES

1. Read the following words to one group: hurdle, tripod, snake, pencil, ladder, yacht, jail, volcano, diamond, key, elephant, chair, truck, fish, bicycle. Read these words to a second group: method, knowledge, thought, adage, amount, belief, chance, excuse, hint, idea, origin, unit, truth, fact, rating.

   Read the words at intervals of one second. When you finish, ask members of the group to write down as many words as they can remember. Count the number correct. Use the appropriate statistic to see if the two true means differ from one another. If they do, why might this be?

2. Take two random samples of ten single-digit numbers from Appendix H. Add 4 to each number in one of the samples. Test the hypothesis $\mu_1 = \mu_2$. Next add 4 more to each number in the sample so that one sample has had 8 added to each score. Again test the hypothesis $\mu_1 = \mu_2$. How has the F value changed? What has happened to the numerator? To the denominator?

1

$$S_x{}^2 = \frac{\sum\limits_1^N x^2}{(?)} \quad \text{or} \quad S_y{}^2 = \frac{\sum\limits_1^N y^2}{(?)}$$

══════ STOP ══════ ══════ STOP ══════

$$N_x - 1, \ N_y - 1$$

2

$$S^2 = \frac{\sum\limits_1^N x^2}{(N-1)}; \quad \sum\limits_1^N x^2 = 100, \quad N_x = 11; \quad \sum\limits_1^N y^2 = 40, \quad N_y = 21.$$

$$S_x{}^2 = \underline{\hspace{1cm}}, \ S_y{}^2 = \underline{\hspace{1cm}}, \text{ and } F = \underline{\hspace{1cm}}.$$

══════ STOP ══════ ══════ STOP ══════

10, 2, 5.00

---

3 $S_x{}^2 = 10, N_x = 11; S_y{}^2 = 2, N_y = 21.\ F = 5.00,\ df = \underline{\hspace{1cm}}$ and $\underline{\hspace{1cm}}$.

══════ STOP ══════ ══════ STOP ══════

10 and 20

---

4 $S_x{}^2 = 10, \ N_x = 11; \ S_y{}^2 = 2, \ N_y = 21; \ F = 5.00,$ evaluate $F.$ $\underline{\hspace{1cm}}$.

══════ STOP ══════ ══════ STOP ══════

Reject H: $\sigma_x{}^2 = \sigma_y{}^2$

---

5 σ_M is called the *true standard error of the mean* and it is the standard deviation of a large number of $\underline{\hspace{2cm}}$.

══════ STOP ══════ ══════ STOP ══════

means

---

6   In a previous section we said that the true standard error of the mean (σ_M) was related to the true standard deviation (σ) by the formula $\sigma_M = \underline{\hspace{2cm}}$.

══════ STOP ══════ ══════ STOP ══════

$\sigma/\sqrt{N}$

---

7   Since S_M (the standard error of the mean) is an estimate of σ_M, we could compute S_M directly by finding the standard deviation of a group of $\underline{\hspace{1cm}}$.

══════ STOP ══════ ══════ STOP ══════

means

---

8   Since S_M is an estimate of σ_M and S is an estimate of σ then S_M may be estimated from S by the formula $S_M =$ _____.

━━━━━━━━ STOP ━━━━━━━━━━━━━━━━━━━━━━━ STOP ━━━━━━━━

$$S/\sqrt{N}$$

---

9   If we square both sides of the equation then $S_M{}^2 =$ _____.

━━━━━━━━ STOP ━━━━━━━━━━━━━━━━━━━━━━━ STOP ━━━━━━━━

$$S^2/N$$

---

10   Furthermore if we multiply both sides of the equation by N we get $NS_M{}^2 = NS^2/N$ and the right side may be simplified to _____.

━━━━━━━━ STOP ━━━━━━━━━━━━━━━━━━━━━━━ STOP ━━━━━━━━

$$NS^2/N = S^2$$

---

11 $S^2 = NS_M{}^2$. This means then that we can estimate the sample variance by direct computation of the variance or by computation of the variance of a group of _____ and then multiplying by N.

━━━━━━━━ STOP ━━━━━━━━━━━━━━━━━━━━━━━ STOP ━━━━━━━━

means

---

12 N of course refers to the _____ of scores in each sample and *not* to the _____ of samples.

━━━━━━━━ STOP ━━━━━━━━━━━━━━━━━━━━━━━ STOP ━━━━━━━━

number, number

---

13 $S^2 = NS_M{}^2$. If we find $S_M{}^2 = 10$ and the number of scores in each of the several samples is equal to 5, then $S^2 =$ _____.

━━━━━━━━ STOP ━━━━━━━━━━━━━━━━━━━━━━━ STOP ━━━━━━━━

50

---

14   Similarly, since $S_M{}^2 = S^2/N$, if we find from a sample of ten scores that $S^2 = 25$, then $S_M{}^2 =$ _____.

━━━━━━━━ STOP ━━━━━━━━━━━━━━━━━━━━━━━ STOP ━━━━━━━━

2.5

---

15 Thus, if we know the variance of a group of means it is obvious we can estimate _____.

═══════ STOP ═══════════════════════════ STOP ═══════

$$S^2$$

---

16 Note however the relationship between S^2 and $S_M{}^2$. $S^2 = NS_M{}^2$ only if $N\sigma_M{}^2 =$ _____.

═══════ STOP ═══════════════════════════ STOP ═══════

$$\sigma^2$$

---

17 If $\sigma^2 = N\sigma_M{}^2$, and we took several samples from the same population, computed S^2 from individual scores and $S_M{}^2$ from the means then on the average $NS^2{}_M/S^2$ should equal _____ (remember $S^2 = NS^2{}_M$).

═══════ STOP ═══════════════════════════ STOP ═══════

$$N\sigma_M{}^2/\sigma^2 = 1$$

---

18 Since S^2 and $S_M{}^2$ are computed independently of one another, we would *not* always expect $NS_M{}^2/S^2$ to equal _____ (what value).

═══════ STOP ═══════════════════════════ STOP ═══════

1.00

---

19 Rather we would expect the ratio $NS_M{}^2/S^2$ to _____ around an average ‑of 1.00.

═══════ STOP ═══════════════════════════ STOP ═══════

vary

---

20 $S^2 = NS_M{}^2$. Remember that the ratio of two variances $S_x{}^2/S_y{}^2$ is _____ distributed.

═══════ STOP ═══════════════════════════ STOP ═══════

$$F$$

---

21 Then it would seem that the ratio $NS_M{}^2/S^2$ would be _____ distributed.

═══════ STOP ═══════════════════════════ STOP ═══════

$$F$$

---

22   Thus the ratio $NS_M{}^2/S^2$ would be F distributed with an average of 1.00 if (and only if) $N\sigma_M{}^2 = $ _____.

═══════ STOP ═══════════════════ STOP ═══════

$$\sigma^2$$

---

23   If $N\sigma_M{}^2$ were greater than σ^2, then the ratio $NS_M{}^2/S^2$ would be F distributed but the average of the ratio would not equal _____.

═══════ STOP ═══════════════════ STOP ═══════

1.00

---

24   We wish to find those conditions under which $N\sigma_M{}^2$ will be greater than σ^2. The next ten frames are apt to be confusing. They attempt to show why $N\sigma_M{}^2$ is greater than σ^2 when the true means differ.

---

25   Suppose we sample from a population with $\mu = 100$ and $\sigma^2 = 20$. If we take two samples of ten scores each we can find the mean of each sample and compute the variance of the two means $(S_M{}^2)$. Since $\sigma_M{}^2 = \sigma^2/N$ then on the average $S_M{}^2$ will equal _____ (what value).

═══════ STOP ═══════════════════ STOP ═══════

$20/10 = 2.00$

---

26   Two samples of ten each taken from a population with $\mu = 100$ and $\sigma^2 = 20$ will yield $S_M{}^2$ which will have a mean value of 2.00. It is obvious that on the average $NS_M{}^2 = N\sigma_M{}^2 = $ _____.

═══════ STOP ═══════════════════ STOP ═══════

$$\sigma^2$$

---

27   Suppose we now take a sample of ten cases from population X and a sample of ten cases from population Y. $\mu_x = 100$, $\sigma_x{}^2 = 20$; $\mu_y = 40$, $\sigma_y{}^2 = 20$. It may not be obvious, but the average variance of the sample means multiplied by N $(NS_M{}^2)$ will be greater than _____.

═══════ STOP ═══════════════════ STOP ═══════

$\sigma^2(S^2)$

---

28 $\mu_x = 100$, $\sigma_x{}^2 = 20$; $\mu_y = 40$, $\sigma_y{}^2 = 20$. Let us compute the variance of the two true means 100 and 40. $\sum_1^N X^2 = $ _____ and $\sum_1^N X = $ _____.

═══════ STOP ═══════════════════ STOP ═══════

11600, 140

29 Since we are finding the variance of two scores then N in the equation $\sum_1^N x^2 = \sum_1^N X^2 - (\sum_1^N X)^2/N$ will equal _____.

═══ STOP ═══ ═══ STOP ═══

2

30 $\sum_1^N X^2 = 11600, \sum_1^N X = 140. \sum_1^N x^2 = \sum_1^N X^2 - (\sum_1^N X)^2/N = $ _____.

$S^2 = \sum_1^N x^2/(N-1) = $ _____.

═══ STOP ═══ ═══ STOP ═══

1800; 1800

31 $\mu_x = 100, \sigma_x^2 = 20; \mu_y = 40, \sigma_y^2 = 20$. The variance of the two true means (σ_M^2) was found to equal 1800. Yet if the equation $\sigma_M^2 = \sigma^2/N$ holds (and if $N = 10$), then σ_M^2 should equal _____.

═══ STOP ═══ ═══ STOP ═══

2.00

32 $\sigma_M^2 = \sigma^2/N$ only when samples are taken from populations that have equal means (that is, $\sigma_M^2 = \sigma^2/N$ if and only if $\mu_1 = \mu_2 = \cdots \mu_N$). However in our example, we calculated σ_M^2 directly from μ_x and μ_y. If $\mu_x = \mu_y$ then by direct calculation, σ_M^2 would equal _____.

═══ STOP ═══ ═══ STOP ═══

0

33 How can the equation $\sigma_M^2 = \sigma^2/N$ ever hold? The answer is that when the true means differ, there are two sources of variability: the variance of the true means (which we just computed) and the variance of sample means. $\sigma_M^2 = \sigma^2/N$ refers to this second source. However, when the true means are equal there is only *one* source of variability: The variance of the _____ _____.

═══ STOP ═══ ═══ STOP ═══

sample means

34 Thus if $\mu_1 \neq \mu_2$ then $N\sigma_M^2$ will be greater than σ^2 and NS_M^2 will, on the average, be greater than _____.

═══ STOP ═══ ═══ STOP ═══

S^2

35 If $N\sigma_M{}^2 = 100$ and $\sigma^2 = 25$, then on the average F will equal _____.

═══════ STOP ═══════════════════════ STOP ═══════

4.00

---

36 If $N\sigma_M{}^2 = 50$ and $\sigma^2 = 10$, then the average value of F will equal _____.

═══════ STOP ═══════════════════════ STOP ═══════

5.00

---

37 In general if the true means are not equal then $N\sigma_M{}^2$ will be greater than

_____.

═══════ STOP ═══════════════════════ STOP ═══════

σ^2

---

38 And if $N\sigma_M{}^2$ is greater than σ^2 then the average value of F will be greater than _____.

═══════ STOP ═══════════════════════ STOP ═══════

1.00

---

39 Thus if $\mu_1 \neq \mu_2$ (if μ_1 does not equal μ_2) then the average value of F will be greater than _____.

═══════ STOP ═══════════════════════ STOP ═══════

1.00

---

40 We now have a method by which we can determine if _____ differ by seeing if variances differ.

═══════ STOP ═══════════════════════ STOP ═══════

means

---

41 The analysis of variance is the statistical method by which differences between two or more means are tested. It is simply computing the F ratio between $NS_M{}^2$ and _____ and then evaluating the obtained F.

═══════ STOP ═══════════════════════ STOP ═══════

S^2

---

42 More simply, the analysis of variance compares variance estimates based on the variability of means with variance estimates based on variability of _____.

individuals (individual scores)

---

43 Suppose we have three groups and wish to evalute the differences observed among the three means. We would compute an analysis of _____.

variance

---

44 Given these three groups to evaluate, we would, in light of the previous discussion, compute the F ratio between _____ and _____. However, rather than computing _____ and _____ directly, computational formulas are available which enable us to compute F more rapidly.

S^2 and $NS_M{}^2$; S^2 and $NS_M{}^2$

---

45 Suppose we have two groups of three scores each: group I: 0, 2, 3; group II: 4, 7, 8. K, the number of groups, = _____ and N the number of individuals within a group = _____.

2, 3

---

46 Group I: 0, 2, 3; group II: 4, 7, 8. We first must get an estimate of the total variability (variance).

$$\sum_1^K \sum_1^N X = \text{_____} \quad \text{and} \quad \sum_1^K \sum_1^N X^2 = \text{_____}.$$

24, 142

---

47 I: 0, 2, 3; II: 4, 7, 8. $\sum_1^K \sum_1^N X = 24$, $\sum_1^K \sum_1^N X^2 = 142$. We next compute the correction factor (CF).

$$KN = \text{_____}, \qquad CF = \left(\sum_1^K \sum_1^N X\right)^2 / KN = \text{_____}.$$

6, 96

---

48   I: 0, 2, 3; II: 4, 7, 8. $CF = 96, \sum_{1}^{K}\sum_{1}^{N} X = 24, \sum_{1}^{K}\sum_{1}^{N} X^2 = 142$. The measure of total variability is called the *total sum of squares* ($tot_{..}$).

$$tot_{..} = \sum_{1}^{K}\sum_{1}^{N} X^2 - CF = \underline{\qquad}.$$

══════ STOP ══════════════════════════ STOP ══════

46

──

49   The formula for the total sum of squares is already quite familiar (in a simpler form). $\sum_{1}^{K}\sum_{1}^{N} X^2 - (\sum_{1}^{K}\sum_{1}^{N} X)^2/KN$ is exactly the same formula as _____.

══════ STOP ══════════════════════════ STOP ══════

$$\sum X^2 - (\sum X)^2/N$$

──

50   and _____ $= \sum X^2 - (\sum X)^2/N$.

══════ STOP ══════════════════════════ STOP ══════

$$\sum x^2$$

──

51   Hence, _____ is simply the value of $\sum x^2$ for all N scores in all K groups.

══════ STOP ══════════════════════════ STOP ══════

$$tot_{..}$$

──

52

$$\text{Hence } tot_{..} = \underline{\qquad} - \frac{\left(\sum_{1}^{K}\sum_{1}^{N} X\right)^2}{KN}.$$

══════ STOP ══════════════════════════ STOP ══════

$$\sum_{1}^{K}\sum_{1}^{N} X^2$$

──

53

$$tot_{..} = \sum_{1}^{K}\sum_{1}^{N} X^2 - \underline{\qquad}$$

══════ STOP ══════════════════════════ STOP ══════

$$\left(\sum_{1}^{K}\sum_{1}^{N} X\right)^2/KN$$

──

54

$$CF = \underline{\qquad}$$

════ STOP ═══════════════════════ STOP ════

$$\frac{\left(\sum\limits_{1}^{K} \sum\limits_{1}^{N} X \right)^{2}}{KN}$$

---

55   Hence $tot_{..} = \sum\limits_{1}^{K} \sum\limits_{1}^{N} X^{2} - \underline{\qquad}$.

════ STOP ═══════════════════════ STOP ════

$$CF$$

---

56   In a single-classification analysis of variance the tot_{ss} may be broken down into the *between-groups sum of squares* and the *within-groups sum of squares*. The between-groups sum of squares (bg_{ss}) is based on the variability of _____.

════ STOP ═══════════════════════ STOP ════

means

---

57   The within-groups sum of squares ($wg_{..}$) is based on the variability of _____ _____.

════ STOP ═══════════════════════ STOP ════

individual scores

---

58   An important relationship between $tot_{..}$, $bg_{..}$, and $wg_{..}$ is that $bg_{..} + wg_{..}$ always equals _____.

════ STOP ═══════════════════════ STOP ════

$$tot_{..}$$

---

59   The formula for computing $tot_{..} = \underline{\qquad}$.

════ STOP ═══════════════════════ STOP ════

$$\sum\limits_{1}^{K} \sum\limits_{1}^{N} X^{2} - CF$$

---

60   Group I: 0, 2, 3; group II: 4, 7, 8. $tot_{..} = 46$, $CF = 96$. The next step is finding the between-groups sum of squares.

$$bg_{..} = \sum_{1}^{K} \left(\sum_{1}^{N} X \right)^2 / N - CF$$

For group I $\left(\sum_{1}^{N} X \right)^2 = $ _____. For group II $\left(\sum_{1}^{N} X \right)^2 = $ _____.

═══════ STOP ═══════════════════════ STOP ═══════

25; 361

---

61   For group I $\left(\sum_{1}^{N} X \right)^2 = 25$; for group II $\left(\sum_{1}^{N} X \right)^2 = 361$. $N = 3$, $CF = 96$.

$$bg_{..} = \sum_{1}^{K} \left(\sum_{1}^{N} X \right)^2 / N - CF. \qquad \sum_{1}^{K} \left(\sum_{1}^{N} X \right)^2 = \text{_____}.$$

═══════ STOP ═══════════════════════ STOP ═══════

361 + 25 = 386

---

62 $N = 3$, $CF = 96$, $\sum_{1}^{K} \left(\sum_{1}^{N} X \right)^2 = 386$. $bg_{..} = \sum_{1}^{K} \left(\sum_{1}^{N} X \right)^2 / N - CF$,

$bg_{..} = $ _____.

═══════ STOP ═══════════════════════ STOP ═══════

32.67

---

63   The next step is the computation of the within-groups sum of squares: group I: 0, 2, 3, $\sum_{1}^{N} X = $ _____, $\sum_{1}^{N} X^2 = $ _____; group II: 4, 7, 8,

$\sum_{1}^{N} X = $ _____, $\sum_{1}^{N} X^2 = $ _____.

═══════ STOP ═══════════════════════ STOP ═══════

5, 13; 19, 129

---

64   Group I: $\sum_{1}^{N} X = 5$, $\sum_{1}^{N} X^2 = 13$; group II: $\sum_{1}^{N} X = 19$, $\sum_{1}^{N} X^2 = 129$.

$N = 3$.   group I $\sum_{1}^{N} x^2 = $ _____ and group II $\sum_{1}^{N} x^2 = $ _____.

═══════ STOP ═══════════════════════ STOP ═══════

4.67, 8.67

65

Group I: $\sum_1^N x^2 = 4.67$; group II: $\sum_1^N x^2 = 8.67$.

$$wg_{..} = \sum_1^K \left[\sum_1^N X^2 - \frac{\left(\sum_1^N X \right)^2}{N} \right]$$

$$\sum_1^N x^2 + \sum_1^N x^2 = \underline{\hspace{1cm}}.$$

$\equiv$ STOP $\equiv$ $\equiv$ STOP $\equiv$

$$4.67 + 8.67 = 13.33$$

---

66   Hence if the $\sum_1^N x^2$ is found for each group, the sum of these will equal $wg_{..}$.

$$wg_{..} = \sum_1^K \left[\sum_1^N X^2 - \frac{\left(\sum_1^N X \right)^2}{N} \right] = 13.33;$$

$$bg_{..} = 32.67; \qquad bg_{..} + wg_{..} = \underline{\hspace{1cm}}.$$

$\equiv$ STOP $\equiv$ $\equiv$ STOP $\equiv$

$$46.00$$

---

67 $tot_{..} = 46.00$, $bg_{..} = 32.67$, $wg_{..} = 13.33$. $bg_{..} + wg_{..} = 46$. Hence we have demonstrated the relation $bg_{..} + wg_{..} = \underline{\hspace{1cm}}$.

$\equiv$ STOP $\equiv$ $\equiv$ STOP $\equiv$

$$tot_{..}$$

---

68   Since $tot_{..} = bg_{..} + wg_{..}$, it is obvious that $wg_{..}$ can be found by subtraction rather than by direct computation. $wg_{..} = \underline{\hspace{1cm}} - \underline{\hspace{1cm}}$.

$\equiv$ STOP $\equiv$ $\equiv$ STOP $\equiv$

$$tot_{..} - bg_{..}$$

---

69   Group I: 0, 2, 3; group II: 4, 7, 8. The total df is equal to $KN - 1$ (total scores $-$ 1). tot df $= \underline{\hspace{1cm}}$. The bg df is equal to $K - 1$ (groups $-$ 1). bg df $= \underline{\hspace{1cm}}$. The wg df is equal to $K(N - 1)$ (number of groups times the number of scores within a group $-$ 1). wg df $= \underline{\hspace{1cm}}$.

$\equiv$ STOP $\equiv$ $\equiv$ STOP $\equiv$

$$5; 1; 4$$

---

70 *tot* df $= 5$, *bg* df $= 1$, *wg* df $= 4$. Note the relationship here. Just as *tot*$_{..}$ $=$ *bg*$_{..}$ $+$ *wg*$_{..}$, so too *tot* df $=$ _____ $+$ _____.

$\equiv$ STOP $\equiv\equiv\equiv\equiv\equiv$ STOP $\equiv$

bg df $+$ wg df

---

71 The next step in the analysis is to divide *bg*$_{..}$ and *wg*$_{..}$ by their respective df. This yields what are known as *mean squares*. Thus if *bg*$_{..}$ $= 32.67$ and df $= 1$ then *bg*$_{m.}$ $=$ _____.

$\equiv$ STOP $\equiv\equiv\equiv\equiv\equiv$ STOP $\equiv$

32.67

---

72 A source table is set up:

Source	ss	df	ms	F
bg	32.67	1	?	
wg	13.33	4	?	
tot	46.00	5		

$\equiv$ STOP $\equiv\equiv\equiv\equiv\equiv$ STOP $\equiv$

$bg_{m.} = 32.67$, $wg_{m.} = 3.33$

---

73 $bg_{m.}$ is exactly equal to $NS_M{}^2$ and $wg_{m.}$ is exactly equal to S^2. Since $NS_M{}^2$ is an estimate of S^2 then the ratio $NS_M{}^2/S^2$ is _____ distributed and if $N\sigma_M{}^2 = \sigma^2$, then the distribution has an average of _____.

$\equiv$ STOP $\equiv\equiv\equiv\equiv\equiv$ STOP $\equiv$

F, 1.00

---

74 $bg_{m.} = 32.67$, $wg_{m.} = 3.33$.

$$F = \frac{NS_M{}^2}{S^2} = \frac{bg_{m.}}{wg_{m.}} = \underline{\qquad}.$$

$\equiv$ STOP $\equiv\equiv\equiv\equiv\equiv$ STOP $\equiv$

9.81

---

75 Finally we evaluate the F. Looking at Appendix D, we find that if $N\sigma_M{}^2 = \sigma^2$ (that is, $\sigma_x{}^2 = \sigma_y{}^2$) with 1 and 4 df, 5 percent of the sample Fs will be larger than 7.71. What do we do with the hypothesis that $N\sigma_M{}^2 = \sigma^2$? _____.

$\equiv$ STOP $\equiv\equiv\equiv\equiv\equiv$ STOP $\equiv$

Reject

76   Suppose we have four groups of ten scores each. $tot_{..} = 50$, $bg_{..} = 20$, $K = $ _____; $N = $ _____, $wg_{..} = $ _____.

═══════ STOP ═══════════════════ STOP ═══════

4; 10, 30

77
Source	ss	df	ms	F
bg	20	?		
wg	30	?		
tot	50	?		

$K = 4$, $N = 10$.

═══════ STOP ═══════════════════ STOP ═══════

bg df $= 3$, wg df $= 36$, tot df $= 39$

78
Source	ss	df	ms	F
bg	20	3	?	
wg	30	36	?	
tot	50	39		

═══════ STOP ═══════════════════ STOP ═══════

$bg_{ms} = 6.67$, $wg_{ms} = 0.83$

79
Source	ss	df	ms	F
bg	20	3	6.67	?
wg	30	36	0.83	
tot	50	39		

═══════ STOP ═══════════════════ STOP ═══════

$F = 8.04$

---

80 $F = 8.04$ with 3 and 36 df. Do you accept or reject H: $N\sigma_M^2 = \sigma^2$? _____

═══════ STOP ═══════════════════ STOP ═══════

Reject

81
Source	ss	df	ms	F
bg	45	?	?	?
wg	?	190	?	
tot	425	199		

Given ten groups of twenty scores each.

═══════ STOP ═══════════════════ STOP ═══════

$wg_{..} = 380$, bg df $= 9$, $bg_{ms} = 5.00$, $wg_{ms} = 2.00$, $F = 2.50$

82 $F = 2.50$ with df $= 9$ and 190. Interpolating in your table, what do you do with H: $N\sigma_M{}^2 = \sigma^2$? _____.

Reject

83
Source	ss	df	ms	F	Six groups of eleven scores each.
bg	38	5	?	?	
wg	150	60	?		
tot	188	65			

$bg_{ms} = 7.60$, $wg_{ms} = 2.50$, $F = 3.04$

---

84 $F = 3.04$ with 5 and 60 df. What do you do with H: $N\sigma_M{}^2 = \sigma^2$? _____.

Reject

---

85   If $N\sigma_M{}^2 = \sigma^2$ then with 5 and 60 df, 5 percent of the sample Fs would be larger than what value? _____.

2.37

---

86   The obtained F of 3.04 is larger than the tabled F of 2.37. Therefore the probability that an F this large (3.04) or larger will be obtained by chance is less than _____.

.05

---

87   Measures were taken on three subjects in each of three groups: I: 0, 1, 1; II: 2, 1, 5; III: 6, 7, 7.

$$\sum_1^K \sum_1^N X = \underline{\hspace{1cm}} , \qquad \sum_1^K \sum_1^N X^2 = \underline{\hspace{1cm}} .$$

30, 166

88

$$K = 3, \quad N = 3, \quad \sum_1^K \sum_1^N X = 30, \quad \sum_1^K \sum_1^N X^2 = 166.$$

$$CF = \left(\sum_1^K \sum_1^N X \right)^2 / KN = \underline{\quad\quad}.$$

═══════ STOP ═══════════════════════════ STOP ═══════

100

89

$$K = 3, \quad N = 3, \quad \sum_1^K \sum_1^N X = 30, \quad \sum_1^K \sum_1^N X^2 = 166, \quad CF = 100.$$

$$tot_{..} = \sum_1^K \sum_1^N X^2 - CF = \underline{\quad\quad}.$$

═══════ STOP ═══════════════════════════ STOP ═══════

66

90

I: 0, 1, 1; II: 2, 1, 5; III: 6, 7, 7.

$$\sum_1^K \sum_1^N X = 30, \quad \sum_1^K \sum_1^N X^2 = 166, \quad CF = 100.$$

$$bg_{..} = \sum_1^K \left(\sum_1^N X \right)^2 / N - CF = \underline{\quad\quad}.$$

═══════ STOP ═══════════════════════════ STOP ═══════

$$156 - 100 = 56$$

91

$$tot_{..} = 66, \, bg_{..} = 56, \, wg_{..} = \underline{\quad\quad}.$$

═══════ STOP ═══════════════════════════ STOP ═══════

10

92

Source	ss	df	ms	F	
bg	56	?			$K = 3, N = 3.$
wg	10	?			
tot	66	?			

═══════ STOP ═══════════════════════════ STOP ═══════

bg df = 2, wg df = 6, tot df = 8

93

Source	ss	df	ms	F
bg	56	2	?	?
wg	10	6	?	
tot	66	8		

═══ STOP ═══════════════════════════════ STOP ═══

$$bg_{m\bullet} = 28, \; wg_{m\bullet} = 1.67, \; F = 16.77$$

---

94 $F = 16.77$ with 2 and 6 df. What do you do with H: $N\sigma_M{}^2 = \sigma^2$? _____.

═══ STOP ═══════════════════════════════ STOP ═══

Reject

---

95 Two groups of four scores each: I: 0, 0, 1, 3; II: 4, 5, 4, 3.

$$\sum_{1}^{K}\sum_{1}^{N} X = \underline{\hspace{1cm}}, \qquad \sum_{1}^{K}\sum_{1}^{N} X^2 = \underline{\hspace{1cm}},$$

$$\left(\sum_{1}^{K}\sum_{1}^{N} X\right)^2 / KN = CF = \underline{\hspace{1cm}}.$$

═══ STOP ═══════════════════════════════ STOP ═══

20, 76, 50

---

96 Group I: 0, 0, 1, 3; group II: 4, 5, 4, 3. $\sum_{1}^{K}\sum_{1}^{N} X = 20$; $\sum_{1}^{K}\sum_{1}^{N} X^2 = 76$. $CF = 50$.

$$tot_{\bullet\bullet} = \sum_{1}^{K}\sum_{1}^{N} X^2 - CF = \underline{\hspace{1cm}},$$

$$bg_{\bullet\bullet} = \sum_{1}^{K}\left(\sum_{1}^{N} X\right)^2 / N - CF = \underline{\hspace{1cm}}.$$

═══ STOP ═══════════════════════════════ STOP ═══

$76 - 50 = 26, \; 68 - 50 = 18$

---

97 $tot_{\bullet\bullet} = 26, \; bg_{\bullet\bullet} = 18; \; wg_{\bullet\bullet} = \underline{\hspace{1cm}}.$

═══ STOP ═══════════════════════════════ STOP ═══

8

98

Source	ss	df	ms	F
bg	18	1		$\underline{?}$
wg	8	6		
tot	26	7		

══════ STOP ═══════════════════════════════ STOP ══════

$$F = 13.53$$

---

99 $F = 13.53$, df $= 1$ and 6, evaluate F. _____.

══════ STOP ═══════════════════════════════ STOP ══════

Reject

---

100   Three groups of four scores each: I: 0, 0, 2, 2; II: 1, 3, 4, 5; III: 4, 5, 6, 8.

$$CF = \left(\sum_{1}^{K} \sum_{1}^{N} X \right)^{2} / KN = \underline{\qquad}.$$

══════ STOP ═══════════════════════════════ STOP ══════

133.33

---

101   Four groups of two scores each: I: 0, 1; II: 2, 4; III: 5, 7; IV: 9, 12.

$$CF = \frac{\left(\sum_{1}^{K} \sum_{1}^{N} X \right)^{2}}{KN} = \underline{\qquad}.$$

══════ STOP ═══════════════════════════════ STOP ══════

200

---

102   Two groups of five scores each: I: 0, 0, 2, 4, 5; II: 4, 6, 6, 7, 8. $CF = $ _____.

══════ STOP ═══════════════════════════════ STOP ══════

$$1764/10 = 176.4$$

---

103   Three groups of four scores each: I: 0, 0, 2, 2; II: 1, 3, 4, 5; III: 4, 5, 6, 8.

$CF = 133.33$. $tot_{..} = \sum_{1}^{K} \sum_{1}^{N} X^{2} - CF$. $tot_{..} = $ _____.

══════ STOP ═══════════════════════════════ STOP ══════

$$200 - 133.33 = 66.67$$

104 Three groups of two scores each: I: 0, 1; II: 2, 4; III: 5, 8. $CF =$ _____,
$tot_{..} =$ _____.

═══ STOP ═══════════════════════════════ STOP ═══

$$66.67, \quad 110 - 66.67 = 43.33$$

---

105 Three groups of two scores each: I: 0, 1; II: 2, 4; III: 5, 8. $CF = 66.67$.

$$bg_{..} = \sum_1^K \left(\sum_1^N X \right)^2 / N - CF. \qquad bg_{..} = \text{_____}.$$

═══ STOP ═══════════════════════════════ STOP ═══

$$36.33$$

---

106 Two groups of three scores each: I: 0, 2, 3; II: 3, 4, 6.

$$CF = \text{_____}; \qquad bg_{..} = \frac{\sum_1^K \left(\sum_1^N X \right)^2}{N} - CF, \qquad bg_{..} = \text{_____}.$$

═══ STOP ═══════════════════════════════ STOP ═══

$$54; \quad 10.67$$

---

107 Four groups of two scores each: I: 0, 1; II: 2, 4; III: 5, 7; IV: 9, 12. $CF = 200$, $bg_{..} =$ _____.

═══ STOP ═══════════════════════════════ STOP ═══

$$311 - 200 = 111$$

---

108 Four groups of two scores each: I: 0, 1; II: 2, 4; III: 5, 7; IV: 9, 12. $CF = 200$, $bg_{..} = 111$, $tot_{..} =$ _____.

═══ STOP ═══════════════════════════════ STOP ═══

$$\sum_1^K \sum_1^N X^2 - CF = 320 - 200 = 120$$

---

109 $CF = 200$, $bg_{..} = 111$, $tot_{..} = 120$, $wg_{..} =$ _____.

═══ STOP ═══════════════════════════════ STOP ═══

$$9.00$$

---

110 I: 0, 2, 4; II: 5, 4, 3. We wish to compute NS_M^2. For group I $M =$ _____.
For group II $M =$ _____.

═══ STOP ═══════════════════════════════ STOP ═══

$$2; \quad 4$$

---

111 I: 0, 2, 4; II: 5, 4, 3. For group I $M = 2$; for group II $M = 4$. We wish to compute $NS_M{}^2$. $S_M{}^2$ may be thought of as the variance of a group of _____.

═══ STOP ═══════════════════════ STOP ═══

means

---

112 I: 0, 2, 4; II: 5, 4, 3. For I: $M = 2$; for II: $M = 4$. We wish to compute $NS_M{}^2$. Since $S_M{}^2$ is also the variance of a group of means, then $S_M{}^2$ would equal the variance of _____ and _____.

═══ STOP ═══════════════════════ STOP ═══

2, 4

---

113 I: 0, 2, 4; II: 5, 4, 3. I: $M = 2$; II: $M = 4$. We wish to compute $NS_M{}^2$ using only the scores 2 and 4 (the two Ms). $K =$ _____, and $\sum_1^K x^2 = \sum_1^K X^2 - (\sum_1^K X)^2/K =$ _____.

═══ STOP ═══════════════════════ STOP ═══

2, 2

---

114 I: 0, 2, 4; II: 5, 4, 3. I: $M = 2$; II: $M = 4$. We wish to compute $NS_M{}^2$ using scores 2 and 4. $\sum_1^K x^2 = 2$, $S^2 = \sum_1^K x^2/(K - 1) =$ _____. But in this case S^2 (the variance of the scores 2 and 4) is equal to (what symbol) _____.

═══ STOP ═══════════════════════ STOP ═══

2; $S_M{}^2$

---

115 I: 0, 2, 4; II: 5, 4, 3. I: $M = 2$; II: M $= 4$. We wish to compute $NS_M{}^2$. The variance of the means 2 and 4 ($S_M{}^2$) is equal to 2. $N =$ _____ and $NS_M{}^2 =$ _____

═══ STOP ═══════════════════════ STOP ═══

3, 6

---

116 I: 0, 2, 4; II: 5, 4, 3. $NS_M{}^2 = 6$. We wish to compute $bg_{m\bullet}$.

$$CF = \left(\sum_1^K \sum_1^N X\right)^2/KN = \underline{\quad\quad}.$$

═══ STOP ═══════════════════════ STOP ═══

54

117   I: 0, 2, 4; II: 5, 4, 3. $NS_M{}^2 = 6$, $CF = 54$.

$$bg_{\bullet\bullet} = \sum_1^K \left(\sum_1^N X \right)^2 / N - CF. \qquad \sum_1^K \left(\sum_1^N X \right)^2 / N = \underline{\qquad}.$$

━━━━━ STOP ━━━━━━━━━━━━━━━━━━━━━━ STOP ━━━━━

60

---

118   I: 0, 2, 4; II: 5, 4, 3, $NS_M{}^2 = 6$, $CF = 54$, $\sum_1^K (\sum_1^N X)^2 / N = 60$. We wish to compute $bg_{m\bullet}$.

$$bg_{\bullet\bullet} = \sum_1^K \left(\sum_1^N X \right)^2 / N - CF = \underline{\qquad};$$

$$bg \text{ df} = \underline{\qquad} \quad \text{and} \quad bg_{m\bullet} = \underline{\qquad}.$$

━━━━━ STOP ━━━━━━━━━━━━━━━━━━━━━━ STOP ━━━━━

6; 1, 6

---

119 $bg_{m\bullet} = 6$, $NS_M{}^2 = 6$. It is apparent that $bg_{m\bullet}$ is simply a convenient way to compute _____ and $bg_{m\bullet}$ is *always* exactly the same as _____.

━━━━━ STOP ━━━━━━━━━━━━━━━━━━━━━━ STOP ━━━━━

$NS_M{}^2$, $NS_M{}^2$

---

120   Furthermore if the hypothesis is true (that is, if $\mu_1 = \mu_2$) then $NS_M{}^2$ is an estimate of _____ as well as $N\sigma_M{}^2$.

━━━━━ STOP ━━━━━━━━━━━━━━━━━━━━━━ STOP ━━━━━

$S^2 (\sigma^2)$

---

121   I: 0, 2, 4; II: 5, 4, 3. We wish to compute the mean variance of the two groups. For group I $\sum_1^N x^2 = \sum_1^N X^2 - (\sum_1^N X)^2 / N = \underline{\qquad}$. For group II

$$\sum_1^N x^2 = \sum_1^N X^2 - (\sum_1^N X)^2 / N = \underline{\qquad}.$$

━━━━━ STOP ━━━━━━━━━━━━━━━━━━━━━━ STOP ━━━━━

8; 2

---

122   I: 0, 2, 4; II: 5, 4, 3. We wish to compute the mean variance S^2 of the two groups. For group I $\sum_1^N x^2 = 8$, $S^2 = 8/2 = 4$. For group II $\sum_1^N x^2 = 2$, $S^2 = \underline{\qquad}$. The mean variance S^2 is equal to _____.

━━━━━ STOP ━━━━━━━━━━━━━━━━━━━━━━ STOP ━━━━━

1; 2.5

---

123 I: 0, 2, 4; II: 5, 4, 3. For I $S^2 = 4$; for II $S^2 = 1$. Mean $S^2 = 2.5$. We wish to find wg_{ms}. We have found from a previous example $CF = 54$,

$$tot_{..} = \sum_1^K \sum_1^N X^2 - CF = \underline{\hspace{1cm}}.$$

══════ STOP ══════════════════════════ STOP ══════

16

---

124 I: 0, 2, 4; II: 5, 4, 3. For I $S^2 = 4$; for II $S^2 = 1$. Mean $S^2 = 2.5$. We wish to find wg_{ms}. $tot_{..} = 16$ and we have found from a previous example $bg_{..} = 6$. Hence $wg_{..} = \underline{\hspace{1cm}}$, wg df $= \underline{\hspace{1cm}}$, and $wg_{ms} = \underline{\hspace{1cm}}$.

══════ STOP ══════════════════════════ STOP ══════

10, 4, 2.5

---

125 I: 0, 2, 4; II: 5, 4, 3. $wg_{ms} = 2.5$. Mean $S^2 = 2.5$. It is apparent that wg_{ms} is simply a convenient way to compute _____ and they are always exactly the same.

══════ STOP ══════════════════════════ STOP ══════

mean S^2

---

126 In summary (and no matter how complex the analysis): bg_{ms} is exactly equal to _____ and wg_{ms} is exactly equal to _____.

══════ STOP ══════════════════════════ STOP ══════

$NS_M{}^2$, mean S^2

Fourteen

Double Classification Analysis of Variance

Let us suppose that you have devised a method for measuring the strength of a person's "brand loyalty," that is, the likelihood that a customer will buy the same product a second time when several choices are available. Your measure can range from 1 (no brand loyalty) to 10 (complete brand loyalty), and you have found that it meets the requirements of an interval scale. Suppose further that you have taken a variety of social science courses and have become interested in a cross-cultural comparison of brand loyalty. That is, you wish to determine if brand loyalty varies or is the same between two quite dissimilar countries. The two cultures (countries) you are going to compare differ considerably in the standard of living their citizens enjoy. For this reason, if you find a difference in degree of brand loyalty from one country to another, it could be due to economic differences rather than to the cultural differences found between the two countries. Hence, you decide to include class differences (middle versus lower) within a country as well as cultural differences (country X versus country Y) in your research.

The topics covered in this chapter are relevant to the analysis of data collected in experiments such as that described above. More specifically, this chapter will outline in some detail the procedures involved in double-classification analysis of variance. In single-classification analysis of variance, presented in Chapter Thirteen, each score was classified once according to the group to which it belonged. In double-classification analysis of variance, each score is classified twice, again according to the groups to which it belongs. This type of experimental design is extremely powerful and widely used.

Returning to our example, by varying both socioeconomic level and country within the same experiment, we can investigate the influence of the two variables. At the same time we can also look at the combined effects of the two variables on our dependent variable, brand loyalty.

To investigate the influence of these variables on brand loyalty, you randomly select customers from those who qualify for each combination of your two independent variables. Next you measure their degree of brand loyalty by observing their buying habits over a period of time. The resulting data are shown in Table 14-1.

Table 14-1   Four groups of brand loyalty scores

Country X		Country Y	
lower class	middle class	lower class	middle class
3	4	4	1
7	9	4	4
7	9	4	5
7	10	8	6
24	32	20	16

SINGLE-CLASSIFICATION ANALYSIS

It would be possible, of course, to carry out a single-classification analysis of variance with these four groups of scores. Suppose, however, that you found a significant F ratio for between-group differences. How would you know whether the differences were due to class, location, or the combination? It might also happen that the single-classification analysis was unable to show significant differences among the four groups, even though one pair of groups did actually differ from the corresponding pair. This might happen if, for instance, a country differed from another country, although the middle and low groups did not differ. As a matter of fact, the data in our example do not yield a significant F ratio across the four groups in a single-classification analysis of variance:

$$\sum_1^K \sum_1^N X = 92; \quad \sum_1^K \sum_1^N X^2 = 624; \quad CF = \frac{\left(\sum_1^K \sum_1^N X\right)^2}{KN} = \frac{92^2}{4 \times 4} = 529$$

$$tot_{ss} = \sum_1^K \sum_1^N X^2 - CF = 624 - 529 = 95$$

$$bg_{ss} = \frac{\sum_1^K \left(\sum_1^N X\right)^2}{N} - CF = \frac{20^2 + 24^2 + 16^2 + 32^2}{4} - 529 = 35$$

$$wg_{ss} = tot_{ss} - bg_{ss} = 95 - 35 = 60$$

Setting up the usual source table, we find the degrees of freedom, mean squares, and the F ratio:

Source	ss	df	ms	F
bg	35	3	11.67	2.33
wg	60	12	5.00	
tot	95	15		

This F ratio is tested with 3 and 12 df by the use of Appendix D, and is found to be less than the value needed for significance at the 0.05 level.

DOUBLE-CLASSIFICATION ANALYSIS

Before we get into the computational method for double-classification analysis, we should be sure to understand the reasoning that led us to this method. If we had been interested in the effects of only one principle of classification—people from four different countries, for instance—then single-classification analysis would be appropriate. In the present example, however, we want to classify simultaneously each score according to two principles of classification. This may be clearer to you after we put the data from our example in a new form, shown in Table 14-2. These are the same scores and group (cell) sums, but now are arranged to show that every person has been classified in two ways at the same time. We have also added the middle- and lower-class scores in Country X to get an overall Country

Table 14-2   Alternate presentation of four groups of brand loyalty scores shown in Table 14-1

Class	Location		
	COUNTRY X	COUNTRY Y	
MIDDLE	4	1	
	9	4	
	9	5	
	10	6	
	32 (cell sums)	16	48
LOWER	3	4	
	7	4	
	7	4	
	7	8	
	24 (cell sums)	20	44
	56 (column sums)	36	92

X sum, and have added similarly to get an overall Country Y sum. The total sum of scores (92) is thus the sum of all of the Country X scores (56) and all of the Country Y scores (36), and is *also* the sum of all the middle-class scores (48) and all the lower-class scores (44).

We will not go into the theory underlying double-classification analysis of variance except to say that in this procedure we are dealing again with variables that are F distributed. These are derived from the ratio of two independent chi-square distributed variables divided by their respective numbers of df. What we will do is break down the between-groups variance estimate into three independent variance estimates: (1) variance due to class, (2) variance due to location, and (3) variance due to the combination of class and location. As we will see later, the within-groups variance estimate in the double-classification analysis is the same as that already computed for the single-classification analysis with four groups. Instead of comparing the between-groups variance estimate to the within-groups variance estimate, we will be able to make separate comparisons for each of the three sources of variance in which we are interested.

Computation of the double-classification analysis differs from that for the single-classification analysis already completed in Chapter Thirteen in only one way: The three parts of the between-groups sum of squares must be separately computed.

The first component of the between-groups sum of squares will be that due to the country or location of the buyer.

$$\text{location}_{ss} = \frac{\sum\limits_{1}^{L} \left(\sum\limits_{1}^{C} \sum\limits_{1}^{N} X \right)^2}{CN} - CF = \frac{(20+16)^2 + (24+32)^2}{2 \times 4} - CF$$

$$= \frac{36^2 + 56^2}{8} - 529 = 25$$

Notice that we squared the column sums and divided by the number of scores in each column.

The second component of the between-groups sum of squares will be that related to the class of the buyer. Here we square each of the two row sums, add, and divide by the number of scores that make up each of these two totals.

$$\text{class}_{ss} = \frac{\sum\limits_{1}^{C} \left(\sum\limits_{1}^{L} \sum\limits_{1}^{N} X \right)^2}{LN} - CF = \frac{(32+16)^2 + (24+20)^2}{2 \times 4} - CF$$

$$= \frac{44^2 + 48^2}{8} - 529 = 1.00$$

The third component of the between-groups sum of squares is due to the combination of class and location. This kind of a combination effect is usually called an *interaction*—in this case, the interaction of class (C) and location (L)—symbolized $(C \times L)$. Since the sum of the three independent components of the between-groups sum of squares must equal the overall between-groups sum of squares, we can obtain this third component by subtraction. That is, $bg_{ss} = C_{ss} + L_{ss} + (C \times L)_{ss}$, and we already have found C_{ss} and L_{ss}. Therefore, $(C \times L)_{ss} = bg_{ss} - L_{ss} = 35 - 25 - 1 = 9$.

We are ready to set up the source table.

Table 14-3   Source table for vocabulary analysis

Source	ss	df	ms	F
Between groups (bg)	35.00	3		
location (L)	25.00	1	25.00	5.00*
class (C)	1.00	1	1.00	0.20
class by location $(C \times L)$	9.00	1	9.00	1.80
Within groups (wg)	60.00	12	5.00	
Total (tot)	95.00	15		

* Pr < 0.05

As before, the mean square (*ms*) for a row in the table is found by dividing the sum of squares for the row by the degrees of freedom for the row. Notice that the overall between-groups df have been broken down into three parts, just as the sum of squares was broken down. The within-groups sum of squares and their df are the same as before. Each of the three mean squares is divided by the estimate of within-groups variance to yield an *F* ratio for that particular component of the between-groups variance. When we enter the *F* table with 1 and 12 df to evaluate these new *F* ratios, we find that the *F* for location (5.00) exceeds the tabled *F* value at the 0.05 level of significance. We cannot reject the null hypothesis for either of the other two comparisons, however.

DEGREES OF FREEDOM

Just as the between-groups sum of squares is broken down into three independent sums of squares, so too is the total number of degrees of freedom for between-groups partitioned. The principle for determining the degrees of freedom (df) for each variance estimate is similar to that for determining df for between-groups. With K groups, we have $K - 1$ df. Now, if there are two "levels" or classifications for class, we have $C - 1$ df for the class component: $2 - 1 = 1$. With two levels on the location dimension, we have $L - 1 = 2 - 1 = 1$ df for location. The interaction of class and location may now be determined by multiplying together the numbers of df for the two variables that interact: $(C - 1)(L - 1) = (1)(1) = 1$, or by subtraction, $(K - 1) - (C - 1) - (L - 1) = 3 - 1 - 1 = 1$.

INTERACTION EFFECTS

The concept of interaction effects is not easy to grasp because there are a number of specific conditions that can produce a significant interaction effect in an analysis of variance. Generally, an interaction reflects variation in the group means, which cannot be accounted for by knowledge of the means of either of the main effects (row and column variables). If, in our previous example, there were large mean differences between middle and lower classes in Country X, but not in Country Y, we would not be able to account for all the variation just by knowing the class or by knowing whether the person came from Country X or Country Y. To account for the brand loyalty scores adequately, we would need the additional information provided by knowing *both* class and location *at the same time*.

It is often helpful to use a graphic respresentation of group means to understand what is implied by an interaction effect. In Figure 14-1 we have presented the results of the previous experiment. Note that the means of both Country X groups are higher than either of the Country Y groups. This is consistent with the significant *F* ratio for location. The means of the two middle-class groups, however, lie on either side of the two lower-class group means, and you will recall that the class *F* ratio was very small. If the

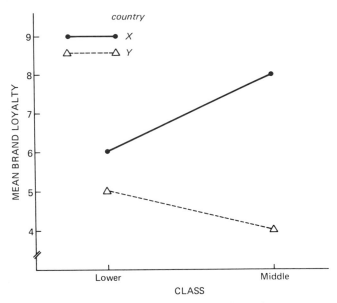

Figure 14-1   *Group means for brand loyalty scores.*

$(C \times L)_{ss}$ had been zero, the two lines in the graph would have been parallel. The fact that they are not parallel does not necessarily mean that there is an interaction between class and location. Because of random sampling, we would rarely expect perfectly parallel lines. But in similar experiments, if we reject the null hypothesis and conclude that there is an interaction, then a graph would show that the lines depart greatly from being parallel.

ANOTHER EXAMPLE   Let us suppose that a drug manufacturer has developed a cure for the common cold, which can be taken in pill form. When the company submits the results of its experiments on the drug to the Food and Drug Administration, however, the application is challenged. One government chemist notices that the drug when used on dog subjects seems to cause considerable pain. The FDA, suspicious of this serious side effect, withholds approval pending further experimental investigation.

The problem confronting the chemists working for the drug manufacturer is not whether the drug is an effective cure for colds, but whether it also causes pain. An experiment is designed to test the hypothesis that the FDA worries are unfounded. Three conditions (levels) are used to represent each of the two dimensions (independent variables) in the design. One of the dimensions is the amount of the new drug given to those people serving in the experiment, and the three levels are designated 0 units, 2 units, and 4 units. Because the effectiveness of the drug had previously been found to increase with the age of the user, a second dimension of age is added to the design, and the levels are set at 20 years, 40 years, and 60 years. A large group of volunteers, all of whom have colds, is selected. The dependent variable, or response measure, of the experiment is the number of times during the week of the experiment a person answers yes to the question, "Did you experience considerable pain today?" Two people are used in each of the nine cells of the design. The 0-drug people were given a dummy sugar pill, and although all people who had served in the experiment had agreed to do so, they were not told the exact nature of the experiment until it was over. The data for the experiment are shown in Table 14-4.

Table 14-4   Scores from the drug experiment

Drug dosage	Age 20	Age 40	Age 60	
4 units	6	4	2	
	4	3	1	
	10	7	3	20
2 units	2	2	2	
	4	3	2	
	6	5	4	15
0 units	1	2	5	
	0	1	4	
	1	3	9	13
	17	15	16	48

There are $N = 2$ scores per cell, where a cell represents a particular combination of the two dimensions of the design. The cell sums (Σ) are also shown in the table. The overall sums ($\Sigma\Sigma$) for each of the drug dosages (D) and for the age levels (A) are also given in the table. Finally, the total sum of scores is represented in the table.

$$\sum_1^A \sum_1^D \sum_1^N X = 48$$

The total sum of squared scores is found by squaring each of the 18 raw scores and summing the results:

$$\sum_1^A \sum_1^D \sum_1^N X^2 = 170$$

The correction factor (CF) is the square of the total sum of scores divided by the total number of scores:

$$CF = \frac{\left(\sum_1^A \sum_1^D \sum_1^N X\right)^2}{ADN} = \frac{48 \times 48}{3 \times 3 \times 2} = 128$$

The total sum of squares is the total sum of squared scores minus the CF:

$$tot_{ss} = \sum_1^A \sum_1^D \sum_1^N X^2 - CF = 170 - 128 = 42$$

We can now break this total sum of squares into two parts, the be-tween-groups and the within-groups sums of squares. The between-groups sum of squares or *cell* sum of squares is found by squaring each of the cell totals, adding these nine values together, dividing by the number of scores within each cell, and subtracting the correction factor:

$$bg_{ss} = \frac{\sum\limits_{1}^{D} \sum\limits_{1}^{A} \left(\sum\limits_{1}^{N} X \right)^2}{N} - CF$$

$$= \frac{10^2 + 7^2 + 3^2 + 6^2 + 5^2 + 4^2 + 1^2 + 3^2 + 9^2}{2} - 128$$

$$= 35$$

The within-groups (cells) sum of squares is the other part of the total sum of squares, and may be obtained by subtraction:

$$wg_{ss} = tot_{ss} - bg_{ss} = 42 - 35 = 7$$

Now we may break down the between-groups sum of squares into three parts, which will be the dosage effect (D), the age effect (A), and the interaction effect $(D \times A)$. The dosage effect is found by squaring the totals for each of the three dosage conditions, dividing the sum of these squared totals by the number of raw scores that are represented by each total, and then subtracting the correction factor:

$$D_{ss} = \frac{\sum\limits_{1}^{D} \left(\sum\limits_{1}^{A} \sum\limits_{1}^{N} X \right)^2}{AN} - CF = \frac{20^2 + 15^2 + 13^2}{3 \times 2} - 128 = 4.33$$

The age effect is found in a similar way, using the sums for each age group. The number of raw scores for each sum remains the same because there is an equal number of levels for each dimension of the design:

$$A_{ss} = \frac{\sum\limits_{1}^{A} \left(\sum\limits_{1}^{D} \sum\limits_{1}^{N} X \right)^2}{DN} - CF = \frac{17^2 + 15^2 + 16^2}{3 \times 2} - 128 = 0.33$$

The interaction effect $(D \times A)$ can be found now by subtraction, since two of the three components of the between-groups sum of squares are known:

$$(D \times A)_{ss} = bg_{ss} - D_{ss} - A_{ss} = 35 - 4.33 - 0.33 = 30.34$$

This completes the calculation of the various sums of squares needed for the analysis. We are now ready to set up the summary table. Since there are 18 raw scores, there are 17 df to be split up among the components of the total variation of these scores. With K cells or groups, there are $K - 1 = 9 - 1 = 8$ df between groups. The number of df for between + within = total, so we can subtract $17 - 8 = 9$ to get the df within groups.

The 8 df between groups are broken up as follows:

$$D \text{ df} = D - 1 = 2$$
$$A \text{ df} = A - 1 = 2$$
$$(D \times A) \text{ df} = (D - 1)(A - 1) = 2 \times 2 = 4$$

Since the three components must equal the between-groups in df as in the sums of squares, we could have obtained $(D \times A)$ df by subtraction: $(D \times A) \text{ df} = BG \text{ df} - D \text{ df} - A \text{ df} = 8 - 2 - 2 = 4$.

The mean squares in the source table (Table 14-5) are derived by dividing each sum of squares by its corresponding number of df. The F

Table 14-5   Source table for drug experiment analysis

Source	ss	df	ms	F
Between	35.00	8		
dosage (D)	4.33	2	2.17	2.78
age (A)	0.33	2	0.17	0.22
$D \times A$	30.33	4	7.58	9.72*
Within	7.00	9	0.78	
Total	42.00	17		

* Pr < 0.01

ratios are obtained by dividing the between-groups-component mean squares by the within-groups mean square, which is the *error term,* or estimate of random variation. You should now be familiar with the entire process for carrying out a double classification analysis of variance. If you have any doubts at this point, go back and review before we get into the problems of interpreting the results.

We can see from Appendix D that with 2 and 9 df, an F of 4.26 is required for significance at the 0.05 level. Neither of the Fs for A or D is larger than 4.26, and we thus cannot reject the null hypothesis.

We must evaluate the $A \times D$ interaction with a different set of df. With 4 and 9 df, an F of 3.63 is required for significance at the 0.05 level. Our F of 9.72 is much larger than is needed to reach significance, and we can reject the null hypothesis and conclude that there is an interaction between the age and drug variables. As an aid to interpreting these results, let us present the cell sums in Figure 14.2.

Interpretation of the meaning of an interaction is often rather difficult. In the present example, we found that neither A (age) nor D (drug dosage) was significant, but since we found that the $A \times D$ interaction was significant, we *cannot* conclude that both age and drug dosage are not effective variables. That they are (at least under certain conditions) is apparent from observation of performance (that is, reported pain) under zero dosage or under age 20. Thus, when we find that two variables interact, we must conclude that both independent variables are related to the dependent variable. This will be true even though the variables themselves (in our exam-

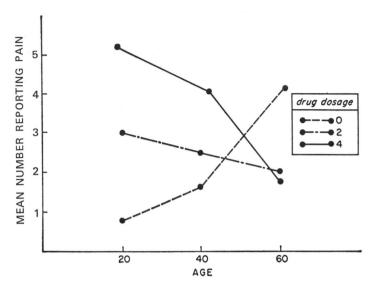

Figure 14-2   *Amount of pain reported as a function of age and dosage.*

ple, A and D) do not reach significance. The fact that they do not reach significance may simply mean that the relationship between the two independent variables and the dependent variable is very complex.

How can the results of our experiment be interpreted? The zero dosage line indicates that amount of reported pain increases with age. That is, older people have more aches and pains when they have a cold. Now, if people with colds are given the drug, it can be seen that the drug increases the pain experienced by younger people, while at the same time it decreases the pain experienced by older people. Thus, even though the age variable did not reach significance, the drug manufacturer might agree to recommend that only people over 50 or so be given the drug.

ONE SCORE PER CELL

Suppose you read in the newspaper about "assertive training" in which people, usually women, are trained to be assertive—to insist on their rights—without being aggressive. At this point in the statistics course you may tend to be a little suspicious of such stories, but in keeping an open mind you decide to determine if assertive training does do what it is supposed to do. To test this, you find that a new training group will begin shortly, and unobstrusively you observe four of the women of the group, each for a 3-hour period, recording the number of assertive responses each makes. (This is commonly called the *pretest.*) Next, the assertive training sessions are given over a period of two weeks, and after the final session you again observe each of the four women in the same work situation as in the pretest. Measuring the number of assertive responses given by each woman over a 3-hour period, you complete the experiment. (This is commonly called the post-test.)

Now you wish to analyze the data, which are presented in Table 14–6. Can this be done, inasmuch as there is only a single score per group (or cell)? There is only one score for Barbara Smith on the pretest, one score for Mary Adams on the post-test, and so on. That is, each score can be uniquely classified as to which woman and as to either the pre- or post-test. The problem this raises is that a variance cannot be computed on a single score (try to, and you will see why), and the within-group mean square, you may recall, is the mean of the cell variances.

Table 14-6 Number of assertive responses given by each of four women before and after training

	Before	After	Total
Kate Jones	0	3	3
Barbara Smith	1	4	5
Nancy Brown	1	8	9
Mary Adams	2	5	7
	4	20	24

Before we go any further, it should be noted that if the rank order of the women's scores would not be expected to be fairly stable from "before" to "after," this design would not be appropriate. If, in fact, the before-and-after scores are not correlated (see Chapter Sixteen), the present analysis would not be the appropriate one to use. In such a case a single classification design using the same eight scores would be more likely to detect a reliable difference. The reason for this is that the df in the denominator of the F would increase from 3 to 6 and hence require a smaller F for significance, whereas the value of the wg_{ms} would, on the average, remain the same. In our present example, we are reasonably sure of a high correlation of pre- and post-test measures, since the rank order of the women's scores is nearly the same in the pre- and post-test conditions.

Let us proceed with the analysis just as we did before, despite the one-score-per-cell situation. In these computations, women $(W) = 4$; conditions $(C) = 2$; scores per cell $(N) = 1$.

$$\sum_1^W \sum_1^C \sum_1^N X = 24; \quad \sum_1^W \sum_1^C \sum_1^N X^2 = 120$$

$$CF = \frac{\left(\sum_1^W \sum_1^C \sum_1^N X\right)^2}{WCN} = \frac{24 \times 24}{4 \times 2 \times 1} = 72.00$$

$$tot_{ss} = \sum_1^W \sum_1^C \sum_1^N X^2 - CF = 120 - 72 = 48.00$$

$$bg_{ss} = \frac{\sum_1^W \sum_1^C \left(\sum_1^N X\right)^2}{N} - CF \text{ [but since } N = 1] = \sum_1^W \sum_1^C X^2 - CF$$

$$= 120 - 72 = 48.00$$

$$wg_{ss} = tot_{ss} - bg_{ss} = 48.00 N - 48.00 N = 0.00$$

$$W = \frac{\sum\limits_{1}^{W}\left(\sum\limits_{1}^{C} X\right)^2}{C} - CF = \frac{3^2 + 5^2 + 9^2 + 7^2}{2} - 72.00 = 10.00$$

$$C_{ss} = \frac{\sum\limits_{1}^{C}\left(\sum\limits_{1}^{W} X\right)^2}{W} - CF = \frac{4^2 + 20^2}{4} - 72.00 = 32.00$$

$$(W \times C)_{ss} = bg_{ss} - W_{ss} - C_{ss} = 48.00 - 10.00 - 32.00 = 6.00$$

Table 14-7 shows that all variation is between groups. This makes sense, since there is only one score per "group," and a variance cannot be computed from a single score. We are, however, without the "within" term we used before to test the mean squares for the main effects and the interaction. The best estimate of random variation in this situation will be provided by the interaction term (women by conditions), and we will use this mean square in the denominator of the F ratio for the conditions effect.

Table 14-7   Source table for analysis of effectiveness of
assertive training

Source	ss	df	ms	F
Between				
women (W)	10.00	3	3.33	
conditions (C)	32.00	1	32.00	16.00*
$W \times C$	6.00	3	2.00	
Within	0.00	0	0.00	
Total	48.00	7		

    * Pr < 0.05

Note that we could have computed an F ratio for differences between the women, but this would have been of no particular interest in the present experiment.

With 1 and 3 df, the F ratio of 16.00 for "conditions"—the pretest versus post-test differences—is larger than the tabled value at the 0.05 level (10.13), and so we are justified in rejecting the null hypothesis. We can conclude that the training had the expected effect (that is, assertiveness increased), and we can conclude that the effect was due to the assertive training.

t TEST FOR PAIRED OBSERVATIONS

A t test, sometimes called the "direct difference t," can be computed when pairs of observations have been taken on the same individuals. This was the case in the preceding example where each person was measured twice, pretest and post-test. As has been noted previously when the analysis of variance has 1 df in the numerator, $F = t^2$ and a t can be obtained by taking the square root of F. In our example, $\sqrt{16.00} = 4.00$. This can be easily verified by subtracting each pretest score from its paired post-test score: 3, 3, 7, and 3. The mean and standard deviation of the difference scores are 4.00 and 2.00, respectively. S_M, which can be found by dividing 2.00 (that is, $\sqrt{4}$), equals 1.00. Next, testing the hypothesis of no change, $\mu = 0$, and $t = (M - \mu)/S_M = 4.00/1.00 = 4.00$, which, of course, is the value obtained by taking the square root of F. Finally, the t of 4.00 is evaluated in Appendix C. With 3 df, a t of 3.18 is the tabled value at 0.05. Hence, the hypothesis of no change is rejected, and the same conclusion is reached.

SUMMARY

This chapter has presented double-classification analysis of variance both when $N = 1$ per cell or group, and when $N > 1$ per cell. The advantage of this statistical test is that two independent variables can be manipulated in the same experiment. Such manipulation can provide us with information that would be extremely difficult to obtain if only a single independent variable could be varied at a time. By varying two independent variables simultaneously, we are able to determine the interaction, if any, between these two variables.

More complex forms of analysis of variance can be easily used but they are beyond the scope of this book. Three, four, five, and even more independent variables can be used in the same experiment, and such data can be analyzed by the analysis-of-variance technique. We have seen only a small portion of this powerful tool of statistical analysis.

PROBLEMS

1. A 2×3 factorial design with three scores per cell is presented below. Complete the source table.

	B_1	B_2	B_3
A_1			
A_2			

Source	ss	df	ms	F
A	20			
B	16			
wg	48			
tot	100			

2. From the following source table, answer these questions:
   (a) How many scores were there in the analysis?
   (b) What was the average (mean) variance of the scores in each of the cells?
   (c) What was S_M for the R variable?
   (d) Is there a significant difference between the means of the R variable?

Source	ss	df	ms	F
R	24	2		
C	18	3		
$R \times C$	6	6		
wg	72	24		

3. From the following factorial design, compute the correction factor (CF) and Q_{ss}.

	Q_1	Q_2
P_1	0	1
	1	2
P_2	2	5
	3	6
P_3	0	2
	0	2

4. Given the following information, complete the analysis.

	C_1	C_2	C_3
R_1	6	9	12
	2	7	8
R_2	8	5	3
	8	3	1

$$\sum_{1}^{R}\sum_{1}^{C}\sum_{1}^{N} X = 72; \qquad \sum_{1}^{R}\sum_{1}^{C}\sum_{1}^{N} X^2 = 550$$

5. In the following 2×2 factorial design, $CF = 128.00$, $A_{ss} = 32.00$, and $B_{ss} = 0.00$. Compute $\sum_{1}^{B}\sum_{1}^{A} \left(\sum_{1}^{N} X\right)^2$ and the $(A \times B)_{ss}$.

	B_1	B_2
A_1	5	0
	3	0
A_2	5	6
	3	10

6. Each of ten people learned a 12-item list of words for five trials. After each trial the number of correct responses was recorded for each person. Set up a source table showing source, df, and what Fs are computed.

7. Given the following information, complete the analysis and set up a source table.

	T_1	T_2	T_3
S_1	0	1	6
S_2	2	4	7
S_3	0	2	2
S_4	2	5	5

$$\sum_{1}^{T}\sum_{1}^{S} X^2 = 168$$

8. Given the following information, complete the analysis.

	B_1	B_2
A_1	5	0
	3	0
A_2	6	5
	10	3

$$\sum_1^A \sum_1^B \sum_1^N X = 32; \quad \sum_1^A \sum_1^B \sum_1^N X^2 = 204; \quad CF = 128$$

9. The concept of interaction is a difficult one to grasp. In Problem 5 the interaction was significant, but in Problem 8 it was not. Graph the results of the experiments in Problems 5 and 8. When are the lines of the graph parallel?

10. Pairs of mothers and daughters were given a "domestic" interest test. A total of 20 persons served in the experiment. Set up a source table showing source and df.

11. Analyze the following data:

	A_1	A_2	A_3
B_1	3	1	0
	2	3	3
B_2	5	5	9
	7	4	6

12. Five boys had three trials each on a certain task. Analyze the following data:

	T_1	T_2	T_3
S_1	0	4	7
S_2	1	2	5
S_3	2	8	10
S_4	1	2	6
S_5	4	2	6

EXERCISES

1. Take the words listed in Exercise 1 of Chapter Thirteen and read them to a group of people. See how many of each kind of word they get correct. Assign to a new line the two scores obtained from each person who participated. Do the appropriate analysis.

2. Give a second trial test to the people from Exercise 1. That is, read the words a second time and have them write down as many as they can remember a second time. In this exercise, write down the total number correct on trial 1 and then the total number correct on trial 2. Do this for each participant. Do the appropriate analysis.

1 There are several types (or classifications) of analysis of variance. In the previous chapter, you learned how to compute a single-classification analysis of _____.

============ STOP ============ STOP ============

variance

---

2 A single-classification analysis of variance is so named because each group is classified in only a _____ way.

============ STOP ============ STOP ============

single

---

3 Similarly a _____-classification analysis of variance is so named because each group is classified in *two* ways.

============ STOP ============ STOP ============

double

---

4 And a triple-classification analysis of variance classifies each group in _____ ways.

============ STOP ============ STOP ============

three

---

5 Suppose an experiment is conducted with four groups. Each group is given a different amount of a drug. The independent variable in this experiment is _____ _____ _____ _____.

============ STOP ============ STOP ============

the amount of drug

---

6 Each group is classified only on the basis of the amount of drug and this would be an example of a _____-classification analysis of variance.

============ STOP ============ STOP ============

single

---

7 One group of men took a small amount of the drug and a second group took a large amount. Two groups of women similarly took either large or small amounts of the same drug. There are a total of _____ groups.

============ STOP ============ STOP ============

four

8   Two groups of men and two groups of women take either large or small amounts of a drug. Each of the four groups can be classified according to the _____ of drug and to the _____ of the group member.

═══════ STOP ═══════════════════════ STOP ═══════

amount, sex

---

9   Two groups of men and two groups of women take either large or small amounts of a drug. There are two independent variables: _____ and _____ .

═══════ STOP ═══════════════════════ STOP ═══════

drug, sex

---

10   Two groups of men and two groups of women take either small or large amounts of a drug. There are two independent variables and the experiment is properly analyzed by a _____-classification analysis of variance.

═══════ STOP ═══════════════════════ STOP ═══════

double

---

11   If an experiment has three independent variables, we would guess that the proper analysis would be in terms of a _____-classification analysis of variance.

═══════ STOP ═══════════════════════ STOP ═══════

triple

---

12   Not all experiments with two independent variables are properly analyzed by a _____-classification analysis of variance. But we will look only at those experiments that are properly analyzed by that method.

═══════ STOP ═══════════════════════ STOP ═══════

double

13

             Drug   In the design at the left we see that all
        Small   Large   people in the bottom row are _____ .

Sex M ————————

      F

═══════ STOP ═══════════════════════ STOP ═══════

female

14

	Drug	
	Small	Large
Sex M		
F		

The column variable is the amount of drug and the row variable is _____.

━━━━ STOP ━━━━━━━━━━━━ STOP ━━━━

sex

15

	Drug	
	Small	Large
Sex M		
F		

Each group (cell) is classified according to _____ and _____.

━━━━ STOP ━━━━━━━━━━━━ STOP ━━━━

sex, drug

16

	Drug	
	Small	Large
Sex M		
F		

The design at the left is properly analyzed by a _____-classification analysis of variance.

━━━━ STOP ━━━━━━━━━━━━ STOP ━━━━

double

17

	Task Difficulty		
	Easy	Med.	Hard
5			
Age 10			
15			

The design at the left has two independent variables: _____ and _____ _____.

━━━━ STOP ━━━━━━━━━━━━ STOP ━━━━

age, task difficulty

---

18 Computation of a double-classification analysis of variance is very similar to the computation of a _____-classification analysis about which we have already learned.

━━━━ STOP ━━━━━━━━━━━━ STOP ━━━━

single

19

	Age	
	5	10
M	**0**	**7**
Sex	**1**	**8**
F	**4**	**0**
	2	**2**

If A = age, S = sex, and N = number of scores per group, then $\sum_{1}^{A}\sum_{1}^{S}\sum_{1}^{N} X =$ _____ and $\sum_{1}^{A}\sum_{1}^{S}\sum_{1}^{N} X^2 =$ _____.

═══ STOP ═══ ═══ STOP ═══

24, 138

20

	Age	
	5	10
M	**0**	**7**
Sex	**1**	**8**
F	**4**	**0**
	2	**2**

$\sum\sum\sum X = 24, \sum\sum\sum X^2 = 138,$ $CF = (\sum\sum\sum X)^2/ASN =$ _____.

═══ STOP ═══ ═══ STOP ═══

$(24 \times 24)/(2 \times 2 \times 2) = 72$

21

	Age	
	5	10
M	**0**	**7**
Sex	**1**	**8**
F	**4**	**0**
	2	**2**

$\sum\sum\sum X = 24, \sum\sum\sum X^2 = 138,$ $CF = 72, tot_{..} = \sum_{1}^{S}\sum_{1}^{A}\sum_{1}^{N} X^2 - CF =$ _____.

═══ STOP ═══ ═══ STOP ═══

$138 - 72 = 66$

22

	Age	
	5	10
M	**0**	**7**
Sex	**1**	**8**
F	**4**	**0**
	2	**2**

$\sum\sum\sum X = 24, \sum\sum\sum X^2 = 138,$ $CF = 72, tot_{..} = 66.$ As in single classification $bg_{..} = \sum_{1}^{S}\sum_{1}^{A}(\sum_{1}^{N} X)^2/N - CF =$ _____.

═══ STOP ═══ ═══ STOP ═══

$(1^2 + 15^2 + 6^2 + 2^2)/2 - CF = 133 - 72 = 61$

23   As in both single- and double-classification analysis of variance $wg_{..} = tot_{..} -$ _____.

══════ STOP ══════════════════════ STOP ══════

$bg_{..}$

──

24

	Age	
	5	10
M	0	7
	1	8
F	4	0
	2	2

(Sex: M / F)

$tot_{..} = 66$, $bg_{..} = 61$, $wg_{..} =$ _____.

══════ STOP ══════════════════════ STOP ══════

$$66 - 61 = 5$$

──

25

	Age	
	5	10
M	0	7
	1	8
F	4	0
	2	2

(Sex: M / F)

$tot_{..} = 66$, $bg_{..} = 61$, $wg_{..} = 5$. However in the double-classification analysis the $bg_{..}$ includes the variability due to sex, _____, and the combination of the two.

══════ STOP ══════════════════════ STOP ══════

age

──

26

	Age	
	5	10
M	0	7
	1	8
F	4	0
	2	2

(Sex: M / F)

$CF = 72$. In double classification the $bg_{..}$ is "broken down" into several parts. For example, variance due to sex may be computed from the formula: $\sum_{1}^{S} (\sum_{1}^{A} \sum_{1}^{N} X)^2/AN - CF$.

For males: $(\sum_{1}^{A} \sum_{1}^{N} X)^2/AN =$ _____.

══════ STOP ══════════════════════ STOP ══════

$$(16 \times 16)/(2 \times 2) = 64$$

──

27

		Age	
		5	10
Sex	M	0	7
		1	8
	F	4	0
		2	2

$CF = 72$

$$\text{sex}_{..} = \sum_{1}^{S}\left(\sum_{1}^{A}\sum_{1}^{N} X\right)^2/AN - CF.$$

$\text{sex}_{..} =$ ______.

STOP STOP

$(16 \times 16)/(2 \times 2) + (8 \times 8)/(2 \times 2) - CF = 64 + 16 - CF = 8.00$

28

		Age	
		5	10
Sex	M	0	7
		1	8
	F	4	0
		2	2

$CF = 72$

$$\text{age}_{..} = \sum_{1}^{A}\left(\sum_{1}^{S}\sum_{1}^{N} X\right)^2/SN - CF.$$

$\text{age}_{..} =$ ______.

STOP STOP

$(7 \times 7)/(2 \times 2) + (17 \times 17)/(2 \times 2) - CF = 84.5 - CF = 12.5$

29

		Age	
		5	10
Sex	M	0	7
		1	8
	F	4	0
		2	2

$bg_{..} = 61$, $\text{age}_{..} = 12.5$, $\text{sex}_{..} = 8.00$. All the variation between the four groups is equal to 61.00. $\text{sex}_{..}$ and $\text{age}_{..}$ account for $8.00 + 12.50 = 20.50$ of the 61. The remaining variability between groups is called the *interaction* $(A \times S)_{..} = bg_{..} - \text{sex}_{..} - \text{age}_{..} =$ ______.

STOP STOP

$(A \times S)_{..} = 61 - 12.5 - 8 = 40.5$

---

30 The variance not accounted for (or remaining) after the sums of squares associated with the two independent variables have been subtracted from the $bg_{..}$ is called the ______.

STOP STOP

interaction

31　Suppose an experiment is conducted with two independent variables A and B. $bg_{..} = 100$, $A_{..} = 25$, $B_{..} = 40$. The interaction $(A \times B)_{..} =$ _____ .

═══ STOP ═══════════════════ STOP ═══

$$100 - 25 - 40 = 35$$

32

	Age	
	5	10
M	0	7
	1	8
Sex F	4	0
	2	2

$tot_{..} = 66$, $age_{..} = 12.5$, $sex_{..} = 8.0$, $(A \times S)_{..} = 40.5$. The number of df are found in much the same way as in the single classification analysis: tot df = _____ , bg df = _____ , and wg df = _____ .

═══ STOP ═══════════════════ STOP ═══

7, 3, 4

33

	Age	
	5	10
M	0	7
	1	8
Sex F	4	0
	2	2

The df for age are found by subtracting 1 from the number of age groups: age df = _____ ; similarly sex df = _____ .

═══ STOP ═══════════════════ STOP ═══

1; 1

34

	Age	
	5	10
M	0	7
	1	8
Sex F	4	0
	2	2

The $(A \times S)$ df are found by multiplying age and sex df together (or by bg df − age df − sex df). Thus $(A \times S)$ df = age df × sex df = _____ .

═══ STOP ═══════════════════ STOP ═══

$$1 \times 1 = 1$$

35

	A_1	A_2	A_3
B_1			
B_2			

A df = _____ , B df = _____ , $(A \times B)$ df = _____ , bg df = _____ .

═══ STOP ═══════════════════ STOP ═══

2, 1, 2, 5

36

$$C_1 \quad C_2 \quad C_3$$

D_1 _____

D_2 _____

D_3

C df = _____, D df = _____,

$(C \times D)$ df = _____, bg df = _____.

════ STOP ════════════════════ STOP ════

2, 2, 4, 8

37

Source	ss	df	ms	F
sex	8.0	1	?	
age	12.5	1	?	
$A \times S$	40.5	1	?	
wg	5.0	4	?	

As in the single-classification analysis df are divided into respective sums of squares to give ms.

════ STOP ════════════════════ STOP ════

$\text{sex}_{ms} = 8.0$, $\text{age}_{ms} = 12.5$, $(A \times S)_{ms} = 40.5$, $wg_{ms} = 1.25$

38

Source	ms	F
sex	8.0	?
age	12.5	?
$A \times S$	40.5	?
wg	1.25	

Several different Fs are computed by dividing wg_{ms} into each of the ms.

════ STOP ════════════════════ STOP ════

$\text{sex}_F = 6.40$, $\text{age}_F = 10.00$, $(A \times S)_F = 32.4$

---

39 From Appendix D we see that with 1 and 4 df an F of _____ is required to reject the null hypothesis. Thus we conclude that in this experiment variables of _____ and _____ vary more than we would expect on the basis of chance.

════ STOP ════════════════════ STOP ════

7.71; age and $A \times S$

40

	A_1	A_2	A_3
B_1	0	2	5
	1	2	3
B_2	10	4	8
	5	6	12

$$\sum_{1}^{A} \sum_{1}^{B} \sum_{1}^{N} X = \underline{\quad},$$

$$CF = \left(\sum_{1}^{A} \sum_{1}^{B} \sum_{1}^{N} X \right)^2 / ABN$$

$$= \underline{\quad}.$$

════ STOP ════════════════════ STOP ════

58, 280.33

41

	C_1	C_2	C_3
D_1	0	4	6
	1	2	4
D_2	0	1	3
	1	1	1

$$\sum_1^C \sum_1^D \sum_1^N X = \underline{\hspace{1cm}},$$

$$CF = \left(\sum_1^C \sum_1^D \sum_1^N X \right)^2 / CDN$$

$$= \underline{\hspace{1cm}},$$

$$\sum_1^C \sum_1^D \sum_1^N X^2 = \underline{\hspace{1cm}}.$$

═══ STOP ═══════════════ STOP ═══

24, 48, 86

42

	C_1	C_2	C_3
D_1	0	4	6
	1	2	4
D_2	0	1	3
	1	1	1

$$\sum_1^C \sum_1^D \sum_1^N X^2 = 86, \qquad CF = 48.$$

$$tot_{..} = \underline{\hspace{1cm}},$$

$$C_{..} = \sum_1^C \left(\sum_1^D \sum_1^N X \right)^2 / DN - CF$$

$$= \underline{\hspace{1cm}}.$$

═══ STOP ═══════════════ STOP ═══

$86 - 48 = 38,\ 66 - 48 = 18$

43

	C_1	C_2	C_3
D_1	0	4	6
	1	2	4
D_2	0	1	3
	1	1	1

$$CF = 48, \quad tot_{..} = 38, \quad C_{..} = 18.$$

$$D_{..} = \sum_1^D \left(\sum_1^C \sum_1^N X \right)^2 / CN - CF$$

$$= \underline{\hspace{1cm}}.$$

═══ STOP ═══════════════ STOP ═══

$56.33 - 48 = 8.33$

44

	C_1	C_2	C_3
D_1	0	4	6
	1	2	4
D_2	0	1	3
	1	1	1

$$CF = 48, \qquad tot_{..} = 38,$$

$$C_{..} = 18, \qquad D_{..} = 8.33.$$

$$bg_{..} = \sum_1^C \sum_1^D \left(\sum_1^N X \right)^2 / N - CF$$

$$= \underline{\hspace{1cm}},$$

$$(C \times D)_{..} = bg_{..} - \underline{\hspace{1cm}}$$

$$- \underline{\hspace{1cm}} = \underline{\hspace{1cm}}.$$

═══ STOP ═══════════════ STOP ═══

$158/2 - CF = 31,\ - 18 - 8.33 = 4.67$

45

	C_1	C_2	C_3
D_1	0	4	6
	1	2	4
D_2	0	1	3
	1	1	1

$CF = 48,$ $tot_{..} = 38,$

$C_{..} = 18,$ $D_{..} = 8.33,$

$(C \times D)_{..} = 4.67,$ $bg_{..} = 31.$

$wg_{..} = $ _____.

════ STOP ════   ════ STOP ════

$$38 - 31 = 7.00$$

46

	C_1	C_2	C_3
D_1	0	4	6
	1	2	4
D_2	0	1	3
	1	1	1

Source	ss	df
C	18	?
D	8.33	?
$C \times D$	4.67	?
wg	7	?
tot	38	?

════ STOP ════   ════ STOP ════

$$C = 2, D = 1, C \times D = 2, wg = 6, tot = 11$$

47

Source	ss	df	ms
C	18	2	?
D	8.33	1	?
$C \times D$	4.67	2	?
wg	7	6	?

════ STOP ════   ════ STOP ════

$$C = 9, D = 8.33, C \times D = 2.33, wg = 1.17$$

48

Source	ss	df	ms	F
C	18.00	2	9.00	?
D	8.33	1	8.33	?
$C \times D$	4.67	2	2.33	?
wg	7.00	6	1.17	

════ STOP ════   ════ STOP ════

$$C = 7.69, D = 7.12, C \times D = 2.00$$

49

Source	df	F
C	2	7.69
D	1	7.12
$C \times D$	2	2.00
wg	6	

With 1 and 6 df an F of _____ and with 2 and 6 df an F of _____ are required to reject the null hypothesis.

════ STOP ════   ════ STOP ════

$$5.99, 5.14$$

50

Source	df	F
C	2	7.69
D	1	7.12
$C \times D$	2	2.00

We reject the null hypothesis for which variables? _____ and _____.

═══ STOP ═══════════════════════════ STOP ═══

C and D

51

$$\begin{array}{c c c} & P_1 & P_2 \\ T_1 & \begin{array}{c} 4 \\ 6 \end{array} & \begin{array}{c} 3 \\ 7 \end{array} \\ \hline T_2 & \begin{array}{c} 0 \\ 0 \end{array} & \begin{array}{c} 7 \\ 5 \end{array} \end{array}$$

$CF = $ _____, $tot_{..} = $ _____.

═══ STOP ═══════════════════════════ STOP ═══

$128, 184 - 128 = 56$

52

$$\begin{array}{c c c} & P_1 & P_2 \\ T_1 & \begin{array}{c} 4 \\ 6 \end{array} & \begin{array}{c} 3 \\ 7 \end{array} \\ \hline T_2 & \begin{array}{c} 0 \\ 0 \end{array} & \begin{array}{c} 7 \\ 5 \end{array} \end{array}$$

$CF = 128$, $tot_{..} = 56$. $P_{..} = $ _____, $T_{..} = $ _____.

═══ STOP ═══════════════════════════ STOP ═══

$(10 \times 10)/(2 \times 2) + (22 \times 22)/(2 \times 2) - CF = 18$,
$(20 \times 20)/(2 \times 2) + (12 \times 12)/(2 \times 2) - CF = 8$

53

$$\begin{array}{c c c} & P_1 & P_2 \\ T_1 & \begin{array}{c} 4 \\ 6 \end{array} & \begin{array}{c} 3 \\ 7 \end{array} \\ \hline T_2 & \begin{array}{c} 0 \\ 0 \end{array} & \begin{array}{c} 7 \\ 5 \end{array} \end{array}$$

$CF = 128$, $tot_{..} = 56$, $P_{..} = 8$, $T_{..} = 18$. $bg_{..} = $ _____, $(P \times T)_{..} = $ _____.

═══ STOP ═══════════════════════════ STOP ═══

$$\frac{10^2 + 10^2 + 0^2 + 12^2}{2} - CF = 44, \qquad 44 - 8 - 18 = 18$$

---

54 $CF = 128$, $tot_{..} = 56$, $P_{..} = 8$, $T_{..} = 18$, $(P \times T)_{..} = 18$. $wg_{..} = $ _____.

═══ STOP ═══════════════════════════ STOP ═══

$56 - 8 - 18 - 18 = 12$

55

Source	ss	df	ms	F
P	8	1	?	?
T	18	1	?	?
$P \times T$	18	1	?	?
wg	12	4	?	
tot	56	7		

=== STOP === === STOP ===

ms	F
8	2.67
18	6.00
18	6.00
3	

---

56 With 1 and 4 df an F of 7.71 is required to reject the null hypothesis. Thus in the preceding experiment we conclude that _____ of the variables are significant.

=== STOP === === STOP ===

57

	T_1	T_2
S_1	0	3
S_2	0	2
S_3	3	5
S_4	1	2

Under certain circumstances there is only *one* score per cell (group). This need not bother us since the procedure again is almost the same as it has been.

$$\sum_1^S \sum_1^T X = \underline{\quad\quad},$$

$$\sum_1^S \sum_1^T X^2 = \underline{\quad\quad},$$

$$CF = \left(\sum_1^S \sum_1^T X \right)^2 / ST = \underline{\quad\quad}.$$

=== STOP === === STOP ===

$16, 52, (16 \times 16)/8 = 32$

58

	T_1	T_2
S_1	0	3
S_2	0	2
S_3	3	5
S_4	1	2

Experiments in which there is only one score per cell usually occur when the same subject is measured two or more times. $\sum_1^S \sum_1^T X^2 = 52$, $CF = 32$, $tot_{..} = \underline{\quad\quad}.$

=== STOP === === STOP ===

$52 - 32 = 20$

59

	T_1	T_2
S_1	0	3
S_2	0	2
S_3	3	5
S_4	1	2

In the experiment at the left four subjects (S) were given two trials (T) each. Column variability would reflect differences between trials while row variability would reflect differences among _____.

═══ STOP ═══ ═══ STOP ═══

subjects

60

	T_1	T_2
S_1	0	3
S_2	0	2
S_3	3	5
S_4	1	2

$CF = 32, \qquad tot_{..} = 20.$

$T_{..} = \sum_1^T \left(\sum_1^S X \right)^2 / S - CF$

$= \underline{\hspace{1cm}}.$

═══ STOP ═══ ═══ STOP ═══

$4^2/4 + 12^2/4 - CF = 4 + 36 - 32 = 8$

61

	T_1	T_2
S_1	0	3
S_2	0	2
S_3	3	5
S_4	1	2

$CF = 32, \qquad tot_{..} = 20, \qquad T_{..} = 8.$

$S_{..} \sum_1^S \left(\sum_1^T X \right)^2 / T - CF = \underline{\hspace{1cm}}.$

═══ STOP ═══ ═══ STOP ═══

$(3^2 + 2^2 + 8^2 + 3^2)/2 - CF = 43 - 32 = 11$

62

	T_1	T_2
S_1	0	3
S_2	0	2
S_3	3	5
S_4	1	2

$CF = 32, tot_{..} = 20, T_{..} = 8, S_{..} = 11.$

$bg_{..} = \sum_1^T \sum_1^S \left(\sum_1^N X \right)^2 / N - CF,$ but

in this type of analysis $N = \underline{\hspace{1cm}}.$

═══ STOP ═══ ═══ STOP ═══

1

63

	T_1	T_2
S_1	0	3
S_2	0	2
S_3	3	5
S_4	1	2

$CF = 32$, $tot_{..} = 20$, $T_{..} = 8$, $S_{..} = 11$, $\sum_1^T \sum_1^S X = 16$, $\sum_1^T \sum_1^S X^2 = 52$. With $N = 1$ then $bg_{..} = \sum_1^T \sum_1^S (\sum_1^N X)^2/N - CF = \sum_1^T \sum_1^S X^2 - CF = $ _____,

and this is *exactly* equal to what other sum of squares? _____.

═══ STOP ═══════════════════════ STOP ═══

20, $tot_{..}$

64

	T_1	T_2
S_1	0	3
S_2	0	2
S_3	3	5
S_4	1	2

$CF = 32$, $tot_{..} = 20$, $T_{..} = 8$, $S_{..} = 11$. *Whenever* an experiment has only one score per cell $tot_{..}$ will always equal _____.

═══ STOP ═══════════════════════ STOP ═══

$bg_{..}$

---

65 Whenever an experiment has only *one* score per cell $tot_{..} = bg_{..}$. But since $wg_{..} = tot_{..} - bg_{..}$ then in this case $wg_{..}$ must equal _____.

═══ STOP ═══════════════════════ STOP ═══

0

66

	T_1	T_2
S_1	0	3
S_2	0	2
S_3	3	5
S_4	1	2

$CF = 32$, $tot_{..} = 20$, $T_{..} = 8$, $S_{..} = 11$. If there were more than one score per cell we could compute the interaction: $(T \times S)_{..} = bg_{..} - T_{..} - S_{..}$. Since there is only one score per cell we can compute the interaction: $(T \times S)_{..} = tot_{..} - $ _____ $- $ _____.

═══ STOP ═══════════════════════ STOP ═══

$T_{..} - S_{..}$

67

	T_1	T_2
S_1	0	3
S_2	0	2
S_3	3	5
S_4	1	2

$CF = 32, tot_{..} = 20, T_{..} = 8, S_{..} = 11.$
$(T \times S)_{..} = tot_{..} - T_{..} - S_{...}$
$(T \times S)_{..} =$ _____.

═══ STOP ═══ ═══ STOP ═══

$20 - 8 - 11 = 1.00$

68

	T_1	T_2
S_1	0	3
S_2	0	2
S_3	3	5
S_4	1	2

$CF = 32, tot_{..} = 20, T_{..} = 8, S_{..} = 11,$
$(T \times S)_{..} = 1.00.$ df are computed as
before: T df = _____, S df = _____. $(T \times S)$ df = T df $\times$ S df = _____, and tot df = _____.

═══ STOP ═══ ═══ STOP ═══

$1, 3; 3, 7$

69

Source	ss	df	ms
T	8	1	?
S	11	3	?
$T \times S$	1	3	?
tot	20	7	

═══ STOP ═══ ═══ STOP ═══

$T = 8.00, S = 3.67, T \times S = 0.33$

70

Source	ss	df	ms	F
T	8	1	8	?
S	11	3	3.67	?
$T \times S$	1	3	0.33	
tot	20	7		

In the absence of wg_{ms} we must make use of what is available. The most appropriate term to use is the $(T \times S)_{ms}$. Compute F.

═══ STOP ═══ ═══ STOP ═══

$T = 24.24, S = 11.12$

---

71 Since subjects differ, the rejection of the null hypothesis with reference to S is of trivial interest. However with 1 and 3 df an F of 10.13 is required to reject the null hypothesis and we conclude that _____ _____.

═══ STOP ═══ ═══ STOP ═══

trials differ

72

	T_1	T_2	T_3
S_1	0	2	5
S_2	1	6	8
S_3	0	1	2
S_4	1	2	4
S_5	1	4	8

$$\sum_1^S \sum_1^T X = 45, \qquad \sum_1^S \sum_1^T X^2 = 237,$$

$$CF = 135, \qquad S_{..} = 32.$$

$tot_{..} = \underline{\hspace{1cm}}, \quad T_{..} = \underline{\hspace{1cm}},$

$(T \times S)_{..} = \underline{\hspace{1cm}}.$

═══ STOP ═══ ═══ STOP ═══

102, 57.6, 12.4

73

	T_1	T_2	T_3
S_1			
S_2			
S_3			
S_4			
S_5			

$tot_{..} = 102, \; T_{..} = 57.6, \; S_{..} = 32,$ $(T \times S)_{..} = 12.4$. Find the F between trials (T): \underline{\hspace{1cm}}. Do the trials differ? \underline{\hspace{1cm}}.

═══ STOP ═══ ═══ STOP ═══

$F = 28.8/1.55 = 18.58$; Yes, reject the null hypothesis

74

	R_1	R_2	R_3	R_4
P_1	0	1	2	3
	1	2	0	1
P_2	2	8	10	15
	3	3	6	7

$CF = 256, \; tot_{..} = 260, \; R_{..} = 52,$ $bg_{..} = 202. \; P_{..} = \underline{\hspace{1cm}}, \; (R \times P)_{..} = \underline{\hspace{1cm}}.$

═══ STOP ═══ ═══ STOP ═══

$121, \; 202 - 52 - 121 = 29$

---

75 $tot_{..} = 260, \; R_{..} = 52, \; P_{..} = 121, \; (R \times P)_{..} = 29, \; bg_{..} = 202. \; wg_{..} = \underline{\hspace{1cm}}.$

═══ STOP ═══ ═══ STOP ═══

$260 - 52 - 121 - 29 = 58$

76

	R_1	R_2	R_3	R_4
P_1				
P_2				

Source	ss	df	ms	F
$R \times P$	29	?	?	?
wg	58	8	?	

Compute the F for the interaction.

═══ STOP ═══ ═══ STOP ═══

	df	ms	F
$R \times P$	3	9.67	1.33
wg	8	7.25	

77

	A_1	A_2	A_3
	0	2	6
B_1	1	1	4
	0	1	2
	7	2	1
B_2	5	2	1
	1	0	0

$\sum \sum \sum X = 36,$

$\sum \sum \sum X^2 = 148,$

$\sum_1^B \left(\sum_1^A \sum_1^N X \right)^2 / AN = 72.22,$

$\sum_1^A \left(\sum_1^B \sum_1^N X \right)^2 / BN = 76.00.$

$bg_{..} = \underline{\qquad}.$

═══ STOP ═══════════════════════════ STOP ═══

$$116.67 - 72 = 44.67$$

78

	A_1	A_2	A_3
	0	2	6
B_1	1	1	4
	0	1	2
	7	2	1
B_2	5	2	1
	1	0	0

$\sum \sum \sum X = 36,$

$\sum \sum \sum X^2 = 148,$

$CF = 72, \; bg_{..} = 44.67,$

$\sum_1^B \left(\sum_1^A \sum_1^N X \right)^2 / AN = 72.22,$

$\sum_1^A \left(\sum_1^B \sum_1^N X \right)^2 / BN = 76.00.$

$(A \times B)_{..} = \underline{\qquad}.$

═══ STOP ═══════════════════════════ STOP ═══

$$44.67 - 4 - 0.22 = 40.45$$

79

	T_1	T_2	T_3
S_1	0	2	3
S_2	2	2	4
S_3	3	2	5
S_4	1	2	6

$\sum \sum \sum X = 32,$

$\sum \sum \sum X^2 = 116,$

$\left(\sum_1^S \sum_1^T X \right)^2 / ST = 85.33,$

$\sum_1^S \left(\sum_1^T X \right)^2 / T = 90,$

$\sum_1^T \left(\sum_1^S X \right)^2 / S = 106.$

$(S \times T)_{..} = \underline{\qquad},$

$wg_{..} = \underline{\qquad}.$

═══ STOP ═══════════════════════════ STOP ═══

$$30.67 - 20.67 - 4.67 = 5.33, \quad 0.00$$

Fifteen

Chi Square

Is the proportion of women students who smoke greater than the proportion of men smokers on your college campus? To find the answer to this question you could ask 50 men and 50 women if they smoked. Unfortunately, you would be unable to determine whether the difference between the proportions was significantly greater than the chance expectation. The reason is that your data would be represented by a "yes" or "no" response from each student, and the techniques of statistical inference we have discussed so far do not apply to *categorical* (nominal) data.

Quite often the research worker is unable to obtain data that can be ordered along some numerical scale. He would find himself in this situation if you asked him your question about smokers. This chapter presents some techniques that may be used when an analysis of categorical data is desired. Is type A blood more frequent than type B? Is preference for a candidate for political office related to social class? Is sickle cell anemia found more often in one race than in another? Is frequency of abortion greater among college-educated women? Investigation of all these questions would yield categorical data.

We will use the statistic called *chi square* to analyze the results of experiments that yield categorical data. We encountered chi square in previous chapters, noting that distributions of sample variances are closely related to this distribution. The statistic introduced in this chapter is also called chi-square because the sampling distribution of the statistic obtained from the formula below closely approximates the chi-square distribution.

$$\chi^2 = \sum_1^K \frac{(o - e)^2}{e}$$

where χ^2 = chi square, K = number of categories or groups, o = observed frequency in a category, and e = expected frequency in a category.

Suppose a neighborhood is rapidly changing, with the original inhabitants moving away and large numbers of college students moving in. The librarian in the neighborhood library decides to try to determine the type of reading material these college students select so that he will be able to reflect their preferences when he orders books in the future. Consequently, he asks each checker to classify the type of book borrowed (or if more than one per person, the first book borrowed) according to these four categories: Fiction, Hobbies and How to Do It, Current Events and History, and Other.

In subsequent research the librarian can be more specific in his hypothesis. However, since he has no prior hypothesis about how book preferences among these college students run, he decides that the hypothesis of chance or random preference would be the best one to test. That is, the librarian wishes to test the hypothesis that selection from the four categories listed above is random and that any differences observed are due to chance.

A survey is taken over a three-day period. Eighty college students borrow books, and these are classified as shown in Table 15-1. If there is simply a random selection of topics, each would have been selected about 20 times. Hence, the value of e (the expected frequency) would be 20 for each of the four categories. The value of o (the observed frequency) would be the number of students selecting a specific topic. The results of the survey are presented in Table 15-1.

Table 15-1   Expected and observed frequencies of topics selected at a neighborhood library by college students

Topic	Expected frequency	Observed frequency
Fiction	20	15
Hobbies and How to Do It	20	40
Current Events and History	20	20
Other	20	5
	80	80

Although there are marked differences among the observed frequencies in the four categories, these differences may be due to chance. For this reason we compute the value of chi square:

$$\chi^2 = \frac{(15-20)^2}{20} + \frac{(40-20)^2}{20} + \frac{(20-20)^2}{20} + \frac{(5-20)^2}{20}$$

$$= \frac{25}{20} + \frac{400}{20} + \frac{0}{20} + \frac{225}{20}$$

$$= 1.25 + 20.00 + 0 + 11.25 = 32.50$$

READING APPENDIX G

To evaluate χ^2 we must consult Appendix G. This table consists of three columns, labeled df, 0.05, and 0.01. The df column indicates that the shape of the chi-square distribution depends upon the number of categories involved. The 0.05 and 0.01 columns indicate that either 5 percent or 1 percent of the values of the chi-square distribution are larger than the tabled value if the null hypothesis is true. The number of df are determined in much the same way that they are in the analysis of variance. With K groups, the number of df is equal to $K - 1$. In our example, we have four groups, and therefore we have $4 - 1 = 3$ df. With 3 df, according to Appendix G, the probability is equal to 0.01 that a chi square will have a value equal to or greater than the tabled value of 11.34. Since the obtained value of chi square is 32.50, we reject the null hypothesis and conclude that the four categories of books are not equally preferred.

CHI SQUARE AS A MEASURE OF GOODNESS OF FIT

The expected frequencies are not required to be equal across categories. For example, an anthropologist studying the diffusion of culture in Iceland found the blood types of a large sample of people. She wished to determine if the distribution of blood types among the people of Iceland was the same as it is in Denmark, which is where the people of Iceland are supposed to have originated. There are four major blood types: A, O, AB, and B; in Denmark, 45 percent of the population has type A and 40 percent has type O, with AB and B occurring about 15 percent of the time between them. Although Icelanders are supposed to have come from Denmark, the observed values are considerably different from the expected values based on the proportions occurring in Denmark. These data are presented in Table 15-2. The expected values were found by multiplying the 80 people measured by the various proportions of blood types.

Table 15-2   Observed and expected blood types in Iceland

Blood type	Expected frequency	Observed frequency
A	36	24
O	32	48
AB or B	12	8
	80	80

The differences between the observed and expected frequencies may be due to chance, and chi square is computed to determine if the hypothesis of chance difference can be rejected.

$$\chi^2 = \frac{(24 - 36)^2}{36} + \frac{(48 - 32)^2}{32} + \frac{(12 - 8)^2}{8}$$

$$= \frac{144}{36} + \frac{256}{32} + \frac{16}{8} = 4.00 + 8.00 + 2.00 = 14.00$$

With 2 df, the tabled value of chi square is 9.21. The anthropologist rejects the hypothesis of chance difference and concludes that the distribution of Icelandic blood types differs from that of the Danes.

This type of chi-square procedure has been called the "test for goodness of fit" and is used when the research worker has some expectation as to how the data should be distributed. If the data of a study were expected to be normally distributed, for example, the expected values could be derived from Appendix B, which includes areas under the normal curve.

RESTRICTIONS ON THE USE OF CHI SQUARE

There are a number of restrictions on the use of the chi-square statistic. Some of these arise because the use of the continuous chi-square distribution as an approximation to the distribution of discrete possible events is inappropriate under certain conditions:

1. Frequency data must be used (counts of persons or events, not scaled scores).
2. The *expected* value in any cell should never be less than 5.
3. The sum of the observed frequencies must equal the sum of the expected frequencies.
4. When df = 1, the correction for continuity must be used (see below).
5. Each score must be independent of every other score (no person or event is allowed to appear more than once in the frequency table).

CORRECTION FOR CONTINUITY   Whenever a chi-square problem has 1 df, the correction for continuity should be used. This simply involves subtracting 0.5 from each absolute difference between an *e* and an *o*. For example, suppose you flipped a coin 100 times and obtained 60 heads. Here the expected numbers of heads and tasks would be 50 and 50, and the observed numbers of heads and tasks are 60 and 40, respectively.

With 1 df the formula for chi square (corrected for continuity)[1] is

$$\chi^2 = \sum \frac{(|o - e| - 0.5)^2}{e}$$

and the obtained chi square is equal to

$$\frac{(|60 - 50| - 0.5)^2}{50} + \frac{(|40 - 50| - 0.5)^2}{50} = \frac{(9.5)^2 + (9.5)^2}{50} = 3.61$$

With 1 df we see from Appendix G that we cannot reject the null hypothesis. Note that if we had not corrected for continuity, we would have obtained a chi square of 4.00. In this case, we would have rejected the null hypothesis and concluded that we had a biased coin, when actually this conclusion would not be justified.[2]

---

[1] The bars bracketing *o* and *e* indicate that the absolute difference between *o* and *e* should be found (change minus to plus if necessary) before subtracting 0.5.

[2] The correction for continuity *does not* make the chi-square value a more accurate estimate of the theoretical chi-square value; it *does* make the use of the chi-square distribution table (Appendix G) more nearly appropriate in terms of the actual probabilities of the discrete events observed. Misunderstanding of this point has led a number of statisticians to challenge the continuity correction. The authors are indebted to Professor Quinn McNemar for clarifying this situation.

CHI SQUARE AS AN INDEX OF ASSOCIATION

Chi square can be used to determine the relationship or association between two categorical variables. To illustrate this, let us look at the case where a zoologist is interested in determining those conditions which favor certain types of fish and those which do not. Suppose he takes three types of fish—carp, river carpsucker, and buffalo—and stocks them in equal numbers in two different water environments. That is, he places all three types of fish into a lake and equal numbers into a relatively swift, clear river. Three years later the zoologist goes to the lake and to the river, and spends a day at each catching fish. He uses a procedure which he feels randomly samples the fish in each place. The fish that he caught are shown in Table 15-3.

Table 15-3   Fish harvest as a function of type and location

Location	Carp	River Carpsucker	Buffalo	Total
Lake	36	12	12	60
River	14	8	18	40
	50	20	30	100

Note that the zoologist is interested in those conditions which are favorable and unfavorable to competing types of fish. If he only wanted to find out which fish was the most successful (that is, the most plentiful fish), then the column totals would be compared. If the better environment was to be identified, then the row totals would be compared. In this case, however, we want to determine if success is related to the environment. Casual observation would seem to say it is, since carp are very successful in the lake whereas buffalo are relatively successful in the river. That is, type of fish and environment are *associated*. We must compute the chi square to determine if the relationship between these two variables is only to chance. The next step is to determine the expected frequencies in each of the fish-environment categories. This is done by dividing the product of the respective row and column frequencies by the total frequency. For example, to find the expected number of carp from the lake, we multiply the total number of fish from the lake (60) by the total number of carp (50) and divide by the total number of fish taken from both lake and river (100). Thus, the expected number of carp from the lake is $(60 \times 50)/100 = 30$. This procedure has been followed for each of the other five cells, and these expected frequencies are represented in Table 15-4.

Table 15-4   Expected frequencies of fish

Environment	Carp	River Carpsucker	Buffalo	Total
Lake	30	12	18	60
River	20	8	12	40
	50	20	30	100

If type of environment is *not* associated with type of fish, then the proportions of fish in the lake should equal their relative proportions in the total sample. The value of chi square is again computed from the formula as follows:

$$\chi^2 = \sum \frac{(o - e)^2}{e}$$

$$= \frac{(36 - 30)^2}{30} + \frac{(12 - 12)^2}{12} + \frac{(12 - 18)^2}{18} + \frac{(14 - 20)^2}{20} + \frac{(8 - 8)^2}{8} + \frac{(18 - 12)^2}{12}$$

$$= \frac{36}{30} + \frac{0}{12} + \frac{36}{18} + \frac{36}{20} + \frac{0}{8} + \frac{36}{12} = 8.00$$

The number of df is computed in a manner similar to that for interaction in the analysis of variance. The number of df is the product of the row and column df. With three names there are $N - 1 = 3 - 1 = 2$ df for the columns, and for the rows there are two locations, with $S - 1 = 2 - 1 = 1$ df. Thus, there are $2 \times 1 = 2$ df in this problem.

A chi square of 5.99 is required to reject the null hypothesis with 2 df at the 0.05 level, and 9.21 is required at the 0.01 level. Our obtained chi square of 8.00 is between these two values, and so we can reject the null hypothesis at the 5 percent level of significance, but not at the 1 percent level.

Remember which null hypothesis we are talking about now; we reject the hypothesis that the two variables are not related. The data of Table 15-3 supports this conclusion. The table shows that carp are more successful in a lake, while buffalo are more successful in a river.

CHI SQUARE WITH A 2 × 2 TABLE

The need may arise to analyze the results of an experiment in which each of two variables has been divided into two categories. An example of this would be to ask a group of males and females a question such as, "Do you smoke?" which can be answered "yes" or "no." This type of problem may be answered like the example for chi square as an index of association. For convenience, however, the following formula can be used instead. The 2 × 2 table is shown in Table 15-5 with letters indicating the *observed* frequencies within each cell.

Table 15-5   Schematic representation of a 2 × 2 chi-square table

	Yes	No	Total
Male	A	B	$A + B$
Female	C	D	$C + D$
	$A + C$	$B + D$	$A + B + C + D = N$

Under these circumstances the formula is as follows:

$$\chi^2 = \frac{N \left(|BC - AD| - \dfrac{N}{2} \right)^2}{(A + B)(C + D)(A + C)(B + D)}$$

The data in Table 15-6 can be analyzed with the following formula.

**Table 15-6   Number of recidivists under
two different parole policies**

	Return to prison?		
	YES	NO	TOTAL
Policy A	40	10	50
Policy B	20	30	50

$$\chi^2 = \frac{100(|200 - 1200| - 50)^2}{50 \times 50 \times 60 \times 40} = \frac{100 \times 950 \times 950}{50 \times 50 \times 60 \times 40} = 15.04$$

The value of the chi square may be calculated by substituting into the formula. The obtained result is exactly the same as would be obtained if the previous procedure had been followed. It should also be noted that there is only 1 df in a 2×2 table and that the formula related to Table 15.5 applies the correction for continuity (by subtracting $N/2$ in the numerator).

It is relatively easy to find chi square in much the same way as it was done with chi square as a measure of association in the preceding section. For example, for the policy A–yes cell in Table 15-6, the expected frequency is equal to the product of the appropriate row and column sums divided by the total frequency: $e = (50 \times 60)/100 = 30$. Correcting for continuity and squaring the difference between observed and expected values: $(|o - e| - 0.5)^2 = (|40 - 30| - 0.5)^2 = 9.5^2 = 90.25$ Dividing by e yields $90.25/30 = 3.01$. For all cells, $\chi^2 = 3.01 + 3.01 + 4.51 + 4.51 = 15.04$. With 1 df, $Pr < 0.01$, and the null hypothesis is rejected. It can be concluded that rate of recidivism is related to parole policy.

MEDIAN TEST

The central tendencies of two or more groups may be compared by use of chi square in situations where the analysis of variance would lead to violation of assumptions. Suppose, for example, that students are asked to report the number of magazines (Table 15-7) to which their parents subscribe. Such data would usually be positively skewed, with most people subscribing to a few magazines. Under such circumstances we would hesitate to analyze our data using the analysis of variance because we would be violating the assumption of normality.

The question to be asked concerns subscription habits of the parents of three groups of students (40 in all): English majors (13), mathematics majors (12), and music majors (15). The null hypothesis to be tested is that the median number of magazines reported will be the same for each group.

Table 15-7   Number of magazines subscribed to by parents of three groups of students

English			Math			Music		
10	3	2	9	2	1	10	3	1
7	3	2	5	2	0	7	2	1
6	3	2	4	2		5	1	1
4	3		3	1		4	1	1
4	3		2	1		3	1	0

A frequency distribution of all the scores is illustrated in Figure 15-1. The median of the distribution is shown to be 2.50.

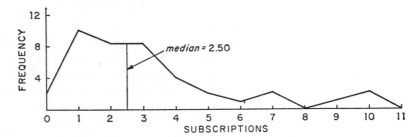

Figure 15-1   *Frequency distribution of magazine subscriptions for parents of 40 students.*

After having found the median of all the scores in all the groups, we may now set up a table of observed frequencies (Table 15-8). This is done by determining the number of scores in each group which fall above and below the overall median. In the group of English majors, for example, only three students reported fewer magazines than the general median.

Table 15-8   Observed and expected frequencies above and below the overall median in the magazine survey

	Observed frequencies			
	ENGLISH	MATH	MUSIC	TOTAL
Above median	10	4	6	20
Below median	3	8	9	20
	13	12	15	40

	Expected frequencies		
	ENGLISH	MATH	MUSIC
Above median	6.5	6.0	7.5
Below median	6.5	6.0	7.5

We next determine the expected frequencies shown in Table 15-8 by finding the product of the respective row and column sums and then dividing by the total number of scores. Thus, the expected frequency above the median for the English majors would be $(20 \times 13)/40 = 6.5$. Next, chi square is computed in the usual manner:

$$\chi^2 = \frac{(10 - 6.5)^2}{6.5} + \frac{(4 - 6.0)^2}{6.0} + \cdots + \frac{(9 - 7.5)^2}{7.5} = 5.70$$

With $(R - 1)(C - 1) = (2 - 1)(3 - 1) = 2$ df, a chi square of 5.70 is not quite large enough to reject the null hypothesis of identical medians for the three groups. These results, although not statistically significant, would probably be encouraging enough to the investigator to lead him to repeat the experiment with larger samples of students. It is interesting to note that the F ratio obtained from an analysis of variance of the same data is not nearly as close to significance as was the chi square computed above ($F = 1.15$). You may want to verify this as an exercise.

CHI SQUARE WITH LARGE NUMBERS OF DF

When df is greater than 30, the chi-square table in Appendix G cannot be used. The following formula may be used in this case to convert the obtained χ^2 to a z score, which may then be used with the table of the areas under the normal curve (Appendix B) to determine its significance.

$$z = \sqrt{2\chi^2} - \sqrt{2 \, df - 1}$$

For instance, suppose we computed chi square for a 15×15 table of event classifications. The df for such a table would be $14 \times 14 = 196$, which is far beyond the limit of Appendix G. Suppose χ^2 was computed to be 225. Then

$$z = \sqrt{2 \times 225} - \sqrt{2 \times 196 - 1} = 1.44$$

In Appendix B the entry for this z-value distance from the mean is 42.51 percent. By subtraction, 7.49 percent of the normal curve would lie beyond a z value this large, and we would conclude that the null hypothesis could not be rejected at the 5 percent level.

SUMMARY

This chapter has presented the essential applications of chi square. The chi-square statistic is a nonparametric statistic (a class of tests about which more will be said later) and, in contrast to earlier tests, is used with nominal or frequency data. The essential feature of the test is the determination of a measure (χ^2), which indicates the size of the difference between the data as they might be found (expected) either on the basis of chance or on the basis of expectations from prior experience.

PROBLEMS

1. Three candidates—Jones, Smith, and Green—were running for local office. A poll found the following results when 120 voters were asked their preference. Is there a difference in the voters' preferences?

$$
\begin{array}{ll}
\text{Jones} & 55 \\
\text{Smith} & 30 \\
\text{Green} & 35
\end{array}
$$

2. Is smoking related to premature birth? In one study of 100 smokers and 100 nonsmokers, the following data were found: Nonsmoking mothers had 5 premature births; smoking mothers had 16 premature births. Are smoking and premature births related?

3. A coin is flipped eight times and not a single head is flipped. Compute the appropriate chi square or tell why chi square should not be computed in this case.

4. *E* was interested in the relationship between age and the solution to the transposition problem. Consequently he used groups of 6-, 12-, and 18-year-olds ($N = 20$ per group). He found that 6 of the 6-year-olds, 12 of the 12-year-olds, and 6 of the 18-year-olds demonstrated transposition. Does transposition vary as a function of age?

5. A poll was taken on the question of whether the sale and use of marijuana should be legalized. Half of the 120 respondents were over 30 years old and the other half were under 30. Of those respondents over 30 years of age, 20 wished to legalize marijuana and 40 did not. Of those respondents under 30 years old, 36 wished to legalize the use of marijuana and 24 did not. Is the desire to legalize the sale of marijuana associated with age?

6. As a demonstration, a professor had each member of his class flip two coins, and then he counted the number of heads. Ten members reported flipping no heads, 30 reported flipping one head (and one tail), and 8 reported flipping two heads. Did the distribution of heads depart from what the professor expected?

7. *E* was interested in the turning bias of customers as a function of sex. Consequently, he set up the entrance to a store so that the customers could turn right or left, or go straight ahead. Of 40 males, 20 turned right, 10 turned left, and 10 went straight. Of 60 females, 45 turned right, 10 went straight, and 5 turned left. Is turning bias related to sex?

8. The five sections of "Introduction to Accounting" are held at 8, 9, 10, 11, and 12 A.M. These sections contain 10, 25, 50, 50, and 15 students, respectively. Test the hypothesis that time is an unimportant factor in choice of section.

9. The enrollment in a certain statistics course included 10 freshmen, 25 sophomores, 10 juniors, and 5 seniors. Professor Snarf took attendance one day

and found that one freshman, eight sophomores, seven juniors, and four seniors were attending class. Is class attendance related to year in school?

10. A poll was taken to determine music preference. Three songs—*A*, *B*, and *C*—were played for several people whose ages ranged widely. The following data indicate the number of people preferring each song. Do the songs differ?

		Preference		
		A	*B*	*C*
	0–20	5	10	5
Age	21–40	20	0	40
	41–60	5	0	35

11. *E* conducted an experiment in which children were given their choice of four different types of food. Each child selected one, and only one, food. Compute x^2 from the following data:

A	*B*	*C*	*D*
10	2	8	12

12. Three candidates run for office. A random sample of 60 men and 40 women yields the following preferences. Compute x^2 to determine if preferences are related to sex.

	A	*B*	*C*
Men	30	15	15
Women	10	15	15

13. An experiment was conducted in turning behavior of right- and left-handed students by noting whether students turn right or left at a choice point. Compute x^2.

	Turn left	Turn right
Right-handed	45	30
Left-handed	5	20

14. Perform a median test to determine whether males and females drink equal amounts of coffee. Each entry represents the number of cups of coffee one person drinks on the test day.

Males		Females	
10	3	7	4
8	3	7	3
6	2	5	3
5	2	5	2
3	1	4	2

15. The numbers below represent the number of mystery novels checked out of the library in one week. Test the hypothesis that equivalent numbers of mysteries are checked out every day. What assumption might you be violating in computing a χ^2 on these data?

Monday	Tuesday	Wednesday	Thursday	Friday	Saturday
40	50	35	45	48	100

16. Chi square cannot be computed for the table below. Why? What could be done to meet the assumptions of chi square? What would be the resulting value?

Grade	Males	Females
A	20	20
B	35	40
C	38	36
D	5	4
F	2	0
	100	100

EXERCISES

1. Have everybody in your class or group raise his hand. Count the number of left hands raised; then count the number of right hands raised. Test the hypothesis that the number of left hands and right hands raised would be the same.
2. If there are both males and females in your group, find the number of left-handed and the number of right-handed women. Do the same for men. Test the hypothesis that handedness and sex are related.
3. Have a friend write down 20 numbers, with those numbers being as random as possible. Count the number of odd numbers: count the number of even numbers. Test the hypothesis of equal numbers of odd and even.

1   We noted in previous chapters that distributions of sample variances are closely related to the _____ distribution.

═══════ STOP ═══════════════════════════════ STOP ═══════

chi-square

---

2   The symbol χ^2 is used to denote _____ _____.

═══════ STOP ═══════════════════════════════ STOP ═══════

chi square

---

3   The formula for the _____ distribution is closely approximated by the formula

$$_____ = \sum \frac{(o - e)^2}{e}.$$

═══════ STOP ═══════════════════════════════ STOP ═══════

χ^2, χ^2

---

4      $$_____ = \sum \frac{(o - e)^2}{e}$$

where o means the observed value and e means the expected value.

═══════ STOP ═══════════════════════════════ STOP ═══════

χ^2

---

5      $$\chi^2 = \sum \frac{(o - e)^2}{e}$$

The _____ means observed value and the _____ means expected value.

═══════ STOP ═══════════════════════════════ STOP ═══════

o, e

---

6   For example, if we flip a coin ten times we would *expect* about _____ heads and about _____ tails.

═══════ STOP ═══════════════════════════════ STOP ═══════

five, five

---

7 Suppose we flip a coin twenty times, the value of e (for heads) in the equation $\sum \frac{(o - e)^2}{e}$ would equal _____.

================ STOP ================ STOP ================

10

---

8 Suppose we flip a coin forty times and find we have twenty-five heads. We suspect that the value of o in the equation $\sum \frac{(o - e)^2}{e}$ is equal to _____.

================ STOP ================ STOP ================

25

---

9 But notice that if we flip twelve coins and observe eight heads we would also observe _____ tails.

================ STOP ================ STOP ================

four

---

10 Thus if we flip a coin twenty times and get twelve heads then the expected number of heads is equal to _____, expected number of tails = _____, observed number of heads = _____ and observed number of tails = _____. Notice that $\sum o = \sum e$.

================ STOP ================ STOP ================

10, 10, 12, 8

---

11 Suppose we flip a coin thirty times and twenty heads come up. (observed number of heads) $o_h =$ _____, (expected number of heads) $e_h =$ _____, and $(o - e)^2 =$ _____.

================ STOP ================ STOP ================

20, 15, 25

---

12 Suppose we flip a coin thirty times and ten tails come up o_t _____, $e_t =$ _____, and $(o - e)^2 =$ _____.

================ STOP ================ STOP ================

10, 15, 25

---

13 Suppose we flip a coin forty times and thirty heads come up. $o_h = $ _____, $e_h = $ _____, and $(o - e)^2/e = $ _____.

══════ STOP ══════════════════════════ STOP ══════

30, 20, 100/20 = 5

---

14 Suppose we flip a coin ten times and six heads come up. For heads $o = $ _____, $e = $ _____, and $(o - e)^2/e = $ _____.

══════ STOP ══════════════════════════ STOP ══════

6, 5, 1/5 = 0.20

---

15 Suppose we flip a coin ten times and six heads come up. For tails $o = $ _____, $e = $ _____, and $(o - e)^2/e = $ _____.

══════ STOP ══════════════════════════ STOP ══════

4, 5, 1/5 = 0.20

---

16 Suppose we flip a coin twenty times and get twelve heads. For heads $(o - e)^2/e = $ _____. For tails $(o - e)^2/e = $ _____.

══════ STOP ══════════════════════════ STOP ══════

$(12 - 10)^2/10 = 0.40$; $(8 - 10)^2/10 = 0.40$

---

17 The formula for chi square is $\chi^2 = \sum \dfrac{(o - e)^2}{e}$ where o means _____ _____ and e means _____ _____.

══════ STOP ══════════════════════════ STOP ══════

observed value, expected value

18
$$\chi^2 = \sum \frac{(o - ?)^2}{e}$$

══════ STOP ══════════════════════════ STOP ══════

e

19
$$\chi^2 = ? \frac{(o - e)^2}{e}$$

══════ STOP ══════════════════════════ STOP ══════

$\sum$

20

$$x^2 = \sum \frac{(o - e)^?}{e}$$

═══ STOP ═══════════════════════════ STOP ═══

$$(o - e)^2$$

21

$$x^2 = \sum \frac{(o - e)^2}{?}$$

═══ STOP ═══════════════════════════ STOP ═══

$$e$$

22

$$x^2 = \underline{\hspace{2cm}}$$

═══ STOP ═══════════════════════════ STOP ═══

$$\sum \frac{(o - e)^2}{e}$$

---

23 If twenty coins are flipped with fifteen heads, then for heads $(o - e)^2/e = \underline{\hspace{2cm}}$ and for tails $(o - e)^2/e = \underline{\hspace{2cm}}$.

═══ STOP ═══════════════════════════ STOP ═══

2.5, 2.5

---

24 Twenty coins are flipped with fifteen heads occurring. For heads $(o-e)^2/e = 2.5$ and for tails $(o - e)^2/e = 2.5$.

$$x^2 = \sum \frac{(o - e)^2}{e} = \underline{\hspace{1.5cm}} + \underline{\hspace{1.5cm}} = \underline{\hspace{1.5cm}}.$$

═══ STOP ═══════════════════════════ STOP ═══

$$2.5 + 2.5 = 5.0$$

---

25 If you flipped ten coins and got eight heads then

$$x^2 = \sum \frac{(o - e)^2}{e} = \underline{\hspace{1.5cm}} + \underline{\hspace{1.5cm}} = \underline{\hspace{1.5cm}}.$$

═══ STOP ═══════════════════════════ STOP ═══

$$\frac{9}{5} + \frac{9}{5} = \frac{18}{5} = 3.6$$

---

26 Thirty coins are flipped and you get twenty heads. $\chi^2 =$ _____ +

_____ = _____.

═══════ STOP ═══════════════════════ STOP ═══════

$$\frac{25}{15} + \frac{25}{15} = \frac{50}{15} = 3.33$$

---

27 You roll a die thirty times. The expected number of times a 6 would appear is _____.

═══════ STOP ═══════════════════════ STOP ═══════

five

---

28 You roll a die thirty times. The expected number of times a 6 would not appear is _____.

═══════ STOP ═══════════════════════ STOP ═══════

twenty-five

---

29 You roll a die thirty times and get nine 6s. $e_1 = 5$, $e_2 = 25$, $o_1 =$ _____ and the number of rolls on which a 6 was not observed: $o_2 =$ _____. Notice that $\sum o = \sum e$.

═══════ STOP ═══════════════════════ STOP ═══════

9, 21

---

30 You roll a die thirty times and get nine 6s. For the times that a 6 occurred $(o - e)^2/e =$ _____. For the times that a 6 did not occur $(o - e)^2/e =$

_____.

═══════ STOP ═══════════════════════ STOP ═══════

$16/5 = 3.20$; $16/25 = 0.64$

---

31 You roll a die thirty times and get nine 6s.

$$\chi^2 = \sum \frac{(o - e)^2}{e} = \text{_____} + \text{_____} = \text{_____}.$$

═══════ STOP ═══════════════════════ STOP ═══════

$3.20 + 0.64 = 3.84$

---

32 Suppose a grocer sells three types of cigarettes A, B, and C. If preference was random and if the grocer sold thirty cartons of cigarettes on a certain day then $e_A =$ _____.

══════ STOP ══════════════════════════ STOP ══════

10

---

33 Of thirty cartons of cigarettes A, B, and C sold by a certain grocer, $e_A = 10$, $e_B =$ _____, and $e_C =$ _____.

══════ STOP ══════════════════════════ STOP ══════

10, 10

---

34 A grocer sold eighteen cartons of A, eight cartons of B and four cartons of C. $o_A =$ _____ and $e_A =$ _____. Hence for A $(o - e)^2/e =$ _____.

══════ STOP ══════════════════════════ STOP ══════

18, 10; $64/10 = 6.4$

---

35 A grocer sells eighteen A, eight B and four C.

$$\chi^2 = \sum \frac{(o - e)^2}{e} = \frac{(18 - 10)^2}{10} + \underline{\quad} + \underline{\quad} = \underline{\quad}.$$

══════ STOP ══════════════════════════ STOP ══════

$(6.4) + 0.4 + 3.6 = 10.4$

---

36 Fifteen cases of beverage X, Y, and Z are sold, again assuming random selection, although $X = 9$, $Y = 4$, $Z = 2$ cases are sold. Then $\chi^2 = (9 - 5)^2/5 +$ _____ $+$ _____ $=$ _____.

══════ STOP ══════════════════════════ STOP ══════

$(16/5) + 1/5 + 9/5 = 26/5 = 5.20$

---

37 Four candidates are running for the presidency of the senior class. A poll taken from a sample of thirty-six finds that four prefer A, twelve B, fourteen C, and six D. $\chi^2 =$ _____ $+$ _____ $+$ _____ $+$ _____ $=$ _____.

══════ STOP ══════════════════════════ STOP ══════

$25/9 + 9/9 + 25/9 + 9/9 = 68/9 = 7.56$

38 Three types of cars are competing for market supremacy. In a sample of twenty-four cars $A = 10$, $B = 12$, $C = 2$. Assuming random buying habits, $\chi^2 =$ _____.

=== STOP === STOP ===

$$4/8 + 16/8 + 36/8 = 56/8 = 7.00$$

---

39 Appendix G contains a table of χ^2. The column on the left is labeled "df", the middle column ".05", and the column on the right ".01." For example with 12 df a value of _____ is entered under the .01 column.

=== STOP === STOP ===

26.22

---

40 In the type of χ^2 problems we have been computing, the number of df is always one less than the number of categories. Thus if we have three types of cars there are _____ df; if there are four candidates there are _____ df.

=== STOP === STOP ===

2; 3

---

41 The entry in Appendix G indicates that 5 percent or 1 percent of the values of χ^2 will exceed the tabled value if events are occurring randomly. Thus if four indistinguishable candidates are running for office, and if a large number of samples were taken, then 5 percent would have χ^2 values larger than _____.

=== STOP === STOP ===

7.82

---

42 If three indistinguishable instant coffees are rated for taste by a large number of samples of subjects then 1 percent of the samples would have χ^2 values larger than _____.

=== STOP === STOP ===

9.21

---

43 The procedure we have been following in previous chapters has been to _____ the null hypothesis whenever the probability that an event will occur by chance is less than _____.

=== STOP === STOP ===

reject, .05

---

44 Thus if there are six groups, there are _____ df and a χ^2 value of _____ or larger would lead us to reject the null hypothesis.

═══ STOP ═══════════════════════════ STOP ═══

5, 11.07

45 If we conducted an experiment with four groups and obtained a χ^2 of 6.00 we would _____ _____ the null hypothesis.

═══ STOP ═══════════════════════════ STOP ═══

not reject

46 If we conducted an experiment with eight groups and obtained a χ^2 of 15,00 we would _____ the null hypothesis.

═══ STOP ═══════════════════════════ STOP ═══

reject

47 When we _____ the null hypothesis after computing a χ^2, we conclude that the frequencies in each category (cell) differ from what was _____.

═══ STOP ═══════════════════════════ STOP ═══

reject, expected

48 One hundred twenty people take Introductory Psychology. Twenty sign up for the 8:00 A.M. section, fifty for the 10:00 A.M. section and fifty for the 11:00 A.M. section. χ^2 = _____, df = _____. We _____ the null hypothesis and conclude that section preference _____ _____ _____.

═══ STOP ═══════════════════════════ STOP ═══

15.00, 2; reject, is not random

49 χ^2 may be computed only on data expressed in frequencies. Number of people preferring brand X would be an example of _____ data but number of seconds to traverse a runway would _____ be an example of frequency data.

═══ STOP ═══════════════════════════ STOP ═══

frequency, not

50 A = number of animals turning right
B = number of years a person spends in prison
The letter _____ represents the example of frequency data.

═══════ STOP ═══════════════════════════════ STOP ═══════

A

---

51 The sum of the observed frequences must equal the sum of the expected frequences. Thus if we flipped one hundred coins and found sixty heads, then $o_h + o_t = e_h + e_t =$ _____.

═══════ STOP ═══════════════════════════════ STOP ═══════

100

---

52 The frequences must be independent. That is, if a single rat turned right ten out of twelve times, the data would not be _____ because the scores were taken from the same rat. If ten out of twelve rats turned right the data would be _____.

═══════ STOP ═══════════════════════════════ STOP ═══════

independent, independent

---

53 The chi-square distribution is a continuous distribution but we apply it to frequency data that are discrete. Because of this we must use a correction for continuity when there is only 1 df. When there is more than 1 df the _____ _____ _____ need not be used.

═══════ STOP ═══════════════════════════════ STOP ═══════

correction for continuity

---

54 Suppose we flip twenty coins and get fifteen heads. The appropriate formula here is $\chi^2 = \sum \frac{(|o - e| - 0.5)^2}{e}$. This simply means that you subtract 0.5 from the absolute difference between o and e. Thus for tails $(|o - e| - 0.5) = |5 - 10| = 5$ and $|5| - 0.5 =$ _____.

═══════ STOP ═══════════════════════════════ STOP ═══════

4.5

---

55 We get fifteen heads in twenty flips. The appropriate formula is corrected for continuity: $\sum \frac{(|o - e| - 0.5)^2}{e}$
$\chi^2 = (4.5)^2/10 + (4.5)^2/10 =$ _____.

═══════ STOP ═══════════════════════════════ STOP ═══════

$2.025 + 2.025 = 4.05$

---

56   We flip fifteen heads in twenty and obtain a χ^2 of 4.05 with _____ df, we _____ the null hypothesis.

══════ STOP ══════════════════════ STOP ══════

---

57   We use the special formula $\chi^2 = \sum \dfrac{(|o - e| - 0.5)^2}{e}$ only when we have

1 df. The reason why we subtract 0.5 from the difference between o and e is that we must correct for _____.

══════ STOP ══════════════════════ STOP ══════

---

58   You notice that of one hundred people entering a movie sixty turn right. Test the hypothesis that people turn right or left randomly. df = _____, $\chi^2 =$ _____. Conclusion: We _____ reject the hypothesis.

══════ STOP ══════════════════════ STOP ══════

---

59   Although the χ^2 distribution is _____ it is applied to discrete data. For this reason the correction for continuity must be used whenever df = _____. For this same reason the expected value of any cell must always be 5 or more.

══════ STOP ══════════════════════ STOP ══════

---

60 χ^2 may also be used to answer questions concerning the comparability of groups. In this case χ^2 is used as an index of association. But as in previous cases each expected cell frequency can never be less than _____.

══════ STOP ══════════════════════ STOP ══════

61

	Yes	No
HS	40	20
JHS	10	30

We ask a group of high school and junior high school students if they date. The results are shown at the left. This is an example of a case when χ^2 may be used as a measure of _____.

══════ STOP ══════════════════════ STOP ══════

62

	Yes	No	Total
HS	40	20	60
JHS	10	30	40
Total	50	50	100

To compute χ^2 in this case the _____ values must be determined. The expected value for the HS-yes cell is equal to $(60 \times 50)/100 =$ _____.

═══ STOP ═══ ═══ STOP ═══

expected; 30

63

	Yes	No	Total
HS	40	20	
JHS	10	30	40
Total	50		100

If we find the product of the respective row and column sums and divide by the sum total we find the expected values. For example to find the expected value for the JHS-yes cell _____ × _____ = _____ and this product divided by _____ = _____.

═══ STOP ═══ ═══ STOP ═══

$50 \times 40 = 2000,\ 2000/100 = 20$

64

	Yes	No	Total
HS	40	20	60
JHS	10	30	40
Total	50	50	100

The observed frequencies are on the left. Compute the expected frequencies for the table on the right. For example $(60 \times 50)/100 = 30$.

	Yes	No
HS	30	
JHS		

═══ STOP ═══ ═══ STOP ═══

	30
20	20

65

		Cigarette	
	A	B	C
	A	10	40	20
Car	B	10	30	50
	C	50	20	10

Cigarette preference may be related to the car you drive. Compute the expected value of cell C-C from the observed data on the left. $(? \times ?)/? =$ _____.

═══ STOP ═══ ═══ STOP ═══

$(80 \times 80)/240 = 26.67$

66

	A	B	C	D	E	F	G
1							
2							
3							
4							
5							

The number of df in this type of problem is computed the same way as in double-classification analysis of variance: (rows − 1) × (columns − 1). The problem at the left has _____ × _____ = _____ df.

══ STOP ══════════════════════════════ STOP ══

$$4 \times 6 = 24$$

---

67 Thus a 2 × 2 table has _____ df and for this reason we must correct for _____. As in previous cases we simply subtract _____ from the difference between expected and observed scores.

══ STOP ══════════════════════════════ STOP ══

1, continuity; 0.5

68

	Observed	Expected	$(o - e)$	$(\lvert o - e \rvert - 0.5)$	$(\lvert o - e \rvert - 0.5)^2$
HS Yes	40	30	10	9.5	90.25
HS No	20	30	?	?	?
JHS Yes	10	20	?	?	?
JHS No	30	20	?	?	?

══ STOP ══════════════════════════════ STOP ══

10	9.5	90.25
10	9.5	90.25
10	9.5	90.25

69

		Expected	$(\lvert o - e \rvert - 0.5)^2$	$(\lvert o - e \rvert - 0.5)^2/e$
HS	Yes	30	90.25	3.008
HS	No	30	90.25	?
JHS	Yes	20	90.25	?
JHS	No	20	90.25	?

$$\chi^2 = \sum \frac{(\lvert o - e \rvert - 0.5)^2}{e} = \underline{\hspace{2cm}}.$$

══ STOP ══════════════════════════════ STOP ══

3.008
4.513
4.513
$$\chi^2 = 15.042$$

70

	Yes	No
HS	40	20
JHS	10	30

$\chi^2 = 15.042$, df = _____. We _____ the _____ hypothesis and conclude that dating patterns for HS and JHS students _____.

═══ STOP ═══════════════════════════ STOP ═══

1; reject, null, differ (are not the same)

71

Candidate

	A	B	C	Total
Male	50	10	20	80
Female	10	10	20	40
Total	60	20	40	

Three candidates run for secretary of the senior class. We wish to determine if voting preference is a function of sex. df = _____ and do we use the correction for continuity? _____.

═══ STOP ═══════════════════════════ STOP ═══

2, No

72

OBSERVED
Candidate

	A	B	C	Total
M	50	20	10	80
F	10	10	20	40
Total	60	30	30	120

EXPECTED
Candidate

	A	B	C
M	40		
F			

$(80 \times 60)/120 = 4800/120 = 40$. Find the other expected values.

═══ STOP ═══════════════════════════ STOP ═══

	20	20
20	10	10

73

Observed

	A	B	C
M	50	20	10
F	10	10	20

Expected

	A	B	C
M	40	20	20
F	20	10	10

Find $(o - e)$ and enter in the table.

	A	B	C
M	10		
F			

═══ STOP ═══════════════════════════ STOP ═══

	0	-10
-10	0	10

74

Expected

	A	B	C
M	40	20	20
F	20	10	10

$\chi^2 = \dfrac{10^2}{40} + \dfrac{0^2}{20} +$ _____
$+$ _____ $+$ _____
$+$ _____ $=$ _____.

═══ STOP ═══════════════════════════ STOP ═══

$$\frac{10^2}{20} + \frac{10^2}{20} + \frac{0^2}{10} + \frac{10^2}{10} = 22.5$$

75

Observed

	A	B	C
M	50	20	10
F	10	10	20

$\chi^2 = 22.5$, df = _____. The null hypothesis is _____ and we conclude that voting preference (is/is not) _____ a function of sex.

═══════ STOP ═══════════════════════ STOP ═══════

2; rejected, is

---

76 To summarize χ^2 we note that we must use _____ data. Percentages (would/would not) _____ be an example of frequency data. In addition, the expected frequency within a cell cannot be less than _____.

═══════ STOP ═══════════════════════ STOP ═══════

frequency; would not; 5

---

77 When df = _____ we must correct for _____. This is done by subtracting _____ from the absolute difference between _____ and _____.

═══════ STOP ═══════════════════════ STOP ═══════

1, continuity; 0.5, o and e

---

78 $\sum o$ must always equal _____.

═══════ STOP ═══════════════════════ STOP ═══════

$$\sum e$$

---

79 The formula for chi square is: _____.

═══════ STOP ═══════════════════════ STOP ═══════

$$\chi^2 = \sum \frac{(o - e)^2}{e}$$

Sixteen

Correlation

Suppose you make an offhand remark to your roommate to the effect that tall people have big feet. You are surprised when your 6-foot 7-inch roommate says in reply that small people have big mouths.

In this chapter we will describe the properties, computation, and uses of a measure of the relationship between two variables. This statistic is known as the *Pearson product-moment coefficient of correlation,* or more simply as *correlation.* You have just expressed your belief that a relationship exists between two variables by the statement that tall people have big feet.

Correlation represents the relationship between two variables. In your case, the two variables were height and foot size, and you suggested that they are *correlated,* that is, tall people have big feet and small people have little feet. This observation may be tested by randomly gathering a group of individuals and measuring their height and shoe size. The essential thing here is that each person's height be compared with his own shoe size. If your observation is correct, then a person who is tall will tend to have larger feet than his shorter friends.

Finally, if you have been able to determine that a person's height and shoe size are related, you can then make predictions from one variable to the other. The value of such predictions cannot be overestimated. Suppose you found an old shoe (size 14) in the street. You can make some statement (a prediction) about the height of the person who wore it (if there is indeed a relationship between height and shoe size). You can predict that the person who wore the big shoe was tall; you can predict that the basketball team will have large feet; you can predict that the Under-Five-Foot Club will have small feet. You may think such examples are trivial, but it is well to remember that extensive use of just such procedures is made in real life. If the police found a large shoe near the scene of a crime, they probably would look for a tall man (wearing one shoe).

Being a student of statistics, you select a sample of five students living in the same rooming house. You measure each student's height and shoe size with the following results:

Student	Shoe size	Height (in inches)
1	9.0	70
2	6.5	62
3	8.0	66
4	7.5	69
5	8.5	72

You can plot the results of this experiment in what is known as a *scatter plot* by entering the shoe size on the x axis (abscissa) and the height on the y axis (ordinate). Thus, each dot represents both a person's shoe size and his height. You may be wondering how you determine which

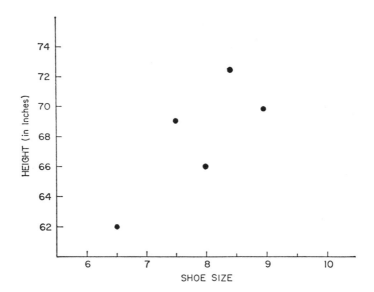

Figure 16-1   *Scatter plot relating shoe size and height measures.*

variable to put on which axis. In general, the independent variable is placed along the x axis and the dependent variable is placed along the y axis. If you predict from one variable to another variable, then the predictor variable is placed along the x axis, and the variable being predicted is placed along the y axis.

It is apparent from a glance at the scatter plot in Figure 16-1 that tall people do tend to have large feet and that small people do tend to have small feet. This relationship, again from the scatter plot, is not perfect, but the two variables (height and shoe size) are correlated. The next problem lies in the development of a technique for describing the degree of correlation between the two variables.

METHOD OF LEAST SQUARES

The problem of determining the degree of relationship between two variables can be approached in a number of ways. The traditional approach, presented here, is through the use of a technique called the *method of least squares* in which the squares of errors made during prediction are kept at a minimum. This technique can be illustrated by observation of the scatter plot in Figure 16-2, where we have passed several straight lines through the data points of Figure 16-1. If the line passes through the point, no error is made, but if the line misses the point, an error equal to the distance between point and line has been made.

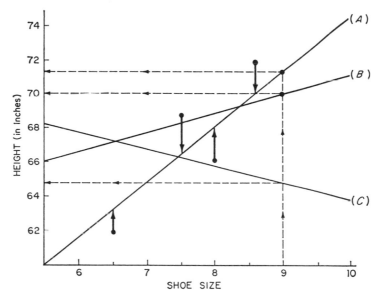

Figure 16-2   *Scatter plot relating shoe size and height measures, with straight lines used to predict height from knowledge of shoe size.*

Predictions are made on the basis of the straight lines passed through the scatter plot. Take, for example, a person who wears a size 9 shoe. If we use line *A*, we go up from 9 until we hit line *A*. Then we follow the horizontal line over until we reach the *y* axis and find that we "predict" a height of 71.5 inches. On the other hand, if we use line *B* going up from size 9 and then move left until we hit the *y* axis, we see that we "predict" a height of 70 inches. Finally, if we use line *C*, we predict a person wearing a size 9 shoe will be 65 inches tall. Casual observation indicates that when all points are considered, less error is made through the use of line *A* than through the use of lines *B* or *C*. It is obvious that we can arbitrarily pass any number of straight lines through the data and use any one of them to predict height from knowledge of shoe size. Which is the best line to use?

As we noted previously, the best line is that line which provides the least error in prediction. Even though line *A* is the best of the three lines we drew, it may not be the best of all possible straight lines. Calculus is required to determine the formula for the best of all possible straight lines.

This best-fitting straight line is called the *regression line,* and the slope of the regression line (when *X* and *Y* are expressed in *z* scores) is equal to the correlation coefficient.[1] Thus we see that prediction and correlation are closely related

POSSIBLE VALUES OF THE CORRELATION COEFFICIENT

The value of a correlation coefficient may vary between a perfect positive correlation and a perfect negative correlation. A perfect positive correlation is illustrated in Figure 16-3.

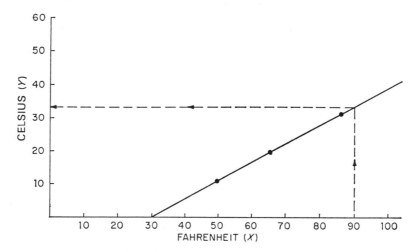

Figure 16-3   *Scatter plot relating Fahrenheit and Celsius temperature scales. All the points fall on the regression line and there are no errors in prediction. Thus, a perfect positive correlation.*

As you can see, all points fall on the regression line. This should not be too surprising, since the relationship denoted is that between the Fahrenheit and Celsius temperature scales. Observation of the scatter plot indicates that a high temperature on the Fahrenheit scale results in a high temperature on the Celsius scale. Prediction is perfect; that is, no errors in prediction are made, and the value of the correlation is symbolized $r_{xy} = 1.00$. This means that the correlation (r) between variables X and Y has a numerical value of 1.00. No correlation can be any larger than 1.00.

---

[1] The slope of a line is defined by the vertical increase $(Y_2 - Y_1)$ divided by the horizontal increase $(X_2 - X_1)$ so that the slope $m = (Y_2 - Y_1)/(X_2 - X_1)$.

Perfect negative correlation is presented in Figure 16-4. This relationship can be illustrated by the distances at which persons X and Y are from the ground while they are teetering on a seesaw. Again the relationship is perfect: Once one person's position is known, the other's can be predicted, and no errors are made in prediction. Observation of the scatter plot indicates that the relationship is quite different from the one shown in Figure 16-3. When person X is high off the ground, person Y is close to the ground;

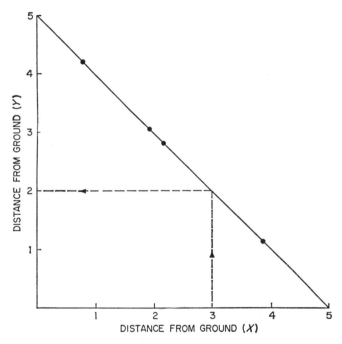

Figure 16-4   *Scatter plot relating distances from the ground of persons X and Y while on a seesaw. All points fall on the regression line. This is a perfect negative correlation*

and when person Y is high off the ground, person X is close to the ground. Each high score is paired with a low score, whereas each high score in the temperature example was paired with another high score. Because the relationship is reversed, it is called a *negative correlation*, and for this reason this perfect correlation is expressed $r_{xy} = -1.00$, since r_{xy} can never be less than -1.00.

Let us look at another relationship. Figure 16-5 is the scatter plot relating height and IQ. In contrast to the data presented in Figures 16-3 and 16-4, all points do not fall on the regression line and no consistent relationship between height and IQ score is apparent. In cases where there is no relationship between the two variables, the correlation coefficient is $r_{xy} = 0.00$. Notice that the regression line parallels the x axis. This means that all predictions of Y (height) are the same and that it makes no difference what the IQ is, since the mean of Y (height) is always predicted.

Finally, let us look at the case where we have a moderate degree of correlation. This would be the case with the relationship between high school grade-point average and college grade-point average. This relationship is plotted in Figure 16-6.

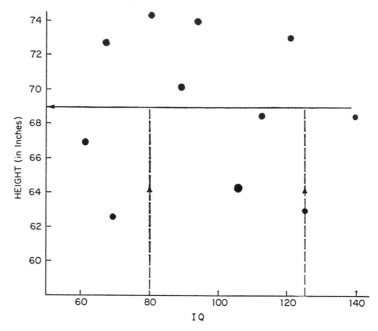

Figure 16-5   *Scatter plot relating height and IQ scores. In this case $r_{xy} = 0.00$, and the best prediction is the mean height (Y) score.*

As can be seen from Figure 16-6, people who did well in high school tend to do well in college, and people who did poorly in high school tend to do poorly in college. However, the relationship is not perfect and some errors are made in prediction, as can be seen from the fact that not all scores fall on the regression line. The value of the correlation for the data presented in this scatter plot is $r_{xy} = 0.62$.

In summary, the value of r_{xy} can vary from $+1.00$ to -1.00. When the value of r_{xy} is 1.00 or -1.00, the relationship between the two variables is perfect and all scores fall on the regression line; but as the relationship between the two variables decreases, the value of r_{xy} also decreases. When the value of $r_{xy} = 0.00$, there is no relationship between the two variables.

FORMULAS FOR CORRELATION

Although any number of straight lines can be passed through a set of data, we are interested only in the line that results in the least error in prediction. As was noted previously, this line is called the *regression line*. Now predictions can be made from Y to X as well as from X to Y. This means that

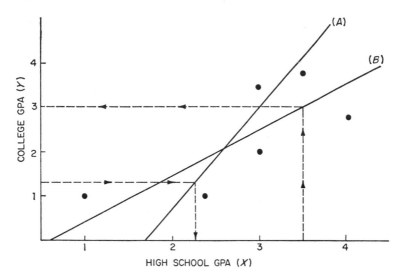

Figure 16-6   *Correlation between high school and college grade-point (GPA) averages. Line A is the best-fitting line for predicting Y from X and Line B is the best-fitting line for predicting X from Y.*

there are two best-fitting regression lines —one computed when predictions are made from X to Y and the other when predictions are made in the opposite direction. Through mathematical manipulation we can show that the formula for the slope of the best-fitting regression line when Y is predicted from X is equal to

$$\frac{\Sigma xy}{\Sigma x^2}$$

where $x = (X - M_x)$, $y = (Y - M_y)$, $xy =$ the product of x and y, and $\Sigma xy =$ the sum of the xy values.

Similarly, it can be shown that the formula for the slope of the best-fitting regression line when X is predicted from Y is equal to

$$\frac{\Sigma xy}{\Sigma y^2}$$

What is called the *geometric mean of these two slopes* is equal to the correlation coefficient r_{xy}. The geometric mean is the square root of the product of the two slopes. Hence,

$$r_{xy} = \sqrt{\frac{\Sigma xy \Sigma xy}{\Sigma x^2 \Sigma y^2}}$$

and this can be simplified to the deviation formula for correlation:

$$r_{xy} = \frac{\Sigma xy}{\sqrt{\Sigma x^2 \Sigma y^2}}$$

As was the case when computation of the mean and standard deviation by deviation-score procedures was considered, the computational difficulties and inconveniences involved in this formula make it almost unusable. As was also the case when computation of the mean and standard deviation was considered, there are also raw-scores formulas available for the correlation coefficient, so that it is not necessary to compute the deviation score for each of the X and Y scores.

If we divide both the numerator and denominator of the deviation-score formula by N, we can derive the raw-score formula:

$$r_{xy} = \frac{\Sigma xy}{\sqrt{\Sigma x^2 \Sigma y^2}} = \frac{\dfrac{\Sigma(X - M_x)(Y - M_y)}{N}}{\dfrac{\sqrt{\Sigma x^2 \Sigma y^2}}{N}} = \frac{\dfrac{\Sigma XY}{N} - M_x M_y}{\sigma_x \sigma_y}$$

You will note that σ_x and σ_y are used in the denominator rather than S_x and S_y. The reason for this is simply that the formula for $\sigma_x = \sqrt{\Sigma x^2/N}$, while the formula for $S_x = \sqrt{\Sigma x^2/(N-1)}$ and we need to find a term equal to $\sqrt{\Sigma x^2/N}$ that we can substitute for it.[2]

In previous chapters we have used S_x rather than σ_x. Use of S_x here would be incorrect because N rather than $N - 1$ was used to derive the raw-score formula. Furthermore, we are interested in describing the variability of our sample rather than in making inferences about population variability. For these reasons we will use the σ_x notation. On the other hand, the value of σ_x computed from $\sigma_x = \sqrt{\Sigma x^2/N}$ will vary from sample to sample. Therefore, it is best to remember that σ_x is used in the raw-score formula for r_{xy} for convenience and description, but that it does not refer to the variability of some larger population.

Finally, let us present a version of the formula that is most easily used when a calculator is available. All three of these formulas are algebraically equivalent, and it makes no difference which of the three you use because each will give you the same answer. The only criterion is that of convenience.

$$r_{xy} = \frac{N\Sigma XY - (\Sigma X)(\Sigma Y)}{\sqrt{N\Sigma X^2 - (\Sigma X)^2}\sqrt{N\Sigma Y^2 - (\Sigma Y)^2}}$$

AN EXAMPLE   Let us compute the correlation between height and shoe size, using the data presented earlier in the chapter. Initially, we find ΣXY, M_x, M_y, σ_x, and σ_y, and then substitute into the formula.

$$r_{xy} = \frac{\dfrac{\Sigma XY}{N} - M_x M_y}{\sigma_x \sigma_y}$$

[2] The raw-score formula for

$$\sigma_x = \sqrt{\frac{\Sigma X^2 - \dfrac{(\Sigma X)^2}{N}}{N}} = \sqrt{\frac{\Sigma X^2}{N} - M_x^2}$$

Let us look once again at the data:

Student	Shoe size	Height (in inches)
	X	Y
1	9.0	70
2	6.5	62
3	8.0	66
4	7.5	69
5	8.5	72

Values of M_x and M_y present no problem, since we have computed these many times before. Computation of σ_x, σ_y, and ΣXY are new, however.

$$\sigma_x = \sqrt{\frac{\Sigma X^2}{N} - M_x^2} = \sqrt{\frac{315.75}{5} - (7.9)^2}$$
$$= \sqrt{63.15 - 62.41} = \sqrt{0.74} = 0.86$$

$$\sigma_y = \sqrt{\frac{\Sigma Y^2}{N} - M_y^2} = \sqrt{\frac{23,045}{5} - (67.8)^2}$$
$$= \sqrt{4609 - 4596.84} = \sqrt{12.16} = 3.49$$

The symbol ΣXY means that each X score is multiplied by its associated or paired Y score and that these products are then summed. For student 1, $9 \times 70 = 630$; for student 2, $6.5 \times 62 = 403$; and so on. Then $\Sigma XY = 630 + 403 + \cdots + 612 = 2690.50$. $\Sigma XY/N = 2690.5/5 = 538.1$. $M_x = 7.9$, $M_y = 67.8$, and $M_x M_y = 535.62$; $\sigma_x = 0.86$, $\sigma_y = 3.49$, and $\sigma_x \sigma_y = 3.00$. Hence,

$$r_{xy} = \frac{\dfrac{\Sigma XY}{N} - M_x M_y}{\sigma_x \sigma_y} = \frac{538.10 - 535.62}{3.00} = \frac{2.48}{3.00} = 0.83$$

SIGNIFICANCE OF A CORRELATION

The coefficient of correlation can be used to describe the relationship between the data obtained from two variables, but, as in other types of research, simple description may not be all that is desired. It may be necessary to make inferences about the relationship between two populations of scores, and the fact that we have found the value of r_{xy} on a sample of pairs does not necessarily answer the question of whether there exists a correlation in the population. Just as we would not expect the sample mean to equal the true mean, we would also not expect the correlation between members of a sample to be exactly equal to the true value in the population. Instead, we would expect that sample correlations would vary around the value of the correlation in the population.

To illustrate this, we selected ten pairs of scores randomly from a population in which $\mu = 50$ and $\sigma = 10$, and found the correlation between the ten pairs of scores. Since these scores were randomly paired, the true correlation between them is zero. We took another ten pairs and found the correlation, then a third set of ten pairs and found the correlation, and so on. We continued to do this until we had computed 10,000 correlations (the computer completed this in 9 minutes). A frequency distribution of the results is presented in Figure 16-7.

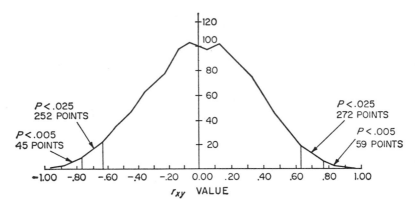

Figure 16-7   *Sampling distribution of 10,000 correlations computed on randomly selected samples of ten pairs each. The true correlation is $r_{xy} = 0.00$.*

As you can see, the values of the correlations varied considerably around the true correlation of zero. In fact, 5 percent of the correlations were larger than 0 ± 0.632. Hence, it is obvious that simply finding a correlation that differs from zero is no indication that the population value of the correlation is also different from zero. Appendix E gives the values of r_{xy} necessary for significance at the 0.05 and 0.01 levels with differing numbers of pairs. As was the case with previous statistics, Appendix E indicates the points beyond which 5 percent or 1 percent of the sample correlations would fall by chance if the true correlation were equal to 0.00. The decision procedure is the same: We reject the null hypothesis ($H: r_{xy} = 0.00$) if the observed value of r_{xy} is greater than the tabled value of r_{xy}. Notice that as N (the number of pairs) increases, the variability of the sample correlations decreases.

Consider your own hypothesis that tall people have big feet. Your correlation of 0.83 in the preceding example seemed to indicate that you were correct in your hypothesis. But are you? Before you can reject the null hypothesis, you must evaluate your correlation from Appendix E. With five pairs of scores, a correlation of 0.88 must be observed before the null hypothesis can be rejected. You found $r_{xy} = 0.83$ and on this basis you can't reject the null hypothesis, and therefore conclude there is a relationship between height and shoe size.

You may have noticed that the word *cause* is rarely mentioned in scientific writing. The reason for this is that causation or causality is difficult if not impossible to define independently of correlation. Yet correlation does not imply causation. The rooster crows and the sun comes up. This is obviously a situation in which a high positive correlation exists between the rooster-crowing and sun-rising variables, but we doubt if the rooster causes the sun to rise. But not all illustrations are so absurd. Suppose we find a positive correlation between frequency of smoking and susceptibility to cancer. This may indicate that smoking causes cancer, that cancer causes smoking, that a third variable (or variables) causes both, or that no cause is involved. Years of cancer research have demonstrated that it is no easy task to determine which of these alternatives is correct. Thus, we are extremely reluctant to use "cause" when we have observed a correlation between two variables, even though we might prefer to believe that one of the variables causes the other.

ASSUMPTIONS OF CORRELATION

There are a number of assumptions that must be made if one wishes to generalize to a population from a correlation based on a sample. For example:

1. Random and independent selection of pairs.
2. Normal distribution of both X and Y.
3. Linear regression lines, that is, the best-fitting lines are straight lines.
4. Homoscedasticity—the assumption that the variances of Y for each X are equal and the assumption that the variances of X for each Y are equal.
5. At least an interval scale is used.

RANK-ORDER CORRELATION

When data are presented in terms of an ordinal scale rather than in an interval scale, then a *rank-order correlation* (rho) is computed. This formula for rho is

$$rho = 1 - \frac{6\Sigma D^2}{N(N^2 - 1)}$$

An example of ranked data may be found in data released by the Austin, Texas, school district. The data indicated the percentage of families in each elementary school district which could be classified as being at or below poverty level. In addition, the median reading achievement, expressed in terms of national percentile norms, was given for each school. Neither of these variables—percentage and median—can meet the assumptions of a Pearson product-moment correlation, but the data of each can be rank-ordered. These data are presented in Table 16-1. (Because of the large number of elementary schools, we have reported only every fourth school. The results reflect closely the correlation found with all 50 schools). Based on

Table 16-1 Rank orders of reading achievement level and percentage of poor families in several elementary school districts

School	Percent Poverty	Reading Achievement	D	D²
Allison	2	13	11	121
Barton Hills	12.5	1	11.5	132.5
Brown	5	9	4	16
Dawson	4	11	7	49
Gullett	12.5	2	10.5	110.25
Maplewood	3	10	7	49
Oak Hill	9	6.5	2.5	6.25
Palm	1	12	11	121
Reilly	6	8	2	4
Summitt	10	3	7	49
Walnut Creek	11	5	6	36
Wooldridge	8	4	4	16
Zilker	7	6.5	0.5	0.25
				710.25

these data, the rank-order correlation between poverty and reading achievement is

$$\text{rho} = 1 - \frac{6 \times 710.25}{13 \times (169 - 1)} = 1 - \frac{4261.50}{2184.00} = 1 - 1.95 = -0.95$$

Although the formula for rho may look entirely different when compared to that for r_{xy}, it is algebraically equivalent to the earlier formulas when used with ranks, and hence is one of the family of Pearson product-moment correlation coefficients. This means that you would obtain an identical result if you applied the earlier formula directly to the rank numbers. The only advantage of the special rho formula is computational convenience. Although the computational formula for rho is the same, the sampling distribution of correlations obtained from ranks is slightly more variable than that upon which the significance levels of r_{xy} are based. Use of a separate table of significant levels for evaluation of rho is necessary (Appendix F). In this case, $N = 13$, and the value must be estimated from the $N = 12$ and $N = 14$ entries. Clearly, the correlation is significant beyond the 0.01 level, and the negative relationship between poverty and reading skills is not due to chance.

As can be seen in the example, tie places will occur in establishing a particular set of rankings. These are easily handled by assigning the mean value of the ranks to each of the tie holders. For example, Barton Hills and Gullett schools tie for the smallest percentage of families living in poverty, and each was assigned a rank of 12.5 (the mean of the 12 and 13 ranks) while the next school, Walnut Creek, was assigned rank 11. The same procedure may be used for ties of three or four cases. If a very large number of ties occurs, however, you would probably be wise to reconsider the use of rho; chi square might be a more appropriate method of showing association between the two variables concerned.

CORRELATION WITH NOMINAL DATA

Rank-order (rho) correlation is only one of several special types of correlation that can be employed when the assumptions of the Pearson product-moment correlation cannot be met. Of these, two others in addition to rho are simply Pearson correlations computed from dichotomized (two-category) data. The first, the point-biserial correlation, has one dichotomized variable, such as male-female, and one normally distributed variable. The second, the phi coefficient, has two dichotomized variables. In both cases one of the two categories of the dichotomized variables is arbitrarily given a value of 0 and the second a value of 1. Thus, for example, with the dichotomized male-female variable, the X score for each male would be 0 and the X score for each female would be 1. If you would take the trouble to use the formula for r_{xy} presented earlier on these types of data, you would compute a correlation equal to the point biserial, if there is one dichotomized variable, or the phi coefficient, if there are two dichotomized variables.

Suppose we take as an example some hypothetical data in which a group of 24 randomly selected students were given a questionnaire measuring interest in intercollegiate athletics. On the basis of the questionnaire, each person was given a score from 0 (low interest) to 4 (high interest). Next, each person was classified on the basis of fraternal affiliation (0 if he belonged to a fraternity and 1 if he did not). The experimenter wishes to find the correlation between fraternal affiliation and interest. The data are shown as a frequency distribution ($N_0 = 10$, $N_1 = 14$, $N = 24$, $M_0 = 2.50$, $M_1 = 1.00$, $S_y = 1.28$).

Interest	Fraternity	Nonfraternity
Y	X_0	X_1
4	2	0
3	4	1
2	2	2
1	1	7
0	1	4

The formula for the point-biserial (pb) correlation is

$$r_{pb} = \frac{M_0 - M_1}{S_y} \sqrt{\frac{N_0 N_1}{N(N-1)}}$$

where M_0 is the mean of the affiliated students on Y; M_1 is the mean of the unaffiliated students on Y; N_0 is the number of affiliated students;

N_1 is the number of unaffiliated students; and S_y is the standard deviation of all the scores on Y and $N_0 + N_1 = N$.

$$r_{pb} = \frac{2.50 - 1.00}{1.28} \sqrt{\frac{10 \times 14}{24 \times 23}} = \frac{1.50}{1.28} \sqrt{0.2536} = 1.17 \times 0.50 = 0.58$$

If we had computed r_{xy}, we would, of course, have gotten the same answer. Appendix E can be used to evaluate r_{pb}, and with $N - 2 = 22$, correlation coefficients of 0.404 and 0.515 are needed for significance at the 0.05 and 0.01 levels, respectively. Thus $r_{pb} = 0.58$ and is significant beyond the 0.01 level.

In contrast to the point-biserial correlation, both variables of the phi coefficient are dichotomized. Such would be the case if we selected a group of students at random and measured fraternal affiliation (yes, no) and sex (male, female). In one college a random sample of 75 students gave these results:

		Affiliation	
		Yes	No
Sex	M	15	40
	F	10	10

Use of the phi coefficient will give an answer equal to that which would be obtained if we used the formula for r_{xy}. The formula for the phi coefficient uses the value for chi square with a 2×2 table:

$$r_\phi = \sqrt{\frac{\chi^2}{N}}$$

where the χ^2 is not corrected for continuity. In the present example,

$$\chi^2 = \frac{75(400 - 150)^2}{55 \times 20 \times 25 \times 50} = \frac{4{,}687{,}500}{1{,}375{,}000} = 3.41$$

and

$$r_\phi = \sqrt{\frac{3.41}{75}} = \sqrt{0.045} = 0.21$$

If the chi square is significant, then the phi coefficient is also. With 1 df, the χ^2 of 3.41 is not significant and therefore neither is r_ϕ.

SUMMARY

This chapter has presented an essentially simple concept—relationship—in all the complexity that comes with converting any concept into numerical measures. Although the basic idea is simple, there are a number of logical traps involved, and it would be well worth the effort to go over the material carefully to be sure you have avoided these traps.

One major fallacy is the idea that correlation implies anything about causation. Correlation is simply a measure of the relationship existing between two variables.

Another point that should be fully understood is that correlation represents the average of the slopes of two regression lines—the slope of the best-fitting line when Y is predicted from X, and the slope of the best-fitting line when X is predicted from Y. When the variables are expressed in terms of z scores, the slopes of both regression lines are the same, and hence the correlation is equal to the slope of the regression line.

Correlation seems to be used primarily as a descriptive statistic, although determining significance levels of correlation is common enough. In this chapter we have been concerned with the development of a statistic which, in contrast to previous statistical tests, describes the similarity between two variables. The stronger the similarity or relationship between the two, the higher the absolute value of the correlation coefficient—up to a maximum of 1.00.

Although we have looked primarily at the development of correlation as a descriptive statistic, correlation as a predictor of subsequent success or failure is where the statistic has its most useful application. Tests you take to get into college, to get a job, to determine your IQ or your personality, all have employed the correlation coefficient as a measure of how closely related test scores were to the subsequent behavior of the individuals involved.

PROBLEMS

1. Find the correlation between the following pairs of scores. Is the correlation significant?

X	Y
0	4
12	5
4	1
8	7
6	3

2. Find the correlation between the following pairs of scores. Notice the relationship between these scores and those in the first problem. Is the correlation significant?

X	Y
0	1
12	7
4	3
8	4
6	5

3. Using the same data, find the correlation again. Is this correlation significant?

X	Y
0	7
12	1
4	5
8	3
6	4

4. An experimenter was interested in the relationship between the number of items remembered and the amount of a certain drug taken over a 24-hour period prior to the test for memory. To do this, 20 subjects learned a long list of words and each was then administered a random amount of drug. Is there a relationship between recall (Y) and amount of drug taken (X)?

$$\Sigma X = 240$$
$$\Sigma Y = 400$$
$$\Sigma X^2 = 3600$$
$$\Sigma Y^2 = 9280$$
$$\Sigma XY = 5280$$

5. An experimenter conducted an experiment and found the correlation between two variables, X and Y. His correlation was based on 25 pairs of scores and $r_{xy} = 0.39$. What can you say about the relationship between the two variables?

6. A professor was interested in seeing how his first hour exam correlated with his second hour exam. Consequently, he rank-ordered the 20 people who took both tests and found their rank differences, which he squared and summed. He found $\Sigma D^2 = 798$. Did those who did well on his first test tend to do well on his second?

7. Professor Snarf found that 30 girls and 10 boys sat in the front while 20 girls and 20 boys sat in the back of the classroom. Find the correlation between sex and seating preference.

8. In computing the correlation between two variables, a formula that is slightly different from the one used previously for the standard deviation is used. Find both S^2 and σ^2 from the following data: 0, 6, 2, 4, 3.

9. A research worker reporting on his work reported a correlation of 1.60 between two variables. What is your comment?

10. What would you expect the correlation between the heights of husbands and wives to be? Why?

11. Think of a stop light. What color is the light on top? How do you think reaction time to this question correlates with color blindness?

EXERCISES

1. Compute the correlation between the height and weight of the members of your group or class.

2. Take ten pairs of numbers from Appendix H. Compute the correlation between these pairs.

3. Compute the correlation between height and shoe size. Do this first for the men; then do it again for the women in your group.

1   The technique used in describing the relationship between two variables is called *correlation*. For example, we would use _____ to study the relationship between height and weight.

========= STOP ===================================== STOP =========

correlation

---

2   The strength of the _____ between two variables, such as height and weight, can be expressed by the correlation coefficient.

========= STOP ===================================== STOP =========

correlation (relationship)

---

3   When the value of the correlation coefficient is low the correlation between two variables is low. But when the value of the _____ _____ is high then the correlation is high.

========= STOP ===================================== STOP =========

correlation coefficient

---

4   The value of a correlation coefficient may vary from -1 to $+1$. When the value of the correlation coefficient (r_{xy}) is equal to 0 there is no correlation (relation) between the two variables. When $r_{xy} = 1.00$ perfect _____ exists between the two variables.

========= STOP ===================================== STOP =========

correlation

---

5   The correlation between two variables enables prediction from one to the other. If $r_{xy} = 1.00$ the correlation is perfect and hence prediction would be _____.

========= STOP ===================================== STOP =========

perfect

---

6   But when $r_{xy} = 0$ no _____ can be made from one variable to the other.

========= STOP ===================================== STOP =========

predictions

---

7 The sign of the correlation coefficient refers not to the degree but to the direction of the relation. Thus the _____ of the relation between X and Y would be the same for $r_{xy} = 0.50$ and $r_{xy} = -0.50$.

═══════ STOP ═══════════════════════════ STOP ═══════

degree

───

8 If the correlation between height and weight is positive (sign is positive) then tall people would tend to be heavy. But if the correlation between height and weight were negative, then tall people would tend to be _____.

═══════ STOP ═══════════════════════════ STOP ═══════

light

───

9 Suppose two tests (X and Y) are positively correlated. Someone who performed well on test X would probably perform _____ on test Y.

═══════ STOP ═══════════════════════════ STOP ═══════

well

───

10 Two other tests are negatively correlated. Someone who performs well on the first test would be expected to perform _____ on the second.

═══════ STOP ═══════════════════════════ STOP ═══════

poorly

───

11 Finally, two other tests have a zero correlation. Could you predict performance on the second test knowing the first test performance? (Yes/No) _____.

═══════ STOP ═══════════════════════════ STOP ═══════

No

───

12 Suppose two children are playing teeter-totter. When child X is high off the ground child Y is _____ _____ _____.

═══════ STOP ═══════════════════════════ STOP ═══════

near (on) the ground

───

13 When child Y is high off the ground child X is on the ground and vice versa. If this were thought of as a test, the correlation between the distance from the ground of child X and the distance from the ground of child Y would be _____.

===== STOP ===== STOP =====

negative

---

14 If child X was high off the ground when child Y was also high off the ground then the correlation would be _____.

===== STOP ===== STOP =====

positive

---

15 Height and weight are positively correlated. This means that a tall man would be _____ and a short man would be _____.

===== STOP ===== STOP =====

heavy, light

---

16 Height and shoe size are positively correlated. If you are tall you probably have _____ feet.

===== STOP ===== STOP =====

big

---

17 Weight and shoe size are positively correlated. If you have big feet you probably are _____.

===== STOP ===== STOP =====

heavy

---

18 Suppose a test and grade point average are negatively correlated. People who have very high grade point averages would tend to get _____ scores on the test.

===== STOP ===== STOP =====

low

---

19 People who do well on a test tend to do well on a second. These two tests are probably _____ correlated.

===== STOP ===== STOP =====

positively

---

20 People who do well on a test tend to do poorly on a second. These two tests are probably _____ correlated.

══════ **STOP** ══════════════════════ **STOP** ══════

negatively

---

21 The value of the correlation coefficient is simply the slope of the best-fitting straight line drawn through the data. Thus if the _____ of the best-fitting straight line is high then the correlation coefficient is high.

══════ **STOP** ══════════════════════ **STOP** ══════

slope

---

22 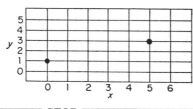 We wish to find the meaning of *slope*. At the left is a graph. Two scores are shown on the graph by use of heavy dots. The X value of the dot on the left side of the graph is $X = 0$, the value of the other dot is $X =$ _____.

══════ **STOP** ══════════════════════ **STOP** ══════

5

---

23 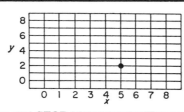 Since there is an x axis and a y axis the position of any point may be described by an X score and a Y score. For example, the position of the dot on the graph is $X =$ _____ and $Y =$ _____.

══════ **STOP** ══════════════════════ **STOP** ══════

5, 2

---

24 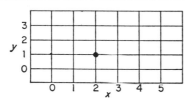 Suppose a person takes two tests (X and Y) and gets scores of $X = 2$ and $Y = 1$. This point is entered on the graph. Suppose a second person takes the tests with the following scores: $X = 5$, $Y = 2$. Place a dot on the graph at the point which represents both of these scores.

══════ **STOP** ══════════════════════ **STOP** ══════

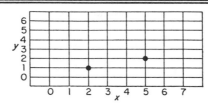

25

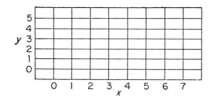

Suppose two other Ss take the same tests. For the first, $X = 1$, $Y = 2$ and for the second, $X = 4$, $Y = 3$. Plot these values on the graph at the left.

═══ STOP ═══════════════════════════ STOP ═══

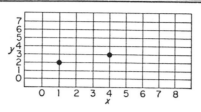

26

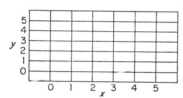

Two other Ss take the same tests. $S1$, $X = 0$, $Y = 2$; $S2$, $X = 5$, $Y = 3$. Plot these results.

═══ STOP ═══════════════════════════ STOP ═══

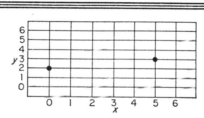

27

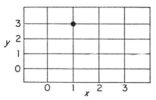

Suppose S took tests X and Y and his results were as plotted. His score on test X = _____. His score on test Y = _____.

═══ STOP ═══════════════════════════ STOP ═══

1; 3

28   Slope may be defined as $(Y_2 - Y_1)/(X_2 - X_1)$. Then if $S1$ got a score of $X = 3$, $Y = 2$ and $S2$ got scores of $X = 8$, $Y = 4$, then $Y_2 =$ _____.

═══ STOP ═══════════════════════════ STOP ═══

4

29 Slope $= (Y_2 - Y_1)/(X_2 - X_1)$. $S1$ got scores of $X = 3$, $Y = 2$ and $S2$ got scores of $X = 8$, $Y = 4$. Slope $= (4 - ?)/(? - 3) = $ _____.

STOP ――――― STOP

$$(4 - 2)/(8 - 3) = 2/5 = 0.40$$

---

30 For $S1$, $X = 0$, $Y = 1$; $S2$, $X = 4$, $Y = 3$. Slope $= $ _____.

STOP ――――― STOP

$$(3 - 1)/(4 - 0) = 2/4 = 0.50$$

---

31 The letter m is commonly used to denote slope. Thus _____ $= (Y_2 - Y_1)/(X_2 - X_1)$.

STOP ――――― STOP

$$m$$

32
$$m = \frac{Y_2 - ?}{X_2 - X_1}$$

STOP ――――― STOP

$$Y_1$$

33
$$m = \frac{Y_2 - Y_1}{?}$$

STOP ――――― STOP

$$X_2 - X_1$$

34
$$m = \frac{?}{?}$$

STOP ――――― STOP

$$\frac{Y_2 - Y_1}{X_2 - X_1}$$

---

35 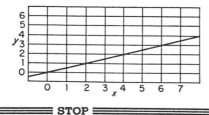 If we wish to find the slope of a line we can select points at "random" from that line. For example $X = 2$, $Y = $ _____, and $X = 6$, $Y = $ _____.

STOP ――――― STOP

$$1, 3$$

36

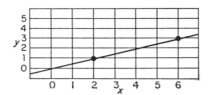

The slope of a straight line may be determined by selecting any two points which the line intersects. For example the straight line on the graph intersects points $X = 2$, $Y = 1$ and $X = 6$, $Y = 3$. Thus the slope of the line equals _____.

═══ STOP ═══ ═══ STOP ═══

$$(3 - 1)/(6 - 2) = 2/4 = 0.50$$

37

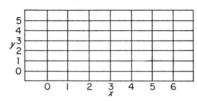

Thus the slope is defined: $m = (Y_2 - Y_1)/(X_2 - X_1)$. Suppose we have three Ss taking two tests (X and Y with X given first). $S1: 1, 2; S2: 3, 2; S3: 5, 5$. Enter the three points on this graph.

═══ STOP ═══ ═══ STOP ═══

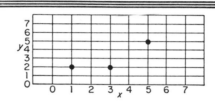

---

38 $S1: 1, 2; S2: 3, 2; S3: 5, 5$ (with the X score given first). It is obvious that we cannot draw a single straight line to connect all three points. But we can draw one best-fitting straight line that has the least amount of error. The formula for the slope of this line is $\sum xy/\sum x^2$. In this example $M_x =$ _____ and $M_y =$ _____.

═══ STOP ═══ ═══ STOP ═══

3, 3

---

39 $x = X - M_x$. Thus if $M_x = 70$ and $X = 76$, $x = 6$. If $M_x = 40$ and $X = 52$, then $x =$ _____.

═══ STOP ═══ ═══ STOP ═══

12

---

40 $S1: 1, 2; S2: 3, 2; S3: 5, 5$. The formula for the slope of the best-fitting straight line is $\sum xy/\sum x^2$ and $x = X - M_x$ and $y = Y - M_y$. $x_1 =$ _____, $x_2 =$ _____, $x_3 =$ _____; $y_1 =$ _____, $y_2 =$ _____, $y_3 =$ _____.

═══ STOP ═══ ═══ STOP ═══

$-2, 0, 2; -1, -1, 2$

41

	S1	S2	S3
x	-2	0	2
y	-1	-1	2

If $x=4$ and $y=3$ then $xy=4 \times 3=12$. Find xy for the values given at the left. $S1 = $ _____, $S2 = $ _____, $S3 = $ _____. $\sum xy = $ _____.

════ STOP ════   ════ STOP ════

2, 0, 4; 6

42

	S1	S2	S3
x	-2	0	2
y	-1	-1	2
xy	2	0	4

$\sum xy = 6.$ $\sum_{1}^{N} x^2 = $ _____,

$\sum_{1}^{N} y^2 = $ _____.

════ STOP ════   ════ STOP ════

8, 6

---

43 $S1: 1, 2; S2: 3, 2; S3: 5, 5.$ $\sum x^2 = 8,$ $\sum xy = 6,$ $\sum y^2 = 6.$ The formula for the slope of the best-fitting straight line is $\sum xy / \sum x^2$. The slope is equal to _____.

════ STOP ════   ════ STOP ════

$6/8 = 0.75$

---

44 $\sum xy / \sum x^2$ is only part of the formula for correlation—the part in which Y is predicted from X. $\sum xy / \sum y^2$ is used to predict X from Y and is also the slope of a best-fitting straight line. Thus the slope of the line when Y is predicted is _____ but the slope is _____ when X is predicted from Y.

════ STOP ════   ════ STOP ════

$\sum xy / \sum x^2, \sum xy / \sum y^2$

---

45   The geometric mean of these two slopes, $\sum xy / \sum x^2$ and $\sum xy / \sum y^2$, is equal to the correlation coefficient. Since the geometric mean of X_1 and X_2 is $\sqrt{X_1 \cdot X_2}$ then the geometric mean of $\sum xy / \sum x^2$ and $\sum xy / \sum y^2$ is equal to _____. (Remember, $A \cdot B = A \times B$)

════ STOP ════   ════ STOP ════

$$\sqrt{\frac{\sum xy}{\sum x^2} \cdot \frac{\sum xy}{\sum y^2}}$$

46

$$\sqrt{\frac{\sum xy}{\sum x^2} \cdot \frac{\sum xy}{\sum y^2}} \text{ can be simplified to } \frac{\sum xy}{\sqrt{\sum x^2 \cdot \sum y^2}}. \text{ Thus in the pre-}$$

vious problem $\sum xy = 6$, $\sum x^2 = 8$, $\sum y^2 = 6$, and the correlation is equal to

$$r_{xy} = \frac{\sum xy}{\sqrt{\sum x^2 \cdot \sum y^2}} = \underline{\qquad}.$$

═══ STOP ═══   ═══ STOP ═══

$$\frac{6}{\sqrt{48}} = \frac{6}{6.9} = 0.88$$

---

47   Thus the correlation coefficient is simply the average of the slopes of two best-fitting straight lines—the line used to predict Y scores and the line used to predict X scores. If $\sum xy = 6$, $\sum x^2 = 3$, $\sum y^2 = 27$, then

$$r_{xy} = \frac{\sum xy}{\sqrt{\sum x^2 \cdot \sum y^2}} = \underline{\qquad}.$$

═══ STOP ═══   ═══ STOP ═══

$$\frac{6}{\sqrt{27 \cdot 3}} = \frac{6}{\sqrt{81}} = \frac{6}{9} = 0.67$$

48

$$r_{xy} = \frac{\sum xy}{\sqrt{\sum x^2 \cdot \sum y^2}}. \qquad \sum xy = 3,$$

$$\sum x^2 = 2, \qquad \sum y^2 = 8, \qquad r_{xy} = \underline{\qquad}.$$

═══ STOP ═══   ═══ STOP ═══

$$\frac{3}{\sqrt{2 \cdot 8}} = \frac{3}{4} = 0.75$$

49

$$r_{xy} = \frac{\sum xy}{\sqrt{\sum x^2 \cdot \sum y^2}}. \qquad \sum x^2 = 6,$$

$$\sum y^2 = 24, \qquad \sum xy = 8, \qquad r_{xy} = \underline{\qquad}.$$

═══ STOP ═══   ═══ STOP ═══

$$\frac{8}{\sqrt{6 \cdot 24}} = \frac{8}{\sqrt{144}} = \frac{8}{12} = 0.67$$

50

$$r_{xy} = \frac{\sum xy}{\sqrt{\sum x^2 \cdot ?}}$$

═══ STOP ═══   ═══ STOP ═══

$$\sum y^2$$

51

$$r_{xy} = \frac{\sum xy}{\sqrt{? \cdot \sum y^2}}$$

═══════ STOP ═══════════════════════════ STOP ═══════

$$\sum x^2$$

52

$$r_{xy} = \frac{?}{\sqrt{\sum x^2 \cdot \sum y^2}}$$

═══════ STOP ═══════════════════════════ STOP ═══════

$$\sum xy$$

53

$$\frac{\sum xy}{\sqrt{\sum x^2 \cdot \sum y^2}} = \underline{\qquad}$$

═══════ STOP ═══════════════════════════ STOP ═══════

$$r_{xy}$$

54

$$r_{xy} = \underline{\qquad}$$

═══════ STOP ═══════════════════════════ STOP ═══════

$$\frac{\sum xy}{\sqrt{\sum x^2 \cdot \sum y^2}}$$

---

55 $\qquad \sum xy = 7, \qquad \sum x^2 = 5, \qquad \sum y^2 = 20, \qquad r_{xy} = \underline{\qquad}.$

═══════ STOP ═══════════════════════════ STOP ═══════

$$\frac{7}{\sqrt{20 \cdot 5}} = \frac{7}{10} = 0.70$$

---

56 $\qquad \sum xy = 6, \qquad \sum x^2 = 4, \qquad \sum y^2 = 16, \qquad r_{xy} = \underline{\qquad}.$

═══════ STOP ═══════════════════════════ STOP ═══════

$$\frac{6}{8} = 0.75$$

---

57 $\qquad \sum xy = 4, \qquad \sum x^2 = 2, \qquad \sum y^2 = 8, \qquad r_{xy} = \underline{\qquad}.$

═══════ STOP ═══════════════════════════ STOP ═══════

$$\frac{4}{4} = 1.00$$

58   We have seen that raw-score formulas are most convenient to use. The raw-score formula for $r_{xy} = \dfrac{\dfrac{\sum XY}{N} - M_x M_y}{\sigma_x \sigma_y}$ and may be derived from the deviation-score formula: $r_{xy} = \underline{\qquad}$.

═══ STOP ═══   ═══ STOP ═══

$$\frac{\sum xy}{\sqrt{\sum x^2 \cdot \sum y^2}}$$

59

$$\underline{\qquad} = \frac{\dfrac{\sum XY}{N} - M_x M_y}{\sigma_x \sigma_y} \text{ and is the } raw\text{-score formula for correlation.}$$

═══ STOP ═══   ═══ STOP ═══

$$r_{xy}$$

---

60 $\dfrac{\dfrac{\sum XY}{N} - M_x M_y}{\sigma_x \sigma_y}$ (Notice that σ_x and σ_y are used) $\sigma_x = \sqrt{\dfrac{\sum x^2}{?}}$

═══ STOP ═══   ═══ STOP ═══

$$N$$

---

61 σ_x and σ_y are used in the raw-score formula because both numerator and denominator of the deviation-score formula were divided by N. Thus

$$\sqrt{\frac{\sum x^2}{N}} = \underline{\qquad}.$$

═══ STOP ═══   ═══ STOP ═══

$$\sigma_x$$

---

62   and $\dfrac{\sqrt{\sum x^2 \cdot \sum y^2}}{N} = \underline{\qquad}.$

═══ STOP ═══   ═══ STOP ═══

$$\sigma_x \sigma_y$$

---

63 $r_{xy} = \dfrac{\dfrac{\sum XY}{N} - M_x M_y}{?}$ When we use σ_x and σ_y in the raw-score formula we are describing the variability of our $\underline{\qquad}$.

═══ STOP ═══   ═══ STOP ═══

$$\sigma_x \sigma_y; \text{ samples}$$

64

$$r_{xy} = \frac{\dfrac{\sum XY}{N} - ?}{\sigma_x \sigma_y}$$

══════ STOP ══════════════════════════ STOP ══════

$$M_x M_y$$

65

$$r_{xy} = \frac{\dfrac{\sum XY}{?} - M_x M_y}{\sigma_x \sigma_y}$$

══════ STOP ══════════════════════════ STOP ══════

$$N$$

---

66 N in this case refers to the _____ of *pairs* and not to the number of scores.

══════ STOP ══════════════════════════ STOP ══════

number

67

$$r_{xy} = \frac{\dfrac{\sum XY}{N} - ?}{?}$$

══════ STOP ══════════════════════════ STOP ══════

$$\frac{M_x M_y}{\sigma_x \sigma_y}$$

---

68   In the raw-score formula for correlation each X score is multiplied by the _____ score paired with it. Then all of these products are _____.

══════ STOP ══════════════════════════ STOP ══════

Y; summed

---

69   After the sum of the XY products is found, then this sum is _____ _____ _____.

══════ STOP ══════════════════════════ STOP ══════

divided by N

---

70   In the numerator of the raw-score formula for the correlation M_x is multiplied by _____.

══════ STOP ══════════════════════════ STOP ══════

$$M_y$$

---

71 In the denominator of the raw-score formula for the correlation σ_y is multiplied by _____.

===== STOP ===== STOP =====

$$\sigma_x$$

72
$$r_{xy} = \text{_____}$$

===== STOP ===== STOP =====

$$\frac{\sum XY/N - M_x M_y}{\sigma_x \sigma_y}$$

---

73 Given scores 3, 6, 0, 4, 2. $\sum x^2 = $ _____, $S^2 = $ _____, $\sigma^2 = $ _____.

===== STOP ===== STOP =====

$$20, \; 20/4 = 5, \; 20/5 = 4$$

---

74 With scores 4, 0, 2, 2. $\sum x^2 = $ _____, $\sigma^2 = $ _____.

===== STOP ===== STOP =====

$$8,2$$

---

75 Suppose $M_x = 5$, and $M_y = 4$, Then _____ would be subtracted from $\sum XY/N$.

===== STOP ===== STOP =====

$$20$$

---

76 Suppose $N = 10$, $\sum XY = 80$, $M_x = 2$, and $M_y = 3$. The numerator of the raw-score formula for correlation would equal _____.

===== STOP ===== STOP =====

$$8 - 6 = 2$$

---

77 If the numerator equals 4 and the denominator equals 8, then $r_{xy} = $ _____.

===== STOP ===== STOP =====

$$0.50$$

---

78 Suppose the numerator equals 16. $\sigma_x = 8$ and $\sigma_y = 5$. $r_{xy} = $ _____.

===== STOP ===== STOP =====

$$16/40 = 0.40$$

79 $\sum XY = 200$, $N = 5$, $M_x = 4$, $M_y = 4$, and the denominator equals 24.
$r_{xy} =$ _____.

══════ STOP ══════   ══════ STOP ══════

1.00

---

80 $\sum XY = 600$, $N = 10$, $M_x = 5$, $M_y = 8$, $\sigma_x = 6$, $\sigma_y = 5$. $r_{xy} =$ _____.

══════ STOP ══════   ══════ STOP ══════

0.67

---

81   The deviation formula for correlation $r_{xy} =$ _____.

══════ STOP ══════   ══════ STOP ══════

$$\frac{\sum xy}{\sqrt{\sum x^2 \cdot \sum y^2}}$$

---

82   The raw-score formula for correlation $r_{xy} =$ _____.

══════ STOP ══════   ══════ STOP ══════

$$\frac{\dfrac{\sum XY}{N} - M_x M_y}{\sigma_x \sigma_y}$$

---

83 $\sum XY = 288$, $M_x = 16$, $M_y = 4$, $N = 4$, $\sigma_x = 4$, $\sigma_y = 3$. $r_{xy} =$ _____.

══════ STOP ══════   ══════ STOP ══════

$(72 - 64)/12 = 0.67$

84

	S1	S2	S3
	4	2	1
	5	8	4
XY	?	?	?

Three people each took two tests with the scores given at the left. Fill in XY.
$\sum XY =$ _____.

══════ STOP ══════   ══════ STOP ══════

$20, 16, 4;\ \sum XY = 40$

85

S1	S2	S3	S4
0	4	1	3
2	6	8	2

Four people each took two tests.
$\sum XY =$ _____.

══════ STOP ══════   ══════ STOP ══════

38

86 Five people each took two tests. If the correlation were computed, $N =$ _____ .

═══════ STOP ═══════════════════════════ STOP ═══════

5

---

87 Suppose three people took two tests with the following scores: $S1$: 4, 5; $S2$: 1, 3; $S3$: 1, 4. $\sum XY =$ _____ .

═══════ STOP ═══════════════════════════ STOP ═══════

27

---

88 $S1 = 4, 5; S2 = 1, 3; S3 = 1, 4.$ $\sum XY = 27, \sigma_x = 1.41, \sigma_y = 0.82.$ $M_x =$ _____ , $M_y =$ _____ .

═══════ STOP ═══════════════════════════ STOP ═══════

2, 4

---

89 $S1 = 4, 5; S2 = 1, 3; S3 = 1, 4.$ $\sum XY = 27, \sigma_x = 1.41, \sigma_y = 0.82.$ $M_x = 2, M_y = 4.$ $r_{xy} =$ _____ .

═══════ STOP ═══════════════════════════ STOP ═══════

$1.00/1.16 = 0.86$

---

90 $S1 = 1, 3; S2 = 5, 8; S3 = 3, 4.$ $\sigma_x = 1.63, \sigma_y = 2.16, M_x =$ _____ , $M_y =$ _____ , $r_{xy} =$ _____ .

═══════ STOP ═══════════════════════════ STOP ═══════

3, 5, 0.95

---

91 $S1 = 3, 5; S2 = 4, 3; S3 = 5, 7.$ $\sum XY =$ _____ , $\sigma_x =$ _____ .

═══════ STOP ═══════════════════════════ STOP ═══════

62, $\sqrt{\sum x^2/N} = 0.82$

---

92 Several assumptions must be made if we wish to generalize to a population from a correlation based on a sample. For example, we must assume that X scores are independent and form a normal distribution. We must assume that the Y scores are also _____ and _____ .

═══════ STOP ═══════════════════════════ STOP ═══════

independent and normal

---

93 In addition to assuming that the X and Y variables are _____ and _____ we must also assume that the X and Y pairs are randomly selected.

═══════ STOP ═══════════════════════════ STOP ═══════

independent and normal

94   We must also assume that the standard deviation of the Y scores for any X is the same as the standard deviation for any other X ($\sigma_{y \cdot x}$ will always be the same for any value of X). This is the assumption of *homoscedasticity*. Write the word. _____.

━━━━━ STOP ━━━━━━━━━━━━━━━━━━━━━━━━ STOP ━━━━━

homoscedasticity

---

95   We also must assume that the standard deviation of the X scores for any Y is the same as the standard deviation of the X scores for any other Y. This, again is the assumption of _____.

━━━━━ STOP ━━━━━━━━━━━━━━━━━━━━━━━━ STOP ━━━━━

homoscedasticity

---

96   We have noted that m is the slope of the line that is used to predict Y scores from X scores. This line is called the *regression line*. The line connecting the mean of the Y scores for each X score is called the _____ _____.

━━━━━ STOP ━━━━━━━━━━━━━━━━━━━━━━━━ STOP ━━━━━

regression line

---

97   We must assume that the _____ _____ is straight and *not* curved.

━━━━━ STOP ━━━━━━━━━━━━━━━━━━━━━━━━ STOP ━━━━━

regression line

---

98   Special correlation formulas have been developed to be used in situations where our _____ cannot be met. Rank-order correlation is an example of this.

━━━━━ STOP ━━━━━━━━━━━━━━━━━━━━━━━━ STOP ━━━━━

assumptions

---

99   Often data are presented in terms of ranks rather than scores. The standings of teams in a league would be an example of _____ data.

━━━━━ STOP ━━━━━━━━━━━━━━━━━━━━━━━━ STOP ━━━━━

ranked (ordinal)

---

100   This section deals with the rank-order correlation (rho). We would use this correlation when data are presented in _____.

━━━━━ STOP ━━━━━━━━━━━━━━━━━━━━━━━━ STOP ━━━━━

ranks

---

101   Suppose you participate in a three-team bowling league. At the end of twelve games the Tigers had won nine games, the Lions had won two games, and the Wolves had won seven. Rank the teams from first to last place: _____, _____, _____.

══════ STOP ══════   ══════ STOP ══════

(1) Tigers, (2) Wolves, (3) Lions

102

	LAST YEAR		THIS YEAR	
	Won	Lost	Won	Lost
Tigers	42	18	40	20
Indians	36	24	26	34
Red Sox	30	30	22	32
Yankees	12	48	32	28

Suppose a four-team league had bowled for some time. The tables show the league's record over a two-year period. Last year the Yankees ranked _____ and this year they rank _____.

══════ STOP ══════   ══════ STOP ══════

fourth, second

103

	Last year	This year	D
Tigers	1	1	0
Yankees	4	2	?
Indians	2	3	?
Red Sox	3	4	?

If we subtract this year's rank from last year's rank for the Tigers we get a score of 0. Do this for the other three teams.

══════ STOP ══════   ══════ STOP ══════

2
−1
−1

---

104   Rank-order correlation uses the *rank difference* in determining the value of the correlation. D is the symbol used to indicate _____ _____.

══════ STOP ══════   ══════ STOP ══════

rank difference

105

	Last year	This year	D	D^2
Tigers	1	1	0	0
Yankees	4	2	2	?
Indians	2	3	−1	?
Red Sox	3	4	−1	?

The square of 0 is equal to 0. Find the other three values of D^2. $\sum D^2 =$ _____.

══════ STOP ══════   ══════ STOP ══════

4
1      $\sum D^2 = 6$
1

106 $\sum D^2$ plays a very important role in rank-order correlation since the formula for it is: rho $= 1 - \dfrac{6 \sum D^2}{N(N^2 - 1)}.$ If $\sum D^2 = 6$, then $6 \sum D^2 = $ _____.

════ STOP ════════════════════════════ STOP ════

$$6 \times 6 = 36$$

---

107   rho $= 1 - \dfrac{6 \sum D^2}{N(N^2 - 1)}.$ $6 \sum D^2 = 36$. N refers to the number of pairs of ranks. In this example $N = 4$. $N(N^2 - 1) = $ _____.

════ STOP ════════════════════════════ STOP ════

$$4(16 - 1) = 4(15) = 60$$

108

$$\text{rho} = 1 - \frac{6 \sum D^2}{N(N^2 - 1)}. \qquad 6 \sum D^2 = 36,$$

$$N(N^2 - 1) = 60. \qquad \frac{6 \sum D^2}{N(N^2 - 1)} = \underline{\qquad}.$$

════ STOP ════════════════════════════ STOP ════

$$\frac{36}{60} = 0.60$$

---

109   rho $= 1 - \dfrac{6 \sum D^2}{N(N^2 - 1)}.$ $\dfrac{6 \sum D^2}{N(N^2 - 1)} = 0.60,$ rho $= $ _____.

════ STOP ════════════════════════════ STOP ════

$$1 - 0.60 = 0.40$$

110

$$\text{rho} = 1 - \frac{?}{N(N^2 - 1)}$$

════ STOP ════════════════════════════ STOP ════

$$6 \sum D^2$$

111

$$\text{rho} = 1 - \frac{6 \sum D^2}{?}$$

════ STOP ════════════════════════════ STOP ════

$$N(N^2 - 1)$$

112

$$\text{rho} = ? \frac{6 \sum D^2}{N(N^2 - 1)}$$

═══ STOP ═══ ═══ STOP ═══

1 −

113

$$? = 1 - \frac{6 \sum D^2}{N(N^2 - 1)}$$

═══ STOP ═══ ═══ STOP ═══

rho

114

$$\text{rho} = \underline{\hspace{1cm}}$$

═══ STOP ═══ ═══ STOP ═══

$$1 - \frac{6 \sum D^2}{N(N^2 - 1)}$$

115

Picture	Judge A	Judge B	D	D²
A	1	2	?	?
B	5	4	?	?
C	2	1	?	?
D	4	5	?	?
E	3	3	?	?

Two judges rate the same five pictures as to beauty. We wish to find the rho between the ratings. $\sum D^2 = \underline{\hspace{1cm}}$.

═══ STOP ═══ ═══ STOP ═══

−1	1
1	1
1	1
−1	1
0	0

$\sum D^2 = 4$

---

116 $\sum D^2 = 4$. Since there were five pictures, $N = \underline{\hspace{1cm}}$.

═══ STOP ═══ ═══ STOP ═══

5

---

117 $N = 5, \sum D^2 = 4$. The formula for the rank-order correlation: rho = $\underline{\hspace{1cm}}$.

═══ STOP ═══ ═══ STOP ═══

$$1 - \frac{6 \sum D^2}{N(N^2 - 1)}$$

118 $N = 5, \sum D^2 = 4.$ Rho = _____.

═════ STOP ═══════════════════════ STOP ═════

$$1 - \frac{24}{120} = 0.80$$

119 Eight teams play in a league. Team A placed 1 last year, 4 this year. B: 2 and 2, C: 3 and 7, D: 4 and 8, E: 5 and 1, F: 6 and 5, G: 7 and 3, H: 8 and 6. $\sum D^2 =$ _____, $N(N^2 - 1) =$ _____, rho = _____.

═════ STOP ═══════════════════════ STOP ═════

78, 504, 0.07

120 Solve the following problem. Two judges made the following rankings of six cigarettes on the basis of "smokeableness." The first rating is always given first. A: 1, 1, B: 2, 3, C: 3, 4, D: 6, 5, E: 5, 2, F: 4, 6. $\sum D^2 =$ _____, $N(N^2 - 1) =$ _____, rho = _____.

═════ STOP ═══════════════════════ STOP ═════

16, 210, 0.54

121 Find rho for the following sets of ranks: A: 1, 5, B: 2, 4, C: 3, 3, D: 4, 2, E: 5, 1. Rho = _____.

═════ STOP ═══════════════════════ STOP ═════

-1.00

Seventeen

Selecting the Right Inferential Method

One of the more frustrating aspects of a course in statistics is that many students often find they spend so much time learning about the particular statistical techniques that they rarely learn how to decide among them when faced with a real research problem. That is, students know how to compute the answers and how to interpret the numbers they compute, but they do not know which statistics to compute. The purpose of this chapter is to show you how to go about selecting the appropriate statistical test for the research question you want to answer.

In deciding which test to use, you must consider the number of groups, the number of times each individual is measured, and the type of data available. In addition, the question you ask or the information you are trying to get is important in determining the appropriate test.

AN EXAMPLE Suppose you administered a test of knowledge and a test of anxiety to a class of 100 freshmen at the beginning and end of their first year in college. Here are only a few of the questions that *could* be asked:

1. Do males and females differ in mean level of anxiety at the beginning of the year?
2. Do males and females differ in variability of knowledge scores at the end of the year?
3. Did the males' anxiety change from the beginning to the end of the year?
4. At the end of the year, what is the relationship between knowledge and anxiety for the males?
5. Were there more males than females present at the beginning of the year?

This list of questions is, of course, not exhaustive, nor does it necessarily represent the most important questions that could be asked. It does, however, give us a starting point for our discussion.

Let us go back over the questions just asked. We will look at each question in turn and show why a certain test is appropriate. The first question was whether males and females differed in mean level of anxiety at the beginning of their freshman year. If we can assume the test which was used to measure the anxiety of the freshmen yielded a distribution of interval-scaled scores, then a single classification analysis of variance would be the appropriate test. But if the test simply identified people as being anxious or not anxious (that is, was a nominal scale), then a chi-square test would be appropriate.

Consider the former situation where a person takes the test and gets a score; the several test scores form a distribution. In other words, each person has a single number that represents anxiety level, and each of these numbers is included in the analysis. There are two groups of numbers—male and female. Since the question concerns the mean levels of anxiety among males and females, the appropriate test is single-classification analysis of variance; this involves a single measurement and a single independent variable, tests differences between means, and assumes the interval scale of measurement. Alternative tests that would *not* be appropriate would include the F test for variances and correlation because we are interested in differences between means; double-classification analysis of variance because we have only one independent variable; z and t because we have more than one group; and chi square because the interval scale of measurement was used.

As we noted earlier, this could be a chi-square problem if a nominal scale had been employed, that is, if the students were classified as being either anxious or not anxious. Such data are frequency data. Any time you count the number of persons, or coins, or trees, and so on, you have frequency (or counting) data. In this case, we would find a number of males and females classified as anxious and a number of males and females classified as not anxious. With such independent frequency data, chi square is the only available test that can be used.

The second question asked is whether males and females differ in variability of knowledge scores at the end of the year. If there is a test given at the end of the year and a wide range of student scores is possible (that is, if we have an interval scale of measurement), then the appropriate test would be the F test for variances, which is the only test designed to compare the variability of one group with another. A single (or double) classification analysis of variance is clearly inappropriate because the question is *not* about differences between means. Similarly, correlation describes the degree of relationship between two variables, and is clearly irrelevant.

The third question asked about a change in the males' anxiety from the beginning to the end of the year. Although the question does not specifically ask whether the means for the beginning and the end of the year differed, this is implied. Each male was tested at the beginning of the year and again at the end. Since only two measurements were made on each person, we know that either correlation or a repeated-measures double-classification analysis of variance would be appropriate because those are the only tests we have available that can be used in the analysis of such data. The question was whether there was a *change* in the (mean) level of anxiety from the beginning to the end of the year. All analysis of variance techniques test differences between means, and the repeated measures type of analysis is the appropriate method here because of the form of the data (that is, repeated measures on the same subjects). Again, correlation would not be appropriate—even though a correlation could be computed with these data—because the question was not about the relationship between the subject order on first and second measures of anxiety.

The fourth question concerns the relationship between knowledge and anxiety among the males: If a male scores high (or low) on one variable, does he score high (or low) on the other? Again, each male has been measured twice—for knowledge at the end of the year and also for anxiety at the end of the year. Although certain of the analysis of variance techniques can be used with repeated measures, these are not appropriate because they would indicate whether the *mean* knowledge score differed from the *mean* anxiety score. Instead, the question concerns the *relationship* between anxiety and knowledge, and computation of the correlation coefficient is clearly the appropriate procedure.

Finally, the fifth question is whether there were more males than females present at the beginning of the year. Since the number of males and females must be counted, the data clearly are frequencies, and chi square is the appropriate test. However, there are several chi-square tests, and the appropriate one depends (as we will see) on the specific question being asked.

In the previous discussion, the choice of statistical technique often depended upon the number of groups and the type of measurement scale employed. Table 17-1 presents the various statistical tests that have been discussed in previous chapters. These tests have been classified according to the number of groups and the scale of measurement used. The purpose of each test has also been included. Knowing this much information will usually (but not always) lead you to the appropriate statistical test.

Let us now turn to specific examples to see what types of mistakes are commonly made and how they can be avoided.

EXAMPLE: INAPPROPRIATE USE OF CHI SQUARE

Consider the following question, which was given in a recent final exam in introductory statistics:

Professor Snarf gave a surprise test composed of 20 true–false questions (scored 1 if correct, 0 if incorrect) to his small seminar. The following results were found:

Student	Score
1	12
2	16
3	12
4	12

Looking at the results, the professor was neither pleased nor surprised with the overall performance of his class and snorted that a chimpanzee could have done just as well. Do everything necessary to test Professor Snarf's assertion.

Most students who attempted to answer the question answered it incorrectly. The reasons they were wrong are part of the focus of this chapter.

Table 17-1   Summary of statistical tests

Method	Groups	Scale of measurement	Purpose
z test	One	Interval	Establish limits or test hypothetical mean
t test	One*	Interval	Establish limits or test hypothetical mean
F test for variances	Two	Interval	Find differences between variances
Analysis of variance:			
single classification	Two or more	Interval	Find differences between means
double classification	Four or more	Interval	Find differences between means
repeated measures	One**	Interval	Find differences between means
Chi square (χ^2):			
simple	One	Nominal	Determine if observed cell frequencies differ from expected
association	Two or more	Both nominal	Determine if two variables are associated
median test	Two	Ordinal or interval	Find differences between groups
Correlation:			
ordinary	One**	X and Y both interval	Determine degree of relationship
rank order	One**	X and Y both ordinal	Determine degree of relationship
point biserial	One**	X interval and Y nominal (dichotomous)	Determine degree of relationship
phi (ϕ)	One**	X and Y both nominal (dichotomous)	Determine degree of relationship

* For purposes of clarity in this discussion, we will assume a *t* test can be computed only on a single group of scores.

** Two measures per individual, or two groups with members of one group paired with the other.

The problem states that four students took a 20-question true-false test and that each got a score. While it is true that each score represents the sum of a series of 0s and 1s (and hence represents repeated measures taken on the same subject), the only score we have to work with is the total correct for each student. That is, we have only a single score for each student.

The problem involves four scores, and they do not appear to be subdivided in any manner, so it seems that there is a single group of scores. Thus, the appropriate test is the z, t, or chi-square test. To determine which of these three tests is appropriate, we must look at those things that make the tests different from one another. The data are not independent frequencies, so chi square can be eliminated, leaving only z and t. The formulas for the z and t tests are

$$z = \frac{M - \mu}{\sigma_M} \quad \text{and} \quad t = \frac{M - \mu}{S_M}$$

The difference between the two tests is immediately apparent. For z, σ_M is in the denominator; for t, S_M is in the denominator; σ_M is a parameter; and S_M is a statistic. We have no way of determining σ_M from the information given in the problem, but we can find S_M simply by finding the standard deviation of the four scores and dividing by the square root of N, which is 2. Therefore, it is apparent that the t test is the appropriate one to use. Before considering why this problem is difficult, let us complete it. Given scores 12, 16, 12, and 12. $\Sigma X = 52$, $\Sigma X^2 = 688$, and

$$M = \frac{\Sigma X}{N} = \frac{52}{4} = 13$$

$$S^2 = \frac{\Sigma X^2 - \frac{(\Sigma X)^2}{N}}{N - 1} = \frac{688 - \frac{52 \times 52}{4}}{3} = \frac{688 - 676}{3} = \frac{12}{3} = 4$$

$$S^2 = 4, \quad S = \sqrt{4} = 2, \quad S_M = \frac{S}{\sqrt{N}} = \frac{2}{\sqrt{4}} = \frac{2}{2} = 1.00$$

Since $t = (M - \mu)/S_M$, and we know M and S_M, we have everything needed to compute t except the value of μ. To determine μ, we must look at the problem again. Professor Snarf said that a chimpanzee could do as well on the examination as the students. From this, we may (reasonably, we hope) assume that the professor was not saying anything about chimpanzees, but rather that he was saying something about his students— namely, that taken as a sample representative of a population, they were only guessing. If they were guessing, they would have correctly answered about half of the true-false questions. Therefore, the professor was saying that the mean of the four examination scores did not differ from the mean that would be obtained by chance. That mean would be $\mu = 10$, since this is half of the 20 true-false questions on the test, and it can be assumed that the chance is one of two that the student can guess each answer correctly. To finish the problem,

$$t = \frac{M - \mu}{S_M} = \frac{13 - 10}{1} = 3.00$$

The number of df $= N - 1 = 3$, and a t of 3.18 or larger is needed to reject (at the 0.05 level) the hypothesis that the observed mean (M) comes from a population that has a true mean equal to the hypothesized mean (μ). Thus, Professor Snarf's assertion that the students were guessing may be true; there is no evidence to suggest that the students, as a randomly selected sample from some population, scored better than chance. (If Professor Snarf's only interest was in the four students in his seminar, then he would not need to use inferential statistics, since he already knows the characteristics of the population.)

It may be helpful to look at the way most of the students taking the final statistics exam attempted to solve the problem. Most of them tried to use chi square. Their apparent reasoning was that there were four groups of right-wrong, or 1-0, responses. This would yield an experimental design consisting of four rows and two columns as shown in Table 17-2.

**Table 17-2   Incorrect design
for problem**

Student	Correct	Incorrect
1	12	8
2	16	4
3	12	8
4	12	8

The expected values were then computed for each cell by multiplying the appropriate row sum by the appropriate column sum and dividing by the total, which was 80. This yields a chi square of 3.45, which with 3 df is not significant. Thus, through an incorrect technique, the proper conclusion was accidentally reached.

The use of chi square here is improper for a number of reasons. The first is that, although the data are frequency data (each score represents the frequency of correct responses), they are not independent, since each score represents several 0 or 1 scores taken from the same person. That is, student 2 got 16 correct of 20, and we can perhaps assume that he had studied. The number of correct responses he got on the first ten items probably is correlated with the number he got correct on the second ten items; that is, if he knew enough to get several items correct on the first half of the test, he probably knew enough to get several items correct on the second half of the test. Chi square should not be used unless the data are independent. If we decide that the data taken from each student are correlated (not independent), we cannot use chi square appropriately.

On the other hand, even if we could assume the data were independent, chi square would still be inappropriate. This is because the chi square that would test the data presented in Table 17-2 would not answer the question originally posed: Does the mean of the correct responses differ from the mean of 10 expected by chance? It can be easily shown that the chi square obtained in Table 17-2 can be increased or decreased by changing the number of correct and incorrect responses while leaving the row and column totals constant. Suppose the scores for the four students were 10, 18, 8, and 16 correct. The chi square would equal **23.71**, and with df = 3, this chi square is highly significant. Since the hypothesis deals with the mean of the four scores, and since the value of chi square can vary while the mean remains constant, it is apparent that chi square does not test the hypothesis. Chi square in this type of problem (when appropriate data are available) tests the interaction between the rows (in the example, students) and columns (correct versus incorrect). Since the original hypothesis implies nothing about this interaction, chi square is clearly inappropriate.

In summary, a *t* test is the appropriate method to test Professor Snarf's hypothesis that the performance of his class on the true-false test was not different from chance. Most students inappropriately used chi square to solve the problem. The apparent reason for this was that they erroneously assumed that the score obtained for a student represented the sum of several *independent* correct responses. When several scores are obtained from the same individual, it is likely that a correlation exists among these scores— that the scores are *not* independent. Chi square, then, is inappropriate because its use assumes independent frequency measures.

EXAMPLE: ANSWERING THE WRONG QUESTION

Another type of question which has given students considerable difficulty is illustrated by the following question about the Müller-Lyer illusion. This

Figure 17-1 *Müller-Lyer illusion.*

illusion is illustrated in Figure 17-1; the task of the subject is to judge when the variable line segment *B* is equal in length to the constant line segment *A*. The following question was asked after the students in the class gained some acquaintance with the illusion.

An experimenter conducted an experiment investigating the Müller-Lyer illusion. To one group of observers, she gave a narrow-angle illusion, and to a second group of observers she gave a wide-angle illusion. The response measure was the amount of error in centimeters. The data are presented below. Was there an illusion?

Wide	Narrow
$M = 6$	$M = 12$
$S = 12$	$S = 12$
$N = 36$	$N = 36$

Again, virtually all students failed to answer this question correctly. The probable reasons for their failure may help to illustrate possible ways of proceeding to answer the question. Most students compared either the means or the variances of the two groups. Either case is inappropriate because it does not answer the question. The student may rightfully say that, with two groups, a number of techniques may be eliminated. Unfortunately, the grouping of the scores is irrelevant to the question. The problem is to decide whether there was an illusion. If there was no illusion, the average observer would be able to match correctly line segment B with line segment A, and the mean error would be 0.00. However, error was observed for both the narrow- and wide-angle illusions, and the question to be answered is whether the observed error represents sampling error or systematic error. That is, the sample could have been taken from a population where the mean error was zero (that is, sampling error), or the sample could have come from a population where the mean error was different from zero (that is, systematic error). We suspect the latter alternative is the correct one, simply because the apparatus employed is called the "Müller-Lyer illusion."

Thus, the task requires a comparison between an observed mean and a hypothetical mean of zero. Clearly, the t test (or the z) is required. The difficult part about the question is that more data are presented than are required to test this hypothesis. A number of ways of attacking the problem present themselves. For example, the data from the two groups could be combined, and the mean and standard deviation could be estimated from the combined data. This would reflect whether the illusion occurs over a wide range of angle sizes. However, our own preference is to test the group with the smaller amount of error (that is, smaller mean). If a significant amount of error is found, it could be assumed that error also occurs in the illusion with the greater amount of error. Thus, with a single group to consider, either the t, z, or chi-square test is appropriate. However, since the problem indicates that $S = 12$ (along with $M = 6$ and $N = 36$), it is again apparent that the t test is the appropriate one.

Thus
$$S_M = \frac{S}{\sqrt{N}} = \frac{12}{\sqrt{36}} = \frac{12}{6} = 2.00$$

and
$$t = \frac{M - \mu}{S_M} = \frac{6 - 0}{2} = 3.00$$

With 35 df, a t of 2.03 is required for significance at the 0.05 level. Hence, with the obtained t greater than the tabled t, the null hypothesis can be rejected, and we can conclude that the sample comes from a population with a mean greater than 0.00. With the mean error greater than zero, we can conclude that an illusion did occur.

In summary, the purpose of this example is to illustrate the importance of being certain about the nature of the question to be answered. Simply counting the number of groups, as may seem to have been suggested earlier, does not invariably lead to the appropriate statistical test.

EXAMPLE: COMPUTATION OF THE *F* RATIO

Here is another test question that the authors have used several times in various forms and which has also proved to be extremely difficult.

A sociologist was reading an old journal from the last quarter of the nineteenth century and found an article which presented an interesting study. Descriptive statistics for each of the four groups of ten respondents were reported as shown below. No tests of significance were computed, however, and the sociologist was frustrated in his desire to know if the groups differed. Can you help him?

I	II	III	IV
$M = 1.00$	$M = 2.00$	$M = 7.00$	$M = 10.00$
$S = 4.00$	$S = 2.00$	$S = 6.00$	$S = 4.00$

With means and standard deviations reported, it would appear that chi square is not the appropriate statistic to use here; the data of the experiment are not independent frequencies. With means greater than 1.00, the scores for many respondents would have to be greater than 1.00. When there are four groups involved, all tests can be eliminated except single- and double-classification analysis of variance. The major difference between single- and double-classification analysis of variance is that each score is classified according to either one or two independent variables. Since there is no reason to suspect that each score is classified in more than one way (that is, in terms of anything other than group), it must be concluded that the appropriate method in this case is a single-classification analysis of variance. Most students attempting to answer this question do, in fact, decide that a single-classification analysis of variance is appropriate. The steps following this decision are what confused them.

Apparently the computing formulas used in finding tot_{ss}, bg_{ss}, wg_{ss}, and so on are so much a part of the analysis of variance that many people forget that these formulas are simply for the convenience of calculating the bg_{ms} and wg_{ms}. If you remember what bg_{ms} and wg_{ms} represent, the problem is easy, since $F = NS_M{}^2/S^2$, where $NS_M{}^2$ is equal to bg_{ms} and S^2 is equal to wg_{ms}.

$$S_M{}^2 = \frac{\Sigma M^2 - \dfrac{(\Sigma M)^2}{K}}{K - 1} = \frac{154 - \dfrac{20 \times 20}{4}}{3} = \frac{154 - 100}{3} = 18$$

where $\Sigma M^2 = 1^2 + 2^2 + 7^2 + 10^2$, $\Sigma M = 1 + 2 + 7 + 10$, and $K = 4$. S^2 in the problem is the mean of the variances of the four groups. Hence,

$$S^2 = \frac{4^2 + 2^2 + 6^2 + 4^2}{4} = \frac{72}{4} = 18$$

Substituting into the formula,

$$F = \frac{NS_M{}^2}{S^2} = \frac{10 \times 18}{18} = 10$$

Since $N = 10$ for each of the four groups, bg_{ms}, or $NS_M{}^2$, has 3 df; wg_{ms}, or S^2, has 36 df. With 3 and 36 df, an F of 10.00 is highly significant. That is, the null hypothesis (the hypothesis that the true means are all equal) can be rejected, and it can be concluded that the means differ in some manner from one another.

EXAMPLE: ARRANGING THE DATA FOR CHI SQUARE

It can happen that the appropriate test is relatively easy to identify, while the proper way to set up the test is more difficult. The following is an example.

An experimenter tells an audience that he wants to conduct an experiment on extrasensory perception. Consequently, he writes the numbers 1, 2, 3, and 4 on the blackboard, and says, "I am thinking of a number. Concentrate. Try to think of the number I'm thinking of. Write it down." Shortly thereafter he says, "Your number is 3." Suppose that the data in Table 17-3

Table 17-3   Frequency of choice of numbers 1 through 4

Chosen number	Number choosing
1	4
2	13
3	40
4	3

were obtained from 60 people. We can ask: Are the choices random? We can also ask: Is the number 3 chosen more often than the other numbers? While these questions may seem similar, there actually is a considerable difference between them.

First, however, let us decide which statistical test is appropriate here. We have a single group of people selecting one of four numbers; the data are the counts of subjects selecting each number. In this case the data are frequencies that appear to be independent. In such a case, chi square is the appropriate test, and it can be seen from Table 17-3 that a simple chi square is the appropriate variant. However, two questions have been asked, and while a simple chi-square test is appropriate to answer both questions, the two questions could lead to different conclusions.

The first question is whether the choices of numbers are random. That is, are the choices distributed randomly over the four numbers? If this is the case, we would expect each number to be chosen 15 times.

The expected and observed values are presented in Table 17–4. The obtained chi square with 3 df is highly significant, and we can conclude

Table 17-4   Expected and observed number of subjects selecting choices 1 through 4

Chosen number	Observed	Expected	$(o - e)^2$
1	4	15	121
2	13	15	4
3	40	15	625
4	3	15	144

$$\chi^2 = \sum \frac{(o - e)^2}{e} = \frac{894}{15} = 59.60$$

that the choices are not distributed randomly over the four numbers. Does this demonstrate the existence of ESP? Probably not, since it makes no difference what number the experimenter is thinking of; the audience will tend to select the number 3 if the instructions previously indicated are given.

The second question was whether the number 3 was selected more often than the others. In this case, the statistician is not interested in the distribution of choices over all four numbers. Rather, he is interested only in comparing the choice of 3 with the others. Again, if choices are random, the expected number of times 3 would be chosen would be 15, and the number of times 3 would *not* be chosen would be 45 (the sum of the expected values for the other three numbers). The chi square based on these data is presented in Table 17-5. Since there is only one degree of freedom, the correction for continuity has been used. Again the obtained chi square is highly significant. Although this chi square is smaller than the one previously obtained, the significance level is higher because only a single degree of freedom was used this time.

Table 17-5   Expected and observed number of people selecting number 3 or some other choice

Chosen number	Observed	Expected	$(|o - e| - 0.5)^2$
3	40	15	600.25
Other	20	45	600.25

$$\chi^2 = \sum \frac{(|o - e| - 0.5)^2}{e} = \frac{600.25}{15} + \frac{600.25}{45} = 53.35$$

Before leaving this topic, it should be pointed out that the question just tested, "Is 3 chosen more often than the others?" can legitimately be asked only *before* the experimenter sees the data. If a statistical question is asked *after* the results of an experiment are known, the experimenter may be guilty of "data snooping." That is, at the start of the experiment, the experimenter may have thought that 2 would be selected more often than the others. After glancing at the data, he sees that 2 is not selected more often, and decides to test 3, as was just done. If hypotheses are formed and tested after the data are collected (especially when there are a large number of groups in a complex experiment), the probability is increased that a significant difference will be found when there is no real difference. In other words, the experimenter has taken advantage of chance variation to reject the null hypothesis. The likelihood of capitalizing on chance variation is reduced if the experimenter forms his hypotheses—states his questions—before he collects the data.

SUMMARY

In this chapter we have discussed the importance of using the information available to decide which test of significance to use. The number of groups, the number of times each individual is measured, the type of measurement scale employed, and the question being asked, all must be used to determine the appropriate test. Table 17-1 presented the various tests discussed in this chapter. This table, used in conjunction with the information given about the problem at hand, should be sufficient to identify the appropriate statistical test in most cases.

PROBLEMS

1. An experimenter did a study measuring ratings of "liking" as a function of perceived similarity of the personality of an opposite-sexed person. To do this, he instructed a female experimenter to act "aggressive" or "passive" in apparently accidental conversations with male subjects. Half the males were "aggressive" and the other half were "passive," as indicated by their scores on a personality test. In the following table each entry represents an "attraction" score taken from a subject. Compute the appropriate statistic and interpret the results of the test.

		Female experimenter	
		Aggressive	Passive
Male Subjects	Aggressive	0	4
		1	3
		2	2
	Passive	6	2
		7	1
		5	3

2. The true mean of a normal distribution was 60 and the true standard deviation was 15. If we took a sample of 25 cases from this population, what is the probability that the mean of this sample would fall below 57?

3. An experimenter conducted an experiment to measure the amount of group interaction in groups that were familiar and groups that were not. To test this, he measured the number of conversations during his lecture in each of his seven classes in each of two semesters. The Group A data were taken at the end of the fall semester (familiar) and the Group B data were taken at the beginning of the spring semester (unfamiliar). The data are presented in the table. Do the two groups differ in the number of conversations they have?

Group A	Group B
6	2
11	7
11	6
13	5
6	7
8	8
8	0

4. Professor Snarf, who has many interests, noticed that girls tend to sit in the front of the class, while boys tend to sit in the back. The semester after he made this observation, he found that in his class 25 girls and 5 boys sat in rows 1–5, 15 girls and 15 boys sat in rows 6–10, and 10 girls and 30 boys sat in rows 11–15. Is his observation correct?

5. The members of Alpha Phalpha fraternity, a small select group, have been criticized recently for having poor scholarship. It has been noted that they have been consistently below the all-campus average of 1.80. The grades for the members are listed below. Are their grades different from the rest of the students on campus?

Grade-point average
0
3
1
0
1

6. The slogan Brand X headache pills has used for years is, "Remove headaches reliably." With truth in advertising being important, a government agency wants to determine if the slogan is correct. To do this, they give one group of subjects Brand X pills and another the leading brand of pills (Brand Y), and measure the time it takes for relief to come. Given the following data, has the truth of the Brand X slogan been demonstrated?

**Time in minutes for
relief to come**

BRAND X		BRAND Y	
	19		5
	17		21
	24		19
	19		8
	18		20
	23		17
ΣX	120	ΣX	90
ΣX^2	2440	ΣX^2	1580

7. An experiment was conducted in which 36 first-born and 36 later-born students were divided into three subgroups each. The three subgroups were low stress (L_1), medium stress (L_2), and high stress (L_3). The response measure used was the rated desire to wait with some other person prior to the time a hypothetical (stressful) experiment was to begin. Set up a source table for the experiment showing source and degrees of freedom.

8. "An elephant never forgets" is a common saying that an experimenter was interested in testing. Consequently, he taught a randomly selected group of six elephants to traverse an elephant-sized maze. Every year for four years after learning the maze, the elephants were tested for retention by having them run the maze. Below are the data, with the response measure being time in minutes to run the maze. Complete the analysis and see if the E's hypothesis is correct. Much of the computation follows:

Year

	1967	1968	1969	1970
E_1	2	5	6	7
E_2	4	3	6	7
E_3	2	4	3	6
E_4	1	2	2	5
E_5	3	2	5	8
E_6	2	4	4	3

Elephant

$$\sum_{E}^{E}\sum_{Y}^{Y} X = 96$$

$$\sum_{E}^{E}\sum_{Y}^{Y} X^2 = 470$$

$$\sum_{E}^{E}\left(\sum_{Y}^{Y} X\right)^2 = 1618$$

$$\sum_{Y}^{Y}\left(\sum_{E}^{E} X\right)^2 = 2568$$

9. Two groups, A and B, were each composed of ten nurses. After an experimental treatment, the mean of Group A was 12 and the mean of Group B

was 20. The variance of the scores in each of the two groups was 10. What sensible hypothesis can you test in this case? Test it.

10. Manley Wainright is considering running for city council. To determine his chances in the race, he samples 100 people who are likely to vote in the election and asks them which of the two likely candidates they prefer. Of those queried, 55 prefer Mr. Wainright, while 45 prefer the other candidate. Compute whichever statistical test is appropriate. On the basis of the results of this test, what advice do you give the candidate? (Explain the results of the test and give your advice, assuming Manley has no knowledge of statistics.)

11. *E* conducted an experiment with ink blots such that each of ten paranoids was shown 40 ink blots. If the paranoid saw a human form in the blot, the experimenter said, "That's good." The experimenter was interested in determining if the frequency of "human form" responses increased as the paranoid was shown more and more blots. The response measure (dependent variable) for each paranoid was the number of "human form" responses given to the first five blots (that is, blots 1–5), the second 5, and so on (blots 6–10, 11–15, 16–20, and upward). Set up a source table for the analysis of variance *E* computed, showing the source of variability and the df associated with each source.

12. Two competing parachute-manufacturing companies have been arguing about the quality of their parachute-opening equipment. The Speedy-Fast Company says that its parachutes open faster than those of any other company. The Never-Die Company says that while the Speedy-Fast Company claims may be correct, its own parachute opens with more consistency, and they claim this is most important. An independent research company randomly selects ten parachutes from each company's stock and tests them with the following results. The response measure is the number of seconds between leaving the airplane and the opening of the parachute. Are the claims of the Never-Die Company justified?

Speedy-Fast	Never-Die
$M = 10.00$	$M = 12.00$
$S = 8.00$	$S = 4.00$
$N = 10$	$N = 10$

13. "This strain of armadillos has a right-turning bias," Professor Snarf told his class one day. To test this hypothesis, three members of his class built an apparatus which would allow an armadillo to turn right, left, go straight ahead, or turn back. They put 100 armadillos in the maze and found the following results: left, 20; right, 50; straight ahead, 20; turn back, 10. Test Professor Snarf's hypothesis.

14. "Bamo Bullets go farther" is a slogan of an ammunition company. To test this statement, a police department randomly selects 36 bullets from Bamo's assembly line and shoots them on their long rifle range. The mean distance (in feet) the bullets travel is 6000 with an *SD* of 240. The mean of all previous tests of bullets is 5900 feet. Is Bamo's slogan correct?

1   An experimenter conducts an experiment in which he collects one score from each person in a single group. The statistical tests he could use include χ^2, t, and _____.

═══════ STOP ═══════════════ STOP ═══════

z

2   If one score is taken from each person in a single group and the data are frequencies, the appropriate statistical test would be _____.

═══════ STOP ═══════════════ STOP ═══════

χ^2

3   In a single-group experiment (one score per person), the standard deviation of the population is known. Most likely the appropriate test would be _____.

═══════ STOP ═══════════════ STOP ═══════

z

4   But if σ was not known in a single group experiment (with a single score per person), the appropriate test would most likely be _____.

═══════ STOP ═══════════════ STOP ═══════

t

5   A group of voters indicate that ten will vote for Candidate A and that twenty will vote for Candidate B. The null hypothesis would be appropriately tested by _____.

═══════ STOP ═══════════════ STOP ═══════

χ^2

6   Chi square is the appropriate test in Frame 5, because the data of the problem are _____ data.

═══════ STOP ═══════════════ STOP ═══════

frequency (or nominal)

7 $M = 10$, $\sigma_M = 5$, and the hypothesized $\mu = 0$. The appropriate test is _____.

═══════ STOP ═══════════════════════ STOP ═══════

z

---

8 $M = 10$, $\sigma_M = 5$, $\mu = 0$, $z = $ _____.

═══════ STOP ═══════════════════════ STOP ═══════

2.00

---

9 An experimenter measures a group of sixteen children who have been given large quantities of Vitamin A. A randomly selected child would see with twenty units of light; the mean of this group is fifteen. What test should be used? _____.

═══════ STOP ═══════════════════════ STOP ═══════

t

---

10 Thus the critical distinction between the use of z or t is that for z, _____ is known, while for t it is *not* known.

═══════ STOP ═══════════════════════ STOP ═══════

σ_x

---

11 A jaded TV football watcher noted that in fifteen out of twenty games he watched, the home team "won the toss" and received the kickoff. What test is appropriate? _____.

═══════ STOP ═══════════════════════ STOP ═══════

χ^2

---

12 _____ $= \dfrac{M - \mu}{S_M}$.

═══════ STOP ═══════════════════════ STOP ═══════

t

---

13 - Which of the tests, appropriate for use with a single group, is also appropriate (under suitable conditions) with two groups? _____.

═══════ STOP ═══════════════════════ STOP ═══════

$$\chi^2$$
(In this chapter we are restricting the use of t to a single group.)

14 Are t or z used in a two-group experiment? _____.

═══════ STOP ═══════════════════════ STOP ═══════

No

15 Two groups of scores (one score per armadillo) are collected. χ^2 and the F test between variances could be used under these conditions. What other test could also be used? _____.

═══════ STOP ═══════════════════════ STOP ═══════

F test between means (also called single classification analysis of variance)

16 Eight of twenty males and sixteen of twenty females turned right upon entering a store. What test would you use? _____.

═══════ STOP ═══════════════════════ STOP ═══════

$$\chi^2$$

17 Enter the observed values of the previous question into the table below.

	R	L
M		
F		

═══════ STOP ═══════════════════════ STOP ═══════

	R	L
M	8	12
F	16	4

18   If the data from Frame 17 were tested, and a significant χ^2 was found, you could conclude that turning bias (is/is not) related to sex.

═══════ STOP ═══════════════════════ STOP ═══════

is

---

19   From the same data, suppose E wished to determine if more people turn right than left. What would be the expected values? _____.

═══════ STOP ═══════════════════════ STOP ═══════

$R = 20, L = 20$

---

20   From the data of the previous question, E wishes to determine if more people turn right than left. The expected values are: $R = 20$ and $L = 20$. What are the observed values?

═══════ STOP ═══════════════════════ STOP ═══════

$R = 24, L = 16$

---

21   Each of several students takes two tests and a score (grade) is found for each. What statistical tests might be used on these data? ___ _____.

═══════ STOP ═══════════════════════ STOP ═══════

r_{xy} and repeated measures analysis of variance

---

22   If the experimenter was interested in determining the relationship between the scores of the first test and the scores of the second, what statistical test should he use? _____.

═══════ STOP ═══════════════════════ STOP ═══════

r_{xy}

---

23   But if the experimenter wanted to know if the mean of the second test differed from the mean of the first, which test would he use? _____.

═══════ STOP ═══════════════════════ STOP ═══════

Double classification analysis of variance (repeated measures)

---

24 The running times of two groups of athletes were measured. If the experimenter wished to compare the variance of one group with the variance of the other, what test would he use? _____.

═══════ STOP ═══════════════════════════ STOP ═══════

F test for variances

---

25 If the experimenter wished to compare the means of the same two groups, he would use _____ _____ _____ _____ _____.

═══════ STOP ═══════════════════════════ STOP ═══════

single classification analysis of variance

---

26 If he wished to determine if one group was more homogeneous in its performance than the second, he would use _____ _____ _____ _____.

═══════ STOP ═══════════════════════════ STOP ═══════

F test for variances

---

27 Is chi square an appropriate test in Frame 24? _____.

═══════ STOP ═══════════════════════════ STOP ═══════

No

---

28 Chi square is not appropriate because the data of the experiment—running times of athletes—is not _____ data.

═══════ STOP ═══════════════════════════ STOP ═══════

frequency (or nominal)

---

29 The number of ears of corn taken from each of twenty corn stalks was used as the response measure in an experiment. Three different types of corn (i.e., three groups of twenty plants each) were used in the experiment. The appropriate statistical test is _____ _____ _____ _____ _____.

═══════ STOP ═══════════════════════════ STOP ═══════

single classification analysis of variance

---

30   Double classification analysis of variance is *not* appropriate for three independent groups because _____.

===== STOP ===================== STOP =====

they are not doubly classified

---

31   The number of ears of corn on a corn stalk is frequency data. Could chi square be used here? _____.

===== STOP ===================== STOP =====

No

---

32   The reason chi square is inappropriate when the number of ears of corn on a stalk is used as the response measure is that the various ears on a stalk are not _____ of each other.

===== STOP ===================== STOP =====

independent

---

33   Chi square is used on _____ data only if these data are _____.

===== STOP ===================== STOP =====

frequency, independent

---

34   The number of cigarettes smoked by each member of a sample is found along with his weight. What statistic is most appropriate? _____.

===== STOP ===================== STOP =====

r_{xy}

---

35   The number of cigarettes smoked and the weight of each member of the sample is found. It is likely that the correlation between the two variables will be the statistical test of interest. Why would analysis of variance be inappropriate? _____.

===== STOP ===================== STOP =====

A difference between means would have no sensible interpretation

---

36   Hemlines of dresses go up and down more often than does trouser length. That is, we are saying that hemlines are more _____ than are pants lengths.

═══════ STOP ═══════════════════════ STOP ═══════

variable

───

37   We want to test the hypothesis that hemlines are more variable than pants lengths. What is the appropriate test? _____.

═══════ STOP ═══════════════════════ STOP ═══════

F test for variances

───

38 E conducts a test for ESP. She draws twelve cards at random and has thirty-six subjects guess which of four suits will appear. The number correct for each subject is the response measure. What is the appropriate test? _____.

═══════ STOP ═══════════════════════ STOP ═══════

t

───

39   From Frame 38, if people randomly guessed which suit would appear, the mean number of correct guesses would be _____.

═══════ STOP ═══════════════════════ STOP ═══════

three

───

40   The mean number of correct guesses would be three because the probability of a correct guess on a *single* card would be _____ and through the _____ law of probability, the sum of the twelve separate probabilities would be three.

═══════ STOP ═══════════════════════ STOP ═══════

$\frac{1}{4}$, additive

───

41   From Frame 38, if the null hypothesis is true, the true mean, μ, would be equal to _____.

═══════ STOP ═══════════════════════ STOP ═══════

3

───

42 From Frame 38, $\mu = 3$ and E finds $M = 4.00$ and $S = 3$. $S_M = $ _____.

=== STOP === === STOP ===

$$\frac{3}{\sqrt{36}} = \frac{3}{6} = .50$$

---

43 From Frame 38, $\mu = 3$, $M = 4.00$, $S = 3.00$ and $S_M = .50$. $t = $ _____.

=== STOP === === STOP ===

$$\frac{4.00 - 3.00}{.50} = 2.00$$

---

44 $t = 2.00$, df = _____. The tabled value of $t = $ _____.

=== STOP === === STOP ===

35; 2.03

---

45 The observed value of $t = 2.00$, the tabled value of $t = 2.03$. What do you do with the null hypothesis? _____. Can E conclude that ESP has been demonstrated; i.e., that more cards have been guessed correctly than would be expected on the basis of chance? _____.

=== STOP === === STOP ===

Do not reject; No

---

46 Suppose an urn had twelve red balls, six white balls, four green balls, and two black balls. What is the probability of drawing a white ball from the urn? _____.

=== STOP === === STOP ===

$$\frac{6}{24} = .25$$

---

47 From Frame 46, what is the probability of drawing two red balls in a row (replacing the first before drawing the second)? _____.

=== STOP === === STOP ===

$$\tfrac{1}{2} \times \tfrac{1}{2} = \tfrac{1}{4}$$

---

48 From Frame 46, with replacement, what is the probability of *not* drawing a white ball in two draws? _____.

═══════ STOP ═══════════════════════════ STOP ═══════

Pr(not white) = ¾
¾ × ¾ = ⁹⁄₁₆

---

49 We use the additive law of probability if we wish to find the probability of Event A _____ Event B. This could be illustrated when we find that the probability of drawing a green _____ a white ball is $\frac{4}{24} + \frac{6}{24} = \frac{10}{24}$.

═══════ STOP ═══════════════════════════ STOP ═══════

or; or

---

50 We use the multiplicative law of probability if we wish to find the probability of Event A _____ Event B. This could be illustrated when we find that the probability of drawing two green balls in a row is $\frac{4}{24} \times \frac{4}{24} = \frac{1}{6} \times \frac{1}{6} = \frac{1}{36}$.

═══════ STOP ═══════════════════════════ STOP ═══════

and then (followed by)

---

51 Given two groups of scores: A: 1, 5, 1, 1. B: 8, 8, 16, 8. E wishes to test the hypothesis that the means of the two groups differ. The appropriate test would be _____ _____ _____ _____ _____.

═══════ STOP ═══════════════════════════ STOP ═══════

single classification analysis of variance

---

52 From Frame 51, $\Sigma X = 48$, $\Sigma X^2 = 476$. $CF =$ _____; $tot_{ss} =$ _____.

═══════ STOP ═══════════════════════════ STOP ═══════

288; 188

---

53   From Frame 51, $bg_{ss} =$ _____; $wg_{ss} =$ _____.

═══════ STOP ═══════════════════════ STOP ═══════

128; 60

---

54   From Frame 51, bg df = _____; $bg_{ms} =$ _____; wg df = _____; $wg_{ms} =$ _____.

═══════ STOP ═══════════════════════ STOP ═══════

1; 128; 6; 10

---

55   From Frame 51, $F =$ _____. What can you conclude about the means of groups A and B? _____.

═══════ STOP ═══════════════════════ STOP ═══════

12.80; The means differ.

---

56   From Frame 51, on the basis of the analysis of variance just completed, we reject the null hypothesis and conclude that the means differ. On the basis of this analysis what can be said about the variances of groups A and B? _____.

═══════ STOP ═══════════════════════ STOP ═══════

Nothing

---

57   What is the appropriate test for the variances? _____.

═══════ STOP ═══════════════════════ STOP ═══════

$$F = \frac{S_A{}^2}{S_B{}^2} \text{ or } \frac{S_B{}^2}{S_A{}^2}$$

---

58   From Frame 51, for group A, $S^2 =$ _____ and for group B, $S^2 =$ _____ $F =$ _____

═══════ STOP ═══════════════════════ STOP ═══════

$S_A{}^2 = 4$, $S_B{}^2 = 16$, $F = \dfrac{16}{4} = 4$. Without a prior hypothesis, the larger variance is placed in the numerator.

59 How many df for this F? _____.

≡ STOP ≡ ≡ STOP ≡

3 and 3

---

60 What do you do with the null hypothesis? _____.
What do you conclude? _____.

≡ STOP ≡ ≡ STOP ≡

Do not reject; There is no evidence that the null hypothesis is false

---

61 An experimenter was interested in determining if a certain diet pill resulted in a weight loss. Consequently he randomly selected a group of subjects from a population of fat people. He weighed each and then had each take one pill a day for thirty days. After thirty days he weighed them again. What is the appropriate test here? _____.

≡ STOP ≡ ≡ STOP ≡

Double classification analysis of variance with repeated measures

---

62 From Frame 61, what are the independent variables in the analysis? _____.

≡ STOP ≡ ≡ STOP ≡

Time of weighing, subjects

---

63 Suppose there were twenty subjects in the experiment in Frame 61. Set up a summary (or source) table showing the source of variance and the number of df associated with each source.

≡ STOP ≡ ≡ STOP ≡

Source	df
time (T)	1
subject (S)	19
$T \times S$	19
tot	39

64 From Frame 61, there were a total of _____ scores, there were _____ subjects, and each was weighed _____ times.

══════ STOP ══════════════════════ STOP ══════

forty, twenty, two

---

65 There is no wg_{ss} in the source table for Frame 61. The reason for this is that each cell contains _____ _____.

══════ STOP ══════════════════════ STOP ══════

one score

---

66 There is no wg_{ss} in the source table for Frame 61. What is used as the denominator in the F test for time? _____.

══════ STOP ══════════════════════ STOP ══════

$T \times S_{ms}$

	T_1	T_2
S_1	0	2
S_2	0	6
S_3	2	6
S_4	1	4
S_5	6	10
S_6	3	8

67 At the right are some scores which were taken from six subjects. The scores are the number of correct responses after Trial 1 and Trial 2.

$\Sigma X = 48, \Sigma X^2 = 306, CF =$ _____.

══════ STOP ══════════════════════ STOP ══════

$$\frac{48 \times 48}{12} = 192$$

---

68 From Frame 67, $CF = 192$, $tot_{ss} =$ _____, $T_{ss} =$ _____.

══════ STOP ══════════════════════ STOP ══════

$$306 - 192 = 114,$$
$$\frac{12^2 + 36^2}{6} - CF = 48$$

69   From Frame 67, $tot_{ss} = 114$, $T_{ss} = 48$, $S_{ss} =$ _____, $T \times S_{ss} =$ _____.

════ STOP ════════════════════════ STOP ════

$$\frac{2^2 + 6^2 + 8^2 + 5^2 + 16^2 + 11^2}{2}$$
$$- CF = 61,$$
$$tot_{ss} - T_{ss} - S_{ss} = 5$$

70
Source	ss	df	ms	F
T	48	?		
S	61	?		
$T \times S$	5	?		
tot	114	?		

════ STOP ════════════════════════ STOP ════

1, 5, 5, 11

71
Source	ss	df	ms	F
T	48	1	?	?
S	61	5		
$T \times S$	5	5	?	
tot	114	11		

════ STOP ════════════════════════ STOP ════

$T_{ms} = 48$, $T \times S_{ms} = 1$, $F = 48$

---

72   Does Trial 1 differ from Trial 2 significantly?_____ ____.

════ STOP ════════════════════════ STOP ════

Yes

---

73   Consider the data of Frame 67. If someone were interested in the consistency of performance for each S from Trial 1 to Trial 2 and were *not* interested in the mean differences between trials or subjects, what would be the appropriate statistical test to use? _____.

════ STOP ════════════════════════ STOP ════

r_{xy}

74 If a person's performance were highly consistent from Trial 1 to Trial 2, then the correlation would be _____, but if a person who did well on Trial 1, did poorly on Trial 2 (or vice versa) then the correlation would be _____.

═══════ STOP ═══════════════════════════ STOP ═══════

high positive, negative

---

75 Suppose a normally distributed population has a mean of 200 and a standard deviation of 20. About ⅔ of the population falls between _____ and _____ (equidistant from the mean).

═══════ STOP ═══════════════════════════ STOP ═══════

180–220

---

76 Suppose also that 25 scores are randomly selected from the population mentioned in frame 75 and the mean of this sample is computed. The probability is about .68 that the mean of the sample falls between _____ and _____ (equidistant from the mean).

═══════ STOP ═══════════════════════════ STOP ═══════

196–204

---

77 Suppose the mean and standard deviation of the sample from frame 76 are found to be 206 and 15, respectively. In testing the hypothesis that the true mean of the population is 200, $t =$ _____.

═══════ STOP ═══════════════════════════ STOP ═══════

2.00

---

78 What do you do with the hypothesis that the true mean is equal to 200?

═══════ STOP ═══════════════════════════ STOP ═══════

Do not reject

---

79 Establish the 95 percent confidence interval for the mean, using the information provided in frames 76 and 77.

═══════ STOP ═══════════════════════════ STOP ═══════

$206 \pm 3 \times 2.06$

80 Two groups, 10 men and 10 women, are each given 5 trials on a learning task. If we wished to know if the men had increased their performance over trials, what test would we use? _____ .

═══════ STOP ═══════════════════ STOP ═══════

Double classification analysis of variance (repeated measures)

---

81 The rows and columns in this analysis would be _____ and _____ .

═══════ STOP ═══════════════════ STOP ═══════

subjects, trials

---

82 Set up a summary table showing source of variance and df for this analysis.

═══════ STOP ═══════════════════ STOP ═══════

Source	df
T	4
S	9
$T \times S$	36
tot	49

---

83 From Frame 80, if we wished to find if the men had more correct answers on the fifth trial than the women, what test would we use? _____ .

═══════ STOP ═══════════════════ STOP ═══════

Single classification analysis of variance

---

84 Set up a summary table showing source of variance and df for this analysis.

═══════ STOP ═══════════════════ STOP ═══════

Source	df
bg	1
wg	18
tot	19

---

85   From Frame 80, past experiences with men and women indicate that most women try to learn the list, while some men don't try very hard. Using N correct per person as the dependent variable, what test would be appropriate here? _____.

===== STOP ================================ STOP =====

$$F = \frac{S^2 \text{ men}}{S^2 \text{ women}}$$

---

86   In the same experiment, eight women learned the list while only three men did likewise. What test would we use if we wanted to know whether more women learned the list than men? _____.

===== STOP ================================ STOP =====

$$\chi^2$$

---

87   From Frame 80, suppose the men and women were married couples. If we wish to determine if a husband's performance is related to his wife's performance, what test would we use? _____.

===== STOP ================================ STOP =====

$$r_{xy}$$

---

88   We could do this because for each couple we have _____ scores, the husband's and the wife's.

===== STOP ================================ STOP =====

two

---

89   From Frame 80, suppose we wanted to know whether the men had learned anything on Trial 1. What test would we use?_____.

===== STOP ================================ STOP =====

$$t$$

---

90 In this case S_x would be determined from the number correct on Trial 1, μ would be equal to _____, and M would be _____

_____ _____ _____ _____ _____.

═══════ STOP ═══════════════════════════ STOP ═══════

0, the mean correct on Trial 1

---

91 Suppose previous research had found that women get 8.3 items correct on Trial 5 with a standard deviation of 3. If we wish to compare the performance of our sample of women with these norms, we would use what test? _____.

═══════ STOP ═══════════════════════════ STOP ═══════

z

---

92 Suppose half the men and half the women in Frame 80 are ages 20–25, while the other half are between 50 and 60. (Assume here that no one is married.) How might the number of correct responses on Trial 5 be analyzed? _____.

═══════ STOP ═══════════════════════════ STOP ═══════

Double classification analysis of variance

---

93 Set up a source table for this analysis showing source and df.

═══════ STOP ═══════════════════════════ STOP ═══════

Source	df
Age	1
Sex	1
$A \times S$	1
Within	16
tot	19

Eighteen

Recent Trends in Statistics

The material presented in Chapters One through Seventeen represents the state of introductory statistics in much the same way as it was presented in the first edition of this text. The topics covered in the first edition at that time represented a departure from most other introductory statistics texts in only one major respect. Until then, the t test for two samples was taught in the introductory statistics course with the analysis of variance covered later in advanced or graduate courses. We rejected this approach on the grounds that the analysis of variance gave the same result in the two-sample case (that is, $t^2 = F$) while at the same time it was both easier to learn and more general than the t test. Over the past 15 years we have not seen any reason to change our minds about this matter. Analysis of variance is still preferable to the t test on the same grounds.

There are, however, some nontraditional topics and issues which are attracting considerable attention even in the relatively stable world of introductory statistics. We have selected three of these topics to be presented in this chapter as a sort of postscript to the major portion of the text. It was something of a problem to decide whether to present the material covered in Chapter Seventeen as the final summary chapter or to put it where it is. We settled on this arrangement because we wanted to emphasize that the area is changing and that the next few years might see considerable changes in the way the introductory statistics course is taught. The reason for this is that the three topics we will cover have already had some impact upon the field, and we do not foresee any return to the more traditional ways of teaching statistics. Rather, we see only a gradually accelerating change in the direction toward which we are already headed.

The three nontraditional topics covered in this chapter are (a) nonparametric statistics, (b) Bayesian statistics, and (c) computers. Computers have probably had a greater impact upon the teaching of introductory statistics than any of the other nontraditional topics, and yet it is the most difficult topic to correlate with this course. We will attempt to explain why this is the case when we deal with the topic. Bayesian statistics has had considerable impact upon statistics because Bayesian principles take a somewhat different point of view of the whole process of statistical inference than does the more traditional approach. We will try to explain the difference in the two points of view. You might ask which is the correct point of view. As in many other controversies, the answer to that will be very clear 100 years from now. Right now, we suspect that portions of both positions are correct, but as with the alpha and beta errors, we don't know which. Finally, nonparametric statistics have been around for as long as traditional statistics. However, it has been only relatively recently that the interest in *distribution-free* (that is, nonparametric) statistical tests was developed. The advantage, however, of not having to make assumptions about the characteristics of the population from which the samples were drawn was seen to be a considerable advantage in many cases where the experimenter did not know the characteristics of the population used or was reluctant to make the necessary assumptions.

NONPARAMETRIC TESTS

You may recall in the development of the one-way analysis of variance that the variance of each group was computed and then the average (mean) of those variances was used as the within-group mean square (wg$_{MS}$). We were able to do this because the assumption had been made that the true variance of Group I was equal to the true variance of Group II, and so on (that is, $\sigma_1^2 = \sigma_2^2 = \sigma_N^2$). In other words, some assumptions about the parameters (in this case the variances) of the population from which the samples were taken is made by the test. Since only rarely are such assumptions tested (or even can be tested), experimenters often are uncomfortable about using such *parametric* tests. Nearly all the tests we have looked at in previous chapters make some such assumption about one or another parameter. Often, the assumption is made that sampling is done from a population which is normally distributed. Again, research workers are often uncomfortable about making such an assumption when it may not be the case, and they worry about what effect failing to meet such an assumption will have upon the statistical decision-making process. Similarly, many of the tests which have been introduced assume that at least an interval scale of measure has been employed. But many times in social science data the attainment of an interval scale is questionable.

For example, if IQ scores met the requirements of an interval scale, the difference between IQs of 70 and 80 would be the same as the IQ difference between 130 and 140. Probably relatively few psychologists would insist that these differences are the same, but if they are not, the requirements of an interval scale have not been met. Similarly, an archeologist at a certain excavation site may believe that the time elapsed between the beginings of cultures A and B was the same as that between the beginnings of cultures C and D. Yet this is based on assumption, such as equal amounts of deposit laid down per year, but the archeologist may not be willing to make such an assumption. Under these conditions the experimenter wishing to analyze the data might prefer to use a nonparametric test, which characteristically makes no assumption about the characteristics of the population from which the data were taken and requires only an ordinal scale instead of an interval scale.

We have previously presented two tests which are commonly thought of as nonparametric statistics. Both are illustrative of the characteristics of nonparametric tests. The median test, presented in Chapter Fifteen, uses chi square to test the difference between the medians—typically an ordinal-scale measure of central tendency—of two groups, and is one nonparametric test to which you have already been introduced. The second nonparametric test is rank-order correlation, presented in Chapter Sixteen. Here the correlation between ranked data—that is, ordered data—was found. Thus, it is apparent that nonparametric statistics are not tests which require a point of view different from the one we have been taking, but rather are additional tools for the analysis of data when the experimenter is not confident that the data at hand meet the criteria necessary for the usual parametric tests.

There are two commonly used nonparametric tests which we will present to illustrate the wide variety of nonparametric tests available. These are, of course, only two of several tests that we could present. Additional tests and the one appropriate for the analysis of the data of your experiments can probably be found in a text dealing with nonparametric statistics.

MANN-WHITNEY U TEST.   The U test was developed as an alternative to the two-group analysis of variance test, or the two-group t test. The U test requires only two independent groups and an ordinal scale of measurement (and special tables if the number of cases per group is not relatively large). If two groups come from two identical populations (whatever the shapes), then the probability that a score taken from one population is larger than a score taken from the second population should be 0.50 if the two populations are the same. That is, half the time a score from the first population will be larger than a score taken from the second population and half the time vice versa. The U test is based on this probability. Essentially, it compares the sum of the ranks from one group with the average rank of two populations expected to be the same. The difference between the two (the observed sum and expected or average sum) is expressed in z-score units. If the difference is greater than $z = 1.96$, the null hypothesis is rejected, and it is concluded that the two populations differ, with more scores from one population being higher than the scores from the second population. The z obtained from the U test will be normally distributed if the number of scores in each group is larger than, say, 15. With smaller numbers of scores, the obtained z will not necessarily be normal, and special tables should be used.

Consider the following example. A dietitian wanted to know if the addition of Vitamin C to children's diets would cut down on the number of days a child was sick during the year. Consequently, she randomly selected 10 children who were given a Vitamin C supplement and then randomly selected 12 children who were given a placebo (no additional food value) supplement. The experiment was conducted for a year, and after a year she asked each family to report on the number of days the child was ill. Since one family might remember that a child was sick while forgetting just how long the illness lasted (especially if prolonged), the dietician decided to use a Mann-Whitney U test because of the way in which the data were reported. Here are the original data in number of days reported sick. These data were then ranked from smallest score to largest score, and then the ranks were placed on the right.

Vit. C	Placebo	Vit. C	Placebo
0	6	1	4
1	16	2	6.5
5	27	3	10.5
15	32	5	13
16	39	6.5	15
18	42	8	16
22	48	9	17
27	52	10.5	18
29	55	12	19
37	60	14	20
	61		21
	63		22
		$\Sigma R_1 = 71$	$\Sigma R_2 = 182$

The sum of the ranks for the Vitamin C group is found to equal 71, while the sum of the ranks for the placebo group is equal to 182. Notice how ties are handled. The tied ranks are added and then divided by the number of ties. For example, the two scores of 16 are then sixth and seventh ranked scores; the mean of these two scores is 6.5, and this is the rank given to each of these two scores.

The U is computed and then substituted into the formula $z = (U - \mu_U)/ \sigma_U$, where $U = N_1 N_2 + N_1 [(N_1 + 1)/2] - \Sigma R_1$. Here, ΣR_1 is the sum of the ranks in the Vitamin C group, N_2 is the number of scores in the placebo group, $\mu_U = (N_1 N_2)/2$ and

$$\sigma_U = \sqrt{\frac{(N_1)(N_2)(N_1 + N_2 + 1)}{12}}.$$

Before computing U, it should be noted that the designation of Group 1 and Group 2 is arbitrary. Whichever one you choose to call Group 1 will give you the same value of z. The signs will differ, however, but this should be ignored. There is a relationship between the two U values (that is, the U computed when Group A is called 1 and B 2, and when B is called 1 and A 2), which can be used to check on the accuracy of your computations. This relationship is

$$U_1 + U_2 = N_1N_2$$

Computing:

$$U = 10 \times 12 + \frac{10(11)}{2} - 71$$

$$= 120 + 55 - 71$$
$$= 104$$

Computing:

$$\mu_U = \frac{10 \times 12}{2} = 60$$

Computing:

$$\sigma_U = \sqrt{\frac{10 \times 12 \times 23}{12}} = \sqrt{230} = 15.16$$

Combining:

$$z = \frac{U - \mu_U}{\sigma_U} = \frac{104 - 60}{15.16} = \frac{44}{15.16} = 2.90$$

Since $z = 2.90$ and is greater than 1.96, the hypothesis of equivalent populations is rejected. Note that the other U is equal to 16; that is,

$$U_2 = N_1N_2 + \frac{N_2(N_2 + 1)}{2} - \Sigma R_2$$

$$= 10 \times 12 + \frac{12 \times 13}{2} - 182$$

$$= 120 + 78 - 182$$
$$= 16$$

and $z = (16 - 60)/15.16 = -44/15.16 = -2.90$. Ignore the sign and again reject the null hypothesis. We should note in passing that N_1 and N_2 are probably too small, and so tables that are constructed to give the probability of obtaining a U of a certain size should be used. These tables can be found in texts dealing with nonparametric statistics.

The hypothesis of equivalent populations is rejected, and it is concluded that reported number of days sick is less for the Vitamin C group than for the placebo group.

FRIEDMAN TWO-WAY ANALYSIS OF VARIANCE BY RANKS   A second important nonparametric test is the Friedman test. This test may be used when a number of attributes are rank-ordered by more than one subject or judge. Suppose several suggested solutions to a union-management problem are ranked for the quality of solution by five judges, as illustrated in the tabulation below. One way of scoring the solutions would be on the basis of a rating from 0–100. These are presented on the left side of the table. Next, these scores are translated into ranks, and these ranks are given on the right side of the table.

Judge	Solution A	B	C	D		Judge	Solution A	B	C	D
1	95	65	50	70		1	1	3	4	2
2	80	50	70	85		2	2	4	3	1
3	90	60	80	50		3	1	3	2	4
4	50	70	20	40		4	2	1	4	3
5	80	50	40	70		5	1	3	4	2
						ΣR	7	14	17	12

It can be seen that the design of this experiment is the same as that found in the double classification analysis of variance with $N = 1$ per cell. Hence, whenever repeated measures on several subjects or scores taken from matched subjects are encountered, then the Friedman two-way analysis of variance by ranks is an appropriate test if at least rank (that is, ordinal) data are employed. The Friedman test was used in this case instead of a double-classification analysis of variance because the experimenter felt that the assumptions of an interval scale and a normal distribution—assumptions of the analysis of variance—would be difficult to justify. Hence, the appropriate nonparametric test was used.

Essentially, the formula for the Friedman test determines the difference between the rank sums expected by chance and the observed rank sums. This difference is then translated into a χ^2 with the formula

$$\chi^2 = \frac{12}{NK(K+1)} \sum_{1}^{K} \left(\sum_{1}^{N} R \right)^2 - 3N(K+1)$$

where ΣR is the sum of the ranks, K is the number of groups (solutions), and N is the number of subjects (judges) in each group. Solving,

$$\chi^2 = \frac{12}{5 \times 4 \times 5} \times (7^2 + 14^2 + 17^2 + 12^2) - 3 \times 5 \times 5$$

$$= \frac{12}{100} \times 49 + 196 + 289 + 144$$

$$= 0.12 \times (678) - 75$$

$$= 81.36 - 75$$

$$= 6.36$$

Using a standard chi-square table with df $= K - 1 = 3$, the tabled value of χ^2 at 0.05 with df $= 3$ is 7.82. Hence, the null hypothesis is not rejected, and there is no evidence to suggest that the solutions differ from one another in quality.

The Mann-Whitney and Friedman tests are two examples of the many nonparametric tests available for use by research workers. Many others have been developed and are often applicable when no suitable parametric test is available. You may wonder why research workers do not simply use nonparametric statistics exclusively so that they would never have to worry about the possible effects of violations of assumptions. There are two reasons for not doing this. First, parametric tests permit much more complex experimental designs. Nonparametric tests can be used in experiments in which about the largest number of independent variables that can be manipulated is two. Parametric tests have no such limitation. In theory, any number of independent variables can be used in an experiment. The second reason is that even with simple experimental designs, the parametric tests tend to be more powerful because they are more likely to detect a difference when there is one. Even so, nonparametric statistical tests have shown themselves to be useful in a wide variety of conditions and will be with us for a long time to come.

BAYESIAN STATISTICS

In contrast to the classical or traditional statistical methods presented up to this point, Bayesian statistics employs a somewhat different approach. As is characteristic of adherents to a progressive new technique or method, the Bayesian statistician has little if anything positive to say about the older classical techniques. For example, Novick and Jackson[1] describe "hypothesis testing as a method of giving a misleading answer to a question nobody is asking." According to these authors, the question nobody is asking is: Is the null hypothesis, $\mu_1 = \mu_2$, correct? If it is found not to be correct (that is, if the null hypothesis is rejected), we would be giving a misleading answer in saying an extremely small difference is significant when such a difference would make no real difference in any event. Pretty strong stuff, and even if exaggerated, the statement by Novick and Jackson does excite interest in Bayesian statistics.

What is it about the approach of Bayesian statistics that differs from that of classical statistics? Perhaps the greatest difference is that the Bayesian system incorporates prior opinion *formally* into its statistical procedures. This contrasts with classical statistics, in which an experimenter may have a prior opinion about the outcome of some event, but does not incorporate it into the classical system. Indeed, classical experiments are set up in such a way that it is usual to reject the null hypothesis—that "question nobody is asking." Hence, since everyone has an opinion about the outcome of an experiment prior to the actual conduct of the experiment, the Bayesian statistician includes such prior opinion into the system. Alternatively, while the classical statistician designs his (or her) experiments to take advantage of prior opinion, he does not make formal use of it.

Basic to the character of Bayesian statistics is the concept of *conditional* probability. Conditional probability is simply the probability that some event, say A, will occur, given that some other event, say B, has al-

[1] Novick, M. R., & Jackson, P. H. *Statistical methods for educational and psychological research.* New York: McGraw-Hill, 1974, p. 245.

ready occurred. This is a new concept for us because in the discussion of probability we had in Chapter Seven our discussion was directed toward events which were independent of one another. The concept of conditional probability is illustrated by the two overlapping circles in Figure 18-1. In

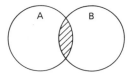

Figure   18-1

this case, if items are sampled randomly, either an A or a B, or as in the cross-hatched area, an AB can be selected. That is, the set of As and Bs are *not* mutually exclusive. A nonmutually exclusive case is illustrated when we consider the sets as face cards and hearts in an ordinary bridge deck. There are face cards (A), hearts (B), and some face cards that are hearts (AB). Now, a conditional probability would be the probability of having a B, given an A. This probability can be shown to be $\Pr(B/A) = \Pr(AB)/\Pr(A)$. Where $\Pr(B/A)$ is the probability of a B, given that an A has occurred, $\Pr(AB)$ is the probability of both A and B occurring. $\Pr(A)$ is the probability of A.

Suppose the population of circles illustrated in Figure 18-2 represents 100 balls in an urn. Forty of the balls are large, 50 are painted blue, and 10

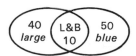

Figure   18-2

are both large and painted blue. Having these data, what is the probability of drawing at random a large ball, given that the ball you have drawn is blue? Substituting into our equation,

$$\Pr(\text{large/blue}) = \frac{\Pr(\text{large and blue})}{\Pr(\text{blue})}$$

Here, $\Pr(\text{large and blue}) = 10/100 = 0.10$, since there are 10 large and blue balls of a total of 100; and $\Pr(\text{blue}) = 60/100 = 0.60$, since 60 of the 100 balls are blue. Hence,

$$\Pr(\text{large/blue}) = \frac{\Pr(\text{large and blue})}{\Pr(\text{blue})} = \frac{0.10}{0.60} = 0.17$$

or 1 of 6.

This, of course, can be checked directly, since 10 of the 60 blue balls are large. Most problems will not be as easy. Similarly,

$$\text{Pr(blue/large)} = \frac{\text{Pr(large and blue)}}{\text{Pr(large)}}$$

or

$$\text{Pr } (B/L) = \frac{0.10}{0.50} = 0.20$$

Now, since $\text{Pr}(L/B) = \text{Pr}(LB)/\text{Pr}(B)$ then $\text{Pr}(LB) = \text{Pr}(L/B) \text{ Pr}(B)$ and Since $\text{Pr}(B/L) = \text{Pr}(LB)/\text{Pr}(L)$, then $\text{Pr}(LB) = \text{Pr}(B/L) \text{ Pr}(L)$. Combining, then $\text{Pr}(B/L) \text{ Pr}(L) = \text{Pr}(L/B) \text{ Pr}(B)$, and

$$\text{Pr } (B/L) = \frac{\text{Pr}(L/B) \text{ Pr}(B)}{\text{Pr}(L)}$$

When in this form, the formula is called the *Bayes' theorem,* or *Bayes' rule.* Using the probability values computed above,

$$\text{Pr } (B/L) = \frac{0.17 \times 0.60}{0.50} = 0.20$$

That is, the probability that a ball will be blue, given that it is large, is 0.20. This is reasonable, since of the 50 large balls there are 10 blue. Note that $\text{Pr}(L/B)$ and $\text{Pr}(B/L)$ differ from one another. In the first case we found that the probability that the ball was large, given that it was blue, was 0.17; in the second case the probability that the ball was blue, given that it was large, was 0.20. Careful attention must be paid to such distinctions because they are used quite often in Bayesian statistics, as can be seen in the next equation.

Bayes' theorem can also be expressed in a slightly different but equivalent form:

$$\text{Pr } (A_1/D) = \frac{\text{Pr}(D/A_1) \text{ Pr}(A_1)}{\text{Pr}(D/A_1) \text{ Pr}(A_1) + \text{Pr}(D/A_2) \text{ Pr}(A_2)}$$

Although the formula appears forbidding, it can be readily understood and used. The following example illustrates this.

Suppose a new, cheap diagnostic technique can correctly identify 70 percent of those people who have cancer before any symptoms of it show up. Unfortunately, it also incorrectly diagnoses 20% of the healthy population as having cancer. Now suppose an advertising campaign is conducted to persuade people to come to a hospital and be tested for cancer. To the surprise of the hospital administrators, 1000 people turn out, and among them are 300 who have positive symptoms. We can use the formula to determine the probability that a person identified as having cancer does in fact have it. The table below shows the various numbers of people involved, but this is for illustrative purposes only and is not an adjunct of the formula. From the discussion above it can be seen that $\text{Pr}(A_1)$—the probability of having cancer—is 0.30, and $\Pr(A_2)$—the probability of not having cancer—is 0.70. These proportions are represented as the row totals in the table. Furthermore, $\Pr(D/A_1)$ for those people who have cancer, the probability of being diagnosed as having cancer is 0.70. Finally, $\Pr(D/A_2)$ for those people who don't have cancer, the probability of being erroneously diagnosed as having cancer is 0.20. These two values are represented in the left column of the table.

Diagnosed as cancer

		Yes	No	
Cancer	Yes	210	90	300
	No	140	560	700
		350	650	1000

Substituting into the formula, we find

$$\Pr(A_1/D) = \frac{0.70 \times 0.30}{0.70 \times 0.30 + 0.20 \times 0.70} = \frac{0.21}{0.21 + 0.14} = \frac{0.24}{0.35} = 0.60$$

That is, the probability that someone diagnosed as having cancer actually has cancer is 0.60. Again this is made more understandable by looking at the table, which shows that of 350 people diagnosed as having cancer, 210 actually had it. The formula expresses each of these numbers as a probability and arrives at the same answer: $\Pr(A_1/D) = 0.60$. In other problems the numbers will not be as obvious and therefore it is best to use the formula.

You will note that of the people who came to be tested, 300 had cancer while 700 did not, but these numbers are given only for comparative purposes. In real life, we could not know which of these people actually had cancer unless we followed up each one several months later to confirm the diagnosis with another test. Nevertheless the example illustrates how Bayesian statistics differs from classical statistics. The Bayesian statistician is willing to make a guess about the nature of the universe and express this opinion as a probability statement. The only requirement is that once the probability has been given (that is, guessed at) by the Bayesian, then the statistician should be indifferent as to whether the probability is greater than or less than his estimate. In other words, the Bayesian statistician is willing to bet on either side of the statement that the probability is equal to some value.

To clarify this further, let us look at a very simple situation, one that allows us to ignore other important considerations that we would have to look at if we were to talk about such things as cancer. Suppose we have two urns, appropriately called urn 1 and urn 2. These are illustrated in Figure 18-3. Urn 1, or u_1, contains 40 white balls and 60 black balls; urn 2, or u_2, has 70 white balls and 30 black balls. One of the two urns is selected (we do not know which) and balls are drawn from it (only from the one selected). Our job as Bayesian statisticians is to express our opinion (by a

urn #1   urn #2

Figure 18-3

probability statement) as to which urn the balls are coming from. The formulas we use are the same as before, but the letters are changed to be a little more meaningful for this problem. For urn 1, the formula is

$$\Pr(u_1/D) = \frac{\Pr(D/u_1)\ \Pr(u_1)}{\Pr(D/u_1)\ \Pr(u_1) + \Pr(D/u_2)\ \Pr(u_2)}$$

and for urn 2, the formula is

$$\Pr(u_2/D) = \frac{\Pr(D/u_2)\ \Pr(u_2)}{\Pr(D/u_1)\ \Pr(u_1) + \Pr(D/u_2)\ \Pr(u_2)}$$

Notice that the denominators are the same and have only two alternatives, $\Pr(u_1/D) + \Pr(u_2/D) = 1.00$.

   If we were asked to bet on which urn had been selected, we would have no information—no *prior* opinion—about the particular urn from which the balls were being taken. It would seem reasonable to say that the chances were even that one urn or the other had been selected. Hence, our prior opinion (that is, opinion prior to the gathering of any data) would be that $\Pr(u_1) = \Pr(u_2) = 0.50$. Suppose now we conduct an "experiment."

   We take a single ball from this unknown urn. It is white (W). Using the two Bayesian equations given above, we can now change our opinion, somewhat, as to the urn from which we are taking balls. This is done in the following manner: The D (data) of our experiment is a single white ball drawn at random from our unknown urn. If we are sampling from u_1, then the probability that the ball would be white would be 40/100; that is, $\Pr(W/u_1) = 0.40$. Similarly, if we were sampling from the second urn, the probability of getting a white ball would be $\Pr(W/u_2) = 70/30 = 0.70$. Since our information is the outcome of the experiment—that is, a white ball (W)—then $\Pr(W/u_1) = \Pr(D/u_1)$ in the basic equation given above. Now all our unknowns are specified, and $\Pr(u_1/D)$ and $\Pr(u_2/D)$ which represent our *posterior* (P) opinion—our opinion after the results of the experiment have been considered—can be determined:

$$\Pr(u_1/D) = \frac{\Pr(D/u_1)\ \Pr(u_1)}{\Pr(D/u_1)\ \Pr(u_1) + \Pr(D/u_2)\ \Pr(u_2)}$$

$$= \frac{0.40 \times 0.50}{0.40 \times 0.50 + 0.70 \times 0.50} = \frac{0.20}{0.20 + 0.35} = \frac{0.20}{0.55} = 0.36$$

and

$$\Pr(u_2/D) = \frac{\Pr(D/u_2)\ \Pr(u_2)}{\Pr(D/u_2)\ \Pr(u_2) + \Pr(D/u_1)\ \Pr(u_1)}$$

$$= \frac{0.70 \times 0.50}{0.70 \times 0.50 + 0.40 \times 0.50} = \frac{0.35}{0.55} = 0.64$$

Hence, after the selection of a single white ball, our opinion that we have selected urn 1 decreased from 0.50 to 0.36. Similarly, our *prior* opinion that urn 2 was selected was 0.50, but our opinion has been changed by the results of the experiment. Therefore our *posterior* opinion that urn 2 was selected is now 0.64.

Suppose we conduct another experiment and select another ball from the unknown urn. Then our *posterior* (P) opinions from the first experiment become the *prior* (Pr) opinions for the second experiments. This is, of course, perfectly reasonable because we have changed our opinion about which urn was selected. Now, after the first experiment but prior to the second, our opinion expressed as a probability is that $\Pr(u_1) = 0.36$ and $\Pr(u_2) = 0.64$. That is, we now think the chances are 64 of 100 that the second urn was selected.

Suppose now that we conduct another experiment and select a second ball from the urn. Again it is white. Using our posterior opinions from the first experiment as the prior opinions for the second experiment, we substitute into the original equations. Hence, for urn 1

$$\Pr(u_1/D) = \frac{0.40 \times 0.36}{0.40 \times 0.36 + 0.70 \times 0.64} = \frac{0.144}{0.144 + 0.448} = \frac{0.144}{0.592} = 0.24$$

and for urn 2,

$$\Pr(u_2/D) = \frac{0.70 \times 0.64}{0.70 \times 0.64 + 0.40 \times 0.36} = \frac{0.448}{0.592} = 0.76.$$

Now our opinion—our posterior opinion—after two experiments is that the probability that urn 2 was selected is 0.76, while the probability that our urn is urn 1 has slipped to 0.24.

What happens if a black ball is now selected? We do essentially what we have been doing before. That is, we substitute into the original equation our posterior opinions from the second experiment, which, of course, are now our prior opinions in the third experiment. The data from the third experiment is a black ball and for urn 1, $\Pr(B/u_1) = 0.60$ and for urn 2, $\Pr(B/u_2) = 0.30$. Substituting once again into the equations, we have for urn 1,

$$\Pr(u_1/D) = \frac{0.60 \times 0.24}{0.60 \times 0.24 + 0.30 \times 0.76} = \frac{0.144}{0.144 + 0.228} = \frac{0.144}{0.371} = 0.39$$

and for urn 2,

$$\Pr(u_2/D) = 1 - P(u_1/D) = 0.61$$

since the two posterior probabilities must sum to 1.00.

The results of the third experiment wherein a black ball was selected has made us less certain that urn 2 was selected. That is, our opinion that urn 2 was selected went from $\Pr(u_2) = 0.76$ after the second experiment to $\Pr(u_2) = 0.61$ after the third. (Remember, $\Pr(u_2/D)$ for one experiment becomes $\Pr(u_2)$ for the next.)

It should be noted that it makes no difference (other than rounding errors) if we modify our opinion after every time a ball is drawn or after a whole group of draws. For example, the probability of the specific sequence of draws W, W, B is equal to $\Pr(W, W, B) = 0.4 \times 0.4 \times 0.6 = .096$ for urn 1 and $\Pr(W, W, B/u_2) = 0.7 \times 0.7 \times 0.3 = 0.147$ for urn 2. Substituting into our equation,

$$\Pr(u_1/D) = \frac{0.096 \times 0.50}{0.096 \times 0.50 + 0.147 \times 0.50} = \frac{0.048}{0.1215} = 0.40$$

and since $\Pr(u_1/D) + \Pr(u_2/D) = 1$, then $\Pr(u_2/D) = 0.60$. These are, of course, the same posterior probabilities as were previously computed for the third experiment. You might wonder why $\Pr(u_1) = \Pr(u_2) = 0.50$. The reason for this is that the results of all three experiments were lumped into one large experiment, and prior to looking at any of the data of the large experiment, our prior opinions were $\Pr(u_1) = P(u_2) = 0.50$.

Now suppose we were to conduct experiment after experiment until we have drawn ten times. These results are presented in Figure 18.4. It can be seen that as the data were collected, our opinion that urn 2 had been selected rose to about 0.95, while $\Pr(u_1)$ dropped to about 0.05. Note that we have plotted the probabilities from the first three experiments and have computed the successive probabilities in exactly the same manner as was done in the first three experiments.

We assumed at the outset that we had no strong opinion as to which urn was being used in the experiment. Does holding a strong opinion influence the posterior probabilities? The answer is yes, or yes *but*. That is, you can hold any opinion that you wish, but as your opinions are modified by data collected in various experiments, your opinions become similar to those of us who held other points of view at the outset of the experiment.

For example, suppose there was a Bayesian statistician who also believed in extrasensory perception. After reading the mind of the experimenter who selected the urn, this person said that his prior opinion was $\Pr(u_1) = 0.80$ and $\Pr(u_2) = 0.20$. That is, he believed strongly that urn 1 was the urn from which the balls would be selected. For purposes of comparison, suppose we drew ten balls from the urn and found the same results as in our draws before: $W, W, B, W, W, W, W, W, B, W$ with $\Pr(u_1) = 0.80$, $\Pr(u_2) = 0.20$, $\Pr(D/u_1) = 0.40$, $\Pr(D/u_2) = 0.70$. When a white ball is drawn, the results of the first experiment are, for urn 1,

$$\Pr(u_1/D) = \frac{0.40 \times 0.80}{0.40 \times 0.80 + 0.70 \times 0.20} = \frac{0.32}{0.32 + 0.14} = \frac{0.32}{0.46} = 0.70$$

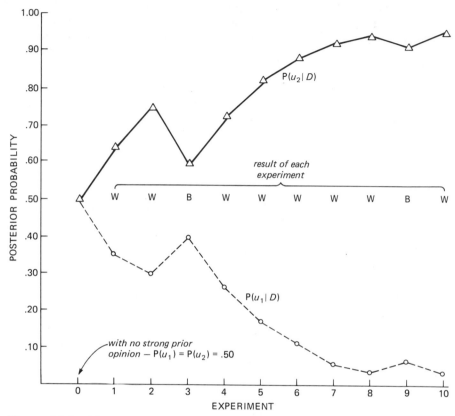

Figure 18-4   *The opinion-posterior probability of a Bayesian statistician changes as new data are provided.*

and for urn 2,

$$\Pr(u_2/D) = \frac{0.70 \times 0.20}{0.70 \times 0.20 + 0.40 \times 0.80} = \frac{0.14}{0.46} = 0.30$$

On the basis of the results of the first experiment, the clairvoyant Bayesian has had to revise his opinion downward for urn 1 and upward for urn 2. If we were to continue the series of experiments with the same results as before, we would see a gradual shifting of his opinion until after the fifth experiment, when he would then believe that we are more likely to be sampling from urn 2 than from urn 1. These data are presented in Figure 18.5. It can be seen from the graph that after ten experiments, the posterior opinion that we were sampling from urn 1 had slipped to 0.15, while the opinion that we were sampling from urn 2 had increased to 0.85—not a great deal different from those people who started with no prior opinion.

Not all examples are so trivial. For example, suppose the sonar in a ship has been found to give a warning "ping" when it encounters a solid object in the sea. When a submarine is in the area, it will "ping" 80% of

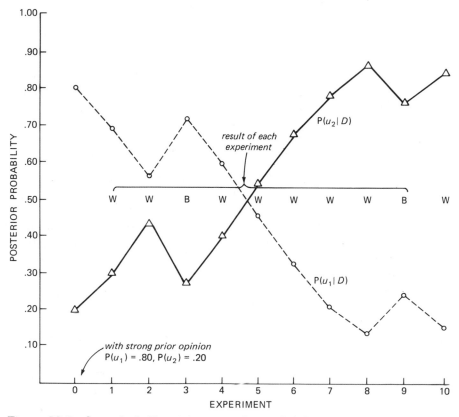

Figure 18-5 *Strongly held opinions are also modified as new data are provided.*

the time, but when a submarine is *not* in the area, the sonar will ping only 30% of the time. It should be obvious that the importance attached to a warning ping will depend upon the captain's assessment of the chances that a submarine is in the area—if the sonar of a ship on Lake Michigan were to give a warning ping, the captain would probably have to adjust his perspective to allow for other causes. Alternatively, a warning ping in a war zone would be given a great deal more of attention. Notice that it makes little difference what the "real" probability of a submarine's being present is. Rather, it is the opinion of the individual in command that is of importance.

Suppose, then, the captain of the ship is in an area where his opinion is that a submarine presence in this area would be 1 in 10, that is $\Pr(S) = 0.10$. Then, if the sonar gave a warning, the captain's posterior opinion that a submarine was in the area would be modified through the use of the formula

$$\Pr(S/D) = \frac{\Pr(D/S)\ \Pr(S)}{\Pr(D/S)\ \Pr(S) + \Pr(D/\bar{S})\ \Pr(\bar{S})}$$

where S = Submarine, $\bar{S}$ = No Submarine, and D = Ping

$$= \frac{0.80 \times 0.10}{0.80 \times 0.10 + 0.30 \times 0.90} = \frac{0.08}{0.8 + 0.27} = \frac{0.08}{0.35} = 0.23$$

On the other hand, if the ship were in a war zone where the captain's prior opinion was that $\Pr(S) = 0.70$, then his opinion would be revised somewhat if he heard the warning of the sonar:

$$\Pr(S/D) = \frac{0.80 \times 0.70}{0.80 \times 0.70 + 0.30 \times 0.30} = \frac{0.56}{0.56 + 0.09} = \frac{0.56}{0.65} = 0.86$$

Given the warning from the sonar, the commander of the ship must now decide what action to take. The commander might decide to take evasive action only when the probability of a hostile submarine presence was greater than 0.50. That is, he would take evasive action only when $\Pr(S/D) > 0.50$. Since the subjective estimate of probability enters into the equations, then such things as anxiety level and experience in battle will all influence the value of $\Pr(S/D)$.

We hope we have given you some insight into what Bayesian statistical procedures can do. As in the case of nonparametric statistics, only a few examples can be given. However, the approach is considerably different from the classical statistics presented in previous chapters. The basic assumption which Bayesian statisticians employ is that subjective probability estimates can be used, and these estimates can be modified in light of the data gathered. It is this assumption that classical and Bayesian statisticians argue about.

TECHNOLOGICAL CHANGES

The use of pocket calculators is already having a considerable impact on the teaching of such courses as introductory statistics. A calculator which will perform the basic arithmetic functions of addition, subtraction, multiplication, and division presently costs only about 3% to 5% of what it cost only a few years ago. Arguments about whether students should be allowed to use them on tests have been heard, and these have tended to be resolved either by allowing only the type that performs the four basic functions or by denying use of all. More complex or sophisticated calculators will find the mean and standard deviation, and can perform or can be programmed to perform such tests as the analysis of variance and correlation. Obviously, if a student is asked to compute a standard deviation or to do an analysis of variance when only a button has to be pushed to find the answer, then this does not test the student's understanding of the procedures being used on the material being tested. Clearly, a reappraisal of what the student should be required to learn in a variety of courses including statistics needs to be made in light of the changes in computer and calculator technology.

Consider now the discussions in early chapters about the methods for the grouping of data. With the availability of pocket calculators, such methods—that is, methods for calculating summary statistics such as the median and SIQR from grouped data—have become largely obsolete. When only paper and pencil were available before the widespread availability of calculators of any size, procedures for grouping data into intervals and then calculating summary statistics from these grouped data were of considerable importance. The small errors which occurred when statistics were computed from grouped data were of minor importance when compared to the ease of computation of these summary statistics from ungrouped data. Now, however, the effort of punching a number into a calculator is little more than that required to assign a number to its proper group. Because of this, there seems to be little justification or advantage in first grouping data and then computing the desired statistics, as compared to simply computing the desired statistics directly from the calculator. If this conclusion is thought to be reasonable, it would then appear that time spent in teaching statistics could be better spent in teaching topics other than how to compute summary statistics from grouped data. This whole argument may seem somewhat unnecessary to the student who has not spent much of his time learning how to group data, but who uses a calculator advantageously. However, dropping this topic of data grouping and computation will cause a considerable change in the content of an introductory statistics course, and any departure from tradition is made only with some degree of hesitation on the part of the authors. However, we feel that the pocket calculator is with us to stay, and since it will never go away, the teaching of statistics must be adjusted to make use of the new technology.

The use of computers seems to be an entirely different situation. There are, of course, computer programs which can do all the statistics presented in this text. However, in contrast to the use of pocket calculators, the use of a computer requires considerable information. In time, of course, programming languages will be made easy enough so that a computer will be no more difficult to use than a pocket calculator. At present, however, we cannot in a one-semester course teach a student how to run a computer as well as we can teach him about introductory statistics. The reason for this is that there is a wide variety of computers as well as a wide variety of computer languages. No very general language now available allows us to teach it as *the* programming language for statistics. Because of this, we have avoided mentioning anything about the computer in this book. To provide only an overview of the computer would most likely leave the student without enough information to use a specific computer. However, in the laboratory section of our statistics course, we have attempted to introduce students to the computer, either to further their understanding of introductory statistics or to introduce them to compute usage.

The future promises a considerable change in the amount of information needed to use a computer. Interactive modes—that is, techniques which allow the operator to talk to the computer via a computer terminal, and which in turn allow the computer (via a program) to guide the operator to do what the operator wishes—are presently available. These techniques are being refined so that, progressively, less and less skill is required on the part of the people who operate the terminals. At present, computer operators can analyze data using statistical techniques with which they are not familiar, and the trend toward the use of computers by less well trained individuals promises that such use will accelerate. What will be the effect of this? No one knows, of course, but it would seem that a statistics course would be forced to change from one that emphasized (at least as far as the students taking it were concerned) computation to one which emphasized the understanding of the meaning and philosophy of the tests being used. Just how that will be done, only time will tell.

Finally, computers are allowing widespread use of statistics which would be prohibitively time consuming if done without the use of computers. Although the theoretical background for linear regression analysis and multivariate analysis has been developed for quite some time, such techniques are cumbersome, to say the least, when only paper and pencil are available. With the increase in computer sophistication and usage, these techniques are being used more often, and it may be that there will be a time in the not so distant future when these techniques will be taught in an introductory statistics course. This would require, it appears, ready access to an intearctive computer terminal. As soon as the problem of a ready supply of computer terminals is solved, we may find that the material presently being covered in introductory statistics has become obsolete. However, before the reader forgets everything learned in this course, he should wait to see what the future will bring. Twenty-five years from now, the topics may be essentially the same as they are today.

SUMMARY

The development of the t test early in this century and the development of the F test and analysis of variance in the thirties had tremendous impact upon hypothesis testing and the content of statistics courses. It would seem that nonparametric statistics, Bayesian statistics, and electronic calculators and computers are likely to have a similar impact. Nonparametric statistics are in demand by research workers who find themselves unable to meet the many and varied assumptions necessary for the use of parametric statistics. Similarly, Bayesian statisticians often feel the need to choose between two alternative courses of action. Since classical statistics provides relatively little direction in many cases (e.g., when the null hypothesis cannot be rejected, it also cannot necessarily be accepted), Bayesian techniques have proved to be attractive. Finally, the advent of high speed, miniaturized calculators and computers presently permit the use of statistical techniques by people who do not understand them. Whether these recent developments will have as great an impact upon the use and teaching of statistical techniques as the development of the t and F tests, only time will tell.

PROBLEMS

1. Ask your friends to name as many of Snow White's dwarfs as they can. Then, after finding the percentage of names ending in "y" (5) or not "y" (2), rank-order recall. Below are typical data. Using a Friedman test, determine if a greater *proportion* of "y" than not "y" names are remembered.

"y"	not "y"
4	1
2	0
1	0
3	1
5	1
4	0

2. Two groups of students were taught by different methods, but were given the same essay test. Believing that the assumption of an interval scale was difficult to justify, the instructor asked you to analyze the following data, using the Mann-Whitney U test. Compute the value of z_μ.

Group I	Group II
17	9
23	11
12	8
14	7
18	14
	5
	6

3. A criminologist found that the probability of a positive identification (I_1) of a burglar was 0.80, while the probability of a positive identification of one of the members of a group of witnesses as a burglar was 0.30. You believe that a man accused of robbing (R) a grocery is innocent (he is your brother). Consequently, you set the probability at 0.20 that he is guilty. After positive identification of your brother by a witness (I_2), your opinion changes somewhat. Find the new probability.

4. Several students were in both Professor Snarf's and Professor Finke's classes. As a consequence, each instructor could be rated by the same students and a comparison made between the two classes. Below are the ratings. Test to see if there is a preference in instructors.

Student	Snarf	Finke
1	90	70
2	50	60
3	60	40
4	30	20
5	70	60
6	50	40
7	80	50

5. $\Pr(D/A_1) = 0.50$, $\Pr(D/A_2) = 0.10$, $\Pr(A_1) = 0.30$, $\Pr(A_2) = 0.70$. Find $\Pr(A_1/D)$.

6. An experimenter selected two groups of people: Group I was composed of males and Group II was composed of females. Below are the data. Compute U.

Group I	Group II
37	16
42	23
19	12
24	22
33	28
36	18
30	14
29	10
	17
	21

7. From Problem 6, compute μ_U.
8. From Problem 6, compute σ_U.
9. From Problem 6, compute z_U.
10. $\Pr(AB) = 0.20$, $\Pr(B) = 0.40$. $\Pr(A/B) =$ _____.

11.

	B_1	B_2	
A_1	60	40	100
A_2	120	180	300

From the information on the left: Find $\Pr(A_1) =$ _____, $\Pr(B_1) =$ _____, $\Pr(A_1/B_1) =$ _____.

12. Each of ten judges ranked a group of four paintings. The rankings are presented below. Is there a difference in preference?

Judge	Painting 1	2	3	4
1	2	1	3	4
2	2	1	4	3
3	1	2	4	3
4	2	1	3	4
5	3	4	2	1
6	2	4	1	3
7	1	2	4	3
8	4	3	2	1
9	2	1	4	3
10	2	1	3	4

13. For two kinds of pupfish—Hubbs (H) and Capps (C)—a zoologist finds that each has two subgroups (R). Of the Hubbs, 30% have 12 rays (or spikes) in their dorsal fin and 70% have 13 rays. Of the Capps, 60% have 12 rays and 40% have 13 rays. The zoologist has reason to believe that a certain lake contains 40% Hubbs and 60% Capps. If he catches 100 fish with 12 rays, how many Hubbs should he find?

1 The Mann-Whitney U test is a nonparametric test used when the assumptions of the analysis of _____ cannot be met.

===== STOP ===== STOP =====

variance

---

2 A nonparametric test which may be used to determine if two independent groups of scores come from the same population is called the _____ test.

===== STOP ===== STOP =====

U, or Mann-Whitney

---

3 The U test is a form of the z test. You may remember $z = (X - ?)/\sigma$.

===== STOP ===== STOP =====

μ

---

4 Similarly, the U test uses a raw score, a mean and a standard deviation: $z = (U - \mu_u)/?$.

===== STOP ===== STOP =====

σ_u

---

5 That is, when $z = (X - \mu)/\sigma$, for the U test, $z = ? - \mu_u/\sigma_u$

===== STOP ===== STOP =====

U

---

6 $z = U - \mu_u/?$

===== STOP ===== STOP =====

σ_u

---

7 Thus, for the U test, a z score is computed using the formula $z =$ _____ .

===== STOP ===== STOP =====

$U - \mu_u/\sigma_u$

---

8   The U test uses ranked scores, and the mean and _____ of these scores in the U test are easily computed.

═══════ STOP ═══════════════════════════ STOP ═══════

standard deviation

---

9   In the computation of the mean of the U from a group of ranked scores, $\mu_u = N_1N_2/2$, where N_1 = the number of scores in Group 1 and N_2 = the number of scores in Group _____.

═══════ STOP ═══════════════════════════ STOP ═══════

2

---

10   With 10 scores in Group 1 and 6 scores in Group 2, and with $\mu_u = N_1N_2/2$, then $\mu_u = $ _____.

═══════ STOP ═══════════════════════════ STOP ═══════

30

---

11   With $N_1 = 6$ and $N_2 = 12$, $\mu_u = $ _____.

═══════ STOP ═══════════════════════════ STOP ═══════

36

---

12   The formula for the mean of $U = N_1N_2/?$

═══════ STOP ═══════════════════════════ STOP ═══════

2

---

13 $\mu_u = ?/2$

═══════ STOP ═══════════════════════════ STOP ═══════

N_1N_2

---

14   Finally, the formula for $\mu_u = $ _____.

═══════ STOP ═══════════════════════════ STOP ═══════

$N_1N_2/2$

---

15 The value in the denominator of z is: $? = \sqrt{\dfrac{(N_1)(N_2)(N_1 + N_2 + 1)}{12}}$

$\equiv\!\equiv\!\equiv$ STOP $\equiv\!\equiv\!\equiv\!\equiv\!\equiv$ STOP $\equiv\!\equiv$

σ_u

---

16 That is, $\sigma_u = \sqrt{\dfrac{(N_1)(N_2)(N_1 + N_2 + 1)}{?}}$

$\equiv\!\equiv\!\equiv$ STOP $\equiv\!\equiv\!\equiv\!\equiv\!\equiv$ STOP $\equiv\!\equiv$

12

---

17 and $\sigma_u = \sqrt{\dfrac{(?)(?)(N_1 + N_2 + 1)}{12}}$

$\equiv\!\equiv\!\equiv$ STOP $\equiv\!\equiv\!\equiv\!\equiv\!\equiv$ STOP $\equiv\!\equiv$

$(N_1)(N_2)$

---

18 $\sigma_u = \sqrt{\dfrac{?}{?}}$

$\equiv\!\equiv\!\equiv$ STOP $\equiv\!\equiv\!\equiv\!\equiv\!\equiv$ STOP $\equiv\!\equiv$

$\dfrac{(N_1)(N_2)(N_1 + N_2 + 1)}{12}$

---

19 $\sigma_u = ?$

$\equiv\!\equiv\!\equiv$ STOP $\equiv\!\equiv\!\equiv\!\equiv\!\equiv$ STOP $\equiv\!\equiv$

$\sqrt{\dfrac{(N_1)(N_2)(N_1 + N_2 + 1)}{12}}$

---

20 $z = \dfrac{U - \mu_u}{\sigma_u}$. Finally, we must calculate _____.

$\equiv\!\equiv\!\equiv$ STOP $\equiv\!\equiv\!\equiv\!\equiv\!\equiv$ STOP $\equiv\!\equiv$

U

---

21 $U = N_1N_2 + \dfrac{N_1(N_1 + 1)}{2} - \Sigma R_1.$ If $N_1 = 5,$ then $\dfrac{N_1(N_1 + 1)}{2} =$

_____

═══ STOP ═══════════════════════════ STOP ═══

15

---

22 $U = N_1N_2 + \ ? - \Sigma R_1,$ where $\Sigma R_1 =$ the sum of the ranks in Group 1.

═══ STOP ═══════════════════════════ STOP ═══

$$\frac{N_1(N_1 + 1)}{2}$$

---

23 $U = N_1N_2 + \dfrac{N_1(N_1 + 1)}{2} - \ ?$

═══ STOP ═══════════════════════════ STOP ═══

ΣR_1

---

24 $U = \ ? - \Sigma R_1$

═══ STOP ═══════════════════════════ STOP ═══

$$N_1N_2 + \frac{N_1(N_1 + 1)}{2}$$

---

25   Finally, $U = \ ?$

═══ STOP ═══════════════════════════ STOP ═══

$$N_1N_2 + \frac{N_1(N_1 + 1)}{2} - \Sigma R_1$$

---

26 $N_1 = 5,$ $N_2 = 8,$ and $\Sigma R_1 = 25.$ Compute $U,$ using the formula $U = N_1N_2 + \dfrac{N_1(N_1 + 1)}{2} - \Sigma R_1$

═══ STOP ═══════════════════════════ STOP ═══

$40 + 15 - 25 = 30$

27 $U = N_1 N_2 + \dfrac{N_1(N_1 + 1)}{2} - \Sigma R_1$. The scores of Group 1 rank 1, 3, 4, 7, 2, and 8. $\Sigma R_1 = $ _____.

STOP ═══ STOP

25

28 If the test scores are ranked from lowest to highest for Group 1 (0, 1, 4, 19, 23) and for Group 2 (5, 8, 25, 30, and 31), the rank of the scores in Group 1 are 1, 2, 3, 6, and 7. For Group 2 they are _____.

STOP ═══ STOP

4, 5, 8, 9, 10

29 Given the following scores

1	2
0	4
1	6
3	8
5	10
	12

Find the ranks of Group 1 and then find ΣR_1. _____

STOP ═══ STOP

1, 2, 3, 5 and $\Sigma R_1 = 11$

30 $U = N_1 N_2 + \dfrac{N_1(N_1 + 1)}{2} - \Sigma R_1$. From 29, find the value of U. _____.

STOP ═══ STOP

$20 + 10 - 11 = 19$

31 $\mu_u = N_1 N_2 / 2$; from 29, μ_u _____.

STOP ═══ STOP

10

32 $\sigma_u = \sqrt{\dfrac{(N_1)(N_2)\,(N_1 + N_2 + 1)}{12}}$. From 29, $\sigma_u =$ _____.

══════ STOP ══════════════════ STOP ══════

$$\sqrt{200/12} = 4.08$$

---

33 From 29, $U = 19$, $\mu_u = 10$, $\sigma_u = 4.08$. Using the formula $z = (U - \mu_u)/\sigma_u = ?$

══════ STOP ══════════════════ STOP ══════

$$(19 - 10)/4.08 = 2.21$$

---

34 Given the following scores:

I	II
2	0
4	1
6	5
	10
	13

Find ΣR_1 _____.

══════ STOP ══════════════════ STOP ══════

13

---

35 Given the following scores:

I	II
1	0
3	4
8	7
10	9
12	11

Find ΣR_1. _____

══════ STOP ══════════════════ STOP ══════

29

---

36 From frame 34, find U. _____

══════ STOP ══════════════════ STOP ══════

$$15 + 6 - 13 = 8$$

---

37 From frame 34, find μ_u. _____

STOP ══════════════ STOP

$$\frac{3 \times 5}{2} = 7.50$$

38 From frame 34, find σ_u. _____

STOP ══════════════ STOP

$$\sqrt{\frac{15 \times 9}{12}} = \sqrt{11.25} = 3.35$$

39 From frame 34, $U = 8$, $\mu_u = 7.5$, $\sigma_u = 3.35$. Find z. _____

STOP ══════════════ STOP

$$(8 - 7.5)/3.35 = 0.15$$

40 Finally, with two larger groups, $N_1 = 20$, $N_2 = 26$, $\Sigma R_1 = 430$, and $\sigma_u = 6.87$. Find z. _____

STOP ══════════════ STOP

$$(300 - 260)/6.87 = 5.82$$

41 Reference to the Appendix B table indicates that if z is larger than 1.96, the null hypothesis is rejected at the 0.05 level. If z is larger than 2.58, the null hypothesis is rejected at the _____ level.

STOP ══════════════ STOP

0.01

42 The z obtained when N_1 and N_2 are small tends not to be normal. However, when $N_1 + N_2 = 30$ or larger, the obtained z is very close to

_____.

STOP ══════════════ STOP

normal

43 The Friedman two-way analysis of variance test is another non _____ test which is used on ranked data.

══════ STOP ════════════════════════════════ STOP ══════

parametric

---

44 Four designs are rank-ordered for quality. The appropriate nonparametric test would be the _____ two-way analysis of variance by ranks.

══════ STOP ════════════════════════════════ STOP ══════

Friedman

---

45 Three candidates are rank-ordered by four voters. The first step in the Friedman test is to find the sum of the ranks. For candidate A, $\Sigma R = 5$; for B, $\Sigma R =$ _____, and for C, $\Sigma R =$ _____.

	A	B	C
V_1	1	2	3
V_2	1	3	2
V_3	2	1	3
V_4	1	2	3

══════ STOP ════════════════════════════════ STOP ══════

8, 11

---

46 The Friedman two-way test determines if the sums of the _____ differ significantly by using a form of chi square.

══════ STOP ════════════════════════════════ STOP ══════

ranks

---

47 The Friedman formula is

$$\chi^2 = \frac{12}{NK(K+1)} \sum^{K} (\Sigma R)^2 - 3N(K+1)$$

Referring to frame 45: K is the number of candidates and N is the number of _____.

══════ STOP ════════════════════════════════ STOP ══════

voters, subjects, or rankers

---

48 $\chi^2 = \dfrac{12}{NK(K+1)} \Sigma(?)^2 - 3N(K+1)$

═════════ STOP ═══════════════════ STOP ═════════

$$\Sigma R$$

---

49 $\chi^2 = \dfrac{?}{NK(K+1)} \Sigma(\Sigma R)^2 - 3N(K+1)$

═════════ STOP ═══════════════════ STOP ═════════

$$12$$

---

50 $\chi^2 = \dfrac{12}{NK(?)} \Sigma(\Sigma R)^2 - 3N(?)$

═════════ STOP ═══════════════════ STOP ═════════

$$K+1$$

---

51 $\chi^2 = \dfrac{12}{?(K+1)} \Sigma(\Sigma R)^2 - 3N(K+1)$

═════════ STOP ═══════════════════ STOP ═════════

$$NK$$

---

52 $\chi^2 = ?\Sigma(\Sigma R)^2 - 3N(K+1)$

═════════ STOP ═══════════════════ STOP ═════════

$$\dfrac{12}{NK(K+1)}$$

---

53 $\chi^2 = \dfrac{12}{NK(K+1)} \Sigma(\Sigma R)^2 - ?$

═════════ STOP ═══════════════════ STOP ═════════

$$3N(K+1)$$

---

54 $\chi^2 = ?\Sigma(\Sigma R)^2 - ?$

══════ STOP ══════════════════════ STOP ══════

$$\frac{12}{NK(K+1)}, \; 3N(K+1)$$

---

55 $\chi^2 = ?$

══════ STOP ══════════════════════ STOP ══════

$$\frac{12}{NK(K+1)} \Sigma(\Sigma R)^2 - 3N(K+1)$$

---

56 $\chi^2 = \dfrac{12}{NK(K+1)} \Sigma(\Sigma R)^2 - 3N(K+1). \; N = 4, K = 3, \; \dfrac{12}{NK(K+1)} = ?$

══════ STOP ══════════════════════ STOP ══════

$$\frac{12}{12 \times 4} = .25$$

---

57 $\chi^2 = \dfrac{12}{NK(K+1)} \Sigma(\Sigma R)^2 - 3N(K+1). \; N = 4, K = 3, \; 3N(K+1) =$

_____.

══════ STOP ══════════════════════ STOP ══════

$$12 \times 4 = 48$$

---

58 $\Sigma R_1 = 5, \; \Sigma R_2 = 8, \; \Sigma R_3 = 11. \; \Sigma(\Sigma R)^2 = ?$

══════ STOP ══════════════════════ STOP ══════

$$25 + 64 + 121 = 210$$

---

59 $\chi^2 = \dfrac{12}{NK(N+1)} \Sigma(\Sigma R)^2 - 3N(K+1)$

$\dfrac{12}{NK(N+1)} = 0.25, \; \Sigma(\Sigma R)^2 = 210, \; 3N(K+1) = 48. \; \chi^2 = ?$

══════ STOP ══════════════════════ STOP ══════

$$.25 \times 210 - 48 = 52.5 - 48 = 4.5$$

---

60 $\chi^2 = 4.5$. For the Friedman test, the number of degrees of freedom equal $K - 1$. df = ?

═══════ STOP ═══════════════════════ STOP ═══════

2

───

61 $\chi^2 = 4.5$, df = 2. Is this χ^2 significant?

═══════ STOP ═══════════════════════ STOP ═══════

No

───

62   Two pictures are rated as follows:

	P_1	P_2
S_1	1	2
S_2	2	1
S_3	1	2
S_4	1	2
S_5	1	2

$K = ?$ _____
$N = ?$ _____

═══════ STOP ═══════════════════════ STOP ═══════

$K = 2; N = 5$

───

63   From frame 62, $\dfrac{12}{NK(K + 1)} = ?$ $\Sigma(\Sigma R)^2 = ?$ $3N(K + 1) = ?$

═══════ STOP ═══════════════════════ STOP ═══════

.40, 117, 45

───

64   From frame 62, $\chi^2 = ?$

═══════ STOP ═══════════════════════ STOP ═══════

$.40 \times 117 - 45 = 1.8$

───

65   The probability of an event is defined as the ratio of the number of favorable events to the total number of _____.

═══════ STOP ═══════════════════════ STOP ═══════

events

───

66 On the other hand, the *conditional* probability of an event is defined as the probability of one event, given the occurrence of another event. $Pr(A/B)$ is used to indicate the conditional probability of A, given _____.

═══════ STOP ═══════════════════════ STOP ═══════

B

---

67 The conditional probability of R, given S, is .50. $Pr(R/S) = $ _____.

═══════ STOP ═══════════════════════ STOP ═══════

.50

---

68 The conditional probability of X, given Y, is symbolized as _____.

═══════ STOP ═══════════════════════ STOP ═══════

$Pr(X/Y)$

---

69 A | 1 | 2 | 3 | 4 | 5 | 6 | 7 |
        | | | | | 8 | 9 | 10 |
        B

From the diagram at the left there are _____ A's and _____ B's and _____ possible events.

═══════ STOP ═══════════════════════ STOP ═══════

7, 6, 10

---

70 From frame 69, there are 7 A's and 6 B's. If a number were drawn at random, $Pr(A) = $ _____.

═══════ STOP ═══════════════════════ STOP ═══════

$7/10 = 0.70$

---

71 From frame 69, with 7 A's in 10 possible events, $Pr(A)$, $= .70$. $Pr(B) = $ _____.

═══════ STOP ═══════════════════════ STOP ═══════

$6/10 = .60$

72   From frame 69, the probability of a randomly selected number being *both* an A and a B is _____.

═══════ STOP ═══════════════════ STOP ═══════

$$Pr(AB) = .30$$

---

73 $A \{$ | 1 | 2 | 3 | 4 | 5 | 6 | 7 |
                          | 8 | 9 | 10 |
                 B

Suppose an A had been selected at random. How many A's are also B? _____

═══════ STOP ═══════════════════ STOP ═══════

3

---

74   Of the 7 A's, 3 are also B; therefore, if an A had been selected at random, the probability that a B had also been selected is _____.

═══════ STOP ═══════════════════ STOP ═══════

$$\tfrac{3}{7} = .43$$

---

75   The conditional probability of B, given A—that is, $Pr(B/A)$—was found to equal .43. Find $Pr(A/B)$. _____

═══════ STOP ═══════════════════ STOP ═══════

$$\tfrac{3}{6} = .50$$

---

76   The formula for conditional probability is given as
$$Pr(A/B) = Pr(AB)/Pr(B)$$
From frame 69, $Pr(B)$ = ?

═══════ STOP ═══════════════════ STOP ═══════

$$\tfrac{6}{10} = .60$$

---

77   And the probability that both A and B would occur on the same draw is $Pr(AB)$ = _____.

═══════ STOP ═══════════════════ STOP ═══════

$$\tfrac{3}{10} = .30$$

---

78 $Pr(AB) = .30$, $P(B) = .60$, $Pr(A/B) =$ _____.

═══════ STOP ═══════════════════════════ STOP ═══════

$$.30/.60 = \frac{1}{2} = .50$$

---

79   That is, the probability of A's occurring, given B, = _____.

═══════ STOP ═══════════════════════════ STOP ═══════

$$.50$$

---

80 $Pr(A/B) = \dfrac{Pr(AB)}{?}$.

═══════ STOP ═══════════════════════════ STOP ═══════

$$Pr(B)$$

---

81 $Pr(A/B) = \dfrac{?}{Pr(B)}$.

═══════ STOP ═══════════════════════════ STOP ═══════

$$Pr(AB)$$

---

82 $? = \dfrac{Pr(AB)}{Pr(B)}$.

═══════ STOP ═══════════════════════════ STOP ═══════

$$Pr(A/B)$$

---

83   Similarly, $Pr(B/A) = \dfrac{?}{Pr(A)}$.

═══════ STOP ═══════════════════════════ STOP ═══════

$$Pr(AB)$$

---

84   And $Pr(B/A) = \dfrac{?}{?}$

════════ STOP ═══════════════════════════ STOP ════════

$$Pr(AB)/Pr(A)$$

---

85 A {

1	2	3	4	5	6	7
				8	9	10

    B

Again from frame 69, $Pr(AB) =$ _____.

════════ STOP ═══════════════════════════ STOP ════════

.30

---

86 $Pr(A) =$ _____ and $Pr(B/A) =$ _____.

════════ STOP ═══════════════════════════ STOP ════════

.70; .43

---

87   Note what happens to conditional probabilities when events are independent of one another. The probability of rolling a 6 in a single roll of a die is $Pr(6) =$ _____.

════════ STOP ═══════════════════════════ STOP ════════

⅙

---

88   Now, given that a 6 has been rolled on the first roll, what is the probability that a 6 will be rolled on the second roll? That is, $Pr(6/6) =$ _____.

════════ STOP ═══════════════════════════ STOP ════════

⅙

---

89   If $Pr(A/B) = Pr(A)$, then A and B are defined as independent of one another. $Pr(H/HHH) = ?$ _____

════════ STOP ═══════════════════════════ STOP ════════

½

---

90   Similarly, if $\Pr(A/B) \neq \Pr(A)$, then events A and B are *not* _____ of one another.

════════ STOP ════════════════════ STOP ════════

independent

---

91 $\Pr(A/B) = \Pr(AB)/\Pr(B)$. Multiplying both sides of the equation by $\Pr(B)$ and canceling, we find that $\Pr(AB) =$ _____.

════════ STOP ════════════════════ STOP ════════

$\Pr(A/B)\,\Pr(B)$

---

92 $\Pr(B/A) = \Pr(AB)/\Pr(A)$. In this case, $\Pr(AB) =$ _____.

════════ STOP ════════════════════ STOP ════════

$\Pr(B/A)\,\Pr(A)$

---

93 $\Pr(AB) = \Pr(B/A)\,\Pr(A)$
$\Pr(AB) = \Pr(A/B)\,\Pr(B)$
Since things equal to the same thing are equal to each other, then $\Pr(B/A)\,\Pr(A) =$ _____.

════════ STOP ════════════════════ STOP ════════

$\Pr(A/B)\,\Pr(B)$

---

94 $\Pr(B/A)\,\Pr(A) = \Pr(A/B)\,\Pr(B)$. Dividing both sides of the equation by $\Pr(B)$, we find that $\Pr(A/B) = ?/\Pr(B)$.

════════ STOP ════════════════════ STOP ════════

$\Pr(B/A)\,\Pr(A)$

---

95   Suppose $\Pr(B) = .60$, $\Pr(A) = .70$, and $\Pr(B/A) = .43$.
$$\Pr(A/B) = \frac{\Pr(B/A)\,\Pr(A)}{\Pr(B)}.$$
Find $\Pr(A/B)$. _____

════════ STOP ════════════════════ STOP ════════

$$\frac{.43 \times .70}{.60} = 0.50$$

---

96 Compare this with frame 69. Similarly, if $\Pr(B/A) = 0.50$, $\Pr(A) = 0.30$, and $\Pr(B) = 0.75$, the $\Pr(A/B) =$ _____.

===== STOP ===== STOP =====

$$\frac{.50 \times .30}{.75} = .20$$

---

97 This new formula is called Bayes' theorem: $\Pr(A/B) = \dfrac{\Pr(B/A)\,\Pr(A)}{?}$.

===== STOP ===== STOP =====

$$\Pr(B)$$

---

98 $\Pr(A/B) = \dfrac{? \times \Pr(A)}{\Pr(B)}$.

===== STOP ===== STOP =====

$$\Pr(B/A)$$

---

99 $\Pr(A/B) = \dfrac{\Pr(B/A) \times ?}{?}$.

===== STOP ===== STOP =====

$$\frac{\Pr(A)}{\Pr(B)}$$

---

100 Bayes' theorem says $\Pr(A/B) = ?$

===== STOP ===== STOP =====

$$\frac{\Pr(B/A)\,\Pr(A)}{\Pr(B)}$$

101

	M	W	
P			60
F			40
	70	30	

Of a group of 100 persons taking a physical test, 70 are men and 30 are women. Only 60 people passed (that is, were said to be physically fit); 50 of the men passed. $\Pr(M) =$ _____; $\Pr(P) =$ _____.

===== STOP ===== STOP =====

$$\Pr(M) = .70$$
$$\Pr(P) = .60$$

---

102 Of the 70 males, 50 passed. Hence, $\Pr(P/M) = $ _____.

════════ STOP ══════════════════════════ STOP ════════

$$0.72$$

---

103 $\Pr(M/P) = \dfrac{\Pr(P/M)\,\Pr(M)}{\Pr(P)}$, where $\Pr(M) = .60$, $\Pr(P) = .70$, and $\Pr(P/M) = .72$. Hence, the probability that someone who is physically fit is a male = _____.

════════ STOP ══════════════════════════ STOP ════════

$$\frac{.72 \times .70}{.60} = 0.83$$

---

104 Notice that $\Pr(P/M) \neq \Pr(M/P)$. The former is the probability that a male is physically fit, while the latter is the probability that a person who is physically fit is a _____.

════════ STOP ══════════════════════════ STOP ════════

male

---

105 Now suppose that you are in a crowded theater and someone shouts fire. Suppose also that last year there were 70 fires and 30 false alarms. What is the probability that there is a fire?

════════ STOP ══════════════════════════ STOP ════════

$$70/_{100} = .70$$

---

106 This is where Bayesian statistics enters in. Bayesian statistics assumes that you have some subjective expectation of the probability of fire at the theater. This is called a *prior* probability. Estimate the probability of a fire when you go to a movie.

════════ STOP ══════════════════════════ STOP ════════

you can choose any value between 0.00 and 1.00

107   In this case, your prior probability is _____.

===== STOP ================================ STOP =====

same as in frame 106.

---

108   But your prior probability, when combined with data, changes and becomes a *posterior* probability. The theorem or rule used to change your
probability is called _____ rule.

===== STOP ================================ STOP =====

Bayes'

---

109   Bayes' rule allows you to combine your subjective prior probability
(opinion) with data to yield a _____ probability or opinion.

===== STOP ================================ STOP =====

posterior

---

110   To illustrate how Bayes' theorem is used to measure opinion change, we
can consider the following example. Suppose we are set down, lost, on a
college campus that looks like a well-known predominantly female
College X. Here, women comprise 80% of the campus population; that
is, $\Pr(W) =$ _____.

===== STOP ================================ STOP =====

.80

---

111   Alternately, it also looks like another campus, College Y, where men
comprise 60% of the population; that is, $\Pr(M) =$ _____.

===== STOP ================================ STOP =====

.60

---

112   Now you wish to guess which campus you are on, but without much
information you can't be sure. It looks a little more like the women's
college. Hence you decide, *prior* to any data, that $\Pr(X) = .60$ and
$\Pr(Y) =$ _____.

===== STOP ================================ STOP =====

.40

113

	M	F
X		
Y		

200

Continuing from above, if 200 students were taken in proportion to our prior opinion from each campus, there would be 120 taken from College X and _____ taken from College Y.

═══════ STOP ═══════════════════════════════ STOP ═══════

80

───

114

	M	F	
X			120
Y			80

200

Continuing from above, if 80% of the students at College X are women, then enter the number of men and women in the group of 120 taken from College X.

═══════ STOP ═══════════════════════════════ STOP ═══════

24, 96

───

115

	M	F	
X	24	96	120
Y			80

200

And enter the appropriate numbers if 60% of the students at College Y are men.

═══════ STOP ═══════════════════════════════ STOP ═══════

48, 32

───

116

	M	F	
X	24	96	120
Y	48	32	80

200

Although the diagrams at the left are not necessary, they will help you in understanding the following frames. But remember, the estimate that the probability that you are on Campus X is .60 represents your _____ opinion.

═══════ STOP ═══════════════════════════════ STOP ═══════

prior

───

117   Now suppose a student is seen on campus—a man. What campus is it? Bayes' theorem can be expressed in a slightly different but equivalent manner as before:

$$\Pr(X/M) = \frac{\Pr(M/X)\,\Pr(X)}{\Pr(M/X)\,\Pr(X) + \Pr(M/Y)\,\Pr(Y)}$$

where $\Pr(X/M)$ means the probability of being on Campus X after a _____ is seen on campus.

══════ STOP ══════════════════════ STOP ══════

man

---

118   On Campus X, with 20% of the students being men, the $\Pr(M/X) =$ _____. Similarly, $\Pr(M/Y) =$ _____.

══════ STOP ══════════════════════ STOP ══════

.20, .60

---

119   While your prior opinions are $\Pr(X) =$ _____ and $\Pr(Y) =$ _____.

══════ STOP ══════════════════════ STOP ══════

.60, .40

---

120   Substituting in the formula (frame 117) for $\Pr(X/M)$, the numerator = _____.

══════ STOP ══════════════════════ STOP ══════

.20 × .60 = .12

---

121   And then $\Pr(X/M) = \dfrac{0.12}{0.12 + ?}$

══════ STOP ══════════════════════ STOP ══════

.24

---

122 $\Pr(X/M) = 0.12/0.36 =$ _____.

══════ STOP ══════════════════════ STOP ══════

.33

123 Similarly, $\Pr(Y/M) = \dfrac{\Pr(M/Y)\,\Pr(Y)}{\Pr(M/Y)\,\Pr(Y) + \Pr(M/X)\,\Pr(X)} = \underline{\qquad}$.

══════ STOP ══════════════════════ STOP ══════

$$.24/.36 = .67$$

---

124 $\Pr(Y/M) = .67$ and $P(X/M) = .33$. Your prior opinion that you were on Campus X was .60. Now your *posterior* opinion that you are on Campus X is _____.

══════ STOP ══════════════════════ STOP ══════

$$\Pr(X/M) = .33$$

---

125 Your posterior opinion becomes a prior opinion if a second "experiment" is conducted. That is, if a woman student is next seen, your opinion as to which campus you are on would again change. In this case $\Pr(X)$ = _____ and $\Pr(Y)$ = _____.

══════ STOP ══════════════════════ STOP ══════

$$.33, .67$$

---

126 And the probabilities of seeing a woman student on the two campuses would be $\Pr(W/X)$ = _____ and $\Pr(W/Y)$ = _____.

══════ STOP ══════════════════════ STOP ══════

$$.80, .40$$

---

127 Solving for $\Pr(X/W)$, the numerator of the fraction = _____.

══════ STOP ══════════════════════ STOP ══════

$$.80 \times .33 = .26$$

---

128 $\Pr(X/W) = \dfrac{.26}{.26 + ?}$

══════ STOP ══════════════════════ STOP ══════

$$.40 \times .67 = .27$$

---

129 $\Pr(X/W) = \dfrac{.26}{.26 + .27} = \underline{\hspace{2cm}}.$

$\equiv\!\!\equiv\!\!\equiv$ STOP $\equiv\!\!\equiv\!\!\equiv\!\!\equiv\!\!\equiv\!\!\equiv\!\!\equiv\!\!\equiv$ STOP $\equiv\!\!\equiv\!\!\equiv$

.49

---

130 Notice that you started with the prior opinion that $\Pr(X) = .60$. That is, the probability that you were on Campus X was .60. However, the posterior probability that you are on Campus X is now _____.

$\equiv\!\!\equiv\!\!\equiv$ STOP $\equiv\!\!\equiv\!\!\equiv\!\!\equiv\!\!\equiv\!\!\equiv\!\!\equiv\!\!\equiv$ STOP $\equiv\!\!\equiv\!\!\equiv$

.49

---

131 After seeing a man and a woman student, your posterior probability that you are on Campus X has decreased from .60 to .49. Notice that it would not make much difference what your prior opinion was. That is, after seeing the two students, the posterior probability that you were on Campus X (with 80% women) would always _____.

$\equiv\!\!\equiv\!\!\equiv$ STOP $\equiv\!\!\equiv\!\!\equiv\!\!\equiv\!\!\equiv\!\!\equiv\!\!\equiv\!\!\equiv$ STOP $\equiv\!\!\equiv\!\!\equiv$

decrease

---

132 Suppose your prior opinion that you were on Campus X was .40 (instead of .60) and $\Pr(Y) = .60$. For Campus X, what is the probability of seeing a man and then a woman? _____

$\equiv\!\!\equiv\!\!\equiv$ STOP $\equiv\!\!\equiv\!\!\equiv\!\!\equiv\!\!\equiv\!\!\equiv\!\!\equiv\!\!\equiv$ STOP $\equiv\!\!\equiv\!\!\equiv$

$.2 \times .8 = .16$

---

133 Let $D = $ Data. Then $\Pr(D/X) = .16$. Similarly, $\Pr(D/Y) = \underline{\hspace{2cm}}.$

$\equiv\!\!\equiv\!\!\equiv$ STOP $\equiv\!\!\equiv\!\!\equiv\!\!\equiv\!\!\equiv\!\!\equiv\!\!\equiv\!\!\equiv$ STOP $\equiv\!\!\equiv\!\!\equiv$

$.4 \times .6 = .24$

---

134 $\Pr(X/D) = \dfrac{\Pr(D/X)\,\Pr(X)}{\Pr(D/X)\,\Pr(X) + \Pr(D/Y)\,\Pr(Y)}$ and the numerator $=$ _____.

$\equiv\!\!\equiv\!\!\equiv$ STOP $\equiv\!\!\equiv\!\!\equiv\!\!\equiv\!\!\equiv\!\!\equiv\!\!\equiv\!\!\equiv$ STOP $\equiv\!\!\equiv\!\!\equiv$

$.16 \times .40 = .064$ (remember! $\Pr(X) = .40$ now.)

135 $Pr(X/D) = \dfrac{.064}{.064 + ?}$ (with $Pr(Y) = .60$).

══════ STOP ══════════════════════════ STOP ══════

$$.24 \times .60 = .144$$

---

136 $Pr(X/D) = $ _____.

══════ STOP ══════════════════════════ STOP ══════

.31

---

137 Again, $Pr(X/D) < Pr(X)$; that is, $.31 < .40$. And again your posterior opinion that this is Campus X is _____ than your prior opinion.

══════ STOP ══════════════════════════ STOP ══════

less

Answers to the Problems

Chapter Two

1. 5

   The smallest score is 32 and the largest is 90. The difference between the two is 58. If 58 is divided by 12 (the number of intervals desired), a value of 4.83 is obtained. This value is then rounded to 5 to give an interval which is both convenient to use and easily understood.

2. 30–34

   The smallest score is 32 and the interval should contain this number. Usually it is most convenient to have the lower limit divisible by the interval size.

3. 60–79

   Again the lower limit is divisible by the interval size.

4. 49.50–59.50

   The apparent limits are simply whole numbers which do not give the exact or real limits of an interval.

5. 20.00–24.99

   Usually the real limits are simply the midpoint between two whole numbers, although in special cases, such as age, a person is, say, 20 even though he will have his twenty-first birthday the next day. That is, he is 20 from the time he is 20.00 until he is 20.999. . . .

6. The type of fertilizer would be the independent variable, since it is under the control of the experimenter, while the dependent variable would be something like the amount of corn grown per acre or the average height of corn.

7. Abscissa, ordinate. This is simply a convention and there does not seem to be any reason for this other than possibly clarity of presentation.

8.

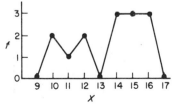

   Notice that it is convenient to construct a frequency distribution before constructing your frequency polygon. Notice also that both 9 and 17 are represented as having a zero frequency in the graph. This is done to indicate there are no scores above 16 or below 10.

9.

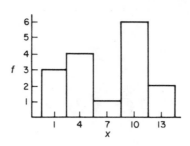

It can be seen that lines are drawn upward from the upper and lower real limits to a height which represents the frequency of each interval.

10.

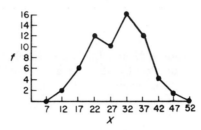

Notice that the lines of the polygon are connected to the points representing the frequency of each interval and that these points are raised over the midpoints of the intervals.

11.

X	f
13	1
12	3
11	2
10	2
9	1

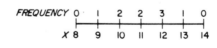

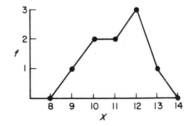

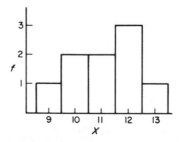

12.

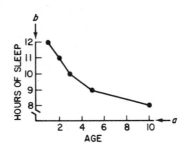

13. From Answer 12:   *a*—abscissa, *x* axis
                         b—ordinate, *y* axis
                         Age—independent variable
                         Hours of sleep—dependent variable

Chapter Three

1. $\displaystyle\sum_{1}^{N} X = 2 + 4 + 8 + 5 + 3 = 22$

2. (a) $\displaystyle\sum_{1}^{4} X = 12 + 0 + 1 + 3 = 16$

   (b) $\displaystyle\sum_{2}^{3} X = 0 + 1 = 1$

   (c) $\displaystyle\sum_{1}^{N} X = 12 + 0 + 1 + 3 + 18 = 34$

3. $\displaystyle\sum_{1}^{N} X^2 = 0^2 = 1^2 + 3^2 + 2^2 = 14$

4. (a) $\displaystyle\sum_{1}^{3} X^2 = 4^2 + 5^2 + 8^2 = 16 + 25 + 64 = 105$

   (b) $\displaystyle\sum_{1}^{N} X^2 = 4^2 + 5^2 + 8^2 + 4^2 + 1^2 + 6^2$
   $$= 16 + 25 + 64 + 16 + 1 + 36 = 158$$

5. (a) $\displaystyle\sum_{1}^{N} X = 1 + 3 + 5 + 2 = 11$

   (b) $\displaystyle\sum_{1}^{N} X^2 = 1^2 + 3^2 + 5^2 + 2^2 = 1 + 9 + 25 + 4 = 39$

   (c) $\displaystyle\sum_{1}^{N} X^2 - \sum_{1}^{N} X = 39 - 11 = 28$

6. (a) $\displaystyle\sum_{1}^{N} X^2 = 2^2 + 0^2 + 4^2 + 6^2 + 1^2 = 4 + 0 + 16 + 36 + 1 = 57$

   (b) $\displaystyle\left(\sum_{1}^{N} X\right)^2 = (2 + 0 + 4 + 6 + 1)^2 = 13^2 = 169$

   (c) $\displaystyle\sum_{1}^{N} X^2 - \left(\sum_{1}^{N} X\right)^2 = 57 - 169 = -112$

7. (a) $\displaystyle\sum_{1}^{N} X^2 = 0^2 + 1^2 + 2^2 + 3^2 + 5^2 + 7^2 = 0 + 1 + 4 + 9 + 25 + 49 = 88$

(b) $\displaystyle\frac{\left(\sum_{1}^{N} X\right)^2}{N} = \frac{(0 + 1 + 2 + 3 + 5 + 7)^2}{6} = \frac{18^2}{6} = 54$

(c) $\displaystyle\sum_{1}^{N} X^2 - \frac{\left(\sum_{1}^{N} X\right)^2}{N} = 88 - 54 = 34$

8. (a) $\displaystyle\sum_{1}^{N} X^2 = 1^2 + 8^2 + 6^2 + 1^2 + 7^2 + 7^2$

$\qquad\qquad = 1 + 64 + 36 + 1 + 49 + 49 = 200$

(b) $\displaystyle\frac{\left(\sum_{1}^{N} X\right)^2}{N} = \frac{(1 + 8 + 6 + 1 + 7 + 7)^2}{6} = \frac{30^2}{6} = 150$

(c) $\displaystyle\sum_{1}^{N} X^2 - \frac{\left(\sum_{1}^{N} X\right)^2}{N} = 200 - 150 = 50$

9. The mean is defined as $\dfrac{\displaystyle\sum_{1}^{N} X}{N}$. In this problem $\displaystyle\sum_{1}^{N} X = 49, N = 7$. The mean $=$ $\dfrac{49}{7} = 7.00$.

10. The variance is defined as $\dfrac{\displaystyle\sum_{1}^{N} X^2 - \dfrac{\left(\sum_{1}^{N} X\right)^2}{N}}{N - 1}$. In this problem $\displaystyle\sum_{1}^{N} X^2 = 65$,

$\displaystyle\sum_{1}^{N} X = 15, N = 5$. The variance $= \dfrac{65 - \dfrac{15 \times 15}{5}}{4} = \dfrac{65 - 45}{4} = \dfrac{20}{4} = 5.00$.

11. 5.00

12. 4.00

13. 4.40

14. 8.50

Chapter Four

1. $\Sigma X = 30, N = 6, M = \dfrac{\Sigma X}{N} = \dfrac{30}{6} = 5.00$

2. $\Sigma X = 5,617,120, N = 5, M = 1,123,424$

A much easier way to compute the mean in this problem would have been to subtract a constant from each of the cases, compute the mean of the remaining scores, and then add the constant back in. For example, if 1,123,420 had been subtracted from each case, this would have given scores of 5, 0, 6, 3, and 6.

Here $\Sigma X = 20$, $N = 5$, and $M = 4$, adding the constant back in $M = 4 + 1,123$, $420 = 1,123,424$.

3. $\Sigma X = 8$, $N = 4$, $M = 2$, $C = 50$, $C = 2 + 50 = 52$

The only difficult part about this question is that subtracting the constant yields a negative case at times. That is, subtracting 50 from each case gives scores of -4, $+6$, $+8$, -2. It is apparent that $\Sigma X = 8$.

4. (a) 5

The median is defined as the point below which 50 percent of the cases fall. If N is uneven, the middle case is the median.

(b) 4.50

In this case 50 percent of the cases (2 and 4) fall below 4.50, while 50 percent (7 and 5) fall above.

(c) 6

Fifty percent of the cases (2 and 4) fall below 4.5 and since there are no cases between 4.5 and 7.5 (the lower real limit of the 8 interval) 50 percent of the cases also fall below 7.5. Thus any of the points between 4.5 and 7.5 satisfy the definition of the median. In cases such as this the midpoint of the (4.5 to 7.5) interval is arbitrarily called the median. Hence, median = 6.00.

5. (a) 4.00, $\dfrac{N}{2} = \dfrac{16}{2} = 8$

The median is the point below which eight cases fall. Seven cases $(1 + 4 + 2)$ fall below the upper limit (i.e., 3.5) of the 3 interval. Hence, with six cases in the 3.5–4.5 interval the median must fall in that interval. The midpoint of the 3.5–4.5 interval is 4.00. Hence, the median is 4.00.

(b) Median $= 6.00$, $\dfrac{24}{2} = 12$

The median is the point below which twelve cases fall. Since eleven cases fall below interval 6, one additional case is needed from the interval. With four cases in the interval, the median must fall in that interval. The midpoint of the interval is 6.00. Hence, the median is 6.00.

6.

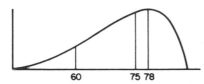

The distribution is a negatively skewed distribution. The effect of this is to pull the mean toward the lower end of the distribution.

7. $\Sigma X = 1176$, $N = 7$, $M = 168$. The median, the middle case, is 175.

8. With such a highly skewed distribution the median would be most appropriate. With $N = 50$, $N/2 = 25$, and the median falls in the 4.50–9.50 interval. The midpoint of that interval is 7.00, and hence the median is 7.00.

9. $\Sigma X = 88$, $N = 8$, $M = 11.00$. The Hutterites probably have the highest birth rate of any identifiable group in the United States or Canada.

	Mean	Median
10. (a)	3.00	2.00
(b)	3.75	3.50
(c)	6.50	6.50
(d)	3.40	3.00
(e)	305.00	306.50
(f)	4.00	4.50
(g)	2.57	3.00
(h)	3.17	4.00

Chapter Five

1. Using the raw-score formula, $\Sigma X^2 = 0^2 + 1^2 + 7^2 + 4^2 + 3^2 + 3^2 = 84$ and $\Sigma X = 18$.

$$S^2 = \frac{\Sigma X^2 - \frac{(\Sigma X)^2}{N}}{N-1} = \frac{84 - \frac{18 \times 18}{6}}{5} = \frac{84 - 54}{5} = \frac{30}{5} = 6$$

Using the deviation-score formula $M = \dfrac{\Sigma X}{N} = \dfrac{18}{6} = 3$

$X - M =$	x	x^2
$0 - 3 = -3$		9
$1 - 3 = -2$		4
$7 - 3 = 4$		16
$4 - 3 = 1$		1
$3 - 3 = 0$		0
$3 - 3 = 0$		0

$$S^2 = \frac{\Sigma x^2}{N-1} = \frac{30}{5} = 6$$

2. $\Sigma X = 54$

$$S^2 = \frac{388 - \frac{54 \times 54}{9}}{8} = \frac{388 - 324}{8} = \frac{64}{8} = 8$$

$\Sigma X^2 = 388$

3. The value of the variance remains the same. Hence, addition (or subtraction) of a constant will not change the value of the variance.

4. $\Sigma X = 12$

$$S^2 = \frac{42 - \frac{12 \times 12}{4}}{3} = \frac{42 - 36}{3} = \frac{6}{3} = 2$$

$\Sigma X^2 = 42$

5. $\Sigma X = 36$

$$S^2 = \frac{378 - \frac{36 \times 36}{4}}{3} = \frac{378 - 324}{3} = \frac{54}{3} = 18$$

$\Sigma X^2 = 378$

The variance is increased by 9 (the square of the constant). Hence multiplication (or division) by a constant will change the value of the variance.

6. $\Sigma X = 1176$, $\Sigma X^2 = 205{,}456$.

$$S^2 = \frac{205{,}456 - \dfrac{1176 \times 1176}{7}}{6} = \frac{205{,}456 - 197{,}568}{6} = \frac{7888}{6} = 1314.67$$

Since the mean equals 168, you could have subtracted 168 from each score and squared the differences. The sum, $\Sigma x^2 = 7888$ and again the variance would equal 1314.67.

7. $\Sigma X = 60$, $\Sigma X^2 = 400$, $N = 12$

$$S^2 = \frac{400 - \dfrac{60 \times 60}{12}}{11} = \frac{400 - 300}{11} = \frac{100}{11} = 9.09$$

8. $SIQR = \dfrac{Q_3 - Q_1}{2} = \dfrac{5.00 - 3.00}{2} = \dfrac{2.00}{2} = 1.00$

For Q_1, $\dfrac{N}{4} = \dfrac{16}{4} = 4$. The point below which four cases fall defines Q_1. Three cases fall below 2.50 and one of three cases is needed from the 2.50–3.50 interval. Hence $Q_1 = 3.00$. Q_3 is defined as the point below which 12 cases fall, 9 cases fall below 4.50 and 3 of the 4 cases are needed from the next interval $Q_3 = 5.00$.

9.

X	f	fX	fX²
7	1	7	49
6	4	24	144
5	7	35	175
4	8	32	128
3	4	12	36
2	4	8	16
1	2	2	2

$\Sigma f = N = 30$
$\Sigma f X = \Sigma X = 120$
$\Sigma f X^2 = \Sigma X^2 = 550$

$$S^2 = \frac{\Sigma X^2 - \dfrac{(\Sigma X)^2}{N}}{N-1} = \frac{550 - \dfrac{120 \times 120}{30}}{29} = \frac{550 - 480}{29} = \frac{70}{29} = 2.41$$

10. Arranging the scores in sequence: 18, 19, 22, 27, 29, 32, 33, 35, 36, 39, 40, 41, 42, 45, 47, 52. With $N = 16$, then Q_1 is the point below which $16 \times 0.25 = 4.00$ scores fall. Similarly, Q_3 is the point below which 12 scores fall. $Q_1 = 28$, $Q_3 = 41.5$, and $SIQR = (Q_3 - Q_1)/2 = 13.5/2 = 6.75$.

11. Since the range is *roughly* five times the value of the standard deviation, the standard deviation should have been approximately 20 and the variance should have been about 400. Instead it was found to be equal to 4.00. While under special conditions the range and variance could equal 100 and 4.00, the probable explanation for the results is that you made a mistake in computing the variance.

12. By the raw-score method: $\Sigma X^2 = 145$, $\Sigma X = 27$, $N = 9$.

$$S^2 = \frac{145 - \dfrac{27 \times 27}{9}}{8} = \frac{145 - 81}{8} = \frac{64}{8} = 8$$

$S = \sqrt{8} = 2.83$

By the deviation-score method:

$X - M = x$

$$
\begin{array}{rcrcr}
0 & - & 3 & = & -3 \\
8 & - & 3 & = & 5 \\
1 & - & 3 & = & -2 \\
7 & - & 3 & = & 4 \\
4 & - & 3 & = & 1 \\
1 & - & 3 & = & -2 \\
3 & - & 3 & = & 0 \\
2 & - & 3 & = & -1 \\
1 & - & 3 & = & -2 \\
\end{array}
$$

$$S^2 = \frac{\Sigma x^2}{N-1} = \frac{64}{8} = 8$$

$$S = \sqrt{8} = 2.83$$

13. Variances:
   (a) 8.67
   (b) 9.20
   (c) 2.50
   (d) 3.33
   (e) 6.67

14. $SIQR$; S^2
   (a) $SIQR = 2.25$, $S^2 = 7.27$
   (b) $SIQR = 0.50$, $S^2 = 1.79$

15. $S^2 = 8.67$

Chapter Six

1. Computation of z, Z, and stanine:
   (a) $z = +0.5$, $Z = 55$, stanine $= 6.0$
   (b) $z = +1.0$, $Z = 60$, stanine $= 7.0$
   (c) $z = -1.5$, $Z = 35$, stanine $= 2.0$
   (d) $z = -2.0$, $Z = 30$, stanine $= 1.0$
   (e) $z = +1.0$, $Z = 60$, stanine $= 7.0$

2. Completion:
   (a) 6
   (b) 2
   (c) 5, 2, 50
   (d) $(X - M)/S$
   (e) interval
   (f) ordinal

3. Answers:
   (a) Refer to Figure 6–1: (1) both equal, and most common; (2) Z much less common; (3) percentile 84 is less common.
   (b) No.
       Raising values to a power constitutes a "non-linear" transformation. Linear transformation is restricted to the operations of addition and multiplication.

(c) Percentiles are easiest to communicate to non-statisticians. Some tests also yield "grade-equivalent" scores, based on national norms. These are non-linear transformed scores.

(d) If the distribution was not symmetrical around the mean (more cases on one side than the other), the median (percentile = 50) will differ from the mean ($Z = 50$).

Chapter Seven

1. Pr (3H) = 1/8
   The probability of flipping a head is 1/2. The multiplicative law is used since this is a problem of the *A and then B* type and $1/2 \times 1/2 \times 1/2 = 1/8$. The probability of three heads in a row is the product of the probabilities of the separate events.

2. Pr (2H) = 3/8
   This problem uses both the multiplicative and the additive laws. The multiplicative law is used to show the probability of any three-coin sequence of events is 1/8. For example, the probability of HTH is $1/2 \times 1/2 \times 1/2 = 1/8$. In addition the additive law is used because this is a problem of the *A or B* type. That is, sequences of HHT, HTH, or THH all satisfy the requirement of two heads. The sum of their separate probabilities is equal to $1/8 + 1/8 + 1/8 = 3/8$.

3. Pr (red then white) = $0.4 \times 0.3 = 0.12$
   The probability is equal to the product of the separate probabilities.

4. Pr (one or more 6s) = $1 - 5/6 \times 5/6 = 11/36$
   Since only zero, one, or two 6s can occur in two rolls of a die, the probability of one or two must equal 1 minus the probability of no 6s. The probability of not rolling a 6 is 5/6 and the probability of not rolling a 6 on either roll is the product of the separate probabilities: $5/6 \times 5/6 = 25/36$. Since this is the probability of no 6s, the probability of one or more must be $1 - 25/36 = 11/36$.

5. Pr (pair) = $52/52 \times 3/51 = 3/51$
   On the first draw the player gets any card, but on the second draw the player must get a card with the same value as the first. There are only three of these cards left in the deck since the fourth was drawn on the first draw and there are only 51 cards left. Hence the probability of drawing a favorable card is 3/51.

6. Pr (three of a kind) = $52/52 \times 3/51 \times 2/50$
   The principle of the last question can be expanded to cover the answer here.

7. Pr (blonde and blue eyed) = $0.20 \times 0.30 = 0.06$
   The probability is the product of the separate probabilities. The assumption, most likely incorrect, which you had to make was that the probability of blonde hair and blue eyes are independent of one another.

8. Pr (6) = $2 \times 1/6 \times 5/6 = 10/36$
   There are two ways to get a single 6 in two rolls of a die: 6 on the first roll and not a 6 on the second or not a 6 on the first and a 6 on the second. As in Problem 2 above, the sum of the separate probabilities yields the answer. The probability of a 6 and then no 6 is $1/6 \times 5/6 = 5/36$. The probability of no 6 and then a 6 is $5/6 \times 1/6 = 5/36$. Hence $5/36 + 5/36 = 10/36$.

9. Pr (red and a club) = $2 \times 1/4 \times 1/2 = 1/4$
   There are two ways to get a red card and a club: a red on the first draw and then a club on the second, or vice versa. The probability of a red and then a club is

$1/2 \times 1/4 = 1/8$. The probability of a club and then a red is $1/4 \times 1/2 = 1/8$. The sum of the two is $1/8 + 1/8 = 1/4$.

10. Using the triangle we get scores of 1, 10, 45, 120, 210, 252, 210, 120, 45, 10, and 1. From left to right this represents the number of ways we can get 0 heads (i.e., 10 tails), 1 head, 2 heads, etc. Thus there are 120 ways to get 7 heads, 45 ways to get 8 heads, 10 ways to get 9 heads, and 1 way to get 10 heads. This sums to 176. Notice that the sum of all the ways to flip the coins is $2^{10} = 1024$. Hence, the probability of flipping 7 or more heads in 10 flips is $176/1024 = 0.17$.

11. (a) $1/6 \times 1/6 = 1/36$
    (b) $5/6 \times 5/6 = 25/36$
    (c) $36/36$
    (d) $1/6 \times 1/6 = 1/36$
    (e) The probability of rolling a 5 or 6 on the first roll is $2/6$ and the probability of rolling the needed number on the second is $1/6$. Hence, $2/6 \times 1/6 = 2/36$
    (f) $6/6 \times 1/6 = 1/6$
    (g) $1/13 \times 1/13 = 1/169$
    (h) $1/4 \times 12/52 = 12/208$
    (i) $1/4 \times 1/4 = 1/16$

Chapter Eight

1. $z = \dfrac{X - \mu}{\sigma} = \dfrac{65 - 50}{10} = \dfrac{15}{10} = 1.50$
2. $X = \mu + \sigma \times z = 120 + 20 \times (-1.50) = 120 - 30 = 90$
3. 44.52 percent

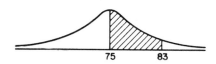

$z = \dfrac{83 - 75}{5} = \dfrac{8}{5} = 1.60$. A score of 83 is 1.60 standard deviations above the mean. From Appendix B, 44.52 percent of the area falls between the mean and $z = 1.60$. Hence, 44.52 percent of the area falls between the mean and 83.

4. 6.68 percent

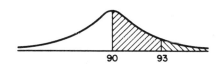

$z = \dfrac{93 - 90}{2} = \dfrac{3}{2} = 1.50$. From Appendix B, Column D indicates that 6.68 percent of the area falls above $z = 1.50$, or 93.

5. 70.11 percent

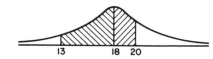

There are two parts to the problem in that the areas above the mean and below the mean must be found. These two areas can then be added to find the area between 13 and 20.

For the area above the mean, $z = \dfrac{20 - 18}{3} = 0.67$ and $z = 0.67$ corresponds

to 24.86 percent in Appendix B. For the area below the mean, $z = \dfrac{13 - 18}{3} =$

-1.67 and $z = 1.67$ corresponds to 45.25 percent in Appendix B. Thus, 24.86 percent of the area falls between the mean and 20; 45.25 percent of the area falls between the mean and 13. To find the area between 13 and 20, the sum of the two areas is computed. Thus, $24.86 + 45.25 = 70.11$ percent of the area falls between 13 and 20.

6. 17.60 percent

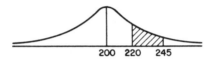

200 220   245

In this case we find the area between the mean and 245 and subtract from this the area between the mean and 220. For 245, $z = \dfrac{245 - 200}{25} = \dfrac{45}{25} = \dfrac{9}{5} = 1.80$,

and for 220, $z = \dfrac{220 - 200}{25} = \dfrac{20}{25} = 0.80$. From Appendix B, a z of 0.80 includes

28.81 percent of the area between the mean, and $z = 0.80$. Subtracting the smaller area from the larger $(46.41 - 28.81)$, the area between 220 and 245 is found to be 17.60 percent. (You could also use Column D to find the areas, but use of the mean as a reference point is recommended.)

7. 86

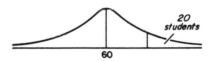

20
students

60

$20/200 = 10$ percent. Ten percent of the class made an "A". Since there are no grades higher than an A then this must be the top 10 percent and reference to Column D of Appendix B shows that 10 percent corresponds to a z of 1.28. That is, all students scoring at or above $z = 1.28$ get an A. The raw score corresponding to $z = 1.28$ is $X = \mu + \sigma \times z = 60 + 20 \times 1.28 = 60 + 25.60 = 85.60$. Rounding, we find that a student had to get at least a score of 86 to get an A on the test.

8. $z = \dfrac{100 - 85}{15} = 1.00$

85   100

From Column D, this z score corresponds to 15.87 percent. That is, only 15.87 percent of the prison population had IQ scores equal to or greater than the mean IQ score in the general population.

9. $z = \dfrac{75 - 85}{15} = 0.67$

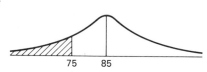

The entry in column D at $z = 0.67$ is 25.14 percent. That is, 25.14 percent of the prison population had IQ scores at or below 75. Nearly one-fourth of the prison population could be considered to be mentally impaired.

10. 69.15 percent

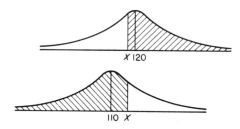

This problem is really two in one. The first is to find the point (X) above which 60 percent of the Drizzly Tech students fall. The second is, given that point (X), to find what percentage of the Flotsam students fall below it. From Appendix B the point above which 60 percent of the area falls is the same as the point which contains 10 percent of the area between the same point and the mean. Hence, $z = -0.25$ and $X = 120 + 10 \times (-0.25) = 117.50$. Thus, 60 percent of the Drizzly students fall above 117.50. Looking at the distribution of Flotsam students, $z = \dfrac{117.50 - 110}{15} = \dfrac{7.50}{15} = 0.50$. A z of 0.50 corresponds to 19.15 percent. Hence, $19.15 + 50.00 = 69.15$ percent of the Flotsam students fall below 60 percent of the Drizzly students.

11. 15.87
12. 69.15
13. $15.54 + 38.49 = 54.03$
14. $110.80 - 189.20$
15. 371–629
16. 206.7
17. 43.84 or 44
18. 53.44 or 53

Chapter Nine

1. The standard deviation of the means is equal to $\sqrt{12} = 3.46$.
   The standard deviation of the means in this problem is computed as an ordinary standard deviation. That is, the sum of the means is equal to 48, the sum of the squared means is 612 and the number of means is 4. Hence:

$$\sqrt{\dfrac{612 - \dfrac{48 \times 48}{4}}{3}} = \sqrt{\dfrac{36}{3}} = \sqrt{12} = 3.46$$

2. $\sigma_M = \dfrac{\sigma}{\sqrt{N}}$

Since $\sigma = 4$, then $\sigma_M = \dfrac{4}{\sqrt{16}} = \dfrac{4}{4} = 1.00$

3. About two-thirds of the area under the normal curve falls between $+1\sigma$ and -1σ of the mean. Hence, $40 \pm 5 = 35 - 45$.

35   45

$\mu = 40$

4. About two-thirds of the means fall between $\pm 1\sigma_M$ of the mean. Hence two-thirds fall between 39–41. σ_M is the appropriate measure of variability to use because σ_M indicates the variability of a group of means. ($N = 25$ from the previous question.)

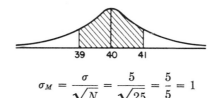

39   40   41

$$\sigma_M = \frac{\sigma}{\sqrt{N}} = \frac{5}{\sqrt{25}} = \frac{5}{5} = 1$$

5. For $N = 4$, $\sigma_M = \dfrac{\sigma}{\sqrt{N}} = \dfrac{24}{\sqrt{4}} = 12$, hence $\mu \pm 12 = 38 - 62$.

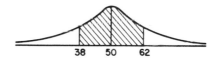

38   50   62

$$N = 9, \sigma_M = \frac{24}{3} = 8, \mu \pm 8 = 42 - 58$$

$$N = 16, \sigma_M = \frac{24}{4} = 6, \mu \pm 6 = 44 - 56$$

$$N = 36, \sigma_M = \frac{24}{6} = 4, \mu \pm 4 = 46 - 54$$

$$N = 64, \sigma_M = \frac{24}{8} = 3, \mu \pm 3 = 47 - 53$$

6. $198 - 202$; $180 - 220$

The z score corresponding to 25 percent is 0.67. Hence, $\mu \pm z = 0.67$ includes 50 percent of the area under the normal curve (and the probability of selecting a score at random from here would be 0.50). The scores corresponding to $\pm z = 0.67$ are found by $X = \mu \pm \sigma_M \times z$. Thus,

$X = 200 + 3 \times 0.67 = 200 + 2.00 = 202.00$
$X = 200 - 3 \times 0.67 = 200 - 2.00 = 198.00$

Notice that σ_M was used in the equation to find the points within which 50 percent of the means fall. The points within which 50 percent of the scores fall are found by substituting σ in the equation. Hence, $X = \mu \pm z \times \sigma$. $X = 200 - 20 = 180$ and $X = 200 + 20 = 220$. $\mu = 200$, $\sigma = 30$, $N = 100$, $\sigma_M = \dfrac{\sigma}{\sqrt{N}} = \dfrac{30}{10} = 3.00$.

7. $45.60 - 104.40$; $69.12 - 80.88$

$\mu = 75$, $\sigma = 15$, $\mu \pm 1.96 \times \sigma$ includes about 95 percent of the *scores*. $\mu \pm 1.96 \times 15 = 75 \pm 29.40$ or $45.60 - 104.40$. Similarly the mean would fall between $\mu \pm 1.96 \times \sigma_M$ (where $\sigma_M = 15/\sqrt{25} = 3$) with a probability of 0.95. $\mu \pm 1.96 \times 3 = 75 \pm 5.88$ or $69.12 - 80.88$.

8. $M = 30$, $N = 25$, $\sigma = 4.00$, $\sigma_M = 4.00/5 = 0.80$. We wish to test the hypothesis that the true mean under the new program is the same as it was under the old program (that is, the null hypothesis). Hence,

$$z = \frac{M - \mu}{\sigma_M} = \frac{30 - 32}{0.80} = \frac{2}{0.80} = 2.50$$

Since the probability of obtaining a z this large or larger is less than 0.05, we reject the hypothesis that $\mu = 32$. This indicates that the true mean is less than 32. We conclude that the mean age of parolees under the new program is less than it was under the old program.

9. $M = 7$, $N = 9$, $\sigma = 6$, $\sigma_M = 6/3 = 2.00$. We want to determine if the last frost date is getting later in the year. To do this, we test the hypothesis that the true mean of the population from which the sample was taken is (March) 3. Hence, $z = (7 - 3)/2 = 2.00$. The probability of obtaining this z by chance is less than 0.05, and we conclude that the true mean must be greater (that is, later) than (March) 3.

10. The hypothesis Captain Snarf is testing is that the obtained sample mean ($M = 29$) comes from a population which has a true mean (μ) of 25. Since $\sigma = 5$ and $N = 100$: $\sigma_M = 5/10 = 0.50$, and $z = (M - \mu)/\sigma_M = (29 - 25)/0.50 = 8.00$. Since 8.00 is larger than 1.96 (or 2.58) the probability of getting a z score this large by chance, from a population with $\mu = 25$, is very small. Hence, the hypothesis that basic training does not help is rejected, and it is concluded that exercise helps.

11. (a) $\sigma_M = 5.00$

    (b) $\sigma_M = 2.00$

    (c) (1) $\sigma_M = 24.00$

        (2) $\sigma_M = 16.00$

        (3) $\sigma_M = 12.00$

        (4) $\sigma_M = 6.00$

    (d) (1) 72–88

        (2) 79–81

    (e) (1) 86.08–93.92

        (2) 84.84–95.16

Chapter Ten

1. When the population standard deviation (σ) is known, the appropriate statistical test to use is z. The reason for this is that $z = (M - \mu)/\sigma_M$, where σ_M is based on the standard deviation of the population.

2. If the population standard deviation is not known and the standard deviation of a sample is known, the t test is the appropriate test to use. $t = (M - \mu / S_M$. Notice that the only difference between the z and the t formulas is in the σ_M or S_M in the denominator.

3. $S_M = \dfrac{S}{\sqrt{N}} = \dfrac{10}{\sqrt{25}} = \dfrac{10}{5} = 2.00$

$S_M = 2.00$, df $= 24$, and 2.06. The number of degrees of freedom is equal to $N - 1$; that is, df $= N - 1 = 25 - 1 = 24$, and the tabled value of t at 24 df $= 2.06$ for the 0.05 level. This means a t of 2.06 or larger is needed to reject the hypothesis that $\mu = 50$.

4. $N = 16$, $M = 40$, $S = 12$; $S_M = \dfrac{S}{\sqrt{N}} = \dfrac{12}{\sqrt{16}} = \dfrac{12}{4} = 3$; $t = \dfrac{M - \mu}{S_M} =$

$\dfrac{40 - 46}{3} = 2.00$

E is testing the hypothesis that $\mu = 46$. The tabled value of t for $N - 1 = $ df $= 15$ is 2.13. The obtained t is smaller than the tabled value of t and the hypothesis that $\mu = 46$ cannot be rejected. Failure to reject this hypothesis merely means that μ could be 46 and that such a hypothesis is tenable. Be sure to remember that failure to reject does not prove $\mu = 46$; μ could just as easily be 45, 44, and so on, as 46 for all we know.

5. $N = 100$, $M = 15$, $S = 3$; $S_M = S/\sqrt{N} = 3/\sqrt{100} = 3/10 = 0.30$
From Appendix C, $t = 1.98$ for 100 (99) df.

$$\mu = M \pm t \times S_M$$
$$= 15 \pm 1.98 \times 0.30$$
$$= 15 \pm 0.594$$
$$= 14.41 - 15.59$$

The 95 percent confidence interval for the true mean extends 0.594 above and below the sample mean.

6. No. The statistician warned him not to buy the machine because while it probably was faster, it may not have been fast enough. Using $t = (M - \mu)/S_M$ where $M = 330$, $\mu = 360$, and $S_M = S/\sqrt{N} = 30/\sqrt{9} = 30/3 = 10$, $t = (330 - 360)/10 = 3.00$. With 8 df a t of 2.31 is required to reject at the 95 percent level the hypothesis that the true mean of the new pickle packer is 360. But the statistician used the 99 percent level, and in this case a t of 3.36 is required to reject the same hypothesis. Because he could not reject at the 99 percent level, it can be concluded that the new machine may be no faster than the old machine.

7. If the advertising campaign has done any good, the number of accidents involving fatalities has been reduced. This means that the true mean would be less than 6.00, which is the usual number of accidents. To see if this is the case, a t test is used to test the hypothesis that $\mu = 6$. The scores obtained were 0, 6, 2, 4, 3. $\Sigma X = 15$, $\Sigma X^2 = 65$, $M = \Sigma X/N = 3.00$.

$$S^2 = \dfrac{\Sigma X^2 - \dfrac{(\Sigma X)^2}{N}}{N - 1} = \dfrac{65 - \dfrac{15 \times 15}{5}}{4} = \dfrac{65 - 45}{4} = \dfrac{20}{4} = 5;$$

$$S_M^2 = \dfrac{S^2}{N} = \dfrac{5}{5} = 1; S_M = 1; t = \dfrac{M - \mu}{S_M} = \dfrac{3 - 6}{1} = 3$$

With 4 df the tabled value of t is 2.78 and the hypothesis that $\mu = 6$ can be rejected. The advertising campaign has reduced the number of fatal accidents.

8. $M = 50$, $N = 4$, $S = 8$, $S_M = 8/2 = 4.00$. We wish to test the hypothesis that the true mean has not been changed by the advertising campaign. That is, we wish to test $H: \mu = 40$. $t = (50 - 40)/4 = 2.50$. With 3 df a t of 3.18 is required at the 0.05 level. Hence, we cannot reject the hypothesis that $\mu = 40$, and so we cannot conclude that advertising increased circulation. We may have violated the assumption of independence, since the same people may have taken books out from one week to the next.

9. $M = 41.00$, $N = 25$, $S = 2$, $S_M = 0.40$. If this sample is the Bigeye Shiner species, then it cannot have more than 40 rows of scales. Hence, we wish to test the hypothesis that $\mu = 40$. $t = (41 - 40)/0.40 = 1/0.40 = 2.50$. With 24 df a t of 2.06 is required to be significant at the 0.05 level. This means that 41 is too far from 40 for 40 to be the true mean. Thus, the fish the zoologist has taken is not the Bigeye Shiner. (Notice that while the last two questions have found $t = 2.50$, the conclusions have been quite different.)

10. $M = 950$, $S = 100$, $N = 16$, $S_M = 100/\sqrt{16} = 25$. The manufacturer claims that $\mu = 1000$. Since $t = (M - \mu)/S_M$, then $t = (950 - 1000)/25 = 50/25 = 2.00$. The obtained $t = 2.00$ and the tabled $t = 2.13$ for 15 df. Since the obtained t is smaller than the tabled t, we cannot reject the manufacturer's claims. Remember, if μ *is* 1000, we would get a t of 2.13 or larger 5 percent of the time by chance alone (and we would get a t of 2.00 or larger even more often). We reject the hypothesis of chance variation only when the probability of such an occurrence is 0.05 or less.

11. $19.75 - 44.25$

12. $S_M = 2.00$; 60.88 to 69.12

13. 106.14 to 133.86

14. 59.40 to 70.60

15. $t = 3.00$. Do not reject the hypothesis that $\mu = 100$.

16. $t = 2.14$. Reject the hypothesis that $\mu = 135$.

Chapter Eleven

1. If $\chi^2 = \dfrac{(N - 1)S^2}{\sigma^2}$, then $\dfrac{\chi^2}{df} = \dfrac{(N - 1)S^2}{(N - 1)\sigma^2}$, which when simplified is equal to S^2/σ^2. This variable has a mean of 1.00.

2. $F = \dfrac{S_1^2/\sigma_1^2}{S_2^2/\sigma_2^2}$

   From the first question, $\chi^2/df = S^2/\sigma^2$; then the ratio of two χ^2 distributions each divided by their respective degrees of freedom yields an F distribution.

3. $F = S_1^2/S_2^2$ is a tabled F distribution only if $\sigma_1^2 = \sigma_2^2$. That is, $\dfrac{S_1^2/\sigma_1^2}{S_2^2/\sigma_2^2}$ is a tabled F distribution and if $\sigma_1^2 = \sigma_2^2$, then these two terms can be canceled out to yield S_1^2/S_2^2.

4. $F = S_1^2/S_2^2$ is not a tabled F distribution if $\sigma_1^2 \neq \sigma_2^2$. That is, the ratio of any two variances will yield an F distribution, but none, except when $\sigma_1^2 = \sigma_2^2$, will be tabled F distributions.

5. $S_1^2 = 20$, $\sigma_1^2 = 40$, $S_2^2 = 10$, $\sigma_2^2 = 5$

$\sigma_1^2/\sigma_2^2 = 40/5 = 8.00$

The average F ratio will be equal to 8.00 because the average S_1^2 will be 40 and the average S_2^2 will be 5. Hence $F = 40/5 = 8.00$ and the ratio of the two true variances will give the average F value.

6. Since $\sigma_x^2 = \sigma_y^2$, the median (average) is equal to 1.00 and 50 percent of the sample F values in the F distribution will fall above 1.00. Hence, $\Pr(F > 1.00) = 0.50$. Similarly, with 14 and 14 df, from Appendix D, the tabled value for the 0.05 level is 2.48. That is, with this F distribution, 5 percent of the sample F values will be greater than 2.48. Hence, $\Pr(F > 2.48) = 0.05$.

7. The probability of rejecting the null hypothesis is 0.50. With $\sigma_x^2 = 3.18$ and $\sigma_y^2 = 1.00$, the average $F = \sigma_x^2/\sigma_y^2 = 3.18/1.00 = 3.18$. The tabled value of F for 9 and 9 df is 3.18. With the tabled F distribution only 5 percent of the area would be larger than 3.18, and the null hypothesis would be rejected only 5 percent of the time. This problem deals, however, with an F distribution which has an average of 3.18. Hence, 50 percent of the area of this F distribution is larger than 3.18, and the null hypothesis would be rejected 50 percent of the time.

8. No. E does not conclude that $\sigma_x^2 > _y^2$.

$$S_x^2 = \frac{\Sigma X^2 - \dfrac{(\Sigma X)^2}{N}}{N - 1} = \frac{198 - \dfrac{30 \times 30}{10}}{9} = \frac{198 - 90}{9} = \frac{108}{9} = 12$$

$$S_y^2 = \frac{\Sigma Y^2 - \dfrac{(\Sigma Y)^2}{N}}{N - 1} = \frac{116 - \dfrac{24 \times 24}{6}}{5} = \frac{116 - 96}{5} = \frac{20}{5} = 4$$

$$F = \frac{S_x^2}{S_y^2} = \frac{12}{4} = 3.00$$

With 9 and 5 df, the tabled value of F is 4.78. However, the obtained $F = 3.00$. Hence the null hypothesis is not rejected and E has no reason to assume that the two true variances are different from one another.

9. Yes, E concludes that $\sigma_x^2 > \sigma_y^2$.

$$S_x^2 = \frac{124 - \dfrac{20 \times 20}{4}}{3} = \frac{124 - 100}{3} = \frac{24}{3} = 8$$

$$S_y^2 = \frac{49 - \dfrac{15 \times 15}{5}}{4} = \frac{49 - 45}{4} = \frac{4}{4} = 1$$

$$F = \frac{S_x^2}{S_y^2} = \frac{8}{1} = 8.00$$

With 3 and 4 df, the tabled value of F is 6.59. Since the obtained F is larger than the tabled value of F, the null hypothesis is rejected and it is concluded that σ_x^2 is larger than σ_y^2.

10. $F = \dfrac{S_x^2}{S_y^2} = \dfrac{9.00}{4.00} = 2.25$, $\Pr < 0.05$, and Bubbly's claims appear justified.

With 24 and 24 df, the tabled value of $F = 1.98$. Since the obtained value of F is larger than the tabled value, the null hypothesis is rejected. The null hypothesis in this case is that there is no difference in the two variances. This is apparent from the question that asks if Bubbly Cola's claims are justified. Bubbly Cola claimed that it was as good as Bouncy Cola (a statement about the two means) and that its quality was more consistent. That is, Bubbly Cola claimed that the variance of the ratings of Bouncy was greater than those of Bubbly. Since there was a significant difference between the two variances, apparently Bubbly's claims about being more consistent in quality are justified.

11. (a) $F = 3.33$. Reject.
    (b) $F = 2.67$. Do not reject.
    (c) $F = 3.00$. Reject.

12. We could test the hypothesis that the variances of the nurses and interns were the same. Without a prior hypothesis we would place the larger variance in the numerator and $F = S_x^2/S_y^2 = 16.00/4.00 = 4.00$. With 9 and 19 df, an F of 3.52 is required for significance at the 0.01 level, and the hypothesis that $\sigma_x^2 = \sigma_y^2$ is rejected. Since we placed the larger variance in the numerator, we could not get an F value less than 1.00. Hence, the significance level must be doubled, and we then reject the null hypothesis at the 0.02 level.

13. $F = 25.00/16.00 = 1.56$ with 199 and 49 df. Although the F for this combination of 199 and 49 df is not tabled, the tabled F values for Fs having similar df make it clear that the null hypothesis is rejected at the 0.05 level, and the sentences given by the older judge are found to be more variable.

Chapter Twelve

1. $N = 2$. There are two scores per cell.
     $AN = 2 \times 2 = 4$
     $BAN = 3 \times 2 \times 2 = 12$. With two levels of A and three levels of B there are a total of 12 scores.

2. $\displaystyle\sum_1^A \sum_1^B \sum_1^N X = 0 + 1 + 3 + \cdots + 4 + 2 + 0 + 0 = 20$

3. $\displaystyle\sum_1^A \sum_1^B \sum_1^N X^2 = 0^2 + 1^2 + 3^2 + \cdots + 4^2 + 2^2 + 0^2 + 0^2 = 52$

4. $\displaystyle\sum_1^A \sum_1^B \left(\sum_1^N X\right)^2 = (0 + 1)^2 + (3 + 1)^2 + \cdots + (4 + 2)^2 + (0 + 0)^2$
     $$= 1^2 + 4^2 + \cdots + 6^2 + 0^2 = 94$$

5. $\displaystyle\sum_1^A \left(\sum_1^B \sum_1^N X\right)^2 = [(0 + 1) + (3 + 1) + (2 + 3)]^2$
     $$+ [(2 + 2) + (4 + 2) + (0 + 0)]^2$$
     $$= 10^2 + 10^2 = 200$$

6. $\displaystyle\sum_1^B \left(\sum_1^A \sum_1^N X\right)^2 = [(0 + 1) + (2 + 2)]^2 + [(3 + 1) + (4 + 2)]^2$
     $$+ [(2 + 3) + (0 + 0)]^2$$
     $$= 5^2 + 10^2 + 5^2 = 25 + 100 + 25 = 150$$

7. $\displaystyle\left(\sum_1^B \sum_1^A \sum_1^N X\right)^2 = (0 + 1 + 3 + \cdots + 4 + 2 + 0 + 0)^2 = 20^2 = 400$

8. $\sum\limits_{1}^{R} \sum\limits_{1}^{C} \left(\sum\limits_{1}^{N} X\right)^2 = (0 + 1 + 1)^2 + (2 + 0 + 0)^2 + \cdots + (0 + 0 + 0)^2$

$$+ (2 + 1 + 2)^2$$
$$= 2^2 + 2^2 + 1^2 + 5^2 + 0^2 + 5^2$$
$$= 4 + 4 + 1 + 25 + 0 + 25 = 59$$

$\sum\limits_{1}^{R} \left(\sum\limits_{1}^{C} \sum\limits_{1}^{N} X\right)^2 = [(0 + 1 + 1) + (3 + 1 + 1)]^2 + \cdots$

$$+ [(1 + 0 + 0) + (2 + 1 + 2)]^2$$
$$= 7^2 + 2^2 + 6^2$$
$$= 49 + 4 + 36 = 89$$

9. $\sum\limits_{1}^{P} \sum\limits_{1}^{Q} \sum\limits_{1}^{N} X^2 = 1^2 + 2^2 + 3^2 + \cdots + 1^2 + 1^2 + 0^2$

$$= 1 + 4 + 9 + \cdots + 1 + 1 + 0$$
$$= 30$$

$\sum\limits_{1}^{P} \left(\sum\limits_{1}^{Q} \sum\limits_{1}^{N} X\right)^2 = [(1 + 2 + 3) + (2 + 0 + 1) + (1 + 2 + 1)]^2$

$$+ [(0 + 0 + 0) + (1 + 1 + 1) + (1 + 1 + 0)]^2$$
$$= 13^2 + 5^2 = 169 + 25 = 194$$

10. $\sum\limits_{1}^{R} \sum\limits_{1}^{C} \sum\limits_{1}^{N} X^2 = 0^2 + 0^2 + \cdots + 2^2 + 1^2$

$$= 0 + 0 + \cdots + 4 + 1$$
$$= 35$$

$\sum\limits_{1}^{R} \sum\limits_{1}^{C} \left(\sum\limits_{1}^{N} X\right)^2 = (0 + 0 + 1 + 1)^2 + \cdots + (1 + 2 + 2 + 1)^2$

$$= 2^2 + 4^2 + 5^2 + 6^2$$
$$= 4 + 16 + 25 + 36 = 81$$

$\sum\limits_{1}^{R} \left(\sum\limits_{1}^{C} \sum\limits_{1}^{N} X\right)^2 = 6^2 + 11^2$

$$= 36 + 121 = 157$$

$\sum\limits_{1}^{C} \left(\sum\limits_{1}^{R} \sum\limits_{1}^{N} X\right)^2 = 7^2 + 10^2$

$$= 49 + 100 = 149$$

Chapter Thirteen

1.

Source	df
bg	3
wg	36
tot	39

With four groups of ten scores each there are a total of 40 scores. The number of df associated with the tot_{ss} is always one less than the total number of scores and *tot* df = 39. There are four groups, and hence *bg* df = 3, since the *bg* df = $K - 1$. The *wg* df are obtained by subtraction.

2.

Source	ss	df	ms	F
bg	24	3	8.00	4.00
wg	16	8	2.00	
tot	40	11		

The tabled value of F with 3 and 8 df is 4.07. The null hypothesis is that there are no differences between the true means, and since the tabled value of F is greater than the obtained value, the null hypothesis cannot be rejected.

3. $CF = \dfrac{\left(\sum\limits_1^K \sum\limits_1^N X\right)^2}{KN} = \dfrac{24 \times 24}{3 \times 4} = 48.00$

$bg_{ss} = \dfrac{\sum\limits_1^K \left(\sum\limits_1^N X\right)^2}{N} - CF = \dfrac{6^2 + 6^2 + 12^2}{4} - CF = 54 - 48 = 6.00$

The correction factor is the square of the sum of all the scores divided by the total number of scores. The bg_{ss} is the sum of the squared sums for each group divided by the number of scores in each group less the CF.

4.

Source	ss	df	ms
bg	450	5	90
wg	540	54	10
tot	990	59	

From the values of the mean squares, we know that $bg_{ms} = NS_M{}^2$. Since $N = 10$ per group, then $S_M{}^2 = 9$ and $S_M = 3.00$. In addition, $wg_{ms} = S^2$ and the average (mean) variance of the scores in each group $= 10.00$.

5. $\sum\limits_1^K \sum\limits_1^N X = 24$

$\sum\limits_1^K \sum\limits_1^N X^2 = 106$

$CF = \dfrac{24 \times 24}{4 \times 2} = 72$

$tot_{ss} = \sum\limits_1^K \sum\limits_1^N X^2 - CF = 106 - 72 = 34$

$bg_{ss} = \dfrac{\sum\limits_1^K \left(\sum\limits_1^N X\right)^2}{N} - CF = \dfrac{200}{2} - 72 = 28$

$wg_{ss} = tot_{ss} - bg_{ss} = 34 - 28 = 6.00$

Source	ss	df	ms	F
bg	28.00	3	9.33	6.22
wg	6.00	4	1.50	
tot	34.00	7		

With 3 and 4 df an F of 6.22 is not significant. The null hypothesis cannot be rejected and there is no reason to conclude that the means differ.

6. $\displaystyle\sum_{1}^{K}\sum_{1}^{N} X = 36$

$\displaystyle\sum_{1}^{K}\sum_{1}^{N} X^2 = 176$

$CF = \dfrac{36 \times 36}{12} = 108$

$tot_{ss} = \displaystyle\sum_{1}^{K}\sum_{1}^{N} X^2 - CF = 68$

$bg_{ss} = \displaystyle\sum_{1}^{K}\left(\sum_{1}^{N} X\right)^2 - CF = \dfrac{4^2 + 12^2 + 20^2}{4} - CF = \dfrac{560}{4} - 108$

$= 140 - 108 = 32$

$wg_{ss} = tot_{ss} - bg_{ss} = 68 - 32 = 36$

Source	ss	df	ms	F
bg	32	2	16	4.00
wg	36	9	4	
tot	68	11		

With 2 and 9 df the tabled value of F is 4.26. Hence the null hypothesis cannot be rejected and there is no reason to conclude that the means differ.

7. Obviously the analysis of variance cannot be computed in the usual manner. Instead the answer can be computed from knowledge of what wg_{ms} and bg_{ms} mean in the computing formula: $F = \dfrac{bg_{ms}}{wg_{ms}} = \dfrac{NS_M^2}{S^2}$. Since there were nine scores per cell, then $N = 9$. With means of 6, 10, 9, and 15, $S_M^2 = \dfrac{\Sigma M^2 - \dfrac{(\Sigma M)^2}{K}}{K - 1} =$

$\dfrac{442 - \dfrac{40 \times 40}{4}}{3} = \dfrac{442 - 400}{3} = 14$. And with variances of 25, 1, 49, and 9, the mean variance is equal to $\dfrac{84}{4} = 21$. Hence $F = \dfrac{NS_M^2}{S^2} = \dfrac{9 \times 14}{21} = 6$ and with 3 and 32 df the null hypothesis is rejected.

8. It looks as if there should be no difference because both groups have the same sum.

$\Sigma X = 48$

$\Sigma X^2 = 230$

$CF = \dfrac{48 \times 48}{12} = 192$

$tot_{ss} = 230 - 192 = 38$

$bg_{ss} = \dfrac{24^2}{4} + \dfrac{24^2}{8} - CF = 216 - 192 = 24$

$wg_{ss} = tot_{ss} - bg_{ss} = 38 - 24 = 14$

Source	ss	df	ms	F
bg	24	1	24	17.14
wg	14	10	1.40	
tot	38	11		

The means of the groups differ with the *Ex* group mean being significantly larger than the Control group mean. The only difficulty this problem has is that the *N*s of the two groups are not equal. To compute the bg_{ss} simply square each group sum and divide by the appropriate *N*. Then sum and subtract *CF*.

9.

Source	ss	df	ms
bg	48	3	16
wg	112	28	4
tot	160	31	

(a) There were four groups, as can be seen from the 3 df for *bg*.
(b) There were eight scores per group, as can be seen by dividing the total number of scores (32) by the number of groups (4).
(c) $S^2 = wg_{ms} = 4$
(d) $bg_{ms} = NS_M^2 = 16$. Since $N = 8$, then $S_M^2 = 2$.

10. $$S_M^2 = \frac{79.25 - \dfrac{12.5 \times 12.5}{2}}{1} = 79.25 - 78.125 = 1.125$$

$$S^2 = \frac{2^2 + 1.5^2}{2} = 3.125$$

$$F = \frac{NS_M^2}{S^2} = \frac{20 \times 1.125}{3.125} = \frac{22.50}{3.125} = 7.20$$

With 1 and 38 df, an *F* of 7.35 is needed for significance at the 0.01 level. Hence, the null hypothesis is rejected at the 0.05 level, and it can be concluded that respiration rates of the two groups differ.

11.

Source	ss	df	ms	F
bg	16	2	8	4.00
wg	24	12	2	
tot	40	14		

With 2 and 12 df, the tabled value of $F = 3.88$. The obtained *F* is larger than the tabled *F*, and we reject the null hypothes s and conclude the groups differ.

$$CF = \frac{\left(\sum_1^K \sum_1^N X \right)^2}{KN} = \frac{45 \times 45}{15} = 135$$

$$tot_{ss} = \sum_1^K \sum_1^N X^2 - CF = 175 - 135 = 40$$

$$bg_{ss} = \frac{\sum_1^K \left(\sum_1^N X \right)^2}{N} - CF = \frac{755}{5} - 135$$

$$= 151 - 135 = 16.$$ The rest of the analysis is completed in the table.

12. $F = 4.00$
13. $F = 2.77$ Do not reject
14. $F = 5.71.$ Reject.

15. $S_M^2 = 4$, $S^2 = \dfrac{1.33 + 2.00 + 3.33 + 5.33}{4} = \dfrac{12}{4} = 3.00$

Chapter Fourteen

Source	ss	df	ms	F
A	20	1	20.00	5.00
B	16	2	8.00	2.00
$A \times B$	16	2	8.00	2.00
wg	48	12	4.00	
tot	100	17		

1. The B_{ss} can be found by subtracting A, $A \times B$, and wg from tot. The reason this can be done is that the total sum of squares has been broken down or partitioned into these various terms. The df can be determined from the description of the experiment.

2. (a) There were a total of 36 scores in the analysis. It can be seen from the df column that the design was a 3×4 factorial with $N = 3$ per cell.
   (b) The mean variance of the scores in each group can be determined from the wg_{ms}. Since wg_{ss} equals 72 and wg df equals 24 then wg_{ms} equals 3.00 which also equals the mean variance for the cells.
   (c) The R_{ms} is equal to $24/2 = 12$. The R_{ms} is equal to CNS_M^2 which corresponds to NS_M^2 in a single-classification analysis of variance. In this example with $C = 4$ and $N = 3$, there are $CN = 12$ scores in each row and $S_M^2 = 1.00$. That is, $R_{ms} = CNS_M^2 = 12$, $S_M^2 = 1$ and $S_M = 1.00$.
   (d) $R_{ms} = 12$, $wg_{ms} = 3$, $F = 4.00$. With 2 and 24 df an F of 3.40 is significant at the 0.05 level. Hence the R_{ms} is significant and there is a difference between the means of the R variable.

3. $CF = 48.00$, $Q_{ss} = 12.00$.

$$CF = \frac{\left(\sum_1^P \sum_1^Q \sum_1^N X\right)^2}{PQN} = \frac{24 \times 24}{3 \times 2 \times 2} = 48.00$$

Remember that the CF is the square of the sum of all the scores divided by the total number of scores.

$$\frac{\sum_1^Q \left(\sum_1^P \sum_1^N X\right)^2}{PN} - CF = \frac{6^2 + 18^2}{6} - CF = \frac{36 + 324}{6} - 48 = 60 - 48 = 12.00$$

The Q_{ss} is found by getting the sum of all the Q_1 scores (6) and the sum of the Q_2 scores (18) squaring each, dividing by 6, and subtracting CF.

4. $$CF = \frac{\left(\sum_1^R \sum_1^C \sum_1^N X\right)^2}{RCN} = \frac{72 \times 72}{2 \times 3 \times 2} = 432$$

$$tot_{ss} = \sum_1^R \sum_1^C \sum_1^N X^2 - CF = 550 - 432 = 118$$

$$R_{ss} = \frac{\sum_1^R \left(\sum_1^C \sum_1^N X\right)^2}{CN} - CF = \frac{44^2 + 28^2}{3 \times 2} - CF = 453.33 - 432 = 21.33$$

$$C_{ss} = \frac{\sum_1^C \left(\sum_1^R \sum_1^N X\right)^2}{RN} - CF = \frac{24^2 + 24^2 + 24^2}{4} - CF = 0.00$$

$$R \times C_{ss} = \frac{\sum\limits_{1}^{R} \sum\limits_{1}^{C} \left(\sum\limits_{1}^{N} X \right)^2}{N} - CF - R_{ss} - C_{ss}$$

$$= \frac{8^2 + 16^2 + 20^2 + 16^2 + 8^2 + 4^2}{2} - CF - R_{ss} - C_{sa}$$

$$= 528 - 432 - 21.33 - 0.00 = 74.67$$

$$wg_{ss} = tot_{ss} - R_{ss} - C_{ss} - R \times C_{ss}$$
$$= 118 - 21.33 - 0 - 74.67 = 22.00$$

Source	ss	df	ms	F
R	21.33	1	21.33	5.81
C	.00	2	.00	
R × C	74.67	2	37.33	10.17*
wg	22.00	6	3.67	
tot	118.00	11		

      * Pr < 0.05

5. $\sum\limits_{1}^{B} \sum\limits_{1}^{A} \left(\sum\limits_{1}^{N} X \right)^2 = 8^2 + 0^2 + 8^2 + 16^2 = 384$

This is the sum of the squares of the cell sums.

$$A \times B_{ss} = \frac{\sum\limits_{1}^{B} \sum\limits_{1}^{A} \left(\sum\limits_{1}^{N} X \right)^2}{N} - CF - A_{ss} - B_{ss}$$

$$= \frac{384}{2} - 128 - 32 - 0 = 32.00$$

The cell sum of squares is simply the two terms on the left of the $A \times B_{ss}$ formula. This sum of squares can be partitioned into A, B, and $A \times B$. The $A \times B_{ss}$ is found by subtracting A and B from the cell sum of squares.

6.

	T_1	T_2	T_3	T_4	T_5		Source	df	F
S_1							T	4	. . .
S_2							S	9	
S_3							T × S	36	
S_4							tol	49	
.									
.									
.									
S_{10}									

Since the number of correct responses for each subject is recorded after each trial, there are a total of 50 scores, which are broken down into T (trial), S (subject), and $T \times S$ (trials by subject interaction). The $T \times S$ interaction mean square is used as the denominator in testing T. S is not usually tested.

7.

	T_1	T_2	T_3
S_1	0	1	6
S_2	2	4	7
S_3	0	2	2
S_4	2	5	5

$$\sum_1^S \sum_1^T X = 36, \ \sum_1^S \sum_1^T X^2 = 168, \ CF = \frac{36 \times 36}{12} = 108$$

$$tot = 60, \ S_{ss} = \frac{7^2 + 13^2 + 4^2 + 12^2}{3} - CF = 18,$$

$$T_{ss} = \frac{4^2 + 12^2 + 20^2}{4} - CF = 32$$

Source	ss	df	ms	F
T	32	2	16	9.60*
S	18	3		
T × S	10	6	1.667	
tot	60	11		

With 2 and 6 df the trials (*T*) differ from one another.
* Pr < 0.05

8.

Source	ss	df	ms	F
A	32	1	32	10.67*
B	32	1	32	10.67*
A × B	0	1	0	
wg	12	4	3	
tot	76	7		

* Pr < 0.05

9.

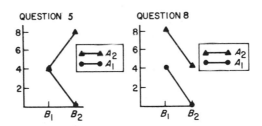

When the lines are parallel or close to parallel, the interaction is small and not significant. When the lines are far from parallel (as in Problem 5) the interaction is more likely to be significant.

10.

	Mother	Daughter	Source	df
Pair 1			Column (*C*)	1
2			(Mother *vs.*	
3			Daughter)	
4				
.			Row (*R*)	9
.			(Pair)	
.				
10			*R* × *C*	9

The *R* × *C* interaction mean square is the denominator of the *F* in the comparison of mothers and daughters.

11. | **Source** | *ms* | *F* |
|---|---|---|
| A | 1.75 | 0.75 |
| B | 48.00 | 20.60* |
| $A \times B$ | 3.25 | 1.39 |
| *wg* | 2.33 | |

* Pr < 0.01

12. | **Source** | *ms* | *F* |
|---|---|---|
| S | 7.50 | |
| T | 34.40 | 12.98* |
| $S \times T$ | 2.65 | |

* Pr < 0.01

Chapter Fifteen

1. $\chi^2 = 8.76$, Pr < 0.05. There are voter preferences.

	o	*e*	*(o − e)*	*(o − e)²*	*(o − e)²/e*
Jones	55	40	15	225	5.63
Smith	30	40	−10	100	2.50
Green	35	40	−5	25	0.63
	120	120	0		8.76

$$\chi^2 = \sum_{1}^{K} \frac{(o - e)^2}{e} = 8.76.$$ Notice that $\Sigma o = \Sigma e$ and that $\Sigma(o - e) = 0$.

2.

	Observed					**Expected**		
	Pre-mature	**Full term**				**Pre-mature**	**Full term**	
Smoke	16	84	100	Smoke		10.5	89.5	100
No Smoke	5	95	100	No Smoke		10.5	89.5	110
Total	21	179	200			21	179	200

$$\chi^2 = \sum \frac{(|o - e| = 0.5)^2}{e} = \frac{25}{10.5} + \frac{25}{89.5} + \frac{25}{10.5} + \frac{25}{89.5}$$
$$= 2.38 + 0.28 + 2.38 + 0.28 = 5.32$$

With 1 df, a chi square of 5.32 is significant at the 0.05 level. On the basis of these data it can be concluded that smoking and premature birth are associated. Note that with 1 df, the correction for continuity had to be used.

3. $\chi^2 = 6.12$, Pr < 0.05 with 1 df.

	o	*e*	*(|o − e| − 0.5)*	*(|o − e| − 0.5)²*	*(|o − e| − 0.5)²/e*
H	0	4	3.5	12.25	3.06
T	8	4	3.5	12.25	3.06
					6.12

In this case the assumption of no expected values of less than 5 has been violated. When this happens the chi-square formula $\Sigma(|o - e| - 0.5)^2/e$ does not closely approximate the χ^2 distribution, and χ^2 should not be used. Alternate statistical tests such as the expansion of the binomial should be used.

4. $\chi^2 = 4.99$, $\mathrm{Pr} < 0.05$ with 2 df.

		Observed				Expected	
AGE	COR-RECT	NOT COR-RECT	TOTAL	AGE	COR-RECT	NOT COR-RECT	TOTAL
6	6	14	20	6	8	12	20
12	12	8	20	12	8	12	20
18	6	14	20	18	8	12	20
	24	36	60		24	36	60

The expected values are found by multiplying by the appropriate row and its column sum and dividing by the total: for age 6, correct: $\dfrac{20 \times 24}{60} = 8.$

$$(o - e)^2$$

AGE	CORRECT	NOT CORRECT
6	4	4
12	16	16
18	4	4

$$\sum \frac{(o-e)^2}{e} = \frac{4}{8} + \frac{4}{12} + \frac{16}{8} + \frac{16}{12} + \frac{4}{8} + \frac{4}{12}$$

$$= 0.50 + 0.33 + 2.00 + 1.33 + 0.50 + 0.33$$

$$\chi^2 = 4.99$$

The null hypothesis cannot be rejected and E cannot conclude that transposition varies with age.

5. $\chi^2 = 7.53$, $\mathrm{Pr} < 0.05$ with 1 df.

	For	Against
over 30	20	40
under 30	36	24

Since the data are independent frequency data, χ^2 is the appropriate statistic.

$$\chi^2 = \frac{N\left(|BC - AD| - \dfrac{N}{2}\right)^2}{(A+B)(C+D)(A+C)(B+D)} = \frac{120(|1440 - 480| - 60)^2}{(60)(60)(56)(64)} = 7.53$$

Note that the problem can be solved by the method used in Problem 4.

	Observed			Expected	
	FOR	AGAINST		FOR	AGAINST
over 30	20	40	over 30	28	32
under 30	36	24	under 30	28	32

	$(|o - e| - 0.5)^2$			$(|o - e| - 0.5)^2/e$	
	56.25	56.25		2.01	1.76
	56.25	56.25		2.01	1.76

6. $\chi^2 = 3.16$, Pr < 0.05 with 2 df.

Number of heads	Frequency	
	OBSERVED	EXPECTED
0	10	12
1	30	24
2	8	12

$$\chi^2 = \sum \frac{(o - e)^2}{e} = \frac{(10 - 12)^2}{12} + \frac{(30 - 24)^2}{24} + \frac{(8 - 12)^2}{12}$$

$$= \frac{4}{12} + \frac{36}{24} + \frac{16}{12} = 3.16$$

The expected frequencies of each number of heads are obtained through expansion of the binomial. Since $\chi^2 = 3.16$, the observed frequencies are not significantly different from the expected. That is, the professor got what he expected.

7. $\chi^2 = 7.58$, Pr < 0.05 with 2 df.

Observed

	L	S	R	
M	10	10	20	40
F	5	10	45	60
	15	20	65	100

Expected

	L	S	R	
M	6	8	26	40
F	9	12	39	60
	15	20	65	100

Multiply appropriate row and column sums and divide by total to get each expected value.

$$\chi^2 = \frac{4^2}{6} + \frac{2^2}{8} + \frac{6^2}{26} + \frac{4^2}{9} + \frac{2^2}{12} + \frac{6^2}{39} = 7.58$$

With 2 df the null hypothesis is rejected, and it can be concluded that turning bias is related to sex.

8. $\chi^2 = 48.33$, Pr < 0.01 with 4 df.

	o	e
8	10	30
9	25	30
10	50	30
11	50	30
12	15	30

If time is unimportant, the students should distribute themselves evenly across the five sections.

$$\chi^2 = \frac{(10 - 30)^2 + (25 - 30)^2 + (50 - 30)^2 + (50 - 30)^2 + (15 - 30)^2}{30}$$

$$= \frac{400 + 25 + 400 + 400 + 225}{30} = \frac{1450}{30} = 48.33$$

This χ^2 is highly significant with 4 df, and apparently the time a class is held does make a difference in the number of students enrolled.

9. $\chi^2 = 4.83$ Pr < 0.05 with 1 df.

	o	e
F	1	4
S	8	10
J	7	4
S	4	2

The expected values are obtained by seeing what proportion of the students in the class are, say, freshmen, and then finding that proportion of the group that attended. That is, 20 percent of the class are freshmen; hence 4 (20 percent of 20) are expected. Since several expected values are less than 5, it probably would be better to collapse over grade levels.

	o	e
F and S	9	14
J and S	11	6

$$\chi^2 = \frac{(4.5)^2}{14} + \frac{(4.5)^2}{6}$$
$$= \frac{20.25}{14} + \frac{20.25}{6}$$
$$= 1.45 + 3.38 = 4.83$$

Class attendance is related to year in school.

10. $\chi^2 = 65.00$, Pr < 0.01 with 2 df.

Song	o	e
A	30	40
B	10	40
C	80	40

$$\chi^2 = \frac{10^2 + 30^2 + 40^2}{40} = \frac{2600}{40} = 65.00$$

With 2 df the χ^2 is highly significant and the songs do differ in how well they are liked. Notice that the question asked nothing about age or whether age and song preference were related.

11. $\chi^2 = 7.00$, Pr < 0.05 with 3 df.
12. $\chi^2 = 6.25$, Pr < 0.05 with 2 df.
13. $\chi^2 = 10.46$, Pr < 0.01 with 1 df.
14. $\chi^2 = 0.20$ with correction for continuity.
15. $\chi^2 = 52.83$, Pr < 0.01 with 5 df. The same person (or persons) could check more than one book out on one day or on different days. Hence, the data may not be independent.
16. $\chi^2 = 1.22$, Pr > 0.05 with 3 df. Combine the D and F categories.

Chapter Sixteen

1. $N = 5$
$\Sigma X = 30$
$\Sigma X^2 = 260$
$\Sigma Y = 20$
$\Sigma Y^2 = 100$
$\Sigma XY = 138$

$$r_{xy} = \frac{\dfrac{\Sigma XY}{N} - M_x M_y}{\sigma_x \sigma_y} = \frac{\dfrac{138}{5} - \dfrac{30}{5} \times \dfrac{20}{5}}{\sigma_x \sigma_y} = \frac{27.6 - 6 \times 4}{4 \times 2} = \frac{3.60}{8} = 0.45$$

$$\sigma_x = \sqrt{\frac{\Sigma X^2 - \dfrac{(\Sigma X)^2}{N}}{N}} = \sqrt{\frac{260 - \dfrac{30 \times 30}{5}}{5}} = \sqrt{\frac{260 - 180}{5}} = \sqrt{\frac{80}{5}}$$
$$= \sqrt{16} = 4$$

$$\sigma_y = \sqrt{\frac{\Sigma Y^2 - \dfrac{(\Sigma Y)^2}{N}}{N}} = \sqrt{\frac{100 - \dfrac{20 \times 20}{5}}{5}} = \sqrt{\frac{100 - 80}{5}} = \sqrt{\frac{20}{5}}$$
$$= \sqrt{4} = 2$$

With $N - 2 = 3$ a correlation of 0.878 is required for significance at the 0.05 level. Thus, this correlation of 0.45 is *not* significant.

2. Using the data from Problem 1:

$$\Sigma XY = 158; \text{ then } r_{xy} = \frac{\dfrac{158}{5} - 6 \times 4}{4 \times 2} = \frac{31.6 - 24}{8} = \frac{7.6}{8} = 0.95$$

Since the obtained correlation coefficient is larger than the tabled value of 0.878, the correlation is significant.

3. Again using the data from Problem 1:

$$\Sigma XY = 80; \text{ then } r_{xy} = \frac{\dfrac{80}{5} - 6 \times 4}{4 \times 2} = \frac{16 - 24}{8} = \frac{-8}{8} = -1.00$$

4. $$r_{xy} = \frac{\dfrac{\Sigma XY}{N} - M_x M_y}{\sigma_x \sigma_y} = \frac{\dfrac{5280}{20} - 12 \times 20}{6 \times 8} = \frac{264 - 240}{48} = \frac{24}{48} = 0.50$$

$$\sigma_x = \sqrt{\frac{3600 - \dfrac{240 \times 240}{20}}{20}} = \sqrt{\frac{3600 - 2880}{20}} = \sqrt{\frac{720}{20}} = \sqrt{36} = 6.00$$

$$\sigma_y = \sqrt{\frac{9280 - \dfrac{400 \times 400}{20}}{20}} = \sqrt{\frac{9280 - 8000}{20}} = \sqrt{\frac{1280}{20}} = \sqrt{64} = 8.00$$

$N - 2 = 18$ and the tabled value of r_{xy} at the 0.05 level is 0.444. Since the obtained correlation coefficient is larger than the tabled value, the null hypothesis is rejected, and it is concluded that there is a relationship between amount recalled and the amount of drug taken.

5. With $r_{xy} = 0.39$ with 25 pairs of scores, you can say very little about the correlation between X and Y. From Appendix E the value required for significance is 0.396 (number of pairs -2). Hence, the correlation from the experiment is not significant and as far as we know has not been shown to be different from zero.

6. rho $= 1 - \dfrac{6\Sigma D^2}{N(N^2 - 1)} = 1 - \dfrac{6 \times 798}{20 \times 399} = 1 - \dfrac{6 \times 2}{20} = 1 - 0.6 = 0.40$

From Appendix F a rank-order correlation of 0.45 is required for significance. The obtained $\rho = 0.40$ and is not significant. Since the correlation is not larger than what would be expected on the basis of chance, we cannot say that people who did well on the first test did well on the second.

7. Both variables are dichotomized and for this reason the phi coefficient is used. The phi coefficient is related to χ^2: $r_\phi = \sqrt{\chi^2/N}$, not correcting for continuity.

$$\chi^2 = \frac{\Sigma(o - e)^2}{e} = \frac{50}{15} + \frac{50}{25} = 3.33 + 2.00 = 5.33$$

$r_\phi = \sqrt{5.33/80} = 0.26$. With 1 df, the χ^2 of 5.33 is significant and therefore the phi coefficient of 0.26 is also.

8. $S^2 = \dfrac{\Sigma X^2 - \dfrac{(\Sigma X)^2}{N}}{N - 1} = \dfrac{65 - \dfrac{15 \times 15}{5}}{4} = \dfrac{65 - 45}{4} = \dfrac{20}{4} = 5$

$\sigma^2 = \dfrac{\Sigma X^2}{N} - M^2 = \dfrac{65}{5} - 3 \times 3 = 13 - 9 = 4$

9. Since the range of values which a correlation can possibly take on is -1.00 to $+1.00$, the research worker obviously made a mistake in his computations.

10. Wives tend to be shorter than their husbands, but this is not always the case. Similarly, tall husbands tend to marry tall wives and short husbands tend to marry short wives, although this, too, is not always the case. The correlation between heights of husbands and wives probably is moderately high, around 0.60 or 0.70.

11. The length of time it takes to answer the question is highly positively correlated with ability to see colors. If you were able to answer the question immediately, you probably are color blind, but if you took a long time to answer or didn't know it, either you have good color vision or have never seen a stop light.

Chapter Seventeen

1. The experiment is a 2×2 factorial double classification analysis of variance.

$$\sum_1^R \sum_1^C \sum_1^N X = 36; \quad \sum_1^R \sum_1^C \sum_1^N X^2 = 158, \quad CF = \frac{\left(\sum_1^R \sum_1^C \sum_1^N X\right)^2}{RCN} = 108$$

$$tot_{ss} = \sum_1^R \sum_1^C \sum_1^N X^2 - CF = 158 - 108 = 50.00$$

$$\text{Male}_{ss} = \frac{\sum\limits_{1}^{R}\left(\sum\limits_{1}^{C}\sum\limits_{1}^{N}X\right)^2}{CN} - CF = \frac{12^2 + 24^2}{6} - 108 = 12.00$$

$$\text{Female}_{ss} = \frac{\sum\limits_{1}^{C}\left(\sum\limits_{1}^{R}\sum\limits_{1}^{N}X\right)^2}{RN} - CF = \frac{21^2 + 15^2}{6} - 108 = 3.00$$

$$M \times F_{ss} = \frac{\sum\limits_{1}^{C}\sum\limits_{j}^{R}\left(\sum\limits_{1}^{N}X\right)^2}{N} - CF - M - F = \frac{3^2 + 9^2 + 18^2 + 6^2}{3}$$
$$- 108 - 12 - 3 = 27.00$$

$$wg_{ss} = tot - M - F - M \times F$$
$$= 50 - 12 - 3 - 27 = 8$$

Source	ss	df	ms	F
M	12	1	12	12*
F	3	1	3	3
M × F	27	1	27	27*
wg	8	8	1	
tot	50	11		

* Pr < 0.01

The results indicate that the passive male subject likes the female more than the aggressive male and that males like females of the opposite personality.

2. $\mu = 60$, $\sigma = 15$, $N = 25$, $\sigma_M = \dfrac{\sigma}{\sqrt{N}} = \dfrac{15}{\sqrt{25}} = \dfrac{15}{5} = 3.00$

$$z = \frac{M - \mu}{\sigma_M} = \frac{57 - 60}{3} = -1.00$$

The probability that a sample mean would be below 57 is equal to 0.16. The reason for this is that 57 is one standard error of the mean below the true mean. About 34 percent of the sample means would fall between 57 and 60 and about 16 percent would fall below 57.

3. This is a single classification analysis of variance.

$$\sum_{1}^{C}\sum_{1}^{N}X = 98; \quad \sum_{1}^{C}\sum_{1}^{N}X^2 = 838; \quad CF = \frac{\left(\sum\limits_{1}^{C}\sum\limits_{1}^{N}X\right)^2}{CN} = \frac{98^2}{14} = 686$$

$$tot_{ss} = \sum_{1}^{C}\sum_{1}^{N}X^2 - CF = 152; \quad bg_{ss} = \frac{\sum\limits_{1}^{C}\left(\sum\limits_{1}^{N}X\right)^2}{N} - CF = \frac{35^2 + 63^2}{7} - 686$$

$$= \frac{5194}{7} - 686 = 742 - 686 = 56; \quad wg_{ss} = tot - bg = 96$$

Source	ss	df	ms	F	
bg	56	1	56	7.00	Pr < 0.05
wg	96	12	8		
tot	152	13			

Familiar groups talk more than unfamiliar groups.

4. Use of χ^2 as a measure of association is appropriate here. (Compare with Problem 7, Chapter Sixteen.)

	Observed				**Expected**				$\dfrac{(o-e)^2}{e}$		

	1–5	Row 6–10	11–15								
M	5	15	30	50	15	15	20	$\dfrac{100}{15}$	0	$\dfrac{100}{20}$	
F	25	15	10	50	15	15	20	$\dfrac{100}{15}$	0	$\dfrac{100}{20}$	
	30	30	40	100							

$$\chi^2 = \sum \frac{(o-e)^2}{e} = 6.67 + 5.00 + 6.67 + 5.00 = 23.34.$$

$\chi^2 = 23.34$. With df $= 2$, Pr < 0.01.

There is some choice in how the problem was set up. For example, E could have used the data in Rows 1–5 and 11–15, and still tested his hypothesis. (But he should decide how he will analyze his data prior to seeing the data.)

5. t is the appropriate test here.

$$\Sigma X = 5,\ \Sigma X^2 = 11,\ S^2 = 1.50,\ S_M{}^2 = 0.30,\ S_M = 0.55$$

$$t = \frac{M - \mu}{S_M} = \frac{1.00 - 1.80}{0.55} = \frac{0.80}{0.55} = 1.45$$

A t of 1.45 is not large enough to reject the null hypothesis, and hence there is no evidence to suggest that their grades differ from the rest of the students on campus.

6. The Brand X slogan says that Brand X pills are *reliable*. This means that Brand X pills will cure headaches in general. Notice that no claim is made about the speed of the cure—just that the headache will be cured. For this reason, the F test for variances is appropriate here.

$$S_x{}^2 = \frac{2440 - \dfrac{120 \times 120}{6}}{5} = \frac{40}{5} = 8 \qquad F = \frac{46}{8} = 5.75,\ \text{Pr} < 0.05 \text{ with 5 df}$$
$$\text{and 5 df.}$$

$$S_y{}^2 = \frac{1580 - \dfrac{90 \times 90}{6}}{5} = \frac{230}{5} = 46$$

$S_x{}^2$ is placed in the denominator because the slogan indicates $S_x{}^2$ will be smaller.

The slogan is found to be true.

7. This is a 2 × 3 double classification analysis of variance design.

		Stress				**Source**		**df**
		L_1	L_2	L_3				
Birth	B_1					Stress	L	2
order	B_2					Birth order	B	1
							$L \times B$	2
							wg	66
							tot	71

8. This is also a double classification design, but is one in which there is only one score per cell.

$$CF = \frac{96 \times 96}{4 \times 6} = 384 \qquad tot = \Sigma X^2 - CF = 470 - 384 = 86$$

$$E_{ss} = \frac{1618}{4} - CF = 404.5 - 384 = 20.50$$

$$Y_{ss} = \frac{2568}{6} - CF = 428 - 384 = 44$$

$$E \times Y_{ss} = tot - E - Y = 86 - 20.50 - 44 = 21.50$$

Source	ss	**df**	ms	F
E	20.50	5		
Y	44.00	3	14.67	10.26*
$E \times Y$	21.50	15	1.43	
tot	86			

* Pr < 0.01

The data indicate that the mean running time for the elephants slows down over the years. This suggests that the elephants are in fact forgetting, although there are other interpretations, such as the elephants are getting older and more feeble. The F for elephants is not computed because they are "subjects" in this experiment.

9.

	A	B
M	12	20
S	Average variance = 10	
N	10	10

We can test the hypothesis that the two true means are equal.

$$F = \frac{NS_M{}^2}{S^2} = \frac{10 \times 32}{10} = 32$$

$$S_M{}^2 = \frac{544 - \dfrac{32 \times 32}{2}}{1} = 32$$

The means of the groups are significantly different with $F = 32.00$ and with 1 df and 18 df, Pr < 0.01.

10. If the likely voters respond randomly, then we would expect a 50:50 split in preferences. Instead there was a 55:45 split. The data are independent frequency data and chi square is appropriate. With 1 df the correction for continuity is used.

$$\chi^2 = \Sigma \frac{(|o - e| - 0.5)^2}{e} = \frac{(|55 - 50| - 0.5)^2}{50} + \frac{(|45 - 50| - 0.5)^2}{50}$$

$$= \frac{40.50}{50} = 0.81$$

With a χ^2 of less than 1.00, it is apparent that at the present time there is no real preference for one candidate or the other. Hence, Mr. Wainright would be starting with no initial advantage over his likely opponent.

11. $B_1 \ldots$ B_8

		Source	df
P_1		B	7
	B = Blocks of 5 blots	P	9
	P = Paranoids	$B \times P$	63
P_{10}		tot	79

12. $S_1 = 8$, $S^2 = 64$
$S_2 = 4$, $S^2 = 16$ $F = \dfrac{64}{16} = 4.00$ Pr < 0.05 with 9 df and 9 df

The Speedy-Fast parachute opening times are significantly more variable than those of the Never-Die parachutes. Hence, the claims of the Never-Die Company are justified.

13.

	o	e
R	50	25
other	50	75

$\chi^2 = \sum \dfrac{(|o - e| - 0.5)^2}{e} = \dfrac{24.5^2}{25} + \dfrac{24.5^2}{75} = 32.01$

Pr < 0.01 with 1 df.

The armadillos do have a right-turning bias. Note that a chi square with 3 df would not have tested Professor Snarf's hypothesis.

14. $M = 6000$, $S = 240$, $N = 36$, $S_M = \dfrac{240}{6} = 40$

$t = \dfrac{M - \mu}{S_M} = \dfrac{6000 - 5900}{40} = \dfrac{100}{40} = 2.50$ Pr < 0.05 with 35 df.

Bamo's slogan is correct.

Chapter Eighteen

1. First find the proportions and then rank-order the recall:

Proportions		Ranking	
"y"	Not "y"	"y"	Not "y"
0.8	0.5	1	2
0.4	0.0	1	2
0.2	0.0	1	2
0.6	0.5	1	2
1.0	0.5	1	2
0.8	0.0	1	2

$\chi^2 = \dfrac{12}{6 \times 2 \times 3} (6^2 + 12^2) - 3 \times 6 \times 3$

$= 0.33 \times 180 - 54$

$= 60 - 54 = 6.00$

df $= K - 1 = 1$. With 1 df, a χ^2 of 6.00 is significant beyond the 5 percent level and confirms your belief that a greater proportion of y than not-y names are remembered.

2. Ranking the data

Group I		Group II	
X	Rank	X	Rank
17	3	9	8
23	1	11	7
12	6	8	9
14	4.5	7	10
18	2	14	4.5
		5	12
$R_1 = 16.5$		6	11
			61.5

$$U_1 = 5 \times 7 + \frac{5 \times 6}{2} - 16.5$$
$$= 35 + 15 - 16.5$$
$$= 33.50$$

$$U_2 = 7 \times 5 + \frac{7 \times 8}{2} - 61.5$$
$$= 35 + 28 - 61.5$$
$$= 1.50$$

$$\mu_U = \frac{7 \times 5}{2} = 17.5$$

$$\sigma_\mu = \frac{7 \times 5 \times 13}{12} = 37.92 = 6.16$$

$$z_\mu = \frac{33.50 - 17.50}{6.16} = 2.60.$$ Since the value of z is greater than 2.58, we would reject the null hypothesis and conclude that the two groups differ.

3. $\Pr(I/R_1) = 0.80$, $\Pr(I/R_2) = 0.30$, $\Pr(R) = 0.2$

$$\Pr(R_1/I) = \frac{0.80 \times 0.20}{0.80 \times 0.20 + 0.30 \times 0.80} = \frac{0.16}{0.16 + 0.24} = \frac{0.16}{0.40} = 0.40$$

You still believe your brother, alias Light Finger Louie, is innocent, but your posterior probability (your belief that he is guilty) has been changed from 0.20 to 0.40.

4.

Student	Ranking Snarf	Finke
1	1	2
2	2	1
3	1	2
4	1	2
5	1	2
6	1	2
7	1	2
	8	13

$$\chi^2 = \frac{12}{7 \times 2 \times 3} \times (8^2 + 13^2) - 3 \times 7 \times 3$$
$$= 0.29 \times 233 - 63$$
$$= 67.57 - 63$$
$$= 4.57$$

With df $= 1$, we can conclude that Professor Snarf is preferred to Professor Finke at the 0.05 level.

5. $$\Pr(A_1/D) = \frac{0.50 \times 0.30}{0.50 \times 0.30 + 0.10 \times 0.70} = \frac{0.15}{0.15 + 0.07} = \frac{0.15}{0.22} = 0.68$$

6.

	Rankings	
Group I		**Group II**

$$
\begin{array}{cc}
2 & 15 \\
1 & 9 \\
12 & 17 \\
8 & 10 \\
4 & 7 \\
3 & 13 \\
5 & 16 \\
6 & 18 \\
R_1 = \overline{41} & 14 \\
& 11 \\
& R_2 = \overline{130}
\end{array}
$$

$$U_1 = 8 \times 10 + \frac{8 \times 9}{2} - 41$$

$$= 80 + 36 - 41$$

$$= 75$$

$$U_2 = 8 \times 10 + \frac{10 \times 11}{2} - 130$$

$$= 80 + 55 - 130$$

$$= 5$$

7. From Problem 6, $N_1 = 8$, $N_2 = 10$. $\mu_U = \dfrac{N_1 \times N_2}{2} = \dfrac{8 \times 10}{2} = 40$.

8. From Problem 6, $\sigma_U = \sqrt{\dfrac{N_1 N_2 (N_1 + N_2 + 1)}{12}} = \sqrt{\dfrac{8 \times 10 \times 19}{12}} = \sqrt{126.67}$.

9. From Problem 6, $z_U = \dfrac{U - \mu_U}{\sigma_\mu} = \dfrac{75 - 40}{11.25} = \dfrac{35}{11.25} = 3.11$

10. $\Pr(A/B) = \dfrac{0.20}{0.40} = 0.50$

11. $\Pr(A_1) = \dfrac{100}{400} = 0.25$, $\Pr(B_1) = \dfrac{180}{400} = 0.45$

$$\Pr(A_1/B_1) = \frac{0.60 \times 0.25}{0.60 \times 0.25 + 0.40 \times 0.75} = \frac{0.15}{0.15 + 0.30} = \frac{0.15}{0.45} = 0.33$$

or using the frequencies themselves $60/180 = 0.33$.

12. $\Sigma R_1 = 21$, $\Sigma R_2 = 20$, $\Sigma R_3 = 30$, $\Sigma R_4 = 29$

$$\chi^2 = \frac{12}{10 \times 4 \times 5} (2582) - 3 \times 10 \times 5$$

$$= 0.06 \times 2582 - 150$$

$$= 154.92 - 150$$

$$= 4.92$$

With df $= 3$, χ^2 is not significant, and it cannot be concluded that a preference for the picture exists.

13. $\Pr(H) = 0.40$, $\Pr(C) = 0.60$, $\Pr(R/H) = 0.30$, $\Pr(R/C) = 0.60$

$$\Pr(H/R) = \frac{0.30 \times 0.40}{0.30 \times 0.40 + 0.60 \times 0.60} = \frac{12}{0.12 + 0.36} = 0.48 = 0.25$$

Hence, the zoologist should find about 25 Hubbs.

Appendixes

Acknowledgments

Appendix A. From Walker and Lev, *Statistical Inference*, 1953; published by Holt, Rinehart and Winston, Inc., and used by permission of the authors and publisher.

Appendix B. Prepared by Richard O. Clements through the courtesy of The University of Texas at Austin Computation Center.

Appendix C. Abridged from Table III of Fisher and Yates, *Statistical Tables for Biological, Agricultural and Medical Research*, published by Oliver & Boyd Ltd., Edinburgh, and by permission of the authors and publishers. Additional entries were taken from Snedecor, *Statistical Methods* (fifth edition), 1956; published by Iowa State University Press, and used by permission of the author and publisher.

Appendix D. From Snedecor, *Statistical Methods* (fifth edition), 1956; published by Iowa State University Press, and used by permission of the author and publisher.

Appendix E. Abridged from Table VI of Fisher and Yates, *Statistical Tables for Biological, Agricultural and Medical Research*, published by Oliver & Boyd Ltd., Edinburgh, and by permission of the authors and publishers. Additional entries were taken from Snedecor, *Statistical Methods* (fifth edition), 1956; published by Iowa State University Press, and used by permission of the author and publisher.

Appendix F. From Underwood et al., *Elementary Statistics*, 1954; published by Appleton-Century-Crofts, Inc. The values in the table were computed from Olds, E. G., "Distributions of the sum of squares of rank differences for small numbers of individuals," *Ann. Math. Statist.*, 1938, 9:133–148, and "The 5% significance levels for sums of squares of rank differences and a correction," *Ann. Math. Statist.*, 1949, 20:117–118, by permission of the author and the Institute of Mathematical Statistics.

Appendix G. Abridged from Table IV of Fisher and Yates, *Statistical Tables for Biological, Agricultural and Medical Research*, published by Oliver & Boyd Ltd., Edinburgh, and by permission of the authors and publishers.

Appendix H. Prepared by Paul C. Jennings through the courtesy of the University of Texas Computation Center.

Appendix A. Squares, square roots, and reciprocals

n	n^2	$\sqrt{n}$	$\sqrt{10n}$	$1/n$	n	n^2	$\sqrt{n}$	$\sqrt{10n}$	$1/n$
1	1	1.000	3.162	1.00000	51	2601	7.141	22.583	.01961
2	4	1.414	4.472	.50000	52	2704	7.211	22.804	.01923
3	9	1.732	5.477	.33333	53	2809	7.280	23.022	.01887
4	16	2.000	6.325	.25000	54	2916	7.348	23.238	.01852
5	25	2.236	7.071	.20000	55	3025	7.416	23.452	.01818
6	36	2.449	7.746	.16667	56	3136	7.483	23.664	.01786
7	49	2.646	8.367	.14286	57	3249	7.550	23.875	.01754
8	64	2.828	8.944	.12500	58	3364	7.616	24.083	.01724
9	81	3.000	9.487	.11111	59	3481	7.681	24.290	.01695
10	100	3.162	10.000	.10000	60	3600	7.746	24.495	.01667
11	121	3.317	10.488	.09091	61	3721	7.810	24.698	.01639
12	144	3.464	10.954	.08333	62	3844	7.874	24.900	.01613
13	169	3.606	11.402	.07692	63	3969	7.937	25.100	.01587
14	196	3.742	11.832	.07143	64	4096	8.000	25.298	.01562
15	225	3.873	12.247	.06667	65	4225	8.062	25.495	.01538
16	256	4.000	12.649	.06250	66	4356	8.124	25.690	.01515
17	289	4.123	13.038	.05882	67	4489	8.185	25.884	.01493
18	324	4.243	13.416	.05556	68	4624	8.246	26.077	.01471
19	361	4.359	13.784	.05263	69	4761	8.307	26.268	.01449
20	400	4.472	14.142	.05000	70	4900	8.367	26.458	.01429
21	441	4.583	14.491	.04762	71	5041	8.426	26.646	.01408
22	484	4.690	14.832	.04545	72	5184	8.485	26.833	.01389
23	529	4.796	15.166	.04348	73	5329	8.544	27.019	.01370
24	576	4.899	15.492	.04167	74	5476	8.602	27.203	.01351
25	625	5.000	15.811	.04000	75	5625	8.660	27.386	.01333
26	676	5.099	16.125	.03846	76	5776	8.718	27.568	.01316
27	729	5.196	16.432	.03704	77	5929	8.775	27.749	.01299
28	784	5.292	16.733	.03571	78	6084	8.832	27.928	.01282
29	841	5.385	17.029	.03448	79	6241	8.888	28.107	.01266
30	900	5.477	17.321	.03333	80	6400	8.944	28.284	.01250
31	961	5.568	17.607	.03226	81	6561	9.000	28.460	.01235
32	1024	5.657	17.889	.03125	82	6724	9.055	28.636	.01220
33	1089	5.745	18.166	.03030	83	6889	9.110	28.810	.01205
34	1156	5.831	18.439	.02941	84	7056	9.165	28.983	.01190
35	1225	5.916	18.708	.02857	85	7225	9.220	29.155	.01176
36	1296	6.000	18.974	.02778	86	7396	9.274	29.326	.01163
37	1369	6.083	19.235	.02703	87	7569	9.327	29.496	.01149
38	1444	6.164	19.494	.02632	88	7744	9.381	29.665	.01136
39	1521	6.245	19.748	.02564	89	7921	9.434	29.833	.01124
40	1600	6.325	20.000	.02500	90	8100	9.487	30.000	.01111
41	1681	6.403	20.248	.02439	91	8281	9.539	30.166	.01099
42	1764	6.481	20.494	.02381	92	8464	9.592	30.332	.01087
43	1849	6.557	20.736	.02326	93	8649	9.644	30.496	.01075
44	1936	6.633	20.976	.02273	94	8836	9.695	30.659	.01064
45	2025	6.708	21.213	.02222	95	9025	9.747	30.822	.01053
46	2116	6.782	21.448	.02174	96	9216	9.798	30.984	.01042
47	2209	6.856	21.679	.02128	97	9409	9.849	31.145	.01031
48	2304	6.928	21.909	.02083	98	9604	9.899	31.305	.01020
49	2401	7.000	22.136	.02041	99	9801	9.950	31.464	.01010
50	2500	7.071	22.361	.02000	100	10000	10.000	31.623	.01000

Appendix B.   Percent of area as a function of the z score

Column A z score	Column B Area between mean and z	Column C Area in larger portion	Column D Area in smaller portion
0.00	0.00	50.00	50.00
.01	.40	50.40	49.60
.02	.80	50.80	49.20
.03	1.20	51.20	48.80
.04	1.60	51.60	48.40
.05	1.99	51.99	48.01
.06	2.39	52.39	47.61
.07	2.79	52.79	47.21
.08	3.19	53.19	46.81
.09	3.59	53.59	46.41
.10	3.98	53.98	46.02
.11	4.38	54.38	45.62
.12	4.78	54.78	45.22
.13	5.17	55.17	44.83
.14	5.57	55.57	44.43
.15	5.96	55.96	44.04
.16	6.36	56.36	43.64
.17	6.75	56.75	43.25
.18	7.14	57.14	42.86
.19	7.53	57.53	42.47
.20	7.93	57.93	42.07
.21	8.32	58.32	41.68
.22	8.71	58.71	41.29
.23	9.10	59.10	40.90
.24	9.48	59.48	40.52
.25	9.87	59.87	40.13
.26	10.26	60.26	39.74
.27	10.64	60.64	39.36
.28	11.03	61.03	38.97
.29	11.41	61.41	38.59
.30	11.79	61.79	38.21
.31	12.17	62.17	37.83
.32	12.55	62.55	37.45
.33	12.93	62.93	37.07
.34	13.31	63.31	36.69
.35	13.68	63.68	36.32
.36	14.06	64.06	35.94
.37	14.43	64.43	35.57
.38	14.80	64.80	35.20
.39	15.17	65.17	34.83
.40	15.54	65.54	34.46
.41	15.91	65.91	34.09
.42	16.28	66.28	33.72
.43	16.64	66.64	33.36
.44	17.00	67.00	33.00
.45	17.36	67.36	32.64
.46	17.72	67.72	32.28
.47	18.08	68.08	31.92
.48	18.44	68.44	31.56
.49	18.79	68.79	31.21
.50	19.15	69.15	30.85
.51	19.50	69.50	30.50
.52	19.85	69.85	30.15
.53	20.19	70.19	29.81

Column A z score	Column B Area between mean and z	Column C Area in larger portion	Column D Area in smaller portion
.54	20.54	70.54	29.46
.55	20.88	70.88	29.12
.56	21.23	71.23	28.77
.57	21.57	71.57	28.43
.58	21.90	71.90	28.10
.59	22.24	72.24	27.76
.60	22.57	72.57	27.43
.61	22.91	72.91	27.09
.62	23.24	73.24	26.76
.63	23.57	73.57	26.43
.64	23.89	73.89	26.11
.65	24.22	74.22	25.78
.66	24.54	74.54	25.46
.67	24.86	74.86	25.14
.68	25.17	75.17	24.83
.69	25.49	75.49	24.51
.70	25.80	75.80	24.20
.71	26.11	76.11	23.89
.72	26.42	76.42	23.58
.73	26.73	76.73	23.27
.74	27.04	77.04	22.96
.75	27.34	77.34	22.66
.76	27.64	77.64	22.36
.77	27.94	77.94	22.06
.78	28.23	78.23	21.77
.79	28.52	78.52	21.48
.80	28.81	78.81	21.19
.81	29.10	79.10	20.90
.82	29.39	79.39	20.61
.83	29.67	79.67	20.33
.84	29.95	79.95	20.05
.85	30.23	80.23	19.77
.86	30.51	80.51	19.49
.87	30.78	80.78	19.22
.88	31.06	81.06	18.94
.89	31.33	81.33	18.67
.90	31.59	81.59	18.41
.91	31.86	81.86	18.14
.92	32.12	82.12	17.88
.93	32.38	82.38	17.62
.94	32.64	82.64	17.36
.95	32.89	82.89	17.11
.96	33.15	83.15	16.85
.97	33.40	83.40	16.60
.98	33.65	83.65	16.35
.99	33.89	83.89	16.11
1.00	34.13	84.13	15.87
1.01	34.38	84.38	15.62
1.02	34.61	84.61	15.39
1.03	34.85	84.85	15.15
1.04	35.08	85.08	14.92
1.05	35.31	85.31	14.69
1.06	35.54	85.54	14.46
1.07	35.77	85.77	14.23
1.08	35.99	85.99	14.01
1.09	36.21	86.21	13.79
1.10	36.43	86.43	13.57
1.11	36.65	66.65	13.35
1.12	36.86	86.86	13.14
1.13	37.08	87.08	12.92
1.14	37.29	87.29	12.71
1.15	37.49	87.49	12.51

Column A z score	Column B Area between mean and z	Column C Area in larger portion	Column D Area in smaller portion
1.16	37.70	87.70	12.30
1.17	37.90	87.90	12.10
1.18	38.10	88.10	11.90
1.19	38.30	88.30	11.70
1.20	38.49	88.49	11.51
1.21	38.69	88.69	11.31
1.22	38.88	88.88	11.12
1.23	39.07	89.07	10.93
1.24	39.25	89.25	10.75
1.25	39.44	89.44	10.56
1.26	39.62	89.62	10.38
1.27	39.80	89.80	10.20
1.28	39.97	89.97	10.03
1.29	40.15	90.15	9.85
1.30	40.32	90.32	9.68
1.31	40.49	90.49	9.51
1.32	40.66	90.66	9.34
1.33	40.82	90.82	9.18
1.34	40.99	90.99	9.01
1.35	41.15	91.15	8.85
1.36	41.31	91.31	8.69
1.37	41.47	91.47	8.53
1.38	41.62	91.62	8.38
1.39	41.77	91.77	8.23
1.40	41.92	91.92	8.08
1.41	42.07	92.07	7.93
1.42	42.22	92.22	7.78
1.43	42.36	92.36	7.64
1.44	42.51	92.51	7.49
1.45	42.65	92.65	7.35
1.46	42.79	92.79	7.21
1.47	42.92	92.92	7.08
1.48	43.06	93.06	6.94
1.49	43.19	93.19	6.81
1.50	43.32	93.32	6.68
1.51	43.45	93.45	6.55
1.52	43.57	93.57	6.43
1.53	43.70	93.70	6.30
1.54	43.82	93.82	6.18
1.55	43.94	93.94	6.06
1.56	44.06	94.06	5.94
1.57	44.18	94.18	5.82
1.58	44.29	94.29	5.71
1.59	44.41	94.41	5.59
1.60	44.52	94.52	5.48
1.61	44.63	94.63	5.37
1.62	44.74	94.74	5.26
1.63	44.84	94.84	5.16
1.64	44.95	94.95	5.05
1.65	45.05	95.05	4.95
1.66	45.15	95.15	4.85
1.67	45.25	95.25	4.75
1.68	45.35	95.35	4.65
1.69	45.45	95.45	4.55
1.70	45.54	95.54	4.46
1.71	45.64	95.64	4.36
1.72	45.73	95.73	4.27
1.73	45.82	95.82	4.18
1.74	45.91	95.91	4.09
1.75	45.99	95.99	4.01
1.76	46.08	96.08	3.92
1.77	46.16	96.16	3.84

Column A z score	Column B Area between mean and z	Column C Area in larger portion	Column D Area in smaller portion
1.78	46.25	96.25	3.75
1.79	46.33	96.33	3.67
1.80	46.41	96.41	3.59
1.81	46.49	96.49	3.51
1.82	46.56	96.56	3.44
1.83	46.64	96.64	3.36
1.84	46.71	96.71	3.29
1.85	46.78	96.78	3.22
1.86	46.86	96.86	3.14
1.87	46.93	96.93	3.07
1.88	46.99	96.99	3.01
1.89	47.06	97.06	2.94
1.90	47.13	97.13	2.87
1.91	47.19	97.19	2.81
1.92	47.26	97.26	2.74
1.93	47.32	97.32	2.68
1.94	47.38	97.38	2.62
1.95	47.44	97.44	2.56
1.96	47.50	97.50	2.50
1.97	47.56	97.56	2.44
1.98	47.61	97.61	2.39
1.99	47.67	97.67	2.33
2.00	47.72	97.72	2.28
2.01	47.78	97.78	2.22
2.02	47.83	97.83	2.17
2.03	47.88	97.88	2.12
2.04	47.93	97.93	2.07
2.05	47.98	97.98	2.02
2.06	48.03	98.03	1.97
2.07	48.08	98.08	1.92
2.08	48.12	98.12	1.88
2.09	48.17	98.17	1.83
2.10	48.21	98.21	1.79
2.11	48.26	98.26	1.74
2.12	48.30	98.30	1.70
2.13	48.34	98.34	1.66
2.14	48.38	98.38	1.62
2.15	48.42	98.42	1.58
2.16	48.46	98.46	1.54
2.17	48.50	98.50	1.50
2.18	48.54	98.54	1.46
2.19	48.57	98.57	1.43
2.20	48.61	98.61	1.39
2.21	48.64	98.64	1.36
2.22	48.68	98.68	1.32
2.23	48.71	98.71	1.29
2.24	48.75	98.75	1.25
2.25	48.78	98.78	1.22
2.26	48.81	98.81	1.19
2.27	48.84	98.84	1.16
2.28	48.87	98.87	1.13
2.29	48.90	98.90	1.10
2.30	48.93	98.93	1.07
2.31	48.96	98.96	1.04
2.32	48.98	98.98	1.02
2.33	49.01	99.01	.99
2.34	49.04	99.04	.96
2.35	49.06	99.06	.94
2.36	49.09	99.09	.91
2.37	49.11	99.11	.89
2.38	49.13	99.13	.87
2.39	49.16	99.16	.84

Column A z score	Column B Area between mean and z	Column C Area in larger portion	Column D Area in smaller portion
2.40	49.18	99.18	.82
2.41	49.20	99.20	.80
2.42	49.22	99.22	.78
2.43	49.25	99.25	.75
2.44	49.27	99.27	.73
2.45	49.29	99.29	.71
2.46	49.31	99.31	.69
2.47	49.32	99.32	.68
2.48	49.34	99.34	.66
2.49	49.36	99.36	.64
2.50	49.38	99.38	.62
2.51	49.40	99.40	.60
2.52	49.41	99.41	.59
2.53	49.43	99.43	.57
2.54	49.45	99.45	.55
2.55	49.46	99.46	.54
2.56	49.48	99.48	.52
2.57	49.49	99.49	.51
2.58	49.51	99.51	.49
2.59	49.52	99.52	.48
2.60	49.53	99.53	.47
2.61	49.55	99.55	.45
2.62	49.56	99.56	.44
2.63	49.57	99.57	.43
2.64	49.59	99.59	.41
2.65	49.60	99.60	.40
2.66	49.61	99.61	.39
2.67	49.62	99.62	.38
2.68	49.63	99.63	.37
2.69	49.64	99.64	.36
2.70	49.65	99.65	.35
2.71	49.66	99.66	.34
2.72	49.67	99.67	.33
2.73	49.68	99.68	.32
2.74	49.69	99.69	.31
2.75	49.70	99.70	.30
2.76	49.71	99.71	.29
2.77	49.72	99.72	.28
2.78	49.73	99.73	.27
2.79	49.74	99.74	.26
2.80	49.74	99.74	.26
2.81	49.75	99.75	.25
2.82	49.76	99.76	.24
2.83	49.77	99.77	.23
2.84	49.77	99.77	.23
2.85	49.78	99.78	.22
2.86	49.79	99.79	.21
2.87	49.79	99.79	.21
2.88	49.80	99.80	.20
2.89	49.81	99.81	.19
2.90	49.81	99.81	.19
2.91	49.82	99.82	.18
2.92	49.82	99.82	.18
2.93	49.83	99.83	.17
2.94	49.84	99.84	.16
2.95	49.84	99.84	.16
2.96	49.85	99.85	.15
2.97	49.85	99.85	.15
2.98	49.86	99.86	.14
2.99	49.86	99.86	.14
3.00	49.87	99.87	.13
3.01	49.87	99.87	.13

Appendix C. Values of t beyond which 5 percent or 1 percent of the area falls

df	.05	.01
1	12.71	63.66
2	4.30	9.93
3	3.18	5.84
4	2.78	4.60
5	2.57	4.03
6	2.45	3.71
7	2.37	3.50
8	2.31	3.36
9	2.26	3.25
10	2.23	3.17
11	2.20	3.11
12	2.18	3.06
13	2.16	3.01
14	2.15	2.98
15	2.13	2.95
16	2.12	2.92
17	2.11	2.90
18	2.10	2.88
19	2.09	2.86
20	2.09	2.85
21	2.08	2.83
22	2.07	2.82
23	2.07	2.81
24	2.06	2.80
25	2.06	2.79
26	2.06	2.78
27	2.05	2.77
28	2.05	2.76
29	2.05	2.76
30	2.04	2.75
35	2.03	2.72
40	2.02	2.70
45	2.01	2.69
50	2.01	2.68
60	2.00	2.66
70	1.99	2.65
100	1.98	2.63
∞	1.96	2.58

Appendix D. Values of F beyond which 5 percent (roman type) or 1 percent (boldface type) of the area falls

n_1 degrees of freedom (for greater mean square)

n_2	1	2	3	4	5	6	7	8	9	10	11	12	14	16	20	24	30	40	50	75	100	200	500	∞
1	161 / **4,052**	200 / **4,999**	216 / **5,403**	225 / **5,625**	230 / **5,764**	234 / **5,859**	237 / **5,928**	239 / **5,981**	241 / **6,022**	242 / **6,056**	243 / **6,082**	244 / **6,106**	245 / **6,142**	246 / **6,169**	248 / **6,208**	249 / **6,234**	250 / **6,258**	251 / **6,286**	252 / **6,302**	253 / **6,323**	253 / **6,334**	254 / **6,352**	254 / **6,361**	254 / **6,366**
2	18.51 / **98.49**	19.00 / **99.00**	19.16 / **99.17**	19.25 / **99.25**	19.30 / **99.30**	19.33 / **99.33**	19.36 / **99.34**	19.37 / **99.36**	19.38 / **99.38**	19.39 / **99.40**	19.40 / **99.41**	19.41 / **99.42**	19.42 / **99.43**	19.43 / **99.44**	19.44 / **99.45**	19.45 / **99.46**	19.46 / **99.47**	19.47 / **99.48**	19.47 / **99.48**	19.48 / **99.49**	19.49 / **99.49**	19.49 / **99.49**	19.50 / **99.50**	19.50 / **99.50**
3	10.13 / **34.12**	9.55 / **30.82**	9.28 / **29.46**	9.12 / **28.71**	9.01 / **28.24**	8.94 / **27.91**	8.88 / **27.67**	8.84 / **27.49**	8.81 / **27.34**	8.78 / **27.23**	8.76 / **27.13**	8.74 / **27.05**	8.71 / **26.92**	8.69 / **26.83**	8.66 / **26.69**	8.64 / **26.60**	8.62 / **26.50**	8.60 / **26.41**	8.58 / **26.35**	8.57 / **26.27**	8.56 / **26.23**	8.54 / **26.18**	8.54 / **26.14**	8.53 / **26.12**
4	7.71 / **21.20**	6.94 / **18.00**	6.59 / **16.69**	6.39 / **15.98**	6.26 / **15.52**	6.16 / **15.21**	6.09 / **14.98**	6.04 / **14.80**	6.00 / **14.66**	5.96 / **14.54**	5.93 / **14.45**	5.91 / **14.37**	5.87 / **14.24**	5.84 / **14.15**	5.80 / **14.02**	5.77 / **13.93**	5.74 / **13.83**	5.71 / **13.74**	5.70 / **13.69**	5.68 / **13.61**	5.66 / **13.57**	5.65 / **13.52**	5.64 / **13.48**	5.63 / **13.46**
5	6.61 / **16.26**	5.79 / **13.27**	5.41 / **12.06**	5.19 / **11.39**	5.05 / **10.97**	4.95 / **10.67**	4.88 / **10.45**	4.82 / **10.27**	4.78 / **10.15**	4.74 / **10.05**	4.70 / **9.96**	4.68 / **9.89**	4.64 / **9.77**	4.60 / **9.68**	4.56 / **9.55**	4.53 / **9.47**	4.50 / **9.38**	4.46 / **9.29**	4.44 / **9.24**	4.42 / **9.17**	4.40 / **9.13**	4.38 / **9.07**	4.37 / **9.04**	4.36 / **9.02**
6	5.99 / **13.74**	5.14 / **10.92**	4.76 / **9.78**	4.53 / **9.15**	4.39 / **8.75**	4.28 / **8.47**	4.21 / **8.26**	4.15 / **8.10**	4.10 / **7.98**	4.06 / **7.87**	4.03 / **7.79**	4.00 / **7.72**	3.96 / **7.60**	3.92 / **7.52**	3.87 / **7.39**	3.84 / **7.31**	3.81 / **7.23**	3.77 / **7.14**	3.75 / **7.09**	3.72 / **7.02**	3.71 / **6.99**	3.69 / **6.94**	3.68 / **6.90**	3.67 / **6.88**
7	5.59 / **12.25**	4.74 / **9.55**	4.35 / **8.45**	4.12 / **7.85**	3.97 / **7.46**	3.87 / **7.19**	3.79 / **7.00**	3.73 / **6.84**	3.68 / **6.71**	3.63 / **6.62**	3.60 / **6.54**	3.57 / **6.47**	3.52 / **6.35**	3.49 / **6.27**	3.44 / **6.15**	3.41 / **6.07**	3.38 / **5.98**	3.34 / **5.90**	3.32 / **5.85**	3.29 / **5.78**	3.28 / **5.75**	3.25 / **5.70**	3.24 / **5.67**	3.23 / **5.65**
8	5.32 / **11.26**	4.46 / **8.65**	4.07 / **7.59**	3.84 / **7.01**	3.69 / **6.63**	3.58 / **6.37**	3.50 / **6.19**	3.44 / **6.03**	3.39 / **5.91**	3.34 / **5.82**	3.31 / **5.74**	3.28 / **5.67**	3.23 / **5.56**	3.20 / **5.48**	3.15 / **5.36**	3.12 / **5.28**	3.08 / **5.20**	3.05 / **5.11**	3.03 / **5.06**	3.00 / **5.00**	2.98 / **4.96**	2.96 / **4.91**	2.94 / **4.88**	2.93 / **4.86**
9	5.12 / **10.56**	4.26 / **8.02**	3.86 / **6.99**	3.63 / **6.42**	3.48 / **6.06**	3.37 / **5.80**	3.29 / **5.62**	3.23 / **5.47**	3.18 / **5.35**	3.13 / **5.26**	3.10 / **5.18**	3.07 / **5.11**	3.02 / **5.00**	2.98 / **4.92**	2.93 / **4.80**	2.90 / **4.73**	2.86 / **4.64**	2.82 / **4.56**	2.80 / **4.51**	2.77 / **4.45**	2.76 / **4.41**	2.73 / **4.36**	2.72 / **4.33**	2.71 / **4.31**
10	4.96 / **10.04**	4.10 / **7.56**	3.71 / **6.55**	3.48 / **5.99**	3.33 / **5.64**	3.22 / **5.39**	3.14 / **5.21**	3.07 / **5.06**	3.02 / **4.95**	2.97 / **4.85**	2.94 / **4.78**	2.91 / **4.71**	2.86 / **4.60**	2.82 / **4.52**	2.77 / **4.41**	2.74 / **4.33**	2.70 / **4.25**	2.67 / **4.17**	2.64 / **4.12**	2.61 / **4.05**	2.59 / **4.01**	2.56 / **3.96**	2.55 / **3.93**	2.54 / **3.91**
11	4.84 / **9.65**	3.98 / **7.20**	3.59 / **6.22**	3.36 / **5.67**	3.20 / **5.32**	3.09 / **5.07**	3.01 / **4.88**	2.95 / **4.74**	2.90 / **4.63**	2.86 / **4.54**	2.82 / **4.46**	2.79 / **4.40**	2.74 / **4.29**	2.70 / **4.21**	2.65 / **4.10**	2.61 / **4.02**	2.57 / **3.94**	2.53 / **3.86**	2.50 / **3.80**	2.47 / **3.74**	2.45 / **3.70**	2.42 / **3.66**	2.41 / **3.62**	2.40 / **3.60**
12	4.75 / **9.33**	3.88 / **6.93**	3.49 / **5.95**	3.26 / **5.41**	3.11 / **5.06**	3.00 / **4.82**	2.92 / **4.65**	2.85 / **4.50**	2.80 / **4.39**	2.76 / **4.30**	2.72 / **4.22**	2.69 / **4.16**	2.64 / **4.05**	2.60 / **3.98**	2.54 / **3.86**	2.50 / **3.78**	2.46 / **3.70**	2.42 / **3.61**	2.40 / **3.56**	2.36 / **3.49**	2.35 / **3.46**	2.32 / **3.41**	2.31 / **3.38**	2.30 / **3.36**
13	4.67 / **9.07**	3.80 / **6.70**	3.41 / **5.74**	3.18 / **5.20**	3.02 / **4.86**	2.92 / **4.62**	2.84 / **4.44**	2.77 / **4.30**	2.72 / **4.19**	2.67 / **4.10**	2.63 / **4.02**	2.60 / **3.96**	2.55 / **3.85**	2.51 / **3.78**	2.46 / **3.67**	2.42 / **3.59**	2.38 / **3.51**	2.34 / **3.42**	2.32 / **3.37**	2.28 / **3.30**	2.26 / **3.27**	2.24 / **3.21**	2.22 / **3.18**	2.21 / **3.16**

n1 degrees of freedom (for greater mean square)

n2	1	2	3	4	5	6	7	8	9	10	11	12	14	16	20	24	30	40	50	75	100	200	500	∞
14	4.60 / 8.86	3.74 / 6.51	3.34 / 5.56	3.11 / 5.03	2.96 / 4.69	2.85 / 4.46	2.77 / 4.28	2.70 / 4.14	2.65 / 4.03	2.60 / 3.94	2.56 / 3.86	2.53 / 3.80	2.48 / 3.70	2.44 / 3.62	2.39 / 3.51	2.35 / 3.43	2.31 / 3.34	2.27 / 3.26	2.24 / 3.21	2.21 / 3.14	2.19 / 3.11	2.16 / 3.06	2.14 / 3.02	2.13 / 3.00
15	4.54 / 8.68	3.68 / 6.36	3.29 / 5.42	3.06 / 4.89	2.90 / 4.56	2.79 / 4.32	2.70 / 4.14	2.64 / 4.00	2.59 / 3.89	2.55 / 3.80	2.51 / 3.73	2.48 / 3.67	2.43 / 3.56	2.39 / 3.48	2.33 / 3.36	2.29 / 3.29	2.25 / 3.20	2.21 / 3.12	2.18 / 3.07	2.15 / 3.00	2.12 / 2.97	2.10 / 2.92	2.08 / 2.89	2.07 / 2.87
16	4.49 / 8.53	3.63 / 6.23	3.24 / 5.29	3.01 / 4.77	2.85 / 4.44	2.74 / 4.20	2.66 / 4.03	2.59 / 3.89	2.54 / 3.78	2.49 / 3.69	2.45 / 3.61	2.42 / 3.55	2.37 / 3.45	2.33 / 3.37	2.28 / 3.25	2.24 / 3.18	2.20 / 3.10	2.16 / 3.01	2.13 / 2.96	2.09 / 2.89	2.07 / 2.86	2.04 / 2.80	2.02 / 2.77	2.01 / 2.75
17	4.45 / 8.40	3.59 / 6.11	3.20 / 5.18	2.96 / 4.67	2.81 / 4.34	2.70 / 4.10	2.62 / 3.93	2.55 / 3.79	2.50 / 3.68	2.45 / 3.59	2.41 / 3.52	2.38 / 3.45	2.33 / 3.35	2.29 / 3.27	2.23 / 3.16	2.19 / 3.08	2.15 / 3.00	2.11 / 2.92	2.08 / 2.86	2.04 / 2.79	2.02 / 2.76	1.99 / 2.70	1.97 / 2.67	1.96 / 2.65
18	4.41 / 8.28	3.55 / 6.01	3.16 / 5.09	2.93 / 4.58	2.77 / 4.25	2.66 / 4.01	2.58 / 3.85	2.51 / 3.71	2.46 / 3.60	2.41 / 3.51	2.37 / 3.44	2.34 / 3.37	2.29 / 3.27	2.25 / 3.19	2.19 / 3.07	2.15 / 3.00	2.11 / 2.91	2.07 / 2.83	2.04 / 2.78	2.00 / 2.71	1.98 / 2.68	1.95 / 2.62	1.93 / 2.59	1.92 / 2.57
19	4.38 / 8.18	3.52 / 5.93	3.13 / 5.01	2.90 / 4.50	2.74 / 4.17	2.63 / 3.94	2.55 / 3.77	2.48 / 3.63	2.43 / 3.52	2.38 / 3.43	2.34 / 3.36	2.31 / 3.30	2.26 / 3.19	2.21 / 3.12	2.15 / 3.00	2.11 / 2.92	2.07 / 2.84	2.02 / 2.76	2.00 / 2.70	1.96 / 2.63	1.94 / 2.60	1.91 / 2.54	1.90 / 2.51	1.88 / 2.49
20	4.35 / 8.10	3.49 / 5.85	3.10 / 4.94	2.87 / 4.43	2.71 / 4.10	2.60 / 3.87	2.52 / 3.71	2.45 / 3.56	2.40 / 3.45	2.35 / 3.37	2.31 / 3.30	2.28 / 3.23	2.23 / 3.13	2.18 / 3.05	2.12 / 2.94	2.08 / 2.86	2.04 / 2.77	1.99 / 2.69	1.96 / 2.63	1.92 / 2.56	1.90 / 2.53	1.87 / 2.47	1.85 / 2.44	1.84 / 2.42
21	4.32 / 8.02	3.47 / 5.78	3.07 / 4.87	2.84 / 4.37	2.68 / 4.04	2.57 / 3.81	2.49 / 3.65	2.42 / 3.51	2.37 / 3.40	2.32 / 3.31	2.28 / 3.24	2.25 / 3.17	2.20 / 3.07	2.15 / 2.99	2.09 / 2.88	2.05 / 2.80	2.00 / 2.72	1.96 / 2.63	1.93 / 2.58	1.89 / 2.51	1.87 / 2.47	1.84 / 2.42	1.82 / 2.38	1.81 / 2.36
22	4.30 / 7.94	3.44 / 5.72	3.05 / 4.82	2.82 / 4.31	2.66 / 3.99	2.55 / 3.76	2.47 / 3.59	2.40 / 3.45	2.35 / 3.35	2.30 / 3.26	2.26 / 3.18	2.23 / 3.12	2.18 / 3.02	2.13 / 2.94	2.07 / 2.83	2.03 / 2.75	1.98 / 2.67	1.93 / 2.58	1.91 / 2.53	1.87 / 2.46	1.84 / 2.42	1.81 / 2.37	1.80 / 2.33	1.78 / 2.31
23	4.28 / 7.88	3.42 / 5.66	3.03 / 4.76	2.80 / 4.26	2.64 / 3.94	2.53 / 3.71	2.45 / 3.54	2.38 / 3.41	2.32 / 3.30	2.28 / 3.21	2.24 / 3.14	2.20 / 3.07	2.14 / 2.97	2.10 / 2.89	2.04 / 2.78	2.00 / 2.70	1.96 / 2.62	1.91 / 2.53	1.88 / 2.48	1.84 / 2.41	1.82 / 2.37	1.79 / 2.32	1.77 / 2.28	1.76 / 2.26
24	4.26 / 7.82	3.40 / 5.61	3.01 / 4.72	2.78 / 4.22	2.62 / 3.90	2.51 / 3.67	2.43 / 3.50	2.36 / 3.36	2.30 / 3.25	2.26 / 3.17	2.22 / 3.09	2.18 / 3.03	2.13 / 2.93	2.09 / 2.85	2.02 / 2.74	1.98 / 2.66	1.94 / 2.58	1.89 / 2.49	1.86 / 2.44	1.82 / 2.36	1.80 / 2.33	1.76 / 2.27	1.74 / 2.23	1.73 / 2.21
25	4.24 / 7.77	3.38 / 5.57	2.99 / 4.68	2.76 / 4.18	2.60 / 3.86	2.49 / 3.63	2.41 / 3.46	2.34 / 3.32	2.28 / 3.21	2.24 / 3.13	2.20 / 3.05	2.16 / 2.99	2.11 / 2.89	2.06 / 2.81	2.00 / 2.70	1.96 / 2.62	1.92 / 2.54	1.87 / 2.45	1.84 / 2.40	1.80 / 2.32	1.77 / 2.29	1.74 / 2.23	1.72 / 2.19	1.71 / 2.17
26	4.22 / 7.72	3.37 / 5.53	2.98 / 4.64	2.74 / 4.14	2.59 / 3.82	2.47 / 3.59	2.39 / 3.42	2.32 / 3.29	2.27 / 3.17	2.22 / 3.09	2.18 / 3.02	2.15 / 2.96	2.10 / 2.86	2.05 / 2.77	1.99 / 2.66	1.95 / 2.58	1.90 / 2.50	1.85 / 2.41	1.82 / 2.36	1.78 / 2.28	1.76 / 2.25	1.72 / 2.19	1.70 / 2.15	1.69 / 2.13

n₁ degrees of freedom (for greater mean square)

n_2	1	2	3	4	5	6	7	8	9	10	11	12	14	16	20	24	30	40	50	75	100	200	500	∞
27	4.21 / 7.68	3.35 / 5.49	2.96 / 4.60	2.73 / 4.11	2.57 / 3.79	2.46 / 3.56	2.37 / 3.39	2.30 / 3.26	2.25 / 3.14	2.20 / 3.06	2.16 / 2.98	2.13 / 2.93	2.08 / 2.83	2.03 / 2.74	1.97 / 2.63	1.93 / 2.55	1.88 / 2.47	1.84 / 2.38	1.80 / 2.33	1.76 / 2.25	1.74 / 2.21	1.71 / 2.16	1.68 / 2.12	1.67 / 2.10
28	4.20 / 7.64	3.34 / 5.45	2.95 / 4.57	2.71 / 4.07	2.56 / 3.76	2.44 / 3.53	2.36 / 3.36	2.29 / 3.23	2.24 / 3.11	2.19 / 3.03	2.15 / 2.95	2.12 / 2.90	2.06 / 2.80	2.02 / 2.71	1.96 / 2.60	1.91 / 2.52	1.87 / 2.44	1.81 / 2.35	1.78 / 2.30	1.75 / 2.22	1.72 / 2.18	1.69 / 2.13	1.67 / 2.09	1.65 / 2.06
29	4.18 / 7.60	3.33 / 5.42	2.93 / 4.54	2.70 / 4.04	2.54 / 3.73	2.43 / 3.50	2.35 / 3.33	2.28 / 3.20	2.22 / 3.08	2.18 / 3.00	2.14 / 2.92	2.10 / 2.87	2.05 / 2.77	2.00 / 2.68	1.94 / 2.57	1.90 / 2.49	1.85 / 2.41	1.80 / 2.32	1.77 / 2.27	1.73 / 2.19	1.71 / 2.15	1.68 / 2.10	1.65 / 2.06	1.64 / 2.03
30	4.17 / 7.56	3.32 / 5.39	2.92 / 4.51	2.69 / 4.02	2.53 / 3.70	2.42 / 3.47	2.34 / 3.30	2.27 / 3.17	2.21 / 3.06	2.16 / 2.98	2.12 / 2.90	2.09 / 2.84	2.04 / 2.74	1.99 / 2.66	1.93 / 2.55	1.89 / 2.47	1.84 / 2.38	1.79 / 2.29	1.76 / 2.24	1.72 / 2.16	1.69 / 2.13	1.66 / 2.07	1.64 / 2.03	1.62 / 2.01
32	4.15 / 7.50	3.30 / 5.34	2.90 / 4.46	2.67 / 3.97	2.51 / 3.66	2.40 / 3.42	2.32 / 3.25	2.25 / 3.12	2.19 / 3.01	2.14 / 2.94	2.10 / 2.86	2.07 / 2.80	2.02 / 2.70	1.97 / 2.62	1.91 / 2.51	1.86 / 2.42	1.82 / 2.34	1.76 / 2.25	1.74 / 2.20	1.69 / 2.12	1.67 / 2.08	1.64 / 2.02	1.61 / 1.98	1.59 / 1.96
34	4.13 / 7.44	3.28 / 5.29	2.88 / 4.42	2.65 / 3.93	2.49 / 3.61	2.38 / 3.38	2.30 / 3.21	2.23 / 3.08	2.17 / 2.97	2.12 / 2.89	2.08 / 2.82	2.05 / 2.76	2.00 / 2.66	1.95 / 2.58	1.89 / 2.47	1.84 / 2.38	1.80 / 2.30	1.74 / 2.21	1.71 / 2.15	1.67 / 2.08	1.64 / 2.04	1.61 / 1.98	1.59 / 1.94	1.57 / 1.91
36	4.11 / 7.39	3.26 / 5.25	2.86 / 4.38	2.63 / 3.89	2.48 / 3.58	2.36 / 3.35	2.28 / 3.18	2.21 / 3.04	2.15 / 2.94	2.10 / 2.86	2.06 / 2.78	2.03 / 2.72	1.98 / 2.62	1.93 / 2.54	1.87 / 2.43	1.82 / 2.35	1.78 / 2.26	1.72 / 2.17	1.69 / 2.12	1.65 / 2.04	1.62 / 2.00	1.59 / 1.94	1.56 / 1.90	1.55 / 1.87
38	4.10 / 7.35	3.25 / 5.21	2.85 / 4.34	2.62 / 3.86	2.46 / 3.54	2.35 / 3.32	2.26 / 3.15	2.19 / 3.02	2.14 / 2.91	2.09 / 2.82	2.05 / 2.75	2.02 / 2.69	1.96 / 2.59	1.92 / 2.51	1.85 / 2.40	1.80 / 2.32	1.76 / 2.22	1.71 / 2.14	1.67 / 2.08	1.63 / 2.00	1.60 / 1.97	1.57 / 1.90	1.54 / 1.86	1.53 / 1.84
40	4.08 / 7.31	3.23 / 5.18	2.84 / 4.31	2.61 / 3.83	2.45 / 3.51	2.34 / 3.29	2.25 / 3.12	2.18 / 2.99	2.12 / 2.88	2.07 / 2.80	2.04 / 2.73	2.00 / 2.66	1.95 / 2.56	1.90 / 2.49	1.84 / 2.37	1.79 / 2.29	1.74 / 2.20	1.69 / 2.11	1.66 / 2.05	1.61 / 1.97	1.59 / 1.94	1.55 / 1.88	1.53 / 1.84	1.51 / 1.81
42	4.07 / 7.27	3.22 / 5.15	2.83 / 4.29	2.59 / 3.80	2.44 / 3.49	2.32 / 3.26	2.24 / 3.10	2.17 / 2.96	2.11 / 2.86	2.06 / 2.77	2.02 / 2.70	1.99 / 2.64	1.94 / 2.54	1.89 / 2.46	1.82 / 2.35	1.78 / 2.26	1.73 / 2.17	1.68 / 2.08	1.64 / 2.02	1.60 / 1.94	1.57 / 1.91	1.54 / 1.85	1.51 / 1.80	1.49 / 1.78
44	4.06 / 7.24	3.21 / 5.12	2.82 / 4.26	2.58 / 3.78	2.43 / 3.46	2.31 / 3.24	2.23 / 3.07	2.16 / 2.94	2.10 / 2.84	2.05 / 2.75	2.01 / 2.68	1.98 / 2.62	1.92 / 2.52	1.88 / 2.44	1.81 / 2.32	1.76 / 2.24	1.72 / 2.15	1.66 / 2.06	1.63 / 2.00	1.58 / 1.92	1.56 / 1.88	1.52 / 1.82	1.50 / 1.78	1.48 / 1.75
46	4.05 / 7.21	3.20 / 5.10	2.81 / 4.24	2.57 / 3.76	2.42 / 3.44	2.30 / 3.22	2.22 / 3.05	2.14 / 2.92	2.09 / 2.82	2.04 / 2.73	2.00 / 2.66	1.97 / 2.60	1.91 / 2.50	1.87 / 2.42	1.80 / 2.30	1.75 / 2.22	1.71 / 2.13	1.65 / 2.04	1.62 / 1.98	1.57 / 1.90	1.54 / 1.86	1.51 / 1.80	1.48 / 1.76	1.46 / 1.72
48	4.04 / 7.19	3.19 / 5.08	2.80 / 4.22	2.56 / 3.74	2.41 / 3.42	2.30 / 3.20	2.21 / 3.04	2.14 / 2.90	2.08 / 2.80	2.03 / 2.71	1.99 / 2.64	1.96 / 2.58	1.90 / 2.48	1.86 / 2.40	1.79 / 2.28	1.74 / 2.20	1.70 / 2.11	1.64 / 2.02	1.61 / 1.96	1.56 / 1.88	1.53 / 1.84	1.50 / 1.78	1.47 / 1.73	1.45 / 1.70

n₁ degrees of freedom (for greater mean square)

n₂	1	2	3	4	5	6	7	8	9	10	11	12	14	16	20	24	30	40	50	75	100	200	500	∞
50	4.03 / 7.17	3.18 / 5.06	2.79 / 4.20	2.56 / 3.72	2.40 / 3.41	2.29 / 3.18	2.20 / 3.02	2.13 / 2.88	2.07 / 2.78	2.02 / 2.70	1.98 / 2.62	1.95 / 2.56	1.90 / 2.46	1.85 / 2.39	1.78 / 2.26	1.74 / 2.18	1.69 / 2.10	1.63 / 2.00	1.60 / 1.94	1.55 / 1.86	1.52 / 1.82	1.48 / 1.76	1.46 / 1.71	1.44 / 1.68
55	4.02 / 7.12	3.17 / 5.01	2.78 / 4.16	2.54 / 3.68	2.38 / 3.37	2.27 / 3.15	2.18 / 2.98	2.11 / 2.85	2.05 / 2.75	2.00 / 2.66	1.97 / 2.59	1.93 / 2.53	1.88 / 2.43	1.83 / 2.35	1.76 / 2.23	1.72 / 2.15	1.67 / 2.06	1.61 / 1.96	1.58 / 1.90	1.52 / 1.82	1.50 / 1.78	1.46 / 1.71	1.43 / 1.66	1.41 / 1.64
60	4.00 / 7.08	3.15 / 4.98	2.76 / 4.13	2.52 / 3.65	2.37 / 3.34	2.25 / 3.12	2.17 / 2.95	2.10 / 2.82	2.04 / 2.72	1.99 / 2.63	1.95 / 2.56	1.92 / 2.50	1.86 / 2.40	1.81 / 2.32	1.75 / 2.20	1.70 / 2.12	1.65 / 2.03	1.59 / 1.93	1.56 / 1.87	1.50 / 1.79	1.48 / 1.74	1.44 / 1.68	1.41 / 1.63	1.39 / 1.60
65	3.99 / 7.04	3.14 / 4.95	2.75 / 4.10	2.51 / 3.62	2.36 / 3.31	2.24 / 3.09	2.15 / 2.93	2.08 / 2.79	2.02 / 2.70	1.98 / 2.61	1.94 / 2.54	1.90 / 2.47	1.85 / 2.37	1.80 / 2.30	1.73 / 2.18	1.68 / 2.09	1.63 / 2.00	1.57 / 1.90	1.54 / 1.84	1.49 / 1.76	1.46 / 1.71	1.42 / 1.64	1.39 / 1.60	1.37 / 1.56
70	3.98 / 7.01	3.13 / 4.92	2.74 / 4.08	2.50 / 3.60	2.35 / 3.29	2.23 / 3.07	2.14 / 2.91	2.07 / 2.77	2.01 / 2.67	1.97 / 2.59	1.93 / 2.51	1.89 / 2.45	1.84 / 2.35	1.79 / 2.28	1.72 / 2.15	1.67 / 2.07	1.62 / 1.98	1.56 / 1.88	1.53 / 1.82	1.47 / 1.74	1.45 / 1.69	1.40 / 1.62	1.37 / 1.56	1.35 / 1.53
80	3.96 / 6.96	3.11 / 4.88	2.72 / 4.04	2.48 / 3.56	2.33 / 3.25	2.21 / 3.04	2.12 / 2.87	2.05 / 2.74	1.99 / 2.64	1.95 / 2.55	1.91 / 2.48	1.88 / 2.41	1.82 / 2.32	1.77 / 2.24	1.70 / 2.11	1.65 / 2.03	1.60 / 1.94	1.54 / 1.84	1.51 / 1.78	1.45 / 1.70	1.42 / 1.65	1.38 / 1.57	1.35 / 1.52	1.32 / 1.49
100	3.94 / 6.90	3.09 / 4.82	2.70 / 3.98	2.46 / 3.51	2.30 / 3.20	2.19 / 2.99	2.10 / 2.82	2.03 / 2.69	1.97 / 2.59	1.92 / 2.51	1.88 / 2.43	1.85 / 2.36	1.79 / 2.26	1.75 / 2.19	1.68 / 2.06	1.63 / 1.98	1.57 / 1.89	1.51 / 1.79	1.48 / 1.73	1.42 / 1.64	1.39 / 1.59	1.34 / 1.51	1.30 / 1.46	1.28 / 1.43
125	3.92 / 6.84	3.07 / 4.78	2.68 / 3.94	2.44 / 3.47	2.29 / 3.17	2.17 / 2.95	2.08 / 2.79	2.01 / 2.65	1.95 / 2.56	1.90 / 2.47	1.86 / 2.40	1.83 / 2.33	1.77 / 2.23	1.72 / 2.15	1.65 / 2.03	1.60 / 1.94	1.55 / 1.85	1.49 / 1.75	1.45 / 1.68	1.39 / 1.59	1.36 / 1.54	1.31 / 1.46	1.27 / 1.40	1.25 / 1.37
150	3.91 / 6.81	3.06 / 4.75	2.67 / 3.91	2.43 / 3.44	2.27 / 3.14	2.16 / 2.92	2.07 / 2.76	2.00 / 2.62	1.94 / 2.53	1.89 / 2.44	1.85 / 2.37	1.82 / 2.30	1.76 / 2.20	1.71 / 2.12	1.64 / 2.00	1.59 / 1.91	1.54 / 1.83	1.47 / 1.72	1.44 / 1.66	1.37 / 1.56	1.34 / 1.51	1.29 / 1.43	1.25 / 1.37	1.22 / 1.33
200	3.89 / 6.76	3.04 / 4.71	2.65 / 3.88	2.41 / 3.41	2.26 / 3.11	2.14 / 2.90	2.05 / 2.73	1.98 / 2.60	1.92 / 2.50	1.87 / 2.41	1.83 / 2.34	1.80 / 2.28	1.74 / 2.17	1.69 / 2.09	1.62 / 1.97	1.57 / 1.88	1.52 / 1.79	1.45 / 1.69	1.42 / 1.62	1.35 / 1.53	1.32 / 1.48	1.26 / 1.39	1.22 / 1.33	1.19 / 1.28
400	3.86 / 6.70	3.02 / 4.66	2.62 / 3.83	2.39 / 3.36	2.23 / 3.06	2.12 / 2.85	2.03 / 2.69	1.96 / 2.55	1.90 / 2.46	1.85 / 2.37	1.81 / 2.29	1.78 / 2.23	1.72 / 2.12	1.67 / 2.04	1.60 / 1.92	1.54 / 1.84	1.49 / 1.74	1.42 / 1.64	1.38 / 1.57	1.32 / 1.47	1.28 / 1.42	1.22 / 1.32	1.16 / 1.24	1.13 / 1.19
1000	3.85 / 6.66	3.00 / 4.62	2.61 / 3.80	2.38 / 3.34	2.22 / 3.04	2.10 / 2.82	2.02 / 2.66	1.95 / 2.53	1.89 / 2.43	1.84 / 2.34	1.80 / 2.26	1.76 / 2.20	1.70 / 2.09	1.65 / 2.01	1.58 / 1.89	1.53 / 1.81	1.47 / 1.71	1.41 / 1.61	1.36 / 1.54	1.30 / 1.44	1.26 / 1.38	1.19 / 1.28	1.13 / 1.19	1.08 / 1.11
∞	3.84 / 6.64	2.99 / 4.60	2.60 / 3.78	2.37 / 3.32	2.21 / 3.02	2.09 / 2.80	2.01 / 2.64	1.94 / 2.51	1.88 / 2.41	1.83 / 2.32	1.79 / 2.24	1.75 / 2.18	1.69 / 2.07	1.64 / 1.99	1.57 / 1.87	1.52 / 1.79	1.46 / 1.69	1.40 / 1.59	1.35 / 1.52	1.28 / 1.41	1.24 / 1.36	1.17 / 1.25	1.11 / 1.15	1.00 / 1.00

Appendix E.   Values of r_{xy} beyond which 5 percent or 1 percent of the area falls

Number of pairs -2	05	01	Number of pairs -2	05	01
1	.997	1.000	24	.388	.496
2	.950	.990	25	.381	.487
3	.878	.959	26	.374	.478
4	.811	.917	27	.367	.470
5	.754	.874	28	.361	.463
6	.707	.834	29	.355	.456
7	.666	.798	30	.349	.449
8	.632	.765	35	.325	.418
9	.602	.735	40	.304	.393
10	.576	.708	45	.288	.372
11	.553	.684	50	.273	.354
12	.532	.661	60	.250	.325
13	.514	.641	70	.232	.302
14	.497	.623	80	.217	.283
15	.482	.606	90	.205	.267
16	.468	.590	100	.195	.254
17	.456	.575	125	.174	.228
18	.444	.561	150	.159	.208
19	.433	.549	200	.138	.181
20	.423	.537	300	.113	.148
21	.413	.526	400	.098	.128
22	.404	.515	500	.088	.115
23	.396	.505	1000	.062	.081

Appendix F. Values of rho (rank-order correlation) beyond which 5 percent or 1 percent of the area falls

N	5%	1%
5	1.000	—
6	.886	1.000
7	.786	.929
8	.738	.881
9	.683	.833
10	.648	.794
12	.591	.777
14	.544	.715
16	.506	.665
18	.475	.625
20	.450	.591
22	.428	.562
24	.409	.537
26	.392	.515
28	.377	.496
30	.364	.478

Appendix G. Values of chi square beyond which 5 percent or 1 percent of the area falls

df	05	01
1	3.84	6.64
2	5.99	9.21
3	7.82	11.34
4	9.49	13.28
5	11.07	15.09
6	12.59	16.81
7	14.07	18.48
8	15.51	20.09
9	16.92	21.67
10	18.31	23.21
11	19.68	24.72
12	21.03	26.22
13	22.36	27.69
14	23.68	29.14
15	25.00	30.58
16	26.30	32.00
17	27.59	33.41
18	28.87	34.80
19	30.14	36.19
20	31.41	37.57
21	32.67	38.93
22	33.92	40.29
23	35.17	41.64
24	36.42	42.98
25	37.65	44.31
26	38.88	45.64
27	40.11	46.96
28	41.34	48.28
29	42.56	49.59
30	43.77	50.89

20011	51629	10880	60108	19473	66946	38309	27159	84905	20766
43449	09018	58448	75959	03586	41174	46382	34679	79351	11492
07454	47812	09510	98643	79121	72663	58803	74887	16073	35333
72389	15876	99430	51024	56440	59275	60936	20645	55435	40152
23619	86072	88572	80964	70908	23874	63145	69818	82800	98812
71506	56265	62299	61329	82889	14325	00794	45269	08532	09179
51416	49318	30975	89981	77746	03490	34247	94863	67995	94115
23711	13096	29964	89785	65844	89235	48868	91462	21553	01484
73756	20462	19630	08605	82545	94422	55019	32932	54569	04151
11914	69280	85336	19303	88214	66915	88065	42135	77407	99978
73549	82413	37446	19744	68214	79579	08370	66935	25431	11830
19024	07726	11325	32792	32910	30276	01298	80196	59005	87570
33705	18082	67335	06310	17665	41871	77212	79783	63492	00062
27953	11344	90840	13162	82842	62227	71475	88558	49257	47170
37133	10378	92205	51212	13806	64209	44453	54385	51663	51758
21610	63045	06793	43324	20920	45679	81508	50128	31073	61689
66745	42211	94968	37361	39549	29502	93004	73652	72852	49827
82905	45739	42093	06187	30055	63542	14306	47819	87364	14036
05451	96493	58533	47667	77803	20661	05776	20493	09971	77180
94748	42336	79651	84663	93156	98725	52779	64539	01039	87123
36160	56133	65811	65040	11478	20597	65679	77820	45930	16728
40050	35746	02376	61661	93133	34140	79838	83200	55008	63859
41783	04041	99711	72390	23485	12218	55622	28542	63638	51379
01721	08880	93179	20091	12898	52696	78393	86711	32182	27153
05229	23127	43145	52627	96734	78436	58516	55569	46005	64045
62671	44647	34970	42863	35359	37303	31353	57982	15471	59918
09410	96302	79021	88662	14135	02160	57270	41812	75943	37635
05810	25957	10659	12888	43427	70872	21630	79924	87785	45061
37235	06475	90250	63405	58598	66301	34796	70180	36361	55060
14048	35721	03157	13076	20013	36312	32133	35446	32035	65495
71614	36558	59743	59766	13034	53769	74004	23584	10169	99229
70296	56849	95373	26337	48026	16535	45772	07459	31129	04128
95457	69459	70409	60654	60352	47474	57802	84934	80278	53054
57463	72251	70217	35581	10376	94449	45458	78874	67979	43871
91676	88089	05159	48981	83462	30325	69103	37141	29597	99444
58460	64838	10600	23718	89974	52966	14100	32599	25495	67635
43179	75359	46903	07656	65276	85234	90815	63113	41037	21308
56197	17518	99432	73658	69730	74995	34610	51546	86587	58328
32877	14178	78550	19589	88702	95111	05849	45762	97508	01559
33584	13203	19623	68312	32554	43446	89896	18624	34165	98863
43680	87457	83012	67691	36651	42865	97115	67997	81920	23104
73531	34803	54083	90589	61357	41231	62869	16744	51138	72147

68856	77183	12729	47523	11407	92034	34871	43198	78106	58499
61093	20418	28816	36440	51258	39047	23401	85722	65228	58948
21723	66208	62868	39984	83647	37760	04041	42019	69034	60243
48610	24915	37750	93518	56438	56740	49656	97451	87388	72746
64616	31904	01324	57407	42494	72734	58110	62384	43153	31822
17607	47538	26456	17014	39681	31883	52266	72180	84193	97835
80446	57181	11008	82703	70860	94550	79988	87203	83372	56147
50996	71198	77844	89838	83897	46100	14139	92818	38553	17124
52122	24951	74828	98782	51655	96896	52585	99388	72601	16128
31687	78805	74824	94900	71997	82301	18188	42276	33379	13524
62554	18123	75696	88554	67788	62680	58812	81848	93751	94676
42589	53269	00307	15109	86890	23397	47320	03465	51580	69946
94653	19606	96522	34929	02169	74814	81578	17492	29731	74699
66612	77500	37203	33377	11488	52296	84448	59294	76068	69299
31329	12312	20215	20817	37710	16207	03794	89233	63453	39109
86668	34408	68422	32613	28699	51911	12481	92673	89751	94493
55492	79151	29687	29129	57319	69770	08371	79978	77825	70815
85665	06904	76874	95728	21434	05149	14328	86513	75540	28439
50052	03032	07847	42775	43907	18413	78217	72640	55759	52728
46515	77898	45469	05632	72603	94923	72901	23723	16345	54047
97919	66866	37605	44664	80385	45045	96244	50127	80164	67758
52127	30300	57117	45234	65117	04142	71109	87215	95077	54226
82003	53563	01871	17707	49663	32577	45361	95350	33950	98815
85412	47019	94729	97446	81885	15716	91863	59897	94646	11888
85216	46033	83556	44814	34649	63920	08479	91220	00028	28809
29279	10967	41215	45176	05818	12555	18073	24681	97961	09942
90646	27186	65570	08896	18256	21983	68508	20645	61308	40650
66639	05053	79484	71336	19826	77569	32648	64476	87933	31298
80663	79933	30822	92862	83393	89677	08358	66537	00700	17249
80402	12188	92448	58836	06819	93669	18500	62193	47472	58867
38178	87183	62224	79622	12969	49911	10938	11806	01114	41516
53478	15281	63015	90585	49707	43765	58537	00741	59488	75559
47543	31847	42685	52088	89896	85596	59160	39361	45460	96360
68778	97243	74097	49494	31401	10767	35671	63031	06892	64283
90046	96834	55115	93168	97084	79642	35934	32114	16648	64692
09211	40984	08603	18473	34809	77585	32812	31973	72593	07950
49137	65056	82425	85773	17442	14959	24169	72973	97590	29422
57687	29414	49444	80432	42844	27129	32869	90478	39502	89471
07726	19422	07525	12814	33880	74457	06775	44850	71194	73461
97118	45443	79530	18281	38414	42309	18753	21454	90529	91755
48725	42842	13324	57199	29310	41047	66664	30654	20370	79717
10412	71982	81771	14931	04430	06036	73373	07813	08583	97712
55042	82227	82734	01841	86640	97639	86745	13296	28030	31103
80480	91941	39076	53292	23803	97639	79641	32465	76576	90253
09589	28487	98663	29648	88782	01219	49927	75685	77083	10527

52288	77417	78319	20467	27142	79441	16273	34357	88229	90232
00900	50440	00732	39123	10770	18645	23530	44023	56531	95275
66727	94016	28287	78759	12467	54256	86785	50524	43757	22579
85132	31010	71347	37240	17597	59139	66399	01723	85270	95010
16480	09284	65277	37430	36524	31158	47738	70486	41435	60431
46133	01704	70441	27191	04612	93175	41164	54674	97614	91705
84457	06131	72201	79388	82224	93056	82042	77153	64173	86697
66814	45431	80922	91884	54724	03663	30735	85786	76474	68270
53568	67467	31968	87543	32476	22861	72608	53437	94881	84288
30084	45103	85703	14230	50843	73513	18023	77969	04759	07614
06725	76202	27489	44807	70190	03483	02345	82246	16470	36114
18854	83628	67691	77139	75880	85635	85937	14133	65380	92649
26835	15246	41673	34089	78277	17832	54164	46492	11850	25085
16033	43918	09799	63522	12745	22939	17388	77188	41246	06284
96811	67509	57432	38300	39647	48819	10974	29084	63931	34111
04533	08882	94935	56288	44348	68336	95286	50044	15268	31430
99562	15902	57674	40349	37210	79353	55687	12932	55622	46103
67262	61431	06011	38347	53598	04735	02541	15612	70357	50995
17998	43335	25311	23147	53876	92345	71212	80948	69835	43971
87132	84475	25936	92611	23407	15047	22063	56802	89421	47892
35029	32717	43251	69603	72556	70705	40459	16039	89479	10624
47053	60924	37620	01988	36685	82182	36763	56524	55372	05030
33566	66960	94407	62629	76159	97343	46339	01118	97465	28973
29934	73688	23974	49389	76341	89050	29550	97687	51120	05318
96519	28972	61687	85133	47595	55168	71761	19094	76702	81928
18685	05677	67908	17725	25286	18561	83335	63203	59574	31668
06797	01665	28002	20027	69776	27092	99636	52878	10100	52400
96218	39801	52332	89905	66429	53626	81028	35981	63645	66989
47311	67949	76261	50221	25610	96025	12874	85378	80571	23298
45441	58971	60155	48839	82682	77154	05539	98932	46243	94191
00972	10733	89376	58624	98008	44876	94385	99507	71024	77533
49266	46097	74289	77438	56954	72056	39778	34966	90278	96186
50689	12928	50257	28147	69881	56556	27080	78173	64369	98014
90603	84088	77643	58612	72155	21241	66655	26992	78660	55882
79372	57443	41812	41699	24138	13975	93081	04287	43516	67652
52360	55856	53128	75271	11195	07621	69081	57921	94300	56190
69932	27190	46953	82192	43690	00043	77659	60759	91376	69357
17450	44310	83653	10326	56986	14104	29965	10663	20108	80020
05278	05078	48590	32535	11446	97670	61364	30499	90859	86040
68781	32360	52129	46685	92436	23602	32218	68129	38993	10282
68322	74018	29633	75639	10318	89766	27892	96417	24875	00610
89264	02916	41466	67260	00456	19025	58750	13228	33867	29887
41972	16919	72992	94413	23214	59242	60155	41425	76333	95977
61810	38890	34575	54961	63956	83282	92471	28871	87638	21744
09140	16692	61577	71767	33046	89008	41062	48430	28145	55052

69328	23190	14364	92697	65847	99284	16292	97967	83218	68764
52735	56247	78299	90613	22723	61973	53524	00345	63220	60745
94728	38727	63745	63379	89037	49941	13122	03427	03516	53857
55668	18493	06067	33860	75155	61049	80450	80078	64468	95965
20921	68411	65628	49918	16439	18162	65871	28162	31442	59933
00849	86342	27792	84417	73253	69145	04750	70541	14801	43624
30816	95152	02922	35222	30961	86859	57451	55080	49907	69902
71187	42704	26383	25196	99927	69171	09336	54643	97126	86631
07325	01861	58538	02203	15514	38942	70770	67093	41822	66675
49593	70487	84752	39107	38087	44037	77116	15295	94356	07897
33356	71447	15387	33771	53009	57320	88739	47111	07897	33161
18978	52604	85807	09059	70643	76654	91002	35406	33161	90095
91821	86821	56377	12835	26354	24904	94268	78043	90095	52271
62251	71963	12460	17963	80505	49932	33902	97887	89400	16844
65630	30893	64420	22307	18461	24603	70268	42800	77637	06914
62322	11475	47621	48729	50584	46781	88720	85648	65169	70346
37691	86573	22426	45095	12239	39329	99648	24292	87359	80002
02102	54051	74199	84268	63790	50111	38390	81598	04571	33747
90917	36772	13304	64111	90599	51991	65319	05430	02170	54444
64501	82600	75105	07488	03032	42832	65798	68650	90519	89957
08216	64398	19965	62263	36451	45499	50191	69122	04981	13150
32428	80032	33249	01301	51221	07855	53861	29711	05921	21887
72500	52363	25319	22464	32705	02764	37173	98280	78702	39030
88795	29257	31541	48616	91267	28089	85491	47693	33688	12446
66678	83577	12277	27622	62271	06695	09177	75814	06242	14996
16511	13187	52891	32344	06080	86445	43597	05506	56729	44544
73660	40950	63747	60648	08059	40203	49112	84633	70513	23956
98487	14730	80209	35395	78571	65833	11089	86060	57957	01093
76356	07392	62641	04451	52980	86198	39889	07649	54424	48820
17632	16798	96407	40679	91649	49163	70877	72264	20279	65002
57677	65812	91869	41943	79943	27590	64417	27770	40886	72500
56856	02299	84392	31107	28225	19345	05873	47030	26607	19180
00533	99122	34341	56034	71860	47290	05607	27908	12808	77905
99071	54145	27077	89587	71209	59289	98985	93267	59851	46539
87834	90232	72966	29632	11639	28206	46369	90972	53100	47946
27186	55246	07371	99032	03511	51906	33124	93886	02920	29989
02490	22051	90656	45649	82191	53251	69613	99873	44674	65532
24111	88511	08184	42349	08041	80105	91200	31796	38726	52439
27412	77490	70320	86995	66426	05333	58249	37520	70439	13574
72756	70320	12426	02451	67710	26798	56124	89908	50177	96800
45508	12426	94868	36580	47255	67363	95188	86825	13305	74982
56032	36851	03008	62247	65427	74893	10805	51132	96800	45982
39691	39459	47211	77314	07588	22251	63339	30696	74982	32665
56848	83177	62839	04647	84102	07302	38153	98379	45982	82895
92869	90868	10258	92108	30334	52908	45613	52045	32665	69535

90449	10648	14557	21080	06933	06646	12561	74831	01423	58116
68501	13984	33780	24919	42242	98404	63870	66921	93647	87953
51555	43983	82578	42494	56699	15970	68899	21892	39164	42744
12474	86009	58814	84169	73166	94708	75512	00108	10837	07853
49122	50425	85352	85307	39507	94982	56573	86933	06530	93643
84364	72595	10055	06272	28587	02156	09944	92731	49108	35479
66061	12884	05788	32428	38270	26594	58491	52865	86433	93724
67080	56654	70414	74138	91418	03659	50076	27699	91370	53742
85282	14270	26797	66767	35896	93715	57564	02597	61781	25896
43533	21095	22801	70678	44568	82126	78817	87922	20532	45551
89695	37494	31289	71234	15297	79255	36702	19039	15486	73070
96633	48829	50126	78801	70947	20467	79079	56312	19506	93817
62210	65466	02174	28741	59382	66126	78815	85103	30457	18155
09291	22766	35299	81418	53465	01594	33771	15777	71201	81450
85738	81095	22363	22196	51061	37236	66813	83697	89604	44063
64415	25816	61231	61439	75033	08415	25804	49228	58528	91359
43188	67292	74765	34510	73244	75496	83608	97733	75838	33702
44918	40889	10831	01774	18560	23842	38087	39575	64397	06456
17470	06968	42291	48594	08843	63817	12107	10119	72068	69984
33708	50895	67010	85334	66957	30784	53531	69729	71717	09650
91496	83032	07851	47357	40766	85108	35223	03767	61206	35818
13696	38744	12678	95027	03134	12151	55046	22598	63399	83852
48174	78394	54355	13709	51924	22279	35865	61586	26160	14115

GLOSSARY OF SYMBOLS

Below are symbols commonly used in this book. They are arranged into three sections. The first deals with arithmetic operations, the second with Greek letters, and the third with English letters.

Symbol	Definition
$a + b$	a plus b $2 + 3 = 5$
$a - b$	a minus b $5 - 2 = 3$
$a \times b$	a times b $5 \times 2 = 10$
$a \cdot b$	a times b $5 \cdot 2 = 10$
ab or $(a)(b)$	a times b $(5)(2) = 10$
$\dfrac{a}{b}$	a divided by b $\frac{10}{5} = 2$
a/b	a divided by b $10/5 = 2$
$a < b$	a is less than b $2 < 5$
$b > a$	b is greater than a $5 > 2$
$\sqrt{}$	Square root $\sqrt{9} = 3$
X^2	X squared $3^2 = 3 \times 3 = 9$
$N!$	N multiplied by all smaller numbers $4! = 4 \times 3 \times 2 \times 1$
$\lvert a \rvert$	Absolute value of a
ΣX^2	Sum of all scores squared
Σx^2	$\Sigma (X - M)^2$
$(\Sigma X)^2$	Square of the sum of all scores

Greek letters

α	Alpha–the probability of rejecting a true null hypothesis, usually .05.
β	Beta–the probability of failing to reject a false null hypothesis.
χ^2	Chi square
μ	Population mean
σ	$\begin{cases}\text{Population standard deviation}\\ \text{Standard deviation used in correlation formula}\end{cases}$
σ^2	Population variance
σ_M	True standard error of the mean (i.e., standard deviation of means of all possible samples of size N)
$\sigma_M{}^2$	True variance of the means

English letters

$(A \times B)$	Interaction of A and B
bg_{ss}	Between-group sum of squares
C	Centile

cf	Cumulative frequency
CF	Correction factor in analysis of variance
D	$\begin{cases} \text{Decile} \\ \text{Rank difference} \end{cases}$
df	Degrees of freedom
e	Expected frequency in a category
F	Ratio of two variances
f	Frequency or number of cases in an interval
fX	Frequency times the score value
fX^2	$fX \cdot X - fX$ times the score value
i	Width of the interval
K	Number of categories or groups
m	Slope
M	Mean
M_x	Mean of X scores
M_y	Mean of Y scores
N	Total number of scores or cases in a group
o	Observed frequencies in a category
P	Probability of event P
$\Pr(A)$	Probability of A
$\Pr(A|B)$	Probability of A given B has occurred
$\Pr(A/B)$	Probability of A given B has occurred
Q	Probability of event Q
Q_1	First quartile or 25th percentile
Q_3	Third quartile or 75th percentile
rho	Rank order correlation
r_{xy}	Pearson product moment correlation coefficient
S	Standard deviation of a sample
S^2	Variance of a sample
S_M	Standard error of the mean or standard deviation of a group of means
$S_M{}^2$	Variance of the means
tot_{ss}	Total sum of squares
t	Value obtained when t test is used
U	Mann-Whitney U test
wg_{ss}	Within-group sum of squares
X	Variable
X_1	The first score
X_i	The ith score
X_N	The last or N^{th} score
x	$X - M$
Y	Variable
Z	Score expressed as a value when $M = 50$, $S = 10$
z	$\begin{cases} \text{Distance from the mean expressed in standard deviation units} \\ \text{Value obtained when } z \text{ test is used} \end{cases}$

Index